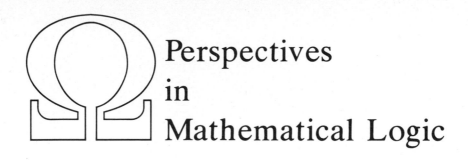
Perspectives
in
Mathematical Logic

Ω-Group:
R. O. Gandy   H. Hermes   A. Levy   G. H. Müller
G. E. Sacks   D. S. Scott

Robert I. Soare

Recursively Enumerable Sets and Degrees

A Study of Computable Functions
and Computably Generated Sets

Springer-Verlag
Berlin Heidelberg New York
London Paris Tokyo

Robert I. Soare
Department of Mathematics
University of Chicago
Chicago, IL 60637, USA

Mathematics Subject Classification (1980): 03 D 25, 03 D 10, 03 D 30

ISBN 3-540-15299-7 Springer-Verlag Berlin Heidelberg New York
ISBN 0-387-15299-7 Springer-Verlag New York Berlin Heidelberg

Library of Congress Cataloging in Publication Data.
Soare, R. I. (Robert Irving), 1940-
Recursively enumerable sets and degrees.
(Perspectives in mathematical logic)
Bibliography: p. Includes indexes.
1. Recursive functions. 2. Computable functions. I. Title. II. Title: Enumerable sets and degrees.
III. Series.
QA9.615.S63 1987 511.3 86-33928
ISBN 0-387-15299-7 (U.S.)

Media conversion: Arbor Text Inc., Ann Arbor
Printing: Druckhaus Beltz, 6944 Hemsbach
Binding: Graphischer Betrieb, Konrad Triltsch, 8700 Würzburg
2141/3140-543210

To my parents Margaret and Irving,
and to my "second parents" Beryl and Lawrence

Only when one's own cup has first
been filled with love abundantly
can one then give love to another

Preface to the Series
Perspectives in Mathematical Logic
(Edited by the Ω-group for "Mathematische Logik" of the
Heidelberger Akademie der Wissenschaften)

On Perspectives. *Mathematical logic arose from a concern with the nature and the limits of rational or mathematical thought, and from a desire to systematise the modes of its expression. The pioneering investigations were diverse and largely autonomous. As time passed, and more particularly in the last two decades, interconnections between different lines of research and links with other branches of mathematics proliferated. The subject is now both rich and varied. It is the aim of the series to provide, as it were, maps of guides to this complex terrain. We shall not aim at encyclopaedic coverage; nor do we wish to prescribe, like Euclid, a definitive version of the elements of the subject. We are not committed to any particular philosophical programme. Nevertheless we have tried by critical discussion to ensure that each book represents a coherent line of thought; and that, by developing certain themes, it will be of greater interest than a mere assemblage of results and techniques.*

The books in the series differ in level: some are introductory, some highly specialised. They also differ in scope: some offer a wide view of an area, others present a single line of thought. Each book is, at its own level, reasonably self-contained. Although no book depends on another as prerequisite, we have encouraged authors to fit their book in with other planned volumes, sometimes deliberately seeking coverage of the same material from different points of view. We have tried to attain a reasonable degree of uniformity of notation and arrangement. However, the books in the series are written by individual authors, not by the group. Plans for books are discussed and argued about at length. Later, encouragement is given and revisions suggested. But it is the authors who do the work; if, as we hope, the series proves of value, the credit will be theirs.

History of the Ω-Group. *During 1968 the idea of an integrated series of monographs on mathematical logic was first mooted. Various discussions led to a meeting at Oberwolfach in the spring of 1969. Here the founding members of the group (R. O. Gandy, A. Levy, G. H. Müller, G. Sacks, D. S. Scott) discussed the project in earnest and decided to go ahead with it.*

Professor F. K. Schmidt and Professor Hans Hermes gave us encouragement and support. Later Hans Hermes joined the group. To begin with all was fluid. How ambitious should we be? Should we write the books ourselves? How long would it take? Plans for authorless books were promoted, savaged and scrapped. Gradually there emerged a form and a method. At the end of an infinite discussion we found our name, and that of the series. We established our centre in Heidelberg. We agreed to meet twice a year together with authors, consultants and assistants, generally in Oberwolfach. We soon found the value of collaboration: on the one hand the permanence of the founding group gave coherence to the overall plans; on the other hand the stimulus of new contributors kept the project alive and flexible. Above all, we found how intensive discussion could modify the authors' ideas and our own. Often the battle ended with a detailed plan for a better book which the author was keen to write and which would indeed contribute a perspective.

Oberwolfach, September 1975

Acknowledgements. *In starting our enterprise we essentially were relying on the personal confidence and understanding of Professor Martin Barner of the Mathematisches Forschungsinstitut Oberwolfach, Dr. Klaus Peters of Springer-Verlag and Dipl.-Ing. Penschuck of the Stiftung Volkswagenwerk. Through the Stiftung Volkswagenwerk we received a generous grant (1970–1973) as an initial help which made our existence as a working group possible.*

Since 1974 the Heidelberger Akademie der Wissenschaften (Mathematisch-Naturwissenschaftliche Klasse) has incorporated our enterprise into its general scientific program. The initiative for this step was taken by the late Professor F. K. Schmidt, and the former President of the Academy, Professor W. Doerr.

Through all the years, the Academy has supported our research project, especially our meetings and the continuous work on the Logic Bibliography, in an outstandingly generous way. We could always rely on their readiness to provide help wherever it was needed.

Assistance in many various respects was provided by Drs. U. Felgner and K. Gloede (till 1975) and Drs. D. Schmidt and H. Zeitler (till 1979). Last but not least, our indefatigable secretary Elfriede Ihrig was and is essential in running our enterprise.

We thank all those concerned.

Heidelberg, September 1982   R. O. Gandy   H. Hermes
                                             A. Levy   G. H. Müller
                                             G. Sacks   D. S. Scott

Author's Preface

One of the fundamental contributions of mathematical logic has been the precise formulation and study of computable functions. This program received an enormous impetus in 1931 with Gödel's Incompleteness Theorem which used the notion of a primitive recursive function and led during the mid-1930's to a variety of definitions of a computable (i.e., *recursive*) function by Church, Gödel, Kleene, Post, and Turing. It was soon proved that these various definitions each gave rise to exactly the same class of mathematical functions, the class now generally accepted (according to Church's Thesis) as containing precisely those functions intuitively regarded as "effectively calculable." Informally, these are the functions which could be calculated by a modern computer if one ignores restrictions on the amount of computing time and storage capacity.

Closely associated is the notion of a computably listable (so-called *recursively enumerable (r.e.)*) set of numbers, namely one which can be generated by a computable procedure. Indeed the notions are, in a sense, interchangeable because one can begin the study of computable functions either: (1) with the notion of a recursive function, and can then define an r.e. set as the range of such a function on the integers; or (2) with the notion of an r.e. set, and can then define a recursive function as one whose graph is r.e. (The latter approach is sometimes preferable in generalized recursion theory.)

Thus although this book is ostensibly about r.e. sets and their degrees, it is intended more generally as an introduction to the theory of computable functions, and indeed it is intended as a replacement for the well-known book by Rogers [1967], which is now both out of date and out of print. This book will serve as an introduction for both mathematicians and computer scientists. The first four chapters cover the basic theory of computable functions and r.e. sets including the Kleene Recursion Theorem and the arithmetical hierarchy. Basic finite injury priority arguments appear in Chapter 7. Well grounded in the fundamentals, the computer scientist can then turn to computational complexity.

In his epochal address to the American Mathematical Society E. L. Post [1944] stripped away the formalism associated with the development of recursive functions in the 1930's and revealed in a clear informal style the essential properties of r.e. sets and their role in Gödel's incompleteness theorem. Recursively enumerable sets have later played an essential role in many other famous undecidability results (such as the Davis-Matijasevič-Putnam-Robinson resolution of Hilbert's tenth problem on the solution of certain Diophantine equations, or the Boone-Novikov theorem on the unsolvability of the word problem for finitely presented groups). This essential role of r.e. sets is because of: (1) the widespread occurrence of r.e. sets in many branches of mathematics; and (2) the fact that there exist r.e. sets which are not computable (i.e., not *recursive*). The first such set (constructed by Church, Rosser, and Kleene jointly) was called by Post *creative* because its existence together with the representability of all r.e. sets even in such a small fragment of mathematics as elementary number theory implied the impossibility of mechanically listing all statements true in such a fragment. Post remarked: "The conclusion is inescapable that even for such a fixed, well-defined body of mathematical propositions, *mathematical thinking is, and must remain, essentially creative*. To the writer's mind, this conclusion must inevitably result in at least a partial reversal of the entire axiomatic trend of the late nineteenth and early twentieth centuries, with a return to meaning and truth as being of the essence of mathematics."

This book represents a kind of progress report over the last forty years on the programs, ideas, and hopes expressed in Post's paper. It is intended to follow the same informal style as Post, but with full mathematical rigor. In doing so, this book is in the style of its principal predecessors on the subject: Kleene [1952a]; Rogers [1967]; and Shoenfield [1971]; to whom the author acknowledges a great debt. It differs from these predecessors in: (1) its emphasis on intuition and pictures of complicated constructions (often accompanied by suggestive terminology intended to create a diagram or image in the reader's mind); and (2) its modular approach of first describing the strategy for meeting each single requirement, and later describing the process by which these various and often conflicting strategies may be fitted together. In this way the book attempts to unveil some of the secrets of classical recursion theory whose seemingly formidable technical obstacles have tended to frighten away the novice from appreciating its considerable intrinsic beauty and elegance.

Classical recursion theory (CRT) is the study of computable functions on ω (the nonnegative integers) as opposed to generalized recursion theory (GRT) which deals with computation in certain ordinals or higher type objects. The beauty of CRT lies in the simplicity of its fundamental notions, just as a classical painting of the Renaissance is characterized by simplicity of line and composition. For example, the notion of an r.e. set as one which can be effectively listed is one of the few fundamental notions of higher mathematics which can easily be explained to the common man. The art and architecture of the Renaissance are characterized by balance, harmony, and a world on a human scale, where the human figure is not dwarfed by his surroundings. In CRT the universe is merely the natural numbers ω

which the human mind can readily grasp and not some more abstract object. Further analogies between CRT and the classical art of the Renaissance may be found in the lectures in Bressanone, Italy (Soare [1981, §7]).

The ideas and methods of CRT (such as the priority method) have been useful not only in GRT but also in many other fields such as recursion theory on sets (so-called E-recursion), recursive model theory, the effective content of mathematics (particularly effective algebra and analysis), theoretical computer science and computational complexity, effective combinatorics (such as the extent to which classical combinatorial results like Ramsey's theorem can be effectivized), and models of formal systems such as Peano arithmetic. It is hoped that workers in these fields may find this book useful (particularly the first twelve chapters). The remainder of the book is written for the genuine devotee of recursion theory who wishes to be initiated into some of its inner mysteries.

Much of the material of this book has been presented in courses and seminars at the University of Chicago, and in short courses at various international mathematical meetings, for instance: the C.I.M.E. conference in Bressanone, Italy during June, 1979 on *Recursive Theory and Computational Complexity;* the British Logic Colloquium in Leeds during August, 1979 on *Recursion Theory: Its Generalizations and Applications;* and the conference in Bielefeld, Germany during July, 1981 on the *Priority Method in Recursion Theory.* I am indebted to my Ph.D. students at the University of Chicago: Victor Bennison, Peter Fejer, Steffen Lempp, David Miller, Steven Schwarz, and Michael Stob; and to the other graduate students Kathy Edwards, Michael Mytilinaios, Nick Reingold, Francesco Ruggeri, Craig Smorynski, and Mitchell Stokes, all of whom have made the course stimulating and exciting to teach, and have contributed substantially to its present form. Special thanks are due for extensive contributions from L. Harrington, C. G. Jockusch, Jr., A. H. Lachlan, S. Lempp, M. Lerman, W. Maass, R. A. Shore, and T. Slaman.   Many other mathematicians have supplied suggestions, corrections, stimulating conversations, or correspondence on the subject including among others F. Abramson, S. Ahmad, D. Alton, K. Ambos-Spies, K. Appel, M. Arslanov, M. Blum, T. Carlson, C. T. Chong, B. Cooper, J. Crossley, M. Davis, J. C. E. Dekker, A. Degtev, R. Downey, B. Dreben, R. Epstein, Y. Ershov, S. Friedman, R. O. Gandy, V. Harizanov, J. Hartmanis, L. Hay, P. Hinman, E. Herrmann, H. Hodes, S. Homer, Huang Wen Qi, I. Kalantari, E. B. Kinber, P. Kolaitis, G. Kreisel, S. Kurtz, R. Ladner, S. MacLane, A. Manaster, D. A. Martin, A. R. D. Mathias, T. McLaughlin, A. Meyer, T. Millar, A. Nerode, P. Odifreddi, J. Owings, D. Pokras, D. Posner, M. B. Pour-El, M. Ramachandran, J. Remmel, R. W. Robinson, J. Rosenstein, J. Royer, G. E. Sacks, L. Sasso, J. Shoenfield, S. Simpson, R. Smith, C. Smorynski, R. Solovay, S. Thomason, S. Wainer, Dong Ping Yang, C. E. M. Yates, and P. Young. Preliminary versions of this book were used in graduate courses by the following mathematicians and computer scientists who supplied very useful suggestions: T. Carlson and L. Harrington, University of California at Berkeley; M. Lerman, University of Connecticut; A. R. D. Mathias, Cambridge University; G. E. Sacks, Harvard University; B. Cooper, University of Leeds; L. Hay, University of Illinois at Chicago; C. G. Jockusch, Jr., University of Illinois at Champaign-Urbana; P. Hinman, University of Michigan; T. Millar, University of Wisconsin; D. Kozen, A. Nerode, and R. A. Shore, Cornell University; M. Stob, Massachusetts Institute of Technology; W. Schnyder, Purdue University; A. H. Lachlan, Simon Fraser University; and Dong Ping Yang, Institute of Software, Academica Sinica, Beijing.

The subject matter of this book includes the contributions of many fine mathematicians, but in particular the unusually innovative discoveries (in historical order) of Stephen C. Kleene, Emil Post, Clifford Spector, Richard Friedberg, A. A. Muchnik, Gerald E. Sacks, and Alistair Lachlan. The book itself reflects the enormous debt which the author owes to his mathematical forbears: Anil Nerode, his thesis advisor, who taught him not only recursion theory but also the enthusiasm and confidence so essential to mathematical success, and to his "mathematical grandfather," Saunders MacLane, whose mathematical vigor, commitment to excellence, and strength of character have deeply influenced the author since his arrival at the University of Chicago in 1975. The author is very grateful to the Heidelberger Akademie der Wissenschaften (Mathematisch-Naturwissenschaftliche Klasse) for its travel support to attend meetings of the Ω-group from 1974 to 1983 to discuss outlines and preliminary versions of the book. These meetings with the editors and other members of the Ω-group were very helpful as the author's view emerged and changed over that period. The Academy also provided support for Steffen Lempp to proofread the entire typescript. A debt of gratitude goes to the author's wife Pegeen for her patience and understanding during the preparation of the book, and for her proofreading parts of the manuscript. The author is indebted to Fred Flowers for typing the first draft of Chapters I–X, to Richard Carnes for typesetting the entire manuscript in TEX, and to Terry Brown for drawing the diagrams and for typing the bibliography.

Chicago   Robert I. Soare
June 18, 1986

Table of Contents

Major Dependencies Diagram

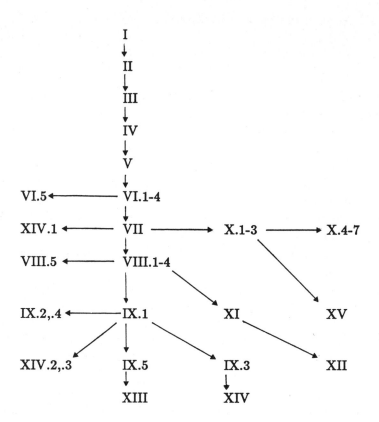

Introduction

An initial segment of this book is intended as an introduction to the theory of recursive functions and r.e. sets. For this purpose, selected portions of Parts A, B, and C could be used in a one-semester or two-quarter course, the exact selection depending upon the lecturer's interests and the length of the course. In most parts of the book no knowledge of logic is necessary, but the reader will find it helpful. The minimum background is the mathematical sophistication normally acquired in an undergraduate course in modern algebra. The more advanced sections and exercises in Parts C and D include an exposition of some of the most important recent results and methods concerning r.e. sets, and are intended to bring the reader to the frontiers of current research.

The first three parts are grouped in three main historical periods, according to the dates of discovery of the main results and techniques. This is only a rough classification because, of course, not all the results in a given part were necessarily proved during the corresponding period. In general, however, they *could* have been proved at that time because the methods were available.

Part A corresponds to the period 1931–1943, beginning with the primitive recursive functions used in Gödel's incompleteness theorem in 1931. It includes various definitions of recursive functions and r.e. sets, their fundamental properties, the Kleene Recursion Theorem and its applications, relative computability and degrees of unsolvability, and the arithmetical hierarchy.

Part B covers the period 1944–1960 beginning with Post's 1944 address to the American Mathematical Society on r.e. sets and their decision problems. It includes Post's problem and his attempts to solve it using simple sets. It continues with the non-r.e. "oracle constructions" of Kleene and Post [1954] and Spector [1956] where a complicated condition such as "A is not computable from B" is decomposed into an infinite sequence $\{\, R_n : n \in \omega \,\}$ of simpler conditions called "requirements," say $A \neq \{\, n \,\}^B$, each one satisfied once and for all at some stage of the construction. Post's problem was solved using the finite injury priority method invented by Friedberg [1957] and independently by Muchnik [1957]. It combines the Kleene-Post type requirements with the effective constructions of Post thereby producing r.e. sets (rather than sets recursive in some oracle). The key innovation is to allow action taken at some stage for a given requirement R_m to be later "injured" by the action of some requirement R_n, $n < m$, of higher priority, so that R_m must be satisfied once again at a still later stage.

Part C covers the period from 1961 to the present, and stresses those constructions where the requirements may be infinitary in nature, for example positive requirements which cause infinitely many elements to enter an r.e. set A being constructed, or negative requirements which tend to restrain infinitely many elements from entering A. This includes the well-known infinite injury priority method and the minimal pair method for studying the r.e. degrees, $\mathbf{R}$, as well as the Friedberg maximal set construction and its extensions by Lachlan and others for studying the lattice $\mathcal{E}$ of r.e. sets under inclusion. After considering $\mathbf{R}$ and $\mathcal{E}$ each separately, special attention is given to certain elegant results relating the algebraic structure of a set $A \in \mathcal{E}$ to its degree in $\mathbf{R}$ (namely to the degree of information which it encodes). Part D is devoted to more complicated results concerning $\mathcal{E}$ and $\mathbf{R}$.

Definitions and Notation. Sections or theorems marked by * are not intended to be studied on a first reading of the book but contain either more difficult or supplementary material. The exercises are divided into three categories. Those unmarked are usually straightforward (at least with the copious hints). More difficult exercises are marked by ⋄ and the most difficult by ⋄⋄.

We deal with sets and functions over the nonnegative integers $\omega = \{0, 1, 2, \ldots\}$, and occasionally for technical convenience we include -1. Lower-case Latin letters $a, b, c, d, e, i, j, k, \ldots, x, y, z$ denote integers; f, g, h (and occasionally other lower-case Latin letters) denote *total* functions from ω^n to ω, for $n \geq 1$; certain upper- and lower-case Greek letters $\Phi, \Psi, \Theta, \varphi, \psi, \theta$ denote (possibly) *partial* functions on ω (functions whose domain is some subset of ω); and upper-case Latin letters $A, B, C, \ldots, X, Y, Z$ denote subsets of ω. The composition of two functions f and g is denoted by $f \circ g$ or simply by fg; $f^n(x)$ denotes the function $f(x)$ composed with itself n times; $\varphi(x) \downarrow$ denotes that $\varphi(x)$ is defined; $\varphi(x) \downarrow = y$ denotes moreover that $\varphi(x)$ has value y; $\varphi(x) \uparrow$ denotes that $\varphi(x)$ is undefined; $\varphi = \psi$ denotes equality of partial functions φ and ψ (which in other books and papers is often written $\varphi \simeq \psi$), namely that for all x, $\varphi(x) \downarrow$ iff $\psi(x) \downarrow$, and if $\varphi(x) \downarrow$ then $\varphi(x) = \psi(x)$; dom φ and ran φ denote the domain and range respectively of φ; χ_A denotes the characteristic function of A which is often identified with A and written simply as $A(x)$; $f \restriction x$ denotes the restriction of f to arguments $y < x$, and $A \restriction x$ denotes $\chi_A \restriction x$; f *majorizes* g if $g(x) \leq f(x)$ for all x, and f *dominates* g if $g(x) \leq f(x)$ for all but finitely many x. We write "1:1" for a function f which is one to one (injective), and "1:1 and onto" if f is bijective.

In addition to the usual set theoretic notation we use $A \subseteq^* B$ to denote that $A \subseteq B$ except for finitely many elements (namely that $A - B$ is finite);

$A \subset B$ to denote that $A \subseteq B$ but $A \neq B$; $\overline{A}$ to denote the complement of A; $A - B$ to denote $A \cap \overline{B}$; $A =^* B$ to denote that $A \subseteq^* B$ and $B \subseteq^* A$; $|A|$ or $\mathrm{card}(A)$ to denote the cardinality of A; $\max(A)$ to denote the maximum element of a finite set A if A is not $\emptyset$, and 0 otherwise; and $A \subset_\infty B$ to denote that $A \subseteq B$ and $|B - A| = \infty$. In II.2.9 we define $W_x \setminus W_y$ and $W_x \searchbslash W_y$. (The former is not to be confused with the set theoretic difference $W_x - W_y$.)

The expression (x, y) denotes the ordered pair consisting of x and y in that order; $\vec{x}$ denotes an n-tuple $(x_1, x_2, \ldots, x_n)$; $\langle x, y \rangle$ denotes the image of (x, y) under the standard pairing function $\frac{1}{2}(x^2 + 2xy + y^2 + 3x + y)$ from $\omega \times \omega$ onto ω; $\langle x_1, x_2, x_3 \rangle$ denotes $\langle \langle x_1, x_2 \rangle, x_3 \rangle$ and similarly for $\langle x_1, x_2, \ldots, x_n \rangle$ as defined in I.3.6; $[n, m]$ denotes the closed interval $\{ x : n \leq x \leq m \}$; and $[n, \infty) = \{ x : n \leq x \}$. A sequence $\{ f_n \}_{n \in \omega}$ is occasionally abbreviated by $\{ f_n \}$, and $\lim_{n \in \omega} f_n(x)$ is abbreviated by $\lim_n f_n(x)$.

We use ω^ω to denote the set of functions from ω to ω; 2^ω denotes the power set of ω; $\omega^{<\omega}$ denotes finite sequences over ω; $2^{<\omega}$ denotes finite sequences of 0's and 1's. In Chapters III, VI, and XIV, we use ρ, σ, τ, ν as finite strings (i.e., ranging over $\omega^{<\omega}$ or $2^{<\omega}$) and the length function $lh(\sigma) = \mu x[x \notin \mathrm{dom}\,\sigma] = \mathrm{card}(\mathrm{dom}\,\sigma)$. We let $|\sigma|$ denote $lh(\sigma)$; $\emptyset$ (or sometimes λ) denote the empty string in $2^{<\omega}$ or $\omega^{<\omega}$; $\sigma\widehat{\ }\tau$ denote the concatenation of string σ followed by τ; and $\sigma \subseteq \tau$ $(\sigma \subseteq A, f)$ denote that σ is an initial segment of τ $(A, f$ respectively). We write $\sigma \subset \tau$ to denote that $\sigma \subseteq \tau$ but $\sigma \neq \tau$. In Chapter X, σ denotes an "e-state" (which can be identified with a string in $2^{<\omega}$).

We form predicates with the usual notation of logic where $\&, \vee, \neg, \Longrightarrow, \exists, \forall, \mu x$ denote respectively: and, or, not, implies, there exists, for all, and the least x. In addition $\exists^\infty x$ denotes "there exist infinitely many x such that," and (a.e. x) is the dual quantifier "for almost every x," namely $(\exists x_0)\,(\forall x \geq x_0)$. We use the usual Church lambda notation for defining partial functions. Namely, let $[\ldots x \ldots]$ be an expression such that for any integer in place of x the expression has at most one corresponding value. Then $\lambda x[\ldots x \ldots]$ denotes the associated partial function. We also use the lambda notation for partial functions of k variables, writing $\lambda x_1 x_2 \ldots x_k$ in place of λx. We use $x, y, z < w$ to abbreviate $x < w$, $y < w$, and $z < w$. In a partially ordered set we let $x \mid y$ denote that x and y are incomparable, namely $x \not\leq y$ and $y \not\leq x$.

Each chapter is divided into sections and the definitions and theorems are numbered consecutively within each section. Thus, Theorem III.2.2 refers to Theorem 2.2 of Chapter III, and §III.2 abbreviates Chapter III Section 2. Certain displayed lines are numbered according to section so that II (3.1) refers to line (3.1) in §3 of Chapter II. The initial Roman numeral will be omitted in reference to a theorem or line within the chapter where it is introduced, but will be retained elsewhere. Thus Theorem 2.2 of Chapter

III will be referred to as Theorem 2.2 within Chapter III but as Theorem III.2.2 elsewhere. These will often be abbreviated 2.2 and III.2.2 respectively. Similarly line (3.1) of Chapter III is referred to as (3.1) within Chapter III and III (3.1) elsewhere.

Part A

*The Fundamental Concepts
of Recursion Theory*

Recursive Functions

Part A contains the basic definitions and methods developed during the beginning period of recursive function theory from 1931 to 1943. We have also included results proved much later if they fit in with the spirit and methods of this period. Gödel's famous Incompleteness Theorem [1931] called attention to the primitive recursive functions which he then extended by 1934 to the (general) recursive functions. Equivalent definitions were given in the mid-1930's by Kleene, Turing, and others, which according to Church's Thesis are commonly accepted as describing the computable functions. In Chapter I we consider the definitions by Kleene and Turing, and base our development on Turing machines, although we refer the reader elsewhere for the formal proofs of the equivalence of these definitions. We sketch proofs of the Enumeration and s-m-n Theorems which are essential to the subject, and we consider the problem of classifying unsolvable problems, particularly those arising in connection with recursively enumerable (r.e.) sets.

After a brief informal description of a computable partial function we give two formal definitions of the class of partial recursive functions and in §3 we state the basic results such as the Enumeration Theorem and s-m-n Theorem which will be essential for most of our work. We give informal sketches of these results, but refer the reader elsewhere for the rather tedious details. We then consider recursively enumerable sets, the existence of unsolvable problems, and the classification of the latter into "degrees" using the s-m-n Theorem.

It should be emphasized that all the results of this book are independent of the particular formal definition of computable function chosen, since the various well-known definitions can all be shown to give rise to the same mathematical class of functions. Once a formal definition is chosen, the results are also independent (by Exercise 5.9) of the particular effective *coding* of that definition.

1. An Informal Description

Informally, an algorithm (for a function f on ω) is a finite set of instructions which, given an input x, yields after a finite number of steps an output

$y = f(x)$. The algorithm must specify how to obtain each step in the calculation from the previous steps and from the input. The algorithm may only yield a partial function. For example, let $\psi(x) = \mu y[p(x, y) = 0]$, where $p(x, y)$ is some polynomial with integer coefficients and where $(\mu x) \; P(x)$ denotes "the least x such that $P(x)$". Now ψ may be defined for some but not all values of x. An algorithmic partial function which is defined on all arguments (i.e., which is *total*) is called *recursive* or *computable*.

For the sake of definiteness we now give two precise mathematical formulations of computable functions, the recursive functions and the Turing computable functions. These and numerous other formal definitions give rise to the same class of functions which is now commonly accepted as corresponding to the informal class of algorithmically computable functions. For our purposes it is immaterial which formal definition is chosen. We shall use the Turing formulation. The reader familiar with some computing model may simply view φ_n as the partial function computed by the nth algorithm (program) P_n under some effective listing of all algorithms, and pass to §3 on basic results.

2. Formal Definitions of Computable Functions

2.1 Primitive Recursive Functions

The name "recursive" comes from definition by recursion where a new function f is defined by specifying each new value $f(x+1)$ in terms of previous values $f(0), f(1), \ldots, f(x)$, and previously defined functions. For example, $f(0) = 1$, and $f(x + 1) = (x + 1)f(x)$ defines $x!$ in terms of multiplication. The primitive recursive functions constitute a very large class of computable functions containing almost all functions on ω commonly found in mathematics.

2.1 Definition. The class of primitive recursive functions is the smallest class $\mathcal{C}$ of functions closed under the following schemata.

(I) The *successor function*, $\lambda x[x + 1]$, is in $\mathcal{C}$.

(II) The *constant functions*, $\lambda x_1 \ldots x_n[m]$, are in $\mathcal{C}$, $0 \le n, m$.

(III) The *identity functions* (also called *projections*), $\lambda x_1 \ldots x_n[x_i]$, are in $\mathcal{C}$, $1 \le n$, $1 \le i \le n$.

(IV) (Composition) If $g_1, g_2, \ldots, g_m, \; h \in \mathcal{C}$, then

$$f(x_1, \ldots, x_n) = h(g_1(x_1, \ldots, x_n), \ldots, g_m(x_1, \ldots, x_n))$$

is in $\mathcal{C}$ where $g_1, \ldots, g_m$ are functions of n variables and h is a function of m variables.

(V) (Primitive Recursion) If $g, h \in C$ and $n \geq 1$ then $f \in C$ where

$$f(0, x_2, \ldots, x_n) = g(x_2, \ldots, x_n)$$

$$f(x_1 + 1, x_2, \ldots, x_n) = h(x_1, f(x_1, x_2, \ldots, x_n), x_2, \ldots, x_n)$$

assuming g and h are functions of $n - 1$ and $n + 1$ variables respectively. (In case $n = 1$, a 0-ary function is a constant function which is in C by schema (II).)

Hence, a function is primitive recursive just if there is a *derivation*, namely a sequence $f_1, f_2, \ldots, f_k = f$ such that each f_i, $i \leq k$, is either an initial function (i.e., is obtained by schemata (I), (II), or (III)), or f_i is obtained from $\{ f_j : j < i \}$, by an application of (IV) or (V). For example, the function $f(x_1, x_2) = x_1 + x_2$ has the following derivation.

$$
\begin{array}{ll}
f_1 = \lambda x[x + 1] & \text{by (I)} \\
f_2 = \lambda x[x] & \text{by (III)} \\
f_3 = \lambda x_1 x_2 x_3[x_2] & \text{by (III)} \\
f_4 = f_1 \circ f_3 & \text{by (IV)} \\
f_5(0, x_2) = f_2(x_2) & \\
f_5(x_1 + 1, x_2) = f_4(x_1, f_5(x_1, x_2), x_2) & \text{by (V)}
\end{array}
$$

Similarly it can be shown (Kleene [1952a, Ch. IX, pp. 217–232]) that all the usual functions on ω are primitive recursive including $x \cdot y$, x^y, $x!$ and

$$
x \dotminus y = \begin{cases} x - y & \text{if } x \geq y, \\ 0 & \text{if } x < y. \end{cases}
$$

A predicate (relation) is *primitive recursive* if its characteristic function is. For example, it can be shown that the relation $R = \{ x : x \text{ is prime} \}$ is primitive recursive. Let $p_0, p_1, \ldots$ be the prime numbers in increasing order. Any $x \in \omega$ has a unique representation

(2.1) $$x = p_0^{x_0} p_1^{x_1} \cdots p_n^{x_n} \cdots ,$$

where finitely many $x_i \neq 0$. It can be shown that the function

(2.2) $$(x)_i = \text{the exponent } x_i \text{ of } p_i \text{ in (2.1)},$$

is a primitive recursive function of x. Thus, for any finite sequence of positive integers $\{ a_0, a_1, \ldots, a_n \}$ there is a unique "code" number $a = p_0^{a_0+1} \cdots p_n^{a_n+1}$ such that each $a_i = (a)_i$ can be obtained primitively recursively from a. The importance of this will become clearer in §3.

2.2 Diagonalization and Partial Recursive Functions

Although the primitive recursive functions include all the usual functions from elementary number theory they fail to include *all* computable functions. Each derivation of a primitive recursive function is a finite string of symbols from a fixed finite alphabet, and thus all derivations can be effectively listed. Let f_n be the function corresponding to the nth derivation in this listing. Then the function $g(x) = f_x(x) + 1$ cannot be primitive recursive because $g \neq f_x$ for all x, but is clearly effectively computable, since to compute $g(n)$ we "call" the nth primitive recursive "program," supply input n and add 1 to the output.

The same argument (known as diagonalization) applies to any effective set of schemata which produces only *total* functions (since the xth function on argument x will always yield an output). Thus to obtain all computable functions we are forced to consider computable *partial* functions, i.e., functions which may not be defined on all arguments. Diagonalization is no longer an obstacle for partial functions. For example, let ψ_n be the partial function computed by the nth algorithm under some effective coding of all algorithms. Suppose $\varphi(x) = \psi_x(x) + 1$ if $\psi_x(x)$ is defined and $\varphi(x)$ is undefined otherwise. Now if φ corresponds to the x_0th algorithm then diagonalization does not imply that $\varphi \neq \psi_{x_0}$ since $\psi_{x_0}(x_0)$ may be undefined.

There are other good reasons to consider partial functions. We wish to include all algorithmic functions, and some algorithms work only for proper subsets of ω. Indeed there are computable partial functions not extendible to any computable total function (see Exercise 4.21), and thus the computable partial functions are not the restriction of the computable total functions to their domains. Finally, we want an enumeration theorem for computable functions (Theorem 3.4) and this is possible only for computable partial functions.

2.2 Definition (Kleene). The class of *partial recursive* (p.r.) functions is the least class obtained by closing under schemata (I) through (V) for the primitive recursive functions and the following schema (VI). A *total recursive* function (abbreviated *recursive* function) is a partial recursive function which is total.

(VI) (Unbounded Search) If $\theta(x_1, \ldots, x_n, y)$ is a partial recursive function of $n + 1$ variables, and

(2.3)
$$\psi(x_1, \ldots, x_n) = \mu y[\theta(x_1, \ldots, x_n, y) \downarrow = 0$$
$$\& \ (\forall z \leq y) \ [\theta(x_1, \ldots, x_n, z) \downarrow]],$$

then ψ is a partial recursive function of n variables.

(It is important to require the second clause in (2.3) since otherwise the p.r. functions are not closed under the μ-operator by Exercise 4.24.)

Every p.r. function obtained by schema (VI) is clearly computable in the intuitive sense since to compute $\psi(\vec{x})$ we simply attempt to compute in order $\theta(\vec{x},0), \theta(\vec{x},1), \ldots$, until, if ever, we find the first y such that $\theta(\vec{x},y) = 0$, and then we output $\psi(\vec{x}) = y$. Thus every p.r. function is intuitively computable. More evidence for the converse will appear in §3. A typical application of the μ operator is to show that $\psi(x) = \mu y[p(x,y) = 0]$ is p.r. where $p(x,y)$ is a polynomial with integer coefficients and hence $p(a,b)$ can be primitively recursively computed for all integers a and b. It is this unbounded search power of schema (VI) which distinguishes the recursive functions from the primitive recursive ones since a function of the form

$$f(x) = \begin{cases} (\mu y < g(x)) \, [h(x,y) = 0] & \text{if such } y \text{ exists,} \\ 0 & \text{otherwise} \end{cases}$$

can be shown to be primitive recursive if g and h are (see Exercise 2.8).

Note that in particular Definition 2.2 applies to θ if θ is a *total* recursive function. However, we do not wish to restrict Definition 2.2 by requiring θ to be total because by Proposition 4.4 and Theorem 4.10 given an index for a partial recursive function θ we cannot determine effectively whether or not θ is total.

2.3 Definition. A relation $R \subseteq \omega^n$, $n \geq 1$, is *recursive* (*primitive recursive, has property* P) if its characteristic function χ_R is recursive, (respectively primitive recursive, has property P) where $\chi_R(x_1, \ldots, x_n) = 1$ if $(x_1, \ldots, x_n) \in R$ and $= 0$ otherwise. Note that a set $A \subseteq \omega$ corresponds to the case $n = 1$ so we have the definition of a set being recursive.

2.3 Turing Computable Functions

A second characterization of the computable partial functions is due to Turing. A *Turing machine M* includes a two-way infinite *tape* divided into *cells*, a *reading head* which scans one cell of the tape at a time, and a finite set of internal *states* $Q = \{ q_0, q_1, \ldots, q_n \}$, $n \geq 1$. Each cell is either blank (B) or has written on it the symbol 1. In a single step the machine may simultaneously: (1) change from one state to another; (2) change the scanned symbol s to another symbol $s' \in S = \{ 1, B \}$; and (3) move the reading head one cell to the right (R) or left (L). The operation of M is controlled by a partial map $\delta : Q \times S \to Q \times S \times \{ R, L \}$ (which may be undefined for some arguments). The interpretation is that if $(q, s, q', s', X) \in \delta$ then the machine M in state q, scanning symbol s changes to state q', replaces s by s', and moves to scan one square to the right if $X = R$ (left if $X = L$). The map δ viewed as a finite set of quintuples is called a *Turing program*. The *input*

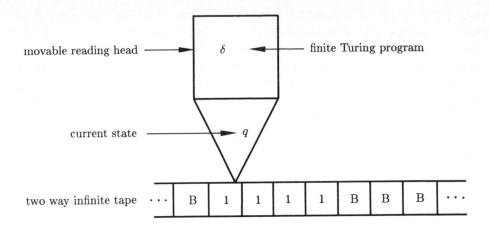

movable reading head ⟶ δ ⟵ finite Turing program

current state ⟶ q

two way infinite tape · · · | B | 1 | 1 | 1 | 1 | B | B | B | · · ·

Diagram 2.1. Turing Machine

integer x is represented by a string of $x + 1$ consecutive 1's (with all other cells blank). The Turing machine can be pictured as follows:

We begin with M in the *starting state* q_1 scanning the left-most cell containing a 1. If the machine ever reaches the *halting state* q_0, then we say M *halts* and the *output* y is the total number of 1's on the tape. (Without loss of generality, we may assume that M never makes any further moves after reaching state q_0; namely that the domain of δ contains no element of the form (q_0, s).) We say that M *computes* the partial function ψ providing that $\psi(x) = y$ if and only if M with input x eventually halts and yields output y. For example, the following machine computes the function $f(x) = x + 3$.

$$
\begin{array}{cccc}
& q_1 & 1 & q_1 & 1 & R \\
(2.4) & q_1 & B & q_2 & 1 & R. \\
& q_2 & B & q_0 & 1 & R
\end{array}
$$

The instantaneous condition at M during each step in a Turing calculation is completely determined by: (1) the current state q_i of the machine; (2) the symbol s_0 being scanned; (3) the symbols on the tape to the right of symbol s_0 up to the last 1, namely $s_1, s_2, \ldots, s_n$; and (4) the symbols to the left of s_0 up to the first 1, namely $s_{-m} \ldots s_{-2} s_{-1}$. This is the (instantaneous) *configuration* of the machine at that step and is written

$$
(2.5) \qquad s_{-m} \ \ldots \ s_{-2} \ s_{-1} \ q_i \ s_0 \ s_1 \ s_2 \ \ldots \ s_n.
$$

For example, the machine of (2.4) in calculating on input $x = 0$ passes through the following configurations, $q_1 1, 1q_1 B, 11q_2 B, 111q_0 B$ and yields output $y = 3$. (Recall that the input x is coded by $x + 1$ consecutive 1's while the output y is coded by the total number of 1's on the tape. Also notice that the tape contains only finitely many non-blank symbols at the beginning of any calculation and that this condition persists at all later stages whether the machine halts or not, so that the integers n and m in (2.5) exist.)

A *Turing computation* according to Turing program P with input x is a sequence of configurations, $c_0, c_1, \ldots, c_n$ such that c_0 represents the machine in the starting state q_1 reading the leftmost symbol of the input x, c_n represents the machine in the halting state q_0, and the transition $c_i \to c_{i+1}$, for all $i < n$, is given by the Turing program P. Thus, from now on a *computation* will always refer to a halting, i.e., *convergent*, calculation. A partial function of n variables is associated with each Turing machine M by representing the input $(x_1, x_2, \ldots, x_n)$ by the following initial configuration of M, $q_1 \alpha_1 B \alpha_2 \ldots B \alpha_n$ where α_i consists of $x_i + 1$ consecutive 1's.

2.4–2.11 Exercises

2.4. Write out primitive recursive derivations as in Definition 2.1 for the following functions: $f(x, y) = x \cdot y$, $g(x, y) = x^y$, $h(x) = x!$ (For example the informal recursion equations for f are $x \cdot 0 = 0$, $x \cdot (y + 1) = x \cdot y + x$.)

2.5. Using $x \mathbin{\dot{-}} y$ show that the function $\overline{sg}(x)$ is primitive recursive where $\overline{sg}(x) = 1$ if $x = 0$, and $\overline{sg}(x) = 0$ otherwise.

2.6. (Definition by cases). If $g_1(x), \ldots, g_n(x)$ are primitive recursive functions and $R_1(x), \ldots, R_n(x)$ are primitive recursive relations which are mutually exclusive and exhaustive show that f is primitive recursive where $f(x) = g_1(x)$ if $R_1(x)$, $f(x) = g_2(x)$ if $R_2(x), \ldots$, and $f(x) = g_n(x)$ if $R_n(x)$.

2.7. Prove that if f is primitive recursive then the finite sum $\sum_{y<z} f(x, y)$ and product $\prod_{y<z} f(x, y)$ are primitive recursive.

2.8. If R is a primitive recursive relation prove that the following are primitive recursive relations on z: $(\forall y < z)R(y)$, $(\exists y < z)R(y)$; and that the function f is primitive recursive, where $f(z) = (\mu y < z)R(y)$ if y exists, and $f(z) = z$ otherwise.

2.9. Write Turing machines which compute the functions $\lambda x[0]$, $\lambda x[k]$, $2x$, $x + y$, $x \cdot y$.

2.10. (a) Use (2.1) to assign a code number to each Turing program. First assign a number to each quintuple in the program, say $(q_i, s_j, q_k, s_l, r_m)$ gets number

$$p_0^{1+i} p_1^{1+j} p_2^{1+k} p_3^{1+l} p_4^{1+m},$$

where $r_0 = R$ and $r_1 = L$. Similarly use (2.1) to assign numbers to all possible finite sequences of quintuples, and hence to all Turing programs. Let P_e be the Turing program with code number e.

(b) Although this map is 1:1 and effective it is not *onto* ω. Devise a coding of Turing programs which is also onto.

2.11.   (a) Show that using (2.1) we can effectively assign a code number, $\#(c)$, to each possible configuration c in (2.5). (For example, let $\#(c) = 2^{1+i}3^{1+\#(s_0)}5^r7^l$ where $r = \prod_{j \geq 1} p_j^{\#(s_j)}$, $l = \prod_{j \leq -1} p_j^{\#(s_j)}$, and $\#(s) = 0$ if $s = B$ and $= 1$ otherwise.) These code numbers are also called *Gödel numbers*.

(b) Define the code number of a Turing computation $c_0, c_1, \ldots, c_n$ according to P_e to be $y = 2^e \cdot \prod_{i \leq n} p_{i+1}^{\#(c_i)}$. Show that the predicate $T(e, x, y)$ which asserts that "y is the code number of a computation according to Turing program P_e on input x" is computable in the intuitive sense. (Indeed it is actually primitive recursive; see Kleene [1952a].)

3. The Basic Results

In this section we state the basic results about the formal definitions which will be necessary for our work. We omit the formal proofs which can be found in Kleene [1952a] and Hermes [1969]. It has been proved that the two formal definitions of Kleene and Turing given in §2 (as well as numerous other formal definitions) give rise to exactly the same class of partial functions. *Church's Thesis* asserts that these functions coincide with the intuitively computable functions. We shall accept Church's Thesis, and from now on shall use the terms "partial recursive (p.r.)," "Turing computable," and "computable" interchangeably.

Since each Turing program is a finite set of quintuples, we can list all Turing programs in such a way that for any program we can effectively find its number on the list and conversely. Fix such an effective listing (say as in Exercise 2.10). (By Exercise 5.9 it will not matter which such listing is chosen.)

3.1 Definition.   Let P_e be the Turing program with code number (Gödel number) e (also called *index* e) in this listing and let $\varphi_e^{(n)}$ be the partial function of n variables computed by P_e, where φ_e abbreviates $\varphi_e^{(1)}$.

3.2 Padding Lemma.   *Each partial recursive (p.r.) function φ_x has $\aleph_0$ indices, and furthermore for each x we can effectively find an infinite set A_x of indices for the same partial function (i.e., $\varphi_y = \varphi_x$ for all $y \in A_x$).*

Proof. For any program P_x mentioning internal states $\{q_0, \ldots, q_n\}$, add extraneous instructions $q_{n+1}B\ q_{n+1}B\ R,\ q_{n+2}B\ q_{n+2}B\ R, \ldots$, to get new programs for the same function.   ▯

3.3 Normal Form Theorem (Kleene).   *There exist a predicate $T(e, x, y)$ (called the* Kleene *T-predicate) and a function $U(y)$ which are recursive (indeed primitive recursive) such that*

$$\varphi_e(x) = U(\mu y\ T(e, x, y)).$$

Proof (sketch). Informally, the predicate $T(e, x, y)$ asserts that y is the code number of some Turing computation (in the sense of Exercise 2.11) according to program P_e with input x. To see whether $T(e, x, y)$ holds we first effectively recover from e the program P_e (see Exercise 2.10); then recover from y the computation $c_0, c_1, \ldots, c_n$ (see Exercise 2.11) if y codes such a computation. Now check whether $c_0, c_1, \ldots, c_n$ is a computation according to P_e with x as the input in c_0. If so $U(y)$ simply outputs the number of 1's in the final configuration c_n. The primitive recursiveness of T and U can be proved by examining the coding of Exercises 2.10 and 2.11 (see Kleene [1952a]). (These are not the original definitions of T and U which were originally stated for deductions in the equational calculus (see Kleene [1952a, p. 281]) but the above definitions are easier to state and are used in an equivalent fashion.)   ▯

It follows from the Normal Form Theorem that every Turing computable partial function is partial recursive. To prove the converse one constructs Turing machines corresponding to the schemata (I)–(VI) (see Kleene [1952a]). The fact that these two formal definitions give the same class of functions is evidence for Church's Thesis. The same analysis as in the proof of Theorem 3.3 can be used to give an informal justification for Church's Thesis (see Shoenfield [1967, p. 120]) where arbitrary algorithms and configurations of steps in their calculations replace Turing programs and Turing configurations. Note that by Theorem 3.3 it follows that every partial recursive function can be obtained from two primitive recursive functions by *one* application of the μ-operator.

From now on we shall describe an algorithm for a recursive function in ordinary mathematical terms and leave it to the reader to convince himself that this algorithm could be translated into one of the formalisms above. Such a demonstration will be called a "proof by Church's Thesis."

3.4 Enumeration Theorem.   *There is a p.r. function of 2 variables $\varphi_z^{(2)}(e, x)$ such that $\varphi_z^{(2)}(e, x) = \varphi_e(x)$. Indeed the Enumeration Theorem holds for p.r. functions of n variables (see Exercise 3.13).*

Proof. By Theorem 3.3 let $\varphi_z^{(2)}(e, x) = U(\mu y T(e, x, y))$. An alternative informal proof is the following. Program P_z given input (e, x) effectively recovers program P_e and applies it to input x until (if ever) an output is obtained. ▯

3.5 Parameter Theorem (s-m-n Theorem). *For every $m, n \geq 1$ there exists a 1:1 recursive function s_n^m of $m+1$ variables such that for all $x, y_1, y_2, \ldots, y_m$*

$$\varphi_{s_n^m(x, y_1, \ldots, y_m)}^{(n)} = \lambda z_1, \ldots, z_n [\varphi_x^{(m+n)}(y_1, \ldots, y_m, z_1, \ldots, z_n)].$$

Proof (informal). For simplicity consider the case $m = n = 1$. The program $P_{s_1^1(x, y)}$ on input z first obtains P_x and then applies P_x to input (y, z). Now $s = s_1^1$ is a recursive function by Church's Thesis since this is an effective procedure in x and y. If s is not already 1:1 it may be replaced by a 1:1 recursive function s' such that $\varphi_{s(x,y)} = \varphi_{s'(x,y)}$ by using the padding lemma, and by defining $s'(x, y)$ in increasing order of $\langle x, y \rangle$, where $\langle x, y \rangle$ is the image of (x, y) under the pairing function of Notation 3.6. ▯

The s-m-n Theorem asserts that y may be treated as a fixed parameter in the program $P_{s(x,y)}$ which operates on z, and furthermore that the index $s(x, y)$ of this program is effective in x and y. A simple application of the s-m-n Theorem is the existence of a recursive function $f(x)$ such that $\varphi_{f(x)} = 2\varphi_x$. Let $\psi(x, y) = 2\varphi_x(y)$. Now by Church's Thesis $\psi(x, y) = \varphi_e^{(2)}(x, y)$ for some e. Let $f(x) = s_1^1(e, x)$. Further applications will occur in §4. (In a very rough sense the Enumeration and Parameter Theorems are inverses to one another since the former "pushes indices up" as variables while the latter "pulls variables down" as indices.)

3.6 Notation. We let $\langle x, y \rangle$ denote the image of (x, y) under the standard pairing function $\frac{1}{2}(x^2 + 2xy + y^2 + 3x + y)$, which is a 1:1 recursive function from $\omega \times \omega$ onto ω. Let π_1 and π_2 denote the inverse functions $\pi_1(\langle x, y \rangle) = x$, and $\pi_2(\langle x, y \rangle) = y$. Let $\langle x_1, x_2, x_3 \rangle$ denote $\langle \langle x_1, x_2 \rangle, x_3 \rangle$ and $\langle x_1, x_2, \ldots, x_n \rangle$ denote $\langle \cdots \langle \langle x_1, x_2 \rangle, x_3 \rangle, \ldots, x_n \rangle$. (All these functions are clearly recursive and indeed are even primitive recursive.)

3.7 Convention. For a relation $R \subseteq \omega^n$, $n > 1$, we say that R has some property P (such as being recursive, r.e., Σ_1, etc.) iff the set $\{ \langle x_1, \ldots, x_n \rangle : R(x_1, x_2, \ldots, x_n) \}$ has property P. (Note that this agrees with the Definition 2.3 of R being recursive iff χ_R is recursive.)

3.8 Definition. We write $\varphi_{e,s}(x) = y$ if $x, y, e < s$ and y is the output of $\varphi_e(x)$ in $< s$ steps of the Turing program P_e. If such a y exists we say $\varphi_{e,s}(x)$ *converges*, which we write as $\varphi_{e,s}(x) \downarrow$, and *diverges* ($\varphi_{e,s}(x) \uparrow$) otherwise.

Similarly, we write $\varphi_e(x)\downarrow$ if $\varphi_{e,s}(x)\downarrow$ for some s and we write $\varphi_e(x)\downarrow = y$ if $\varphi_e(x)\downarrow$ and $\varphi_e(x) = y$, and similarly for $\varphi_{e,s}(x)\downarrow = y$.

We use this simpler definition of $\varphi_{e,s}$ and $W_{e,s}$ only for Chapter I and Chapter II §1 in order to allow the reader to become accustomed to finite approximations. From Chapter II on we assume Convention II.2.6, which asserts that $\varphi_{e,s}$ and $W_{e,s}$ have been defined using Exercise 3.11 in place of Definition 3.8, so that from then on we may assume that condition (3.3) holds for $W_{e,s}$, where $W_{e,s}$ is defined in Definition 4.1.

3.9 Theorem. (i) *The set* $\{\,\langle e, x, s\rangle : \varphi_{e,s}(x)\downarrow\,\}$ *is recursive.*
  (ii) *The set* $\{\,\langle e, x, y, s\rangle : \varphi_{e,s}(x) = y\,\}$ *is recursive.*

Proof. We get (i) and (ii) from Church's Thesis since we calculate until an output is found or until the first s steps have been completed. ▯

3.10–3.13 Exercises

3.10. Prove that if $R(x, y)$ is recursive then $(\exists y \le z); R(x, y)$ is a recursive relation on x and z.

3.11. Prove that the following alternative definition of $\varphi_{e,s}(x) = y$ also satisfies Theorem 3.9 as well as the convenient properties:

(3.1) $\varphi_{e,s}(x) = y \Longrightarrow e, x, y < s;$

and

(3.2) $(\forall s)\,(\exists \text{ at most one } \langle e, x, y\rangle)\,[\varphi_{e,s}(x) = y \ \& \ \varphi_{e,s-1}(x)\uparrow\,],$

and hence

(3.3) $(\forall s)\,(\exists \text{ at most one } \langle e, x\rangle)\,[x \in W_{e,s+1} - W_{e,s}],$

where $W_{e,s}$ is defined from $\varphi_{e,s}$ in Definition 4.1. (In Convention II.2.6 and from then on we assume that Exercise 3.11 rather than Definition 3.8 has been used to define $\varphi_{e,s}$ and $W_{e,s}$ so (3.3) holds for $W_{e,s}$.)
  Define $\varphi_{e,s}(x) = y$ by recursion on s as follows. Let $\varphi_{e,0}(x)\uparrow$ for all x. Let $\varphi_{e,s+1}(x) = y$ iff $\varphi_{e,s}(x) = y$, or $s = \langle e, x, y, t\rangle$ for some $t > 0$ and y is the output of $\varphi_e(x)$ in $\le t$ steps of the Turing program P_e.

3.12. Show that the following alternative definition of $\varphi_{e,s}$ also satisfies Theorem 3.9, (3.1), (3.2), and (3.3):

$$\varphi_{e,s}(x) = \begin{cases} U((\mu z < s)T(e, x, z)) & \text{if } (\exists z < s)\, T(e, x, z), \\ \text{undefined} & \text{otherwise.} \end{cases}$$

3.13 (Enumeration Theorem). Modify either proof of Theorem 3.4 to show that for every $n \geq 1$ there exists a p.r. function $\varphi_{z_n}(e, x_1, \ldots, x_n)$ of $n+1$ variables such that $\varphi_{z_n}(e, x_1, \ldots, x_n) = \varphi_e^{(n)}(x_1, \ldots, x_n)$ for all e and $x_1, \ldots, x_n$.

4. Recursively Enumerable Sets and Unsolvable Problems

4.1 Definition. (i) A set A is *recursively enumerable* (r.e.) if A is the domain of some p.r. function.

(ii) Let the eth r.e. set be denoted by

$$W_e = \text{dom } \varphi_e = \{\, x : \varphi_e(x)\downarrow \,\} = \{\, x : (\exists y)T(e, x, y) \,\}.$$

(iii) $W_{e,s} = \text{dom } \varphi_{e,s}$. (Recall that if $x \in W_{e,s}$ then $x, e < s$.)

Note that $\varphi_e(x) = y$ iff $(\exists s)\, [\varphi_{e,s}(x) = y]$ and $x \in W_e$ iff $(\exists s)\, [x \in W_{e,s}]$. Note also that any recursive set is r.e. since if A is recursive then $A = \text{dom } \psi$, where $\psi(x) = 1$ if $\chi_A(x) = 1$ and $\psi(x) \uparrow$ otherwise. We shall show (Theorem II.1.8) that a nonempty set is r.e. iff A is the range of a recursive function (i.e., iff there is an algorithm for listing the members of A). The frequent occurrence of r.e. sets in other branches of mathematics and the existence of nonrecursive r.e. sets such as K below have yielded numerous undecidability results, such as the Davis-Matijasevič-Putnam-Robinson resolution of Hilbert's tenth problem on the solution of certain Diophantine equations, and the Boone-Novikov theorem on the unsolvability of the word problem for finitely presented groups (see Davis [1973] and Boone [1955–57], [1959] and Novikov [1955]).

4.2 Definition. Let $K = \{\, x : \varphi_x(x) \text{ converges} \,\} = \{\, x : x \in W_x \,\}$.

4.3 Proposition. K *is r.e.*

Proof. K is the domain of the following p.r. function

$$\psi(x) = \begin{cases} x & \text{if } \varphi_x(x) \text{ converges,} \\ \text{undefined} & \text{otherwise.} \end{cases}$$

Now ψ is p.r. by Church's Thesis since $\psi(x)$ can be computed by applying program P_x to input x and giving output x only if $\varphi_x(x)$ converges. Alternatively and more formally, $K = \text{dom } \theta$ where $\theta(x) = \varphi_z^{(2)}(x, x)$, for $\varphi_z^{(2)}$ the p.r. function defined in the Enumeration Theorem 3.4.

4.4 Proposition. K *is not recursive.*

Proof. If K had a recursive characteristic function χ_K then the following function would be recursive,

$$f(x) = \begin{cases} \varphi_x(x) + 1 & \text{if } x \in K, \\ 0 & \text{if } x \notin K. \end{cases}$$

However, f cannot be recursive since $f \neq \varphi_x$ for any x.   ▯

Thus, there is no algorithm for deciding given x whether $x \in K$. This is our first example of an unsolvable problem.

4.5 Definition. $K_0 = \{ \langle x, y \rangle : x \in W_y \}$.

Note that K_0 is also r.e. Indeed, $K_0 = \mathrm{dom}\ \theta$, where $\theta(\langle x, y \rangle) = \varphi_z^{(2)}(y, x)$ for $\varphi_z^{(2)}$ as in Theorem 3.4. ¿

4.6 Corollary. K_0 *is not recursive.*

Proof. Note that $x \in K$ iff $\langle x, x \rangle \in K_0$. Thus if K_0 has a recursive characteristic function, so does K, contrary to Proposition 4.4.   ▯

The *halting problem* is to decide for arbitrary x and y whether $\varphi_x(y)$ converges, i.e., whether program P_x with input y ever halts. Corollary 4.6 asserts the unsolvability of the halting problem. The proof of the corollary suggests an indirect method for proving unsolvability of new problems by reducing K to them.

4.7 Definition. (i) A is a *many-one reducible* (*m-reducible*) to B (written $A \leq_m B$) if there is a recursive function f such that $f(A) \subseteq B$ and $f(\overline{A}) \subseteq \overline{B}$, i.e., $x \in A$ iff $f(x) \in B$.

(ii) A is *one-one reducible* (*1-reducible*) to B ($A \leq_1 B$) if $A \leq_m B$ by a 1:1 recursive function.

For example, the proof of Corollary 4.6 established that $K \leq_1 K_0$ via the function $f(x) = \langle x, x \rangle$. Note that if $A \leq_m B$ via f then $\overline{A} \leq_m \overline{B}$ via f also. It is obvious that $\leq_m$ and $\leq_1$ are reflexive and transitive, and hence induce the following equivalence relations.

4.8 Definition. (i) $A \equiv_m B$ if $A \leq_m B$ and $B \leq_m A$.
(ii) $A \equiv_1 B$ if $A \leq_1 B$ and $B \leq_1 A$.
(iii) $\deg_m(A) = \{ B : A \equiv_m B \}$.
(iv) $\deg_1(A) = \{ B : A \equiv_1 B \}$.

The equivalence classes under $\equiv_m$ and $\equiv_1$ are called the *m-degrees* and *1-degrees* respectively.

4.9 Proposition. *If $A \leq_m B$ and B is recursive then A is recursive.*

Proof. If $A \leq_m B$ via f, then $\chi_A(x) = \chi_B(f(x))$, so χ_A is recursive if B is recursive. ▯

Proposition 4.9 provides a technique for proving the unsolvability of numerous problems such as that of deciding given x whether φ_x is a constant function, a total function, whether dom $\varphi_x \neq \emptyset$, whether φ_x is extendible to a total recursive function, and so on. If we can reduce one unsolvable problem A to another one B, then B is also unsolvable.

4.10 Theorem. $K \leq_1 \text{Tot} =_{\text{dfn}} \{ x : \varphi_x \text{ is a total function} \}.$

Proof. Define the function

$$\psi(x,y) = \begin{cases} 1 & \text{if } x \in K, \\ \text{undefined} & \text{otherwise.} \end{cases}$$

Clearly ψ is p.r. because the program to compute $\psi(x,y)$ says: first attempt to compute $\varphi_x(x)$; if this fails to converge then output nothing; if it converges then output 1 for every argument y. By the *s-m-n* Theorem there is a 1:1 recursive function f such that $\varphi_{f(x)}(y) = \psi(x,y)$. Namely, choose e such that $\varphi_e(x,y) = \psi(x,y)$, and define $f(x) = \lambda x[s_1^1(e,x)]$. Now f is 1:1 because s_1^1 is 1:1. Note that

$$x \in K \Longrightarrow \varphi_{f(x)} = \lambda y[1] \Longrightarrow \varphi_{f(x)} \text{ total} \Longrightarrow f(x) \in \text{Tot},$$
$$x \notin K \Longrightarrow \varphi_{f(x)} = \lambda y \text{ [undefined]} \Longrightarrow \varphi_{f(x)} \text{ not total} \Longrightarrow f(x) \notin \text{Tot}. \quad ▯$$

Notice that this proof shows that the problem of deciding, given x, whether φ_x is constant, or even whether dom $\varphi_x \neq \emptyset$ is an unsolvable problem. Notice also that in this proof K could be replaced by an arbitrary r.e. set A. Suppose $A = \text{dom } \theta$ for θ p.r. Then the former program begins by attempting to compute $\theta(x)$ in place of $\varphi_x(x)$. However, we could not replace K by an arbitrary *non*-r.e. set A because there would be no computable counterpart to this first step in the program for ψ, and ψ must be computable.

Applications of the *s-m-n* Theorem such as that above will occur often. The reader should verify in each case that the instructions for computing $\psi(x,y)$ are effective. From now on we can simply write $\varphi_{f(x)}(y)$ for $\psi(x,y)$ without explicitly mentioning the *s-m-n* Theorem. The method in the above proof applies to numerous other sets A in place of Tot so long as the property defining A is a property of functions and not of certain indices for them, i.e., if A is an "index set."

4.11 Definition.   A set $A \subseteq \omega$ is an *index set* if for all x and y

$$[x \in A \ \& \ \varphi_x = \varphi_y] \Longrightarrow y \in A.$$

4.12 Theorem.   *If A is a nontrivial index set, i.e., $A \neq \emptyset, \omega$, then either $K \leq_1 A$ or $K \leq_1 \overline{A}$.*

4.13 Corollary (Rice's Theorem).   *Let C be any class of partial recursive functions. Then $\{n : \varphi_n \in C\}$ is recursive iff $C = \emptyset$ or C is the class of all partial recursive functions.*   ⬚

Proof (of Theorem 4.12). Choose e_0 such that $\varphi_{e_0}(y)$ is undefined for all y. If $e_0 \in \overline{A}$, then $K \leq_1 A$ as follows. (If $e_0 \in A$, then $K \leq_1 \overline{A}$ similarly.) Since $A \neq \emptyset$ we can choose $e_1 \in A$. Now $\varphi_{e_1} \neq \varphi_{e_0}$ because A is an index set. By the *s-m-n* Theorem define a 1:1 recursive function f such that

$$\varphi_{f(x)}(y) = \begin{cases} \varphi_{e_1}(y) & \text{if } x \in K, \\ \text{undefined} & \text{if } x \notin K. \end{cases}$$

Now

$$x \in K \Longrightarrow \varphi_{f(x)} = \varphi_{e_1} \Longrightarrow f(x) \in A,$$

$$x \in \overline{K} \Longrightarrow \varphi_{f(x)} = \varphi_{e_0} \Longrightarrow f(x) \in \overline{A}.$$

The last implication in each line follows because A is an index set.   ⬚

It is possible that both $K \leq_1 A$ and $K \leq_1 \overline{A}$ for an index set A, for example if $A = \text{Tot}$. (See the exercises below.)

4.14 Definition.   In addition to K and K_0 we shall use the following index sets which correspond to the natural unsolvable problems mentioned above.

> $K_1 = \{x : W_x \neq \emptyset\};$
> Fin $= \{x : W_x \text{ is finite}\};$
> Inf $= \omega - \text{Fin} = \{x : W_x \text{ is infinite}\};$
> Tot $= \{x : \varphi_x \text{ is total}\} = \{x : W_x = \omega\};$
> Con $= \{x : \varphi_x \text{ is total and constant}\};$
> Cof $= \{x : W_x \text{ is cofinite}\};$
> Rec $= \{x : W_x \text{ is recursive}\};$
> Ext $= \{x : \varphi_x \text{ is extendible to a total recursive function}\}.$

Of the above, K, Fin, Inf, Tot, and Cof turn out to be the most important in later work. Each of the above is a nontrivial index set and hence nonrecursive by Rice's Theorem. Only K, K_0 and K_1 are r.e. sets, and these all have the same 1-degree (see Exercise 4.18). Hence, by Theorem 5.4 these three are all recursively isomorphic and may be regarded as interchangeable for all our purposes.

4.15 Definition.   An r.e. set A is *1-complete* if $W_e \leq_1 A$ for every r.e. set W_e.

Clearly K_0 is 1-complete because $x \in W_e$ iff $\langle x, e \rangle \in K_0$. Thus, K and K_1 are 1-complete also by Exercise 4.18. The remaining index sets above (and their complements) are not r.e. as we shall see in Chapter II §1. However, in Chapter IV we shall show that

$$\mathrm{Inf} \equiv_1 \mathrm{Tot} \equiv_1 \mathrm{Con}, \quad \text{and} \quad \mathrm{Cof} \equiv_1 \mathrm{Rec} \equiv_1 \mathrm{Ext}.$$

A major goal of recursion theory is to classify exactly how unsolvable a problem is by computing its "degree" of unsolvability relative to other problems. For example, if $A \equiv_r B$, where $r = 1, m$ or the more general Turing (T) reducibility of Chapter III, then A and B have the same r-degree and intuitively they "code the same information." If $A <_r B$, then B has strictly higher r-degree than A and codes more information.

In Chapter IV we shall give another characterization of these sets in terms of the arithmetical hierarchy. For each $n \in \omega$, we shall have a level in the hierarchy consisting of sets of the same T-degree. For example, the recursive sets lie at level 0; the sets K, K_0, and K_1 at level 1; Inf, Con and Tot at level 2; Cof, Rec, and Ext at level 3; and so on. The sets at higher levels have strictly higher T-degree than those at lower levels.

4.16 Definition.   Let A *join* B, written $A \oplus B$, be

$$\{\, 2x : x \in A \,\} \cup \{\, 2x + 1 : x \in B \,\}.$$

4.17–4.27 Exercises

4.17.   Prove that the m-degree of $A \oplus B$ is the least upper bound (l.u.b.) of the m-degrees of A and B, namely

(a) $A \leq_m A \oplus B$ and $B \leq_m A \oplus B$; and
(b) if $A \leq_m C$ and $B \leq_m C$ then $A \oplus B \leq_m C$.

(It is false that 1-degrees, even of r.e. sets, always have a l.u.b. or greatest lower bound (g.l.b.).)

4.18.   Prove that $K \equiv_1 K_0 \equiv_1 K_1$. (Note that the proof of Theorem 4.10 automatically shows that $K \leq_1 A$ for $A = K_1$, Con, or Inf. Use the same method with K_0 in place of K to show that $K_0 \leq_1 K$ and hence that $K_0 \leq_1 K \leq_1 K_1$, so K and K_1 are 1-complete.)

4.19.   Prove directly (without Rice's Theorem) that $K \leq_1 \mathrm{Fin}$. *Hint.* Let $\varphi_{f(x)}(s)$ be undefined if $x \in K_s$, and $\varphi_{f(x)}(s) = 0$ if $x \notin K_s$, where $K_s = W_{e,s}$ for some e such that $K = W_e$.

4.20. For any x show that $\overline{K} \leq_1 \{\, y : \varphi_x = \varphi_y \,\}$, and $\overline{K} \leq_1 \{\, y : W_x = W_y \,\}$. *Hint.* Consider separately the cases W_x finite or infinite, and use the method of Exercise 4.19. (Also see Exercise II.4.10(b) to see that this reduction cannot be made uniform.)

4.21. Show that $\mathrm{Ext} \neq \omega$. (Hence, not every partial recursive function is extendible to a total recursive function.)

4.22. (a) Disjoint sets A and B are *recursively inseparable* if there is no recursive set C such that $A \subseteq C$ and $C \cap B = \emptyset$. Show that there exist disjoint r.e. sets which are recursively inseparable. The existence of such sets is very useful, for example in Exercise 5.9 and in model theory. *Hint.* Let $A = \{\, x : \varphi_x(x) = 0 \,\}$, $B = \{\, x : \varphi_x(x) = 1 \,\}$ and by considering $\varphi_x(x)$ show that no φ_x can be the characteristic function of a recursive separating set C.
  (b) Give an alternative proof that $\mathrm{Ext} \neq \omega$.
  (c) For A and B as in part (a), prove that $K \equiv_1 A$ and $K \equiv_1 B$.

4.23. A set A is a *cylinder* if $(\forall B)\,[B \leq_m A \Longrightarrow B \leq_1 A]$.

  (a) Show that any index set is a cylinder.
  (b) Show that any set of the form $A \times \omega$ is a cylinder.
  (c) Show that A is a cylinder iff $A \equiv_1 B \times \omega$ for some set B.

4.24. Show that the partial recursive functions are not closed under μ, i.e., there is a p.r. function ψ such that $\lambda x[\mu y[\psi(x,y) = 0]]$ is not p.r. *Hint.* Define $\psi(x,y) = 0$ if $y = 1$, or if $y = 0$ and $\varphi_x(x)\downarrow$.

4.25. If A is recursive and $B, \overline{B}$ are each $\neq \emptyset$, then $A \leq_m B$. (Hence, neglecting the trivial sets $\emptyset$ and ω, there is a least m-degree.)

4.26.$^\diamond$ Prove that $\mathrm{Inf} \equiv_1 \mathrm{Tot} \equiv_1 \mathrm{Con}$. *Hint.* Use the s-m-n Theorem as in Theorem 4.10 or Exercise 4.18. For example, to show that $\mathrm{Inf} \leq_1 \mathrm{Con}$, define

$$\psi(e,x) = \begin{cases} 0 & \text{if } (\exists y > x)\,[\varphi_e(y)\downarrow\,], \\ \uparrow & \text{otherwise.} \end{cases}$$

By the s-m-n Theorem choose a recursive function f such that $\varphi_{f(e)}(x) = \psi(e,x)$, and show that $e \in \mathrm{Inf}$ iff $f(e) \in \mathrm{Con}$.

4.27. Prove that $\mathrm{Fin} \leq_1 \mathrm{Cof}$. *Hint.* Show that $\mathrm{Fin} \leq_1 \mathrm{Cof}$ via f defined as follows. Define

$$\varphi_{f(e)}(s) = \begin{cases} \uparrow & \text{if } W_{e,s+1} - W_{e,s} \neq \emptyset, \\ 0 & \text{otherwise.} \end{cases}$$

5. Recursive Permutations and Myhill's Isomorphism Theorem

5.1 Definition. (i) A *recursive permutation* is a 1:1, recursive function from ω onto ω.

(ii) A property of sets is *recursively invariant* if it is invariant under all recursive permutations.

Examples of recursively invariant properties are the following:

(i) A is r.e.;
(ii) A has cardinality n (written $|A| = n$);
(iii) A is recursive.

The following properties are not recursively invariant:

(i) $2 \in A$;
(ii) A contains the even integers;
(iii) A is an index set.

5.2 Definition. A is *recursively isomorphic* to B (written $A \equiv B$) if there is a recursive permutation p such that $p(A) = B$.

5.3 Definition. The equivalence classes under $\equiv$ are called *recursive isomorphism types*.

We shall attempt to classify sets up to recursive isomorphism just as the algebraist classifies structures up to isomorphism. One reason for introducing $\leq_1$ in §4 (where $\leq_m$ would have sufficed for undecidability results) is the following theorem, which is an effective analogue of the classical Schröder-Bernstein Theorem for cardinal numbers.

5.4 Myhill Isomorphism Theorem. $A \equiv B \iff A \equiv_1 B$.

Proof. ($\Longrightarrow$). Trivial.

($\Longleftarrow$). Let $A \leq_1 B$ via f and $B \leq_1 A$ via g. We define a recursive permutation h by stages so that $h(A) = B$. We let $h = \bigcup_s h_s$, where $h_0 = \emptyset$ and h_s is that portion of h defined by the end of stage s. Assume h_s is given so that in particular we can effectively check for membership in dom h_s and rng h_s, which we assume are both finite.

Stage $s+1 = 2x+1$. (We shall define $h(x)$.) Assume that h_s is 1:1, dom h_s is finite and $y \in A$ iff $h_s(y) \in B$ for all $y \in$ dom h_s. If $h_s(x)$ is defined, do nothing. Otherwise, enumerate the set $\{ f(x), f(h_s^{-1}f(x)), \ldots, f(h_s^{-1}f)^n(x), \ldots \}$, until the first element y not yet in rng h_s is found. Define $h_{s+1}(x) = y$. Note that y must exist since f and h_s are 1:1 and $x \notin$ dom h_s. Furthermore, $x \in A$ iff $y \in B$ by the hypotheses on f and h_s.

Stage $s + 1 = 2x + 2$. Define $h^{-1}(x)$ similarly, with f, h_s, dom and rng replaced by g, h_s^{-1}, rng and dom, respectively.   ▯

The following definition will be used in Exercise 5.7 and later in Chapter XI. See also Definition V.2.2.

5.5 Definition.   A function f *dominates* a function g if (a.e. x) $[f(x) \geq g(x)]$, namely if $f(x) \geq g(x)$ for almost every (all but finitely many) $x \in \omega$.

The notion of an *acceptable numbering* of the partial recursive functions defined in Exercise 5.9 below will be very important. Since most natural numberings are acceptable and any two acceptable numberings differ merely by a recursive permutation, it will not matter exactly *which* acceptable numbering we chose originally.

5.6–5.11 Exercises

5.6.   Prove that the recursive permutations form a group under composition.

5.7.   Prove that the primitive recursive permutations do not form a group under composition. *Hint.* First construct, using the ideas in Definition 2.2, an effective and hence recursive function φ_e which dominates every primitive recursive function. Define $g(x) = \mu y\, T(e, x, y)$ where $T(e, x, y)$ and $U(y)$ are the primitive recursive predicate and function of Theorem 3.3. Note that g also dominates all primitive recursive functions because $U(y) \leq y$ for all y. Construct a primitive recursive permutation f such that $f(g(x)) = x$ if x is even. Note that given y we can primitively recursively compute whether there is an x such that $g(x) = y$.

5.8.   Let $\omega = \bigcup_n A_n = \bigcup_n B_n$ where the sequences $\{A_n\}_{n \in \omega}$ and $\{B_n\}_{n \in \omega}$ are each pairwise disjoint. Let f and g be 1:1 recursive functions such that $f(A_n) \subseteq B_n$ and $g(B_n) \subseteq A_n$ for all n. Show that the construction of Theorem 5.4 produces a recursive permutation h such that $h(A_n) = B_n$ for all n.

5.9 (Rogers).   Let $\mathcal{P}$ be the class of partial recursive functions of one variable. A *numbering* of the p.r. functions is a map π from ω onto $\mathcal{P}$. The numbering $\{\varphi_e\}_{e \in \omega}$ of §3 is called the *standard* numbering. Let $\hat{\pi}$ be another numbering and let ψ_e denote $\hat{\pi}(e)$. Then $\hat{\pi}$ is an *acceptable* numbering if there are recursive functions f and g such that: (i) $\varphi_{f(x)} = \psi_x$; and (ii) $\psi_{g(x)} = \varphi_x$. Show that for any acceptable numbering $\hat{\pi}$, there is a recursive permutation p of ω such that $\varphi_x = \psi_{p(x)}$ for all x. *Hint.* (Jockusch) By Exercise 5.8 with appropriate definitions of A_n and B_n it suffices to convert f and g to 1:1 recursive functions f_1 and g_1 satisfying (i) and (ii). To define f_1 from f use the Padding Lemma 3.2. To define $g_1(x)$ we must be able (uniformly in x) to effectively generate an infinite set S_x of indices such that for each $y \in S_x$,

$\psi_y = \psi_{g(x)}$. Take any two recursively inseparable r.e. sets A and B, such as those of Exercise 4.22, and define

$$\varphi_{k(x,y)}(z) = \begin{cases} \varphi_x(z) & \text{if } y \in A, \\ 0 & \text{if } y \in B, \\ \text{undefined} & \text{otherwise.} \end{cases}$$

and similarly $\varphi_{l(x,y)}$ with 1 in place of 0. Let $C_x = \{ k(x,y) : y \in A \}$ and $D_x = \{ l(x,y) : y \in A \}$. If $\varphi_x \neq \lambda z[0]$, then $g(C_x)$ cannot be finite or else A and B are recursively separable. Hence, $S_x = g(C_x) \cup g(D_x)$ is infinite. Note that we cannot effectively tell which of $\varphi_x \neq \lambda z[0]$ or $\varphi_x \neq \lambda z[1]$ holds, but we do not have to know this in order to see that S_x is infinite. (For a proof using the Recursion Theorem see Exercise II.4.11.)

5.10. Extending the terminology of Exercise 5.9 we say that ψ is an *effective numbering* of the p.r. functions P iff ψ is p.r. and $\{ \psi_e : e \in \omega \} = P$ where $\psi_e = \lambda x[\psi(\langle e, x \rangle)]$. An effective numbering ψ is *acceptable* if $\{ \psi_e : e \in \omega \}$ is acceptable in the sense of Exercise 5.9, i.e., iff there is a recursive function g satisfying Exercise 5.9(ii) since the existence of f satisfying Exercise 5.9(i) follows at once from ψ and the *s-m-n* Theorem for $\{ \varphi_e : e \in \omega \}$.

(a) Show that an effective numbering ψ is acceptable iff for every other effective numbering θ there is a recursive function t such that for all e, $\theta_e = \psi_{t(e)}$.

(b) (Rogers). Show that an effective numbering ψ is acceptable iff it satisfies the *s-m-n* Theorem for $m = n = 1$, namely there is a recursive function s such that $\psi_{s(e,x)} = \lambda y[\psi_e(\langle x, y \rangle)]$ for all e, x and y.

(c) (Machtey and Young). Show that an effective numbering ψ is acceptable iff ψ has a composition function, namely a recursive function $c(p, q)$ such that $\psi_{c(p,q)} = \psi_p \circ \psi_q$. *Hint.* Choose i and j such that $\psi_i = \lambda y[\langle 0, y \rangle]$, and $\psi_j = \lambda xy[\langle x + 1, y \rangle]$. Define an s-1-1 function $s(p, x)$ for ψ by composing $\psi_p \circ \psi_j \cdots \psi_j \circ \psi_i$ where ψ_j is here composed with itself x times.

5.11. Obtain an effective numbering ψ which is not acceptable as follows. Define

$$\psi_{\langle 0,q \rangle}(0) = \text{undefined}$$
$$\psi_{\langle p+1,q \rangle}(0) = p$$
$$\psi_{\langle p,q \rangle}(x) = \varphi_q(x), \text{ if } x > 0.$$

Show that $\{ \psi_n \}_{n \in \omega} = P$ but that ψ is not acceptable. *Hint.* Show that if there is a recursive function g such that $\psi_{g(x)} = \varphi_x$ then we can decide the halting problem. (For an alternative example define a recursive function f such that $\{ \varphi_{f(p)} : p \in \omega \}$ consists of exactly the partial functions with nonempty domain and let $\psi_{p+1} = \varphi_{f(p)}$, $\psi_0(y) = \lambda y$ [undefined].)

Chapter II

Fundamentals of Recursively Enumerable Sets and the Recursion Theorem

This chapter is intended to give the reader a more intuitive feeling for r.e. sets, their alternative characterizations, and their most useful basic properties. We also prove the Recursion Theorem, which will be a very useful tool, and we apply it to prove Myhill's Theorem that all creative sets are isomorphic.

1. Equivalent Definitions of Recursively Enumerable Sets and Their Basic Properties

In Chapter I §4 we defined a set A to be r.e. if A is the domain of some p.r. function. We now show that this definition is equivalent to the more intuitive definition that there is an algorithm for enumerating the members of A. (See the Listing Theorem 1.8.) A third very useful equivalent definition is that A is Σ_1.

Recall our Convention I.3.7 that a relation $R \subseteq \omega^n$ is said to have some property P (such as being recursive, r.e., or Σ_1) iff the set $A = \{\,\langle x_1, x_2, \ldots, x_n \rangle : R(x_1, \ldots, x_n)\,\}$ has property P.

1.1 Definition. (i) A set A is a *projection* of some relation $R \subseteq \omega \times \omega$ if $A = \{\,x : (\exists y)\ R(x,y)\,\}$ (i.e., geometrically A is the projection of the two-dimensional relation R onto the x-axis).

(ii) A set A is in Σ_1-*form* (abbreviated "A is Σ_1") if A is the projection of some recursive relation $R \subseteq \omega \times \omega$. (The reason for the notation Σ_1 is that the corresponding predicate for A has the form of one $\exists$ quantifier followed by a recursive predicate. The forms Σ_n, Π_n, and Δ_n, $n \geq 0$, will be formally defined in Chapter IV on the arithmetical hierarchy. In the literature these forms are usually written Σ_n^0, Π_n^0, and Δ_n^0, to distinguish them from the *analytical hierarchy* Σ_n^1, Π_n^1, and Δ_n^1 where we count function quantifiers. See Rogers [1967] or Hinman [1978].)

The following is the equivalent theorem for r.e. sets of the Normal Form Theorem I.3.3 for partial recursive functions.

1.2 Theorem (Normal Form Theorem for r.e. sets). *A set A is r.e. iff A is Σ_1.*

Proof. ($\Longrightarrow$). If A is r.e. then $A = W_e =_{\text{dfn}}$ dom φ_e for some e. Hence,

$$x \in W_e \iff (\exists s)\,[x \in W_{e,s}] \iff (\exists s)\,T(e, x, s).$$

By Theorem I.3.9(i), the relation $\{\langle e, x, s\rangle : x \in W_{e,s}\}$ is recursive, or alternatively the predicate $T(e, x, s)$ is even primitive recursive.

($\Longleftarrow$). Let $A = \{x : (\exists y)\,R(x, y)\}$, where R is recursive. Then $A = $ dom ψ, where $\psi(x) = (\mu y)R(x, y)$.   ▯

1.3 Quantifier Contraction Theorem. *If there is a recursive relation*

$$R \subseteq \omega^{n+1}$$

and

$$A = \{\,x : (\exists y_1)\cdots(\exists y_n)\,R(x, y_1, \ldots, y_n)\,\}$$

then A is Σ_1.

Proof. Define the recursive relation $S \subseteq \omega^2$ by

$$S(x, z) \iff_{\text{dfn}} R(x, (z)_1, (z)_2, \ldots, (z)_n),$$

where $z = p_1^{(z)_1} \cdots p_k^{(z)_k}$ is the prime decomposition of z as in (2.1) of Chapter I.

$$(\exists z)\,S(x, z) \iff (\exists z)\,R(x, (z)_1, (z)_2, \ldots, (z)_n)$$
$$\iff (\exists y_1)(\exists y_2)\cdots(\exists y_n)\,R(x, y_1, y_2, \ldots, y_n).   ▯$$

1.4 Corollary. *The projection of an r.e. relation is r.e.*   ▯

1.5 Definition. The *graph* of a (partial) function ψ is the relation

$$(x, y) \in \text{graph } \psi \iff_{\text{dfn}} \psi(x) = y.$$

It follows from Theorem 1.3 or Corollary 1.4 that a set A is r.e. if A is of the form

$$\{\,x : (\exists y_1)\ldots(\exists y_n)\,R(x, y_1, \ldots, y_n)\,\},$$

where $R \subseteq \omega^{n+1}$ is recursive. This is very useful for showing various sets to be r.e. For example, using Theorem I.3.9 the following sets and relations are r.e.:

(1.1) $K = \{\,e : e \in W_e\,\} = \{\,e : (\exists s)\,(\exists y)\,[\varphi_{e,s}(e) = y]\,\}$,
(1.2) $K_0 = \{\,\langle x, e\rangle : x \in W_e\,\} = \{\,\langle x, e\rangle : (\exists s)\,(\exists y)\,[\varphi_{e,s}(x) = y]\,\}$,
(1.3) $K_1 = \{\,e : W_e \neq \emptyset\,\} = \{\,e : (\exists s)\,(\exists x)\,[x \in W_{e,s}]\,\}$,
(1.4)   rng $\varphi_e = \{\,y : (\exists s)\,(\exists x)\,[\varphi_{e,s}(x) = y]\,\}$,
(1.5)   graph $\varphi_e = \{\,(x, y) : (\exists s)\,[\varphi_{e,s}(x) = y]\,\}$.

1.6 Uniformization Theorem.   *If $R \subseteq \omega^2$ is an r.e. relation, then there is a p.r. function ψ (called a* selector function *for R) such that*

$$\psi(x) \text{ is defined} \iff (\exists y)R(x, y),$$

and in this case $(x, \psi(x)) \in R$. (Furthermore, an index for ψ can be found recursively from an r.e. index for R.)

Proof. Since R is r.e. and hence Σ_1, there is a recursive relation S such that $R(x, y)$ holds iff $(\exists z)\, S(x, y, z)$. Define the p.r. function

$$\theta(x) = (\mu u)\, S(x, (u)_1, (u)_2)$$

and set $\psi(x) = (\theta(x))_1$.   ▯

The important thing to grasp about the Uniformization Theorem is not that ψ *exists* (which is obvious) but that ψ is partial *recursive*.

1.7 Graph Theorem.   *A partial function ψ is partial recursive iff its graph is r.e.*

Proof. ($\Longrightarrow$). The graph of φ_e is r.e. by Theorem 1.2 and (1.5).
   ($\Longleftarrow$). If the graph of ψ is r.e., then ψ is its own p.r. selector function in Definition 1.5, since $R = \text{graph } \psi$ can have only ψ as its selector function.   ▯

The following is the basic theorem on r.e. sets, and justifies the earlier intuitive description of an r.e. set A as one whose members can be effectively listed, $A = \{a_0, a_1, a_2, \dots\}$. (Note that the listing may have repetitions and need not be in increasing order.)

1.8 Listing Theorem.   *A set A is r.e. iff $A = \emptyset$ or A is the range of a total recursive function f. Furthermore, f can be found uniformly in an index for A as explained in Exercise 1.25.*

Proof. ($\Longleftarrow$). If $A = \emptyset$, A is r.e. Now suppose $A = \text{rng } f$, where f is a total recursive function. Then A is r.e. by (1.4).
   ($\Longrightarrow$). Let $A = W_e \neq \emptyset$. Find the least integer $\langle a, t \rangle$ such that $a \in W_{e,t}$. Define the recursive function f by

$$f(\langle s, x \rangle) = \begin{cases} x & \text{if } x \in W_{e,s+1} - W_{e,s}; \\ a & \text{otherwise.} \end{cases}$$

(Note that each $x \in W_e$, $x \neq a$, is listed exactly once.) Clearly $A = \text{rng } f$, since if $x \in W_e$, choose the least s such that $x \in W_{e,s+1}$. Then $f(\langle s, x \rangle) = x$ so $x \in \text{rng } f$.   ▯

1.9 Union Theorem. *The r.e. sets are closed under union and intersection uniformly effectively, namely there are recursive functions f and g such that $W_{f(x,y)} = W_x \cup W_y$, and $W_{g(x,y)} = W_x \cap W_y$.*

Proof. Using the *s-m-n* Theorem define $f(x, y)$ by enumerating $z \in W_{f(x,y)}$ if $(\exists s) [z \in W_{x,s} \cup W_{y,s}]$, and similarly for g with $\cap$ in place of $\cup$. ▯

1.10 Corollary (Reduction Principle for r.e. sets). *Given any two r.e. sets A and B, there exist r.e. sets $A_1 \subseteq A$ and $B_1 \subseteq B$ such that $A_1 \cap B_1 = \emptyset$ and $A_1 \cup B_1 = A \cup B$.*

Proof. Define the relation $R =_{\mathrm{dfn}} A \times \{0\} \cup B \times \{1\}$ which is r.e. by Theorem 1.9. By the Uniformization Theorem 1.6, let ψ be the p.r. selector function for R. Let $A_1 = \{x : \psi(x) = 0\}$, $B_1 = \{x : \psi(x) = 1\}$. (See also Exercise 2.10(d).) ▯

1.11 Definition. A set A is in Δ_1-*form* (abbreviated "A is Δ_1") if both A and $\overline{A}$ are Σ_1. The following is the basic theorem of Post relating r.e. sets to recursive sets.

1.12 Complementation Theorem. *A set A is recursive iff both A and $\overline{A}$ are r.e. (i.e., iff $A \in \Delta_1$).*

Proof. ($\Longrightarrow$). If A is recursive, then $\overline{A}$ is recursive so A and $\overline{A}$ are both r.e.
($\Longleftarrow$). Let $A = W_e$, $\overline{A} = W_i$. Define the recursive function

$$f(x) = (\mu s) [x \in W_{e,s} \text{ or } x \in W_{i,s}].$$

Then $x \in A$ iff $x \in W_{e,f(x)}$, so A is recursive. ▯

1.13 Corollary. $\overline{K}$ *is not r.e.*

Proof. By Theorems I.4.3 and I.4.4 K is r.e. and not recursive. ▯

It is easily shown that if $\overline{K} \leq_m A$ then A is not r.e., so we can conclude that many sets such as Tot, Fin, and Cof are not r.e. (See Exercise 1.16.)

1.14 Definition. (i) A *lattice* $\mathcal{L} = (L; \leq, \vee, \wedge)$ is a partially ordered set (poset) in which any two elements have a least upper bound (also called supremum or join) and greatest lower bound (also called infimum or meet). If a and b are elements of a lattice $\mathcal{L}$, $a \vee b$ denotes the least upper bound (lub) of a and b, and $a \wedge b$ the greatest lower bound (glb). If $\mathcal{L}$ contains a least element and greatest element these are called the *zero* element 0 and *unit* element 1, respectively. In such a lattice a is the *complement* of b if $a \vee b = 1$ and $a \wedge b = 0$.

(ii) A lattice is *distributive* if all its elements satisfy the distributive laws $(a \vee b) \wedge c = (a \wedge c) \vee (b \wedge c)$, and $(a \wedge b) \vee c = (a \vee c) \wedge (b \vee c)$.

(iii) A lattice is *complemented* if every element has a complement.

(iv) A distributive complemented lattice containing at least the zero element and unit element is a *Boolean algebra*.

(v) A poset closed under suprema but not necessarily under infima is an *upper semi-lattice*.

(vi) $\mathcal{M} = \langle M; \leq, \vee, \wedge \rangle$ is a *sublattice* of $\mathcal{L}$ if $M \subseteq L$ and M is closed under the operations $\vee$ and $\wedge$ in $\mathcal{L}$.

(vii) A nonempty subset $I \subseteq L$ forms an *ideal* $\mathcal{I} = (I; \leq, \wedge, \vee)$ of $\mathcal{L}$ if I satisfies the conditions:

(1) $[a \in L \ \& \ a \leq b \in I] \implies a \in I$, and
(2) $[a \in I \ \& \ b \in I] \implies a \vee b \in I$.

(viii) Similarly a subset $D \subseteq L$ forms a *filter* (also called a *dual ideal*) $\mathcal{D} = (D; \leq, \vee, \wedge)$ of $\mathcal{L}$ if it satisfies the dual conditions:

(3) $[a \in L \ \& \ a \geq b \in D] \implies a \in D$, and
(4) $[a \in D \ \& \ b \in D] \implies a \wedge b \in D$.

(Clearly, an ideal or filter of $\mathcal{L}$ is a sublattice of $\mathcal{L}$.)

(ix) Let $\mathcal{L}$ be an upper semi-lattice. The definitions of ideal and filter are the same except that we require (4) only when $a \wedge b$ exists. Furthermore, we say $\mathcal{D}$ is a *strong filter* in $\mathcal{L}$ if $\mathcal{D}$ satisfies (3) and also:

(5) $[a \in D \ \& \ b \in D] \implies (\exists c \in D) [c \leq a \ \& \ c \leq b]$.

(Clearly, any strong filter is a filter.)

For example, the collection of all subsets of ω forms a Boolean algebra, $\mathcal{N} = (2^{\omega}; \subseteq, \cup, \cap)$ with $\emptyset$ as least element and ω as the greatest element. The finite sets form an ideal $\mathcal{F}$ of $\mathcal{N}$ and the cofinite sets A (i.e., those whose complement $\overline{A}$ is finite) form a filter $\mathcal{C}$ in $\mathcal{N}$.

1.15 Definition. (i) By Theorem 1.9 the r.e. sets form a distributive lattice $\mathcal{E}$ under inclusion with greatest element ω and least element $\emptyset$.

(ii) By Theorem 1.12 an r.e. set $A \in \mathcal{E}$ is recursive iff $\overline{A} \in \mathcal{E}$. Hence the recursive sets form a Boolean algebra $\mathcal{R} \subseteq \mathcal{E}$ consisting precisely of the complemented members of $\mathcal{E}$. (By Corollary 1.13 $\mathcal{R} \neq \mathcal{E}$.)

The Turing degrees $(\mathbf{D}, \leq)$ and the r.e. degrees $(\mathbf{R}, \leq)$ under the partial order induced by Turing reducibility $\leq_T$ to be defined in Definition III.2.1 will form only upper semi-lattices. Ideals and filters of $\mathbf{R}$ will be studied in Chapters IX and XIII, and those of $\mathcal{E}$ will be studied in Chapters X and XI.

1.16–1.25 Exercises

1.16. (a) Prove that $A \leq_m B$ and B r.e. imply A r.e.
(b) Show that Fin and Tot are not r.e.
(c) Show that Cof is not r.e.

1.17. Prove that if A is r.e. and ψ is p.r. then $\psi(A)$ is r.e. and $\psi^{-1}(A)$ is r.e. *Hint.* Write $\psi(A)$ in Σ_1-form.

1.18. Prove that if f is (total) recursive, then graph f is recursive.

1.19. A function f is *increasing* if $f(x) < f(x+1)$ for all x. Show that an infinite set A is recursive iff A is the range of an increasing recursive function.

1.20. Prove that any infinite r.e. set is the range of a 1:1 recursive function.

1.21. Prove that every infinite r.e. set contains an infinite recursive subset.

1.22. A set A is *co-r.e.* (or equivalently Π_1) if $\overline{A}$ is r.e. Use Exercise I.4.22 to prove that the reduction principle Corollary 1.10 fails for Π_1 sets.

1.23. The *separation principle* holds for a class C of sets if for every $A, B \in C$ such that $A \cap B = \emptyset$ there exists C such that $C, \overline{C} \in C$, $A \subseteq C$, and $B \subseteq \overline{C}$. By Exercise I.4.22, the separation principle fails for r.e. sets. Use Corollary 1.10 to show that the separation principle holds for co-r.e. sets.

1.24. Prove that if $A \leq_1 B$ and A and B are r.e. and A is infinite then $A \leq_1 B$ via some f such that $f(A) = B$.

1.25. Show that the proof of Theorem 1.8 is uniform in e in the sense that there is a p.r. function $\psi(e, y)$ such that if $W_e \neq \emptyset$ then $\lambda y \psi(e, y)$ is total and $W_e = \{\, \psi(e, y) : y \in \omega \,\}$. This uniformity is needed, for example, in Exercise 2.11.

2. *Uniformity and Indices for Recursive and Finite Sets*

In the Union Theorem 1.9 we gave an effective procedure f which given indices x, y for the sets mentioned in the hypothesis produced an index $f(x, y)$ for the set mentioned in the conclusion. A theorem will be said to hold *uniformly* if such an effective procedure exists. Procedures may also be uniform for some proper subset S of ω, for example the set of x such that $W_x \neq \emptyset$ (see Exercise 1.25) or such that W_x is recursive (see Theorem 2.2). Formal proofs of uniformity usually follow by the *s-m-n* Theorem. In case we are dealing with an r.e. set A which happens to be recursive or finite it is convenient to introduce certain stronger indices which give more information about A.

2.1 Definition. (i) We say that e is a Σ_1-index (r.e. index) for a set A if $A = W_e = \{ x : (\exists y) \, T(e, x, y) \}$.

(ii) $\langle e, i \rangle$ is a Δ_1-index for a recursive set A if $A = W_e$ and $\overline{A} = W_i$.

(iii) e is a Δ_0-index (*characteristic index*) for A if φ_e is the characteristic function for A.

Clearly, from the Δ_0 index for a recursive set A we can effectively obtain the Δ_1 index for A. The proof of the Complementation Theorem 1.12 allows us to pass uniformly effectively from a Δ_1-index for A to a Δ_0-index for A. However, we cannot, in general, pass effectively from a Σ_1-index for a recursive set A to a Δ_1-index for A, as the following theorem establishes.

2.2 Theorem. *There is no p.r. function ψ such that if $W_x = A$ and A is recursive then $\psi(x)$ converges and $W_{\psi(x)} = \overline{A}$. (There is no uniformly effective way to pass from Σ_1-indices to Δ_0-indices for recursive sets.)*

Proof. Define the recursive function f by

$$W_{f(x)} = \begin{cases} \omega & \text{if } x \in K, \\ \emptyset & \text{otherwise.} \end{cases}$$

Now

$$x \in K \implies W_{f(x)} = \omega \implies W_{\psi f(x)} = \emptyset,$$

and

$$x \notin K \implies W_{f(x)} = \emptyset \implies W_{\psi f(x)} = \omega.$$

Hence

$$x \in \overline{K} \iff W_{\psi f(x)} \neq \emptyset \iff (\exists y) \, (\exists s) \, [y \in W_{\psi f(x), s}],$$

so $\overline{K}$ is Σ_1 and hence r.e., contradicting Corollary 1.13. Therefore no such ψ exists. ∎

2.3 Corollary. *The recursive sets are closed under $\cup$, $\cap$, and complementation. The closure under $\cup$ and $\cap$ is uniformly effective with respect to both Σ_1 and Δ_1-indices. The closure under complementation is uniformly effective with respect to Δ_1-indices only.* ∎

A finite set, being recursive, has both a Σ_1-index and a Δ_0-index. Yet even the latter does not effectively specify the maximum element of A or even the cardinality of A. Thus, for each finite set we introduce a third index which explicitly specifies all the elements of A.

2.4 Definition. (i) Given a finite set $A = \{ x_1, x_2, \ldots, x_k \}$, where $x_1 < x_2 < \cdots < x_k$, the number $y = 2^{x_1} + 2^{x_2} + \cdots + 2^{x_k}$ is the *canonical index* of A. Let D_y denote finite set with canonical index y, and D_0 denote $\emptyset$.

(ii) A sequence $\{ D_{f(x)} \}_{x \in \omega}$ for some recursive function f is called a *recursive sequence* or a *strong array* of finite sets.

If y is written in binary expansion then the elements of D_y are the positions where the digit 1 occurs. For example, the binary expansion of 5 is 101 and $D_5 = \{0, 2\}$. Clearly, there are recursive functions f and g such that $f(y) = \max\{z : z \in D_y\}$ and $g(y) = |D_y|$. However, there is no p.r. function ψ such that if φ_x is the characteristic function of D_y, then $\psi(x)$ converges and $\psi(x) = |D_y|$. (If ψ exists, define $\varphi_{f(x)}(s) = 1$ if $x \in K_{s+1} - K_s$, and $\varphi_{f(x)}(s) = 0$ otherwise. Thus $\psi \circ f$ is actually the characteristic function of K.) It follows that there is no effective way of passing from the Δ_0-index of a finite set to its canonical index.

2.5 Definition. (i) A sequence $\{V_n\}_{n\in\omega}$ of r.e. sets is *uniformly r.e. (u.r.e.)*, also called *simultaneously r.e. (s.r.e.)*, if there is a recursive function f such that $V_n = W_{f(n)}$, for all n.

(ii) A sequence $\{V_n\}_{n\in\omega}$ of recursive sets is *uniformly recursive* if there is a recursive function $g(x, n)$ such that $\lambda x[g(x, n)]$ is the characteristic function of V_n, for all n. (See also Definition III.2.4 for the relativized version of these definitions.)

2.6 Convention. From now on we assume that we have defined $\varphi_{e,s}$ and $W_{e,s}$ using Exercise I.3.11 in place of Definition I.3.8 so that we have condition (3.3) of Chapter I, namely, that at most one set W_e receives a new element at any stage s. (This will be used often from now on. It could also be achieved by defining $W_{e,s}$ using Exercise 2.10(d).)

2.7 Definition. A *recursive enumeration* (usually called simply an *enumeration*) of an r.e. set A consists of a strong array $\{A_s\}_{s\in\omega}$ (of finite sets) such that $A_s \subseteq A_{s+1}$ and $A = \bigcup_s A_s$.

For example, $\{W_{e,s}\}_{s\in\omega}$ is an enumeration of W_e, as we shall see in Exercise 2.12, although an r.e. set A may have an enumeration $\{A_s\}_{s\in\omega}$ which is not of the form $W_{i,s}$ for any i.

2.8 Definition. (i) A *simultaneous (recursive) enumeration* of a u.r.e. sequence $\{V_n\}_{n\in\omega}$ of r.e. sets is a strong array $\{V_{n,s}\}_{n,s\in\omega}$ such that for all s and $n \in \omega$,

(1) $V_{n,s} \subseteq V_{n,s+1}$,
(2) $|V_{n,s+1} - V_{n,s}| \le 1$, and
(3) $V_n = \bigcup_{s\in\omega} V_{n,s}$.

(ii) A *standard enumeration* (of the r.e. sets) is a simultaneous enumeration of $\{V_n\}_{n\in\omega}$ where $\{V_n\}_{n\in\omega}$ is some acceptable numbering of the r.e. sets as defined in Exercise I.5.9.

For example, an easy way to give a simultaneous enumeration of any u.r.e. sequence $\{V_n\}_{n\in\omega}$ is to choose a 1:1 recursive function f with range $\{\langle x, n\rangle : x \in V_n\}$ and to define

$$V_{n,s} = \{\, x : (\exists t < s)\, [f(t) = \langle x, n\rangle]\,\}.$$

It is not strictly necessary to adopt condition (2) of Definition 2.8(i) in order for Definition 2.9 to make sense and for later constructions to work, but it is very helpful, and so we have adopted it in Convention 2.6 and in Definition 2.8.

2.9 Definition. Let $\{\, X_s\,\}_{s\in\omega}$ and $\{\, Y_s\,\}_{s\in\omega}$ be recursive enumerations of r.e. sets X and Y.

(i) Define $X \setminus Y = \{\, z : (\exists s)\, [z \in X_s - Y_s]\,\}$, the elements enumerated in X before (if ever) being enumerated in Y.

(ii) Define $X \searrow Y = (X \setminus Y) \cap Y$, the elements enumerated in X and later in Y.

(Do not confuse $X \setminus Y$ with $X - Y$, which denotes $X \cap \overline{Y}$. Notice that the definitions $X \setminus Y$ and $X \searrow Y$ depend not only upon the *sets* X and Y but also upon the particular recursive *enumerations* $\{\, X_s\,\}_{s\in\omega}$ and $\{\, Y_s\,\}_{s\in\omega}$.

We shall use this notation in Exercise 2.10, in the proof of Theorem X.2.1, and in Chapter XV on automorphisms.

2.10–2.13 Exercises

2.10.   (a) Given recursive enumerations $\{\, X_s\,\}_{s\in\omega}$ and $\{\, Y_s\,\}_{s\in\omega}$ of r.e. sets X and Y prove that both $X \setminus Y$ and $X \searrow Y$ are r.e. sets.

(b) Prove that $X \setminus Y = (X - Y) \cup (X \searrow Y)$.

(c) Prove that if $X - Y$ is nonrecursive then $X \searrow Y$ is infinite.

(d) Give an alternative proof of Corollary 1.10 by letting $A_1 = W_x \setminus W_y$ and $B_1 = W_y \setminus W_x$ where $W_x = A$ and $W_y = B$.

(e) Let f be a 1:1 recursive function from ω onto $K_0 = \{\, \langle x, e\rangle : x \in W_e\,\}$. Define

(2.1) $W_{e,s} = \{\, x : (\exists t \le s)\, [f(t) = \langle x, e\rangle]\,\}.$

Show that $\{W_{e,s} : e, s \in \omega\}$ satisfies condition I (3.3).

2.11.   Prove that there is a recursive function f such that $\{W_{f(n)}\}_{n\in\omega}$ consists precisely of the recursive sets. (Hence we can give an effective list of Σ_1-indices for the recursive sets but not of Δ_1-indices for them.)   *Hint.* Obtain $W_{f(n)} \subseteq W_n$ by enumerating W_n, placing in $W_{f(n)}$ only those elements enumerated in increasing order, and applying Exercise 1.19. Note that we are using the uniformity shown in Exercise 1.25. (See also Exercise 3.17.)

2.12.   Prove that there is a recursive function $f(e, s)$ such that $D_{f(e,s)} = W_{e,s}$ and hence that $W_e = \bigcup_s D_{f(e,s)}$.

2.13.   Prove that there are recursive functions f and g such that $D_x \cup D_y = D_{f(x,y)}$ and $D_x \cap D_y = D_{g(x,y)}$.

3. The Recursion Theorem

The following theorem of Kleene known as the *Recursion Theorem* or *Fixed Point Theorem* is one of the most elegant and important results in the subject. In spirit it is a strong extension of schema (V) for the primitive recursive functions because it allows us to conclude that certain implicitly defined functions are actually recursive. The proof, which uses only the *s-m-n* Theorem, is very short but seems at first somewhat mysterious.

3.1 Recursion Theorem (Kleene).   *For every recursive function f there exists an n (called a* fixed point *of f) such that $\varphi_n = \varphi_{f(n)}$.*

Proof. Define the recursive "diagonal" function $d(u)$ by

$$(3.1) \qquad \varphi_{d(u)}(z) = \begin{cases} \varphi_{\varphi_u(u)}(z) & \text{if } \varphi_u(u) \text{ converges;} \\ \text{undefined} & \text{otherwise.} \end{cases}$$

Note that d is 1:1 and total by the *s-m-n* Theorem. Note also that d is independent of f.

Given f, choose an index v such that

$$(3.2) \qquad\qquad\qquad \varphi_v = f \circ d$$

We claim that $n = d(v)$ is a fixed point for f. First note that f total implies $fd = \varphi_v$ is total, so $\varphi_v(v)$ converges and $\varphi_{d(v)} = \varphi_{\varphi_v(v)}$. Now

$$(3.3) \qquad\qquad \varphi_n = \varphi_{d(v)} = \varphi_{\varphi_v(v)} = \varphi_{fd(v)} = \varphi_{f(n)}.$$

The second equality follows from (3.1) and the third equality follows from (3.2).

3.2 Corollary.   *For every recursive function f, there exists n such that $W_n = W_{f(n)}$.*   ▯

The above proof is best visualized as a diagonal argument that fails [Owings, 1973]. In a typical diagonal argument there is a square array of objects $\{\,\alpha_{x,u}\,\}_{x,u\in\omega}$ and one constructs a sequence $D' = \{\,\alpha'_x\,\}_{x\in\omega}$ such that $\alpha'_x \neq \alpha_{x,x}$ where $D = \{\,\alpha_{x,x}\,\}_{x\in\omega}$ is the diagonal sequence, and hence D' is *not* one of the rows, $R_u = \{\,\alpha_{x,u}\,\}_{x\in\omega}$. Now consider the matrix where $\alpha_{x,u} =$

$\varphi_{\varphi_u(x)}$, and where it is understood that $\alpha_{x,u}$ and $\varphi_{\varphi_u(x)}$ denote the totally undefined function if $\varphi_u(x)$ diverges. Here the strong closure properties of the partial recursive functions under the s-m-n Theorem guarantee that the diagonal sequence $D = \{\alpha_{x,x}\}_{x \in \omega}$ is one of the rows, namely the eth row, $R_e = \{\varphi_{\varphi_e(x)}\}_{x \in \omega}$, where $\varphi_e = d$. Now any recursive function f induces a transformation on the rows $R_u = \{\varphi_{\varphi_u(x)}\}_{x \in \omega}$ of this matrix, mapping R_u to the row $\{\varphi_{f\varphi_u(x)}\}_{x \in \omega}$. In particular, f maps the "diagonal" row $R_e = \{\varphi_{d(x)}\}_{x \in \omega}$ to $R_v = \{\varphi_{fd(x)}\}_{x \in \omega}$. Since R_e is the diagonal sequence, the vth element of the sequence, namely $\varphi_{d(v)} = \varphi_{\varphi_v(v)}$, must be unchanged by this action of f, and hence $\varphi_{d(v)} = \varphi_{fd(v)}$.

A typical application of the Recursion Theorem is that there exists n such that $W_n = \{n\}$. (By the s-m-n Theorem define $W_{f(x)} = \{x\}$, and by the Recursion Theorem choose n such that $W_n = W_{f(n)} = \{n\}$.) Similarly, one can show that there exists n such that $\varphi_n = \lambda x[n]$. We stated the Recursion Theorem above in its simplest form although the proof actually yields considerably more information which will be useful and which we now state explicitly.

3.3 Proposition. *In the Recursion Theorem, n can be computed from an index for f by a 1:1 recursive function g.*

Proof. Let $v(x)$ be a recursive function such that $\varphi_{v(x)} = \varphi_x \circ d$. Let $g(x) = d(v(x))$. Both d and v are 1:1 by the s-m-n Theorem. ▯

3.4 Proposition. *In the Recursion Theorem, there is an infinite r.e. set of fixed points for f.*

Proof. By the Padding Lemma I.3.2 there is an infinite r.e. set V of indices v such that $\varphi_v = f \circ d$, but d is 1:1 so $\{d(v)\}_{v \in V}$ is infinite and r.e. ▯

For more sophisticated applications such as Myhill's Theorem in §4 we need the following form of the Recursion Theorem with parameters. (By coding finite sequences it suffices to consider a single parameter.) The proof exploits the uniformity of the earlier proof and is otherwise identical.

3.5 Recursion Theorem with Parameters (Kleene). *If $f(x, y)$ is a recursive function, then there is a recursive function $n(y)$ such that $\varphi_{n(y)} = \varphi_{f(n(y),y)}$.*

Proof. Define a recursive function d by

$$\varphi_{d(x,y)}(z) = \begin{cases} \varphi_{\varphi_x(x,y)}(z) & \text{if } \varphi_x(x, y) \text{ converges,} \\ \text{undefined} & \text{otherwise.} \end{cases}$$

Choose v such that $\varphi_v(x, y) = f(d(x, y), y)$. Then $n(y) = d(v, y)$ is a fixed point, since $\varphi_{d(v,y)} = \varphi_{\varphi_v(v,y)} = \varphi_{f(d(v,y),y)}$. The first equality follows from the definition of d, and the second from the definition of v. ▯

Notice that the uniformities of Propositions 3.3 and 3.4 apply to Theorem 3.5 as well. Informally, the Recursion Theorem allows us to define a p.r. function φ_n (or an r.e. set W_n) using its own index n in advance as part of the algorithm, $\varphi_n(z) : \dots n \dots$. This circularity is removed by the Recursion Theorem because we are really using the s-m-n Theorem to define a function $f(x)$, $\varphi_{f(x)}(z) : \dots x \dots$, and then taking a fixed point, $\varphi_n(z) = \varphi_{f(n)}(z) :$ $\dots n \dots$. The only restriction on this informal method is that we cannot use in the program any special properties of φ_n (such as φ_n being total, or $W_n \neq \emptyset$). For example, if for all x the function $\varphi_{f(x)}$ being defined is total, then, of course, the fixed point $\varphi_{f(n)} = \varphi_n$ will be total. However, the instructions for $\varphi_{f(x)}$ must not say "wait until $\varphi_x(z)$ converges, take the value $v = \varphi_x(z)$ and do $\dots$". Rather, the instructions must apply to an arbitrary φ_x.

This informal use of the Recursion Theorem is very common in the literature, and considerably clarifies an otherwise notationally cumbersome proof. We illustrate this method by giving an alternative proof of Theorem 2.2. (Theorem 2.2 is equivalent to Theorem 3.6 by the remark preceding the former.)

3.6 Theorem.   There is no p.r. function ψ such that if W_x is recursive then $\psi(x)$ converges and $\varphi_{\psi(x)}$ is the characteristic function for W_x. (There is no uniformly effective way to pass from Σ_1 to Δ_0-indices for recursive sets.)

Proof. Using the Recursion Theorem define a recursive set

$$W_n = \begin{cases} \{0\} & \text{if } \psi(n)\downarrow \text{ and } \varphi_{\psi(n)}(0)\downarrow = 0, \\ \emptyset & \text{otherwise.} \end{cases}$$

Now $\varphi_{\psi(n)}$ cannot be the characteristic function of W_n because $0 \in W_n$ iff $\varphi_{\psi(n)}(0) = 0$.   ▯

The following theorem strengthens Theorem 3.5 by replacing the total function $f(x, y)$ by a partial function $\psi(x, y)$ using the same proof. This form will be useful later.

3.7 Theorem.   If $\psi(x, y)$ is a partial recursive function, then there is a recursive function $n(y)$ such that

$$(\forall y) \; [\psi(n(y), y)\downarrow \implies \varphi_{n(y)} = \varphi_{\psi(n(y),y)}].$$

Proof. Use the same proof as for Theorem 3.5.   ▯

3.8–3.18 Exercises

3.8.   A set A is *self-dual* if $A \leq_m \overline{A}$. For example if $A = B \oplus \overline{B}$ then A is self-dual.

(a) Use the Recursion Theorem to prove that no index set A can be self-dual.

(b) Give a short proof of Rice's Theorem I.4.13.

3.9. Show that for any p.r. function $\psi(x, y)$ there is an n such that $\varphi_n(y) = \psi(n, y)$.

3.10. Show that Corollary 3.2 is equivalent to: For every r.e. set A, $(\exists n)\,[W_n = \{\, x : \langle x, n \rangle \in A \,\}]$. *Hint.* Define the r.e. set

$$A_n = \{\, x : \langle x, n \rangle \in A \,\}$$

and compare the sequence $\{\, A_n \,\}_{n \in \omega}$ with $\{\, W_n \,\}_{n \in \omega}$.

3.11. Use the informal technique in Theorem 3.6 to show that there is no p.r. function φ_e such that if φ_x is the characteristic function of a finite set F, then $\varphi_e(x) \downarrow = \max(F)$. *Hint.* Define $\varphi_n(t + 1) = 1$ if $t = (\mu s)\,[\varphi_{e,s}(n) \downarrow\,]$, and $\varphi_n(x) = 0$ otherwise and apply line (3.1) of Chapter I.

3.12. Show that there is no recursive function $f(x, s)$ such that for all x, $\hat{f}(x) = \lim_s f(x, s)$ exists, and $\hat{f}(x)$ is the characteristic function for Tot. (See Definition III.3.1.) *Hint.* Assume otherwise and define $\varphi_n(s)$ using $\{\, f(n, t) \,\}_{t \geq s}$ so that $\varphi_n(s) \downarrow$ iff $f(n, t) = 0$ for some $t \geq s$.

3.13. Let f be a recursive function. Show that there is an n such that: (i) W_n is recursive; and (ii) $(\mu y)\,[W_y = \overline{W}_n] > f(n)$. *Hint.* Enumerate a finite set W_n by first computing $f(n)$ and then defining $W_n = K \upharpoonright (f(n) + 1)$, where $A \upharpoonright x$ denotes the restriction of A to elements $y < x$.

3.14. (a) Use the recursive permutation p of Exercise I.5.9 to prove that the Recursion Theorem holds for an acceptable numbering $\{\, \psi_e \,\}_{e \in \omega}$ iff it holds for the standard numbering $\{\, \varphi_e \,\}_{e \in \omega}$.

(b) Let ψ_e be the partial recursive function whose graph is obtained by applying the Uniformization Theorem 1.6 to the relation $R_e = \{\, (x, y) : \langle x, y \rangle \in W_e \,\}$, and let $W_{h(e)} = \{\, \langle x, y \rangle : \psi_e(x) = y \,\}$. (We say that W_e is *single-valued* if R_e is a single-valued relation.) Prove that $\{\, \psi_e \,\}_{e \in \omega}$ is an acceptable numbering.

(c) Use (a) and (b) to prove that Corollary 3.2 implies Theorem 3.1. *Hint.* Given f, apply Corollary 3.2 to the function $h \circ f \circ h$ to obtain n such that $\psi_n = \psi_{f(n)}$. (Note that we cannot apply Corollary 3.2 merely to f and then apply h because $W_e = W_i$ does not necessarily imply that $W_{h(e)} = W_{h(i)}$.)

3.15 (Double Recursion Theorem—Smullyan).

(a) Prove that for any recursive functions $f(x, y)$ and $g(x, y)$ there exist a and b such that $\varphi_a = \varphi_{f(a,b)}$ and $\varphi_b = \varphi_{g(a,b)}$. *Hint.* Apply Theorem 3.5

to get a recursive function $\hat{a}(y)$ such that $\varphi_{\hat{a}(y)} = \varphi_{f(\hat{a}(y),y)}$. Let b be a fixed point for the recursive function $h(y)$, where $\varphi_{h(y)} = \varphi_{g(\hat{a}(y),y)}$. Let $a = \hat{a}(b)$.

(b) Use exactly the same method to prove the Double Recursion Theorem with Parameters, namely for any recursive functions $f(x,y,z)$ and $g(x,y,z)$ there exist recursive functions $a(z)$ and $b(z)$ such that

$$\varphi_{a(z)} = \varphi_{f(a(z),b(z),z)} \text{ and } \varphi_{b(z)} = \varphi_{g(a(z),b(z),z)}.$$

3.16. Use the Recursion Theorem to prove that K is not an index set.

3.17 (Lachlan).   Let $\{\, Z_i^s \,\}_{i,s\in\omega}$ be a standard enumeration of some acceptable numbering $\{\, Z_i \,\}_{i\in\omega}$ of the r.e. sets as defined in Definition 2.8. Given the usual enumeration $\{W_{i,s}\}_{i,s\in\omega}$ satisfying Convention 2.6, prove that there is a recursive function h such that for all i, $W_i = Z_{h(i)}$ and the distinct elements of W_i are enumerated in the same order under the enumeration $\{W_{i,s}\}_{i,s\in\omega}$ as under the enumeration $\{\, Z_{h(i),s} \,\}_{i,s\in\omega}$. *Hint.* Since $\{\, Z_i \,\}_{i\in\omega}$ is an acceptable numbering of the r.e. sets we have by Exercise I.5.9 a recursive function g such that $W_i = Z_{g(i)}$ for all i. Use the Recursion Theorem with Parameters to define the recursive function $f(i)$ as follows. If $W_i = \{\, a_1, a_2, \dots \,\}$ enumerated in that order (under $\{W_{i,s}\}_{s\in\omega}$) then enumerate a_1 in $W_{f(i)}$. Later enumerate a_{n+1} in $W_{f(i)}$ iff there is a stage s such that $Z_{g(f(i)),s} = \{\, a_1, a_2, \dots, a_n \,\}$ and the a_i, $i \leq n$, were enumerated in $Z_{g(f(i))}$ in that order (under $\{\, Z_{g(f(i)),s} \,\}_{s\in\omega}$). Let $h(i) = g(f(i))$.

3.18 (Lachlan).   Let $\{\, Z_i^s \,\}_{i,s\in\omega}$ be as in Exercise 3.17. Define recursive sets V_i by:

$$V_i = \{\, n : (\exists s)\, [n \in Z_i^{s+1} - Z_i^s \ \& \ n > \max(Z_i^s)]\,\}.$$

(See also Exercises 1.19 and 2.11.)

(a) Using Exercise 3.17 prove that $\{V_i\}_{i\in\omega}$ is unique in the following sense. If $\{\, \tilde{Z}_i^s \,\}_{i,s\in\omega}$ is another s.r.e. array satisfying (1)–(3) of Exercise 3.17, and $\{\tilde{V}_i\}_{i\in\omega}$ is the corresponding sequence of recursive sets, then there is a recursive isomorphism h of ω such that $V_i = \tilde{V}_{h(i)}$.

(b) Prove that if f is any recursive function, then for some n, $V_n = V_{f(n)}$, namely, the Recursion Theorem holds for the recursive sets $\{V_n\}_{n\in\omega}$.

4. Complete Sets, Productive Sets, and Creative Sets

In Chapter I §3 we defined the reducibilities $\leq_m$ and $\leq_1$. In Chapter III we shall define the more general Turing reducibility $\leq_T$.

4.1 Definition. Let $r = 1$, m or T. A set A is *r-complete* if A is r.e. and $W \leq_r A$ for every r.e. set W.

4.2 Theorem. *The sets K, K_0, and K_1 are all 1-complete (and hence $K_0 \equiv K_1 \equiv K$ by Myhill's Theorem I.5.4).*

Proof. Each is r.e. by (1.1), (1.2), and (1.3). Furthermore, $K_0 = \{ \langle x, y \rangle : x \in W_y \}$ is clearly 1-complete, since $x \in W_y$ iff $\langle x, y \rangle \in K_0$. Let W be any r.e. set. Now as in Theorem I.4.10 define a 1:1 recursive function f by

$$W_{f(x)} = \begin{cases} \omega & \text{if } x \in W, \\ \emptyset & \text{otherwise.} \end{cases}$$

Now if $x \in W$, then $W_{f(x)} = \omega$ and $f(x) \in K \cap K_1$. If $x \in \overline{W}$ then $W_{f(x)} = \emptyset$, so $f(x) \notin (K \cup K_1)$. Thus, f witnesses that $W \leq_1 K$ and $W \leq_1 K_1$.

4.3 Definition. (i) A set P is *productive* if there is a p.r. function $\psi(x)$, called a *productive function* for P, such that

$$(\forall x) \, [W_x \subseteq P \implies [\psi(x)\!\downarrow \quad \& \quad \psi(x) \in P - W_x]].$$

(ii) An r.e. set C is *creative* if $\overline{C}$ is productive.

For example, the set K is creative since $\overline{K}$ is productive via the identity function $\psi(x) = x$. Since $K \equiv K_0 \equiv K_1$ by Theorem 4.2, we know that K_0 and K_1 are also creative. (Note that if $W_x \subset \overline{K}$ then $x \notin W_x$ and $x \notin K$.) A creative set C is "effectively nonrecursive" in the sense that for any candidate W_x for $\overline{C}$, $\psi(x)$ is an effective counterexample; namely $\psi(x) \in \overline{C} - W_x$. These sets were called creative by Post [1944] because their existence justifies "the generalization that every symbolic logic is incomplete and extendible relative to the class of propositions" necessary to code the relation $\{ (x, y) : x \in W_y \}$. Post remarks: "The conclusion is inescapable that even for such a fixed, well-defined body of mathematical propositions, *mathematical thinking is, and must remain, essentially creative.*"

4.4 Theorem. *Any productive set P has a 1:1 total recursive productive function p.*

Proof. Let P be productive via ψ. First obtain a *total* productive function q for P as follows. Define a recursive function g such that

$$W_{g(x)} = \begin{cases} W_x & \text{if } \psi(x)\!\downarrow, \\ \emptyset & \text{otherwise.} \end{cases}$$

Define $q(x)$ to be either $\psi(x)$ or $\psi(g(x))$, whichever converges first. Now if $W_x \subseteq P$, then $\psi(x)$ converges and $W_{g(x)} = W_x$ so both $\psi(g(x))$ and $\psi(x)$ are in $P - W_x$.

Now convert q to a 1:1 productive function p. Let $W_{h(x)} = W_x \cup \{ q(x) \}$. Note that

(4.1) $W_x \subseteq P \implies W_{h(x)} \subseteq P.$

Define $p(0) = q(0)$. To compute $p(x+1)$, enumerate the set $\{ q(x+1), qh(x+1), qh^2(x+1), \dots \}$ until either: some y not in $\{ p(0), \dots, p(x) \}$ is found; or a repetition occurs. In the former case set $p(x+1) = y$. In the latter case, $W_{x+1} \not\subseteq P$ by (4.1), and we can set $p(x+1) = (\mu y)\, [y \notin \{ p(0), \dots, p(x) \}].$ ◻

4.5 Theorem. (i) *If P is productive then P is not r.e.*
   (ii) *If P is productive then P contains an infinite r.e. set W.*
   (iii) *If P is productive and $P \leq_m A$ then A is productive.*

Proof. (i) Immediate.
   (ii) Let $W_n = \emptyset$, and $W_{h(x)} = W_x \cup \{ p(x) \}$. Define

$$W = \{ p(n), ph(n), ph^2(n), \dots \}.$$

   (iii) Let $P \leq_m A$ via f, and p a productive function for P. Let $W_{g(x)} = f^{-1}(W_x)$. Then fpg is a productive function for A.   ◻

Using the fact that $\overline{K}$ is productive, Theorem 4.5(iii) gives us a method for exhibiting new productive sets P by showing that $\overline{K} \leq_m P$. We now use the Recursion Theorem to prove that *every* productive set has this property. An immediate corollary is that all creative sets are 1-complete and hence recursively isomorphic to K.

4.6 Theorem (Myhill [1955]). (i) *If P is productive then $\overline{K} \leq_1 P$.*
   (ii) *If C is creative then C is 1-complete and $C \equiv K$.*

Proof. (i) Let p be a total 1:1 productive function for P. Define the recursive function f by

$$W_{f(x,y)} = \begin{cases} \{ p(x) \} & \text{if } y \in K, \\ \emptyset & \text{otherwise.} \end{cases}$$

By the Recursion Theorem with Parameters 3.5, there is a 1:1 recursive function $n(y)$ such that

(4.2) $W_{n(y)} = W_{f(n(y),y)} = \begin{cases} \{ p(n(y)) \} & \text{if } y \in K, \\ \emptyset & \text{otherwise.} \end{cases}$

Now

(4.3) $y \in K \implies W_{n(y)} = \{ pn(y) \} \implies W_{n(y)} \not\subseteq P \implies pn(y) \in \overline{P},$

and

(4.4) $y \in \overline{K} \implies W_{n(y)} = \emptyset \implies W_{n(y)} \subseteq P \implies pn(y) \in P.$

In both lines the first implication follows from (4.2). The fact that p is a productive function for P yields the second implication in (4.3) and the third implication in (4.4). But (4.3) and (4.4) yield $\overline{K} \leq_1 P$ via the function $\lambda y[pn(y)]$.

Part (ii) follows at once by (i) and the Isomorphism Theorem I.5.4.   ▯

4.7 Corollary.   *The following are equivalent:*

(i) P *is productive;*
(ii) $\overline{K} \leq_1 P$;
(iii) $\overline{K} \leq_m P$.   ▯

4.8 Corollary.   *The following are equivalent:*

(i) C *is creative;*
(ii) C *is 1-complete;*
(iii) C *is m-complete.*   ▯

The following definition will be useful in Exercise 4.15 below and in Chapter IV.

4.9 Definition.   Let (A_1, A_2) and (B_1, B_2) be two pairs of sets such that $A_1 \cap A_2 = \emptyset = B_1 \cap B_2$. Then $(B_1, B_2) \leq_m (A_1, A_2)$ if there is a recursive function f such that $f(B_1) \subseteq A_1$, $f(B_2) \subseteq A_2$ and $f(\overline{B_1 \cup B_2}) \subseteq \overline{A_1 \cup A_2}$. We write "$\leq_1$" if f is 1:1.

4.10–4.17 Exercises

4.10.   Let $A_x = \{\, y : \varphi_y = \varphi_x \,\}$.

(a) Prove that for each x, A_x is productive. *Hint.* Combine Corollary 4.7 and Exercise I.4.20.

(b) (Fejer). Show that this reduction $\overline{K} \leq_m A_x$ is not uniform in x, namely there is no recursive function $f(x, y)$ such that for all x, $y \in \overline{K}$ iff $f(x, y) \in A_x$. *Hint.* Choose $y_0 \in K$, consider $\lambda x[f(x, y_0)]$, and use the Recursion Theorem on f.

4.11.   Use the Recursion Theorem to give an alternative proof of Exercise I.5.9 (that any two acceptable numberings of the p.r. functions differ by an appropriate recursive permutation) as follows. To achieve this prove that for every recursive function g which satisfies the hypotheses of Exercise I.5.9 there is a recursive function g_1 such that for all x, $\psi_{g(x)} = \psi_{g_1(x)}$ and $g_1(x + 1) > g_1(x)$. *Hint.* To obtain g_1 first define a recursive function h such that

(i) $(\forall x)\, [\varphi_{h(e,x)} = \varphi_e]$; and
(ii) $(\forall x)\, (\forall y)\, [x \neq y \implies gh(e, x) \neq gh(e, y)]$.

Fixing e define $h(e, k)$ by induction on k. Define $h(e, 0) = e$. Set $B_k = \{ gh(e, 0), \ldots, gh(e, k) \}$. Use the Recursion Theorem to define

$$\varphi_n(z) = \begin{cases} \varphi_e(z) & \text{if } g(n) \notin B_k \\ \text{undefined} & \text{otherwise.} \end{cases}$$

If $g(n) \notin B_k$, define $h(e, k+1) = n$. Otherwise, $\varphi_e = \varphi_n = \lambda z[\uparrow]$, and $h(e, k+1)$ can be defined using Exercise 4.10(a).

4.12. Prove that if $K \leq_m A$ via f and A is not r.e. then there are $2^{\aleph_0}$ sets B such that $K \leq_m B$ via f. Hence there are $2^{\aleph_0}$ productive sets. *Hint.* If A is not r.e. then $S = A - f(K)$ is infinite.

4.13. A disjoint pair (A, B) of r.e. sets is *effectively inseparable* if there is a partial recursive function ψ, called a *productive* function for (A, B), such that for all x and y

$$[A \subseteq W_x \ \& \ B \subseteq W_y \ \& \ W_x \cap W_y = \emptyset]$$
$$\implies [\psi(\langle x, y \rangle) \downarrow \ \& \ \psi(\langle x, y \rangle) \notin W_x \cup W_y].$$

(a) Use the method of Theorem 4.4 to prove that we may assume ψ is total and 1:1.

(b) Prove that the following r.e. sets are effectively inseparable, $A = \{ x : \varphi_x(x) = 0 \}$ and $B = \{ x : \varphi_x(x) = 1 \}$. *Hint.* Define ψ by

$$\varphi_{\psi(\langle x, y \rangle)}(z) = \begin{cases} 1 & \text{if } z \in W_x \setminus W_y \\ 0 & \text{if } z \in W_y \setminus W_x \\ \uparrow & \text{otherwise.} \end{cases}$$

Recall that $W_x \setminus W_y$ denotes those elements enumerated in W_x before W_y as defined in Definition 2.9.

(c) Show that if (A, B) is a pair of effectively inseparable r.e. sets then each of A and B is creative.

4.14 (Hay). Define A_x as in Exercise 4.10.

(a) Prove that if φ_a and φ_b are distinct p.r. functions then A_a and A_b are effectively inseparable. *Hint.* Use the Recursion Theorem to define $\varphi_{f(x,y)}(z) = \varphi_b(z)$ if $f(x, y) \in W_x \setminus W_y$ (using the notation of Definition 2.9), and $\varphi_{f(x,y)}$ is defined appropriately in the remaining cases.

(b) Suppose that φ_b properly extends φ_a. Prove that there is an r.e. set $W \supseteq A_b$, $W \cap A_a = \emptyset$, but that there is no r.e. set $V \supseteq A_a$, such that $V \cap A_b = \emptyset$.

4.15 (Smullyan). Assume that (A_1, A_2) is a pair of effectively inseparable r.e. sets with productive function p. Prove that if (B_1, B_2) is a pair of disjoint r.e. sets then $(B_1, B_2) \leq_1 (A_1, A_2)$ as defined in Definition 4.9. *Hint.* By the

Double Recursion Theorem with one parameter 3.15(b) define 1:1 recursive functions $g(z)$, $h(z)$ such that

$$W_{g(z)} = \begin{cases} A_1 \cup \{\, p(\langle g(z), h(z)\rangle)\,\} & \text{if } z \in B_2 \\ A_1 & \text{otherwise,} \end{cases}$$

and

$$W_{h(z)} = \begin{cases} A_2 \cup \{\, p(\langle g(z), h(z)\rangle)\,\} & \text{if } z \in B_1, \\ A_2 & \text{otherwise.} \end{cases}$$

Consider $f(z) = p(\langle g(z), h(z)\rangle)$, and show that $(B_1, B_2) \leq_1 (A_1, A_2)$ via f.

4.16° (Lachlan).   (a) Prove that if $K \leq_m A \times B$, and one of the sets is r.e. then $K \leq_m A$ or $K \leq_m B$.

(b) Conclude that there is no pair of incomparable r.e. m-degrees whose l.u.b. is the m-degree of K. (*Hint.* Note that $A \oplus B \leq_m A \times B$.)

Hint for (a). Assume that $K \leq_m A \times B$ and B is r.e. Construct an r.e. set D such that $D \leq_m A \times B$ and either $K \leq_m A$ or $K \leq_m D \leq_m B$. By the Recursion Theorem we may assume that at the start of the construction we have an index for D and hence recursive functions f, g such that $x \in D$ iff $f(x) \in A$ & $g(x) \in B$. Call x *good at stage* s if $x \notin D_s$ and $g(x) \in B_s$. Hence $x \in D$ iff $f(x) \in A$. Let $G = \{\, x_0, x_1, \ldots \,\}$ be a recursive enumeration of the elements good at some stage (called *good* elements). If G is infinite we shall arrange that $i \in K$ iff $x_i \in D$ so $i \in K$ iff $f(x_i) \in A$, and hence $K \leq_m A$. At stage s if $i \in K_s$ and x_i exists put x_i into D. In addition if $i \in K_s$ and i is not good at s, put i into D. If G is infinite then $i \in K$ iff $x_i \in D$, and $x_i \in D$ iff $f(x_i) \in A$ so $K \leq_m A$. If G is finite then for almost every x, $x \in D$ iff $g(x) \in B$, and $D =^* K$ (i.e., the symmetric difference of D and K is finite) so $K \leq_m D \leq_m B$. (This is our first example of a proof where we pursue two strategies simultaneously. Strategy σ_1 attempts to arrange $K \leq_m D \leq_m A$ whenever a good element appears. Meanwhile strategy σ_2 attempts to arrange $K =^* D \leq_m B$. The action of σ_2 to build a specific h reducing K to B is "injured" whenever a new good element appears, and σ_2 begins all over again on a new candidate for the function witnessing $D \leq_m B$.)

4.17.   (Jockusch-Mohrherr). Let A be any r.e. set except ω. Prove that A is creative iff

$$(\forall \text{ r.e. } B)\, [A \cap B = \emptyset \implies A \equiv A \cup B].$$

Chapter III

Turing Reducibility and the Jump Operator

The earlier reducibilities we have studied are now extended to the more general notion of Turing reducibility, which is intended to capture the notion of a set B being computable relative to another set A, and which will be one of our main objects of study. Most of the basic results of Chapter I §3 can be extended to this relativized form by virtually the same proofs. In §2 we consider the upper semi-lattice formed by the Turing degrees and introduce the jump operator, which is merely the relativization of Definition I.4.2 of the nonrecursive r.e. set K. These concepts are then related to one another in §3 where we prove the Modulus Lemma and Limit Lemma, which will be very useful in studying the r.e. degrees and the degrees below $\mathbf{0}'$.

1. Definitions of Relative Computability

Informally a set B is *recursive in (computable in)* a set A, written $B \leq_T A$, if there is an algorithm for deciding whether $x \in B$ given answers to any questions of the form "Is $y \in A$?". For example, if $B \leq_m A$ via f then $x \in B$ iff $f(x) \in A$. Although m-reducibility is the first and most natural reducibility, it is clearly too restrictive. For example, for any set A, $\overline{A}$ is obviously computable in A because $x \in \overline{A}$ iff $x \notin A$, yet no nonrecursive r.e. set satisfies $\overline{A} \leq_m A$. This example led Post [1944] to introduce certain intermediate reducibilities, such as $\leq_{btt}$ and $\leq_{tt}$, which we study in the exercises of Chapter V §2, along with certain others which have later proved useful. We turn now instead to the most important reducibility, $\leq_T$, for which we give two equivalent formal definitions by relativizing the two formal definitions from Chapter I §2.

1.1 Definition. Let $A \subseteq \omega$. A partial function ψ is *partial recursive in A* (*A-partial recursive*) if there is a derivation of ψ using schemata (I)–(VI) of Chapter I §2 with χ_A added to schemata (I)–(III) as a new initial function.

Perhaps more intuitive is the modified Turing machine definition, upon which our formal development will be based. An *oracle Turing machine* is simply a Turing machine with an extra "read only" tape, called the *oracle tape*, upon which is written the characteristic function of some set A (called the *oracle*), and whose symbols cannot be printed over. The old tape is called the *work tape* and operates just as before. The reading head moves along both tapes simultaneously. As before, let Q be a finite set of states, Σ_1 the oracle tape alphabet $\{\,B, 0, 1\,\}$, Σ_2 the work tape alphabet $\{\,B, 1\,\}$, and $\{\,R, L\,\}$ the head moving operations right and left. A *Turing program* is now simply a partial map

$$\delta : Q \times \Sigma_1 \times \Sigma_2 \longrightarrow Q \times \Sigma_2 \times \{\,R, L\,\},$$

where $\delta(q, a, b) = (p, c, X)$ indicates that the machine in state q reading symbol a on the oracle tape and symbol b on the work tape passes to state p, prints "c" over "b" on the work tape and moves one space right (left) on both tapes if $X = R$ $(X = L)$. The other details are just as in Chapter I §2 except that the reading head also starts on that square of the oracle tape which codes $\chi_A(0)$. See Diagram 1.1.

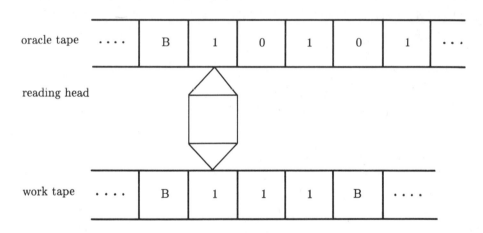

Diagram 1.1. Oracle Turing Machine

Let $y + 1$ be the number of nonblank cells on the oracle tape which are scanned by the reading head during the computation. (Namely, y is the maximum integer which is tested for membership in A.) We say that the elements $z \leq y$ are *used* in the computation. (See Definition 1.7.) (Note that the Turing machine can move to the left of its starting point so the total number of cells on the work tape scanned during a computation may be much greater than y.)

These new oracle Turing programs being finite sets of 6-tuples of the above symbols can be effectively coded as in Exercise I.2.11(b). Let $\hat{P}_e$ denote the eth such program under some effective coding. Note that $\hat{P}_e$ is *independent* of the oracle A.

1.2 Definition. (i) A partial function ψ is *Turing computable in A* (*A-Turing computable*), written $\psi \leq_T A$, if there is a program $\hat{P}_e$ such that if the machine has χ_A written on the oracle tape, then for all x and y, $\psi(x) = y$ iff $\hat{P}_e$ on input x halts and yields output y. In this case we write $\psi = \{e\}^A$, or equivalently $\psi = \Phi_e^A$, or $\psi = \Phi_e(A)$. We say $\psi(x)$ *diverges* (written $\psi(x)\!\uparrow$) iff $\hat{P}_e$ on input x never halts.

(ii) We also allow (total) functions as oracles by defining $\{e\}^f$ to be $\{e\}^A$ where $A = \text{graph}(f)$.

1.3 Theorem. *A partial function ψ is A-partial recursive iff ψ is A-Turing computable.*

We omit the proof of this theorem, which may be found in Kleene [1952a]. The coincidence of these formal definitions and numerous others is evidence for the relativized version of Church's Thesis, which asserts that ψ is *A-partial recursive iff ψ is computable in A* in the intuitive sense. From now on, we shall freely appeal to Church's Thesis, and we shall use the terms "*A-partial recursive*," "*A-Turing computable*," and "*A-computable*" synonymously.

The results of Chapter I easily relativize from partial recursive functions to *A*-partial recursive functions by the same proofs. For example,

1.4 Relativized Enumeration Theorem. *There exists $z \in \omega$ such that for all sets $A \subseteq \omega$ and for all $x, y \in \omega$ the A-partial recursive function $\Phi_z^A(x, y)$ satisfies $\Phi_z^A(x, y) = \Phi_x^A(y)$.*   ∎

1.5 Relativized s-m-n Theorem. *For every $m, n \geq 1$ there exists a 1:1 recursive function s_n^m of $m + 1$ variables such that for all sets $A \subseteq \omega$ and for all $x, y_1, y_2, \ldots, y_m \in \omega$,*

$$\Phi_{s_n^m(x,y_1,\ldots,y_m)}^A = \lambda z_1, \ldots, z_n[\Phi_x^A(y_1, \ldots, y_m, z_1, \ldots, z_n)].$$

(Note that s_n^m is a recursive function, rather than merely A-recursive.)

Proof. Take $m = n = 1$, and let s denote s_n^m. The new program $\hat{P}_{s(x,y)}$ on input z consists of applying program $\hat{P}_x$ to input (y, z). This defines $s(x, y)$ as a *recursive* function since the program $\hat{P}_{s(x,y)}$ is independent of the oracle A. Of course, s can be made 1:1 by padding as in Lemma I.3.2.   ∎

1.6 Relativized Recursion Theorem (Kleene). (i) *For all sets $A \subseteq \omega$ and all $x, y \in \omega$, if $f(x, y)$ is an A-recursive function, then there is a recursive function $n(y)$ such that $\Phi_{n(y)}^A = \Phi_{f(n(y),y)}^A$.*

(ii) *Furthermore, $n(y)$ does not depend upon the oracle A, namely if*

$$f(x, y) = \{e\}^A(x, y)$$

then the recursive function $n(y)$ can be found uniformly in e.

Proof. Apply the proof of Theorem II.3.5 but notice that $n(y)$ is a recursive function (not merely A-recursive) because the s-m-n function $s(x, y)$ in Theorem 1.5 is recursive, and hence the function $d(x, y)$ is recursive where $d(x, y)$ is obtained as in the proof of Theorem II.3.5 but with all partial functions relativized to A.   ☐

Note that the strings σ of 0's and 1's, namely $\sigma \in 2^{<\omega}$, are to be viewed as finite initial segments of characteristic functions. We identify a set A with its characteristic function and write $\sigma \subset A$ if $\sigma \subset \chi_A$ as a partial function, namely, $\sigma(x) = \chi_A(x)$ for all $x \in \text{dom } \sigma$. The *length* of σ, written $lh(\sigma)$, is $|\text{dom } \sigma|$. Note that $lh(\sigma) = \mu x[\sigma(x) \uparrow]$. Fix some effective coding of the strings $\sigma \in 2^{<\omega}$ and identify σ with its code number. Recall that $A \restriction x$ denotes the restriction of A to arguments $y < x$, and likewise for $\sigma \restriction x$. Note that if $n = lh(\sigma)$, then $\sigma = \sigma \restriction n$.

1.7 Definition. (i) We write $\{e\}_s^A(x) = y$ if $x, y, e < s$, $s > 0$, $\{e\}^A(x) = y$ in $< s$ steps according to program $\hat{P}_e$, and only numbers $z < s$ are used in the computation. (Compare this with Definition I.3.8.)

(ii) The *use function* $u(A; e, x, s)$ is $1 +$ the maximum number used in the computation if $\{e\}_s^A(x) \downarrow$, and $= 0$ otherwise. The use function $u(A; e, x)$ is $u(A; e, x, s)$ if $\{e\}_s^A(x) \downarrow$ for some s, and is undefined if $\{e\}^A(x) \uparrow$.

(iii) We write $\{e\}_s^\sigma(x) = y$, if $\{e\}_s^A(x) = y$ for some $A \supset \sigma$, and only elements $z < lh(\sigma)$ are used in the computation. If such $\sigma = A \restriction u$ we also write $\{e\}_s^{A \restriction u}(x) = y$. (The definition in (ii) was arranged so that if $\{e\}_s^A(x) = y$ then $\{e\}_s^\sigma(x) = y$ where $\sigma = A \restriction u(A; e, x, s)$.)

(iv) $\{e\}^\sigma(x) = y$ if $(\exists s) [\{e\}_s^\sigma(x) = y]$.

Note that this definition guarantees that

(1.1) $\{e\}_s^A(x) = y \implies x, y, e < s; \quad u(A; e, x, s) \leq s;$

and

(1.2) $\{e\}_s^A(x) = y \implies (\forall t \geq s) [\{e\}_t^A(x) = y$
                                        $\& \ u(A; e, x, t) = u(A; e, x, s)];$

so that the definition of $u(A; e, x)$ in Definition 1.7(ii) is independent of s.

Note that if A is recursive, then $u(A; e, x, s)$ is a recursive function and its index may be found uniformly in a Δ_0-index for A. In later chapters we define an r.e. set $A = \bigcup_s A_s$ where $\{A_s\}_{s \in \omega}$ is a recursive sequence of finite sets (specified, say, by canonical indices). In this case note that $\lambda x s [u(A_s; e, x, s)]$ is a recursive function.

1.8 Master Enumeration Theorem. (i) $\{\langle e, \sigma, x, s\rangle : \{e\}_s^\sigma(x) \downarrow\}$ *is recursive;*

(ii) $L =_{\text{dfn}} \{\langle e, \sigma, x\rangle : \{e\}^\sigma(x) \downarrow\}$ *is r.e.*

Proof. (i) Perform the Turing computation according to $\hat{P}_e$ on input x with σ written on the oracle tape until an output occurs or the first s steps have been completed.

(ii) $\{e\}^\sigma(x)\downarrow \iff (\exists s)\,[\{e\}^\sigma_s(x) = y]$, so the set L is Σ_1 and hence r.e.   ▯

1.9 Theorem (Use Principle).

(i) $\{e\}^A(x) = y \implies (\exists s)\,(\exists \sigma \subset A)\,[\{e\}^\sigma_s(x) = y]$;

(ii) $\{e\}^\sigma_s(x) = y \implies (\forall t \geq s)\,(\forall \tau \supseteq \sigma)\,[\{e\}^\tau_t(x) = y]$

(iii) $\{e\}^\sigma(x) = y \implies (\forall A \supset \sigma)\,[\{e\}^A(x) = y]$.

Proof. For (i) any computation which converges does so at some finite stage, having used only finitely many elements. Now (ii) and (iii) follow at once from the definition of oracle computation.   ▯

The Use Principle is crucial for most of our later theorems because (i) asserts that if $\{e\}^A(x) = y$ then $\{e\}^\sigma(x) = y$ for some $\sigma \subset A$, indeed $\sigma = A \upharpoonright u$ where $u = u(A; e, x, s)$ is the use function of Definition 1.7(ii). Furthermore, (iii) asserts that $\{e\}^B(x) = y$ for any $B \supseteq \sigma$. (The Use Principle asserts that Φ_e is continuous. See Exercise 1.18.) It follows from (1.1) and the Use Principle that

$$(1.3) \qquad [\{e\}^A_s(x) = y \ \& \ A \upharpoonright u = B \upharpoonright u] \implies \{e\}^B_s(x) = y$$

where $u = u(A; e, x, s)$ because (1.1) asserts that only elements $z < u$ are used in the computation.

1.10 Notation. (i) Let $W^A_e = \text{dom}\,\{e\}^A$ and similarly define $W^A_{e,s}$, W^σ_e, and $W^\sigma_{e,s}$ from $\{e\}^A_s$, $\{e\}^\sigma$, and $\{e\}^\sigma_s$.

(ii) Let $\Phi^A_e(x) = \Phi_e(A; x) = \{e\}^A(x)$ and $\Phi^A_{e,s}(x) = \Phi_{e,s}(A; x) = \{e\}^A_s(x)$.

(iii) Let $\{e\}(x)$ denote $\{e\}^\emptyset(x)$. (See Exercise 1.14.)

1.11 Definition. (i) B is *recursive in* (*Turing reducible to*) A, written $B \leq_T A$, if $B = \{e\}^A$ for some e. (We identify sets with their characteristic functions.) Let $B <_T A$ denote that $B \leq_T A$ but $A \not\leq_T B$.

(ii) B is *recursively enumerable* in A if $B = W^A_e$ for some e.

(iii) B is in Σ^A_1-form (abbreviated B is Σ^A_1) if $B = \{x : (\exists \vec{y})\, R^A(x, \vec{y})\}$ for some A-recursive predicate $R^A(x, \vec{y})$. (By the Quantifier Contraction Theorem II.1.3, this is equivalent to asserting that $B = \{x : (\exists y)\, R^A(x, y)\}$ for some such R^A.)

All the results of Chapter II §1 on r.e. sets relativize to A-r.e. sets by virtually the same proofs, where we replace "r.e." and "recursive" by "A-r.e." and "A-recursive". For example:

1.12 Complementation Theorem. $B \leq_T A$ *iff* B *and* $\overline{B}$ *are r.e. in* A.

1.13 Theorem. *The following are equivalent:*

(i) B *is r.e. in* A;

(ii) $B = \emptyset$ *or* B *is the range of some* A-*recursive total function;*

(iii) B *is* Σ_1^A.

Proof. The proofs of (i) $\Longleftrightarrow$ (ii) and (iii) $\Longleftrightarrow$ (i) are the relativizations to A of the proofs in Chapter II §1. E.g., to prove (i) $\Longrightarrow$ (iii), let $B = W_e^A$. Hence by the Use Principle 1.9,

$$(1.4) \qquad\qquad x \in B \iff (\exists s)\,(\exists \sigma)\,[\sigma \subset A \ \& \ x \in W_{e,s}^\sigma].$$

Now $x \in W_{e,s}^\sigma$ is a recursive relation on (e, σ, x, s) by (1.8)(i), and $\sigma \subset A$ is an A-recursive relation of σ because $\sigma \subset A$ iff $(\forall y < lh(\sigma))\,[\sigma(y) = A(y)]$. Hence (1.4) is of the form $(\exists s)\,(\exists \sigma)\, R(e, \sigma, x, s)$ where R is an A-recursive relation. ▯

1.14–1.20 Exercises

1.14. Give an effective coding of the programs $\{\,\hat{P}_e\,\}_{e \in \omega}$ using the method of Exercise I.2.10 and prove that under this coding $\hat{\varphi}_e =_{\mathrm{dfn}} \{\,e\,\}^\emptyset$ is an acceptable numbering of the partial recursive functions as defined in Exercise I.5.9. Thus from now on we shall often use the functions $\{\,e\,\}(x)$, $e \in \omega$, in place of $\varphi_e(x)$, $e \in \omega$.

1.15. (a) Prove that $\{\,\langle e, \sigma, x, y, s \rangle : \{\,e\,\}_s^\sigma(x) = y\,\}$ is recursive.

(b) Prove that $\{\,\langle e, \sigma, x, y \rangle : \{\,e\,\}^\sigma(x) = y\,\}$ is r.e., and that this set and the set L of Theorem 1.8(ii) are each 1-complete and hence recursively isomorphic to K.

1.16. Given sets B and A prove that $B \leq_T A$ iff there are recursive functions f and g such that

$$x \in B \iff (\exists \sigma)\,[\sigma \in W_{f(x)} \ \& \ \sigma \subset A],$$

$$x \in \overline{B} \iff (\exists \sigma)\,[\sigma \in W_{g(x)} \ \& \ \sigma \subset A].$$

Hint. If $B = \{\,e\,\}^A$ use Exercise 1.15(b) with $y = 1$ and $y = 0$ to obtain f and g, respectively. In the other direction use the Complementation Theorem 1.12.

1.17. Given r.e. sets A and B prove that $B \leq_T A$ iff there is a recursive function h such that

$$x \in \overline{B} \iff (\exists v)\,[v \in W_{h(x)} \ \& \ D_v \subseteq \overline{A}].$$

Hint. Suppose $B = \{e\}^A$. Let g be as in Exercise 1.16. Let

$$W_{h(x)} = \{v : (\exists \sigma) [\sigma \in W_{g(x)} \ \& \ D_v = \{y : \sigma(y) = 0\}$$
$$\& \ \{y : \sigma(y) = 1\} \subseteq A]\},$$

with D_v as defined in Definition II.2.4.

1.18. Consider the Baire space ω^ω with the usual topology where the basic open sets are $U_\sigma = \{f : f \in \omega^\omega \ \& \ \sigma \subseteq f\}$ for $\sigma \in \omega^{<\omega}$. Use the Use Principle 1.9 to show that the functional Φ_e as partial map from ω^ω to ω^ω is continuous and indeed effectively continuous in the sense that $\Phi_e^{-1}(U_\sigma) = \bigcup\{U_\tau : \tau \in V\}$ where V is an r.e. set of code numbers for strings $\tau \in \omega^{<\omega}$.

1.19. In the Relativized *s-m-n* Theorem we obtained a recursive (rather than an *A*-recursive) function s. Usually relativization of the theorems of Chapter I or II produces only an *A*-recursive function. For example, exhibit a set A and an *A*-recursive relation $R^A(x, y)$ which cannot be uniformized by any partial recursive function.

1.20. Prove that if ψ is partial recursive in $Y \subseteq \omega$ then there is a *recursive* function f such that

$$(\forall x \in \text{dom } \psi) \ [W^Y_{f(x)} = W^Y_{\psi(x)}].$$

Furthermore, an index for f can be found uniformly recursively from an e such that $\psi = \{e\}^Y$. *Hint.* Define f such that for all $x \in \omega$

$$W^Y_{f(x)} = \begin{cases} W^Y_{\psi(x)} & \text{if } \psi(x)\downarrow, \\ \emptyset & \text{otherwise.} \end{cases}$$

Namely, the oracle Turing program $\hat{P}_{f(x)}$ uses oracle Y to attempt to compute $\psi(x)$ and outputs nothing until $\psi(x) = z$ say. Then $\hat{P}_{f(x)}$ enumerates W^Y_z.

2. Turing Degrees and the Jump Operator

2.1 Definition. (i) $A \equiv_T B$ if $A \leq_T B$ and $B \leq_T A$. (Note that $\leq_T$ is reflexive and transitive, so $\equiv_T$ is an equivalence relation.)

(ii) The *Turing degree* (also called *degree of unsolvability*) of A is $\deg(A) = \{B : B \equiv_T A\}$.

(iii) $\deg(A) \cup \deg(B) = \deg(A \oplus B)$.

(iv) Lower-case boldface letters **a**, **b**, **c** denote degrees, and **D** denotes the class of all degrees.

(v) The degrees $\mathbf{D}$ form a partially ordered set under the relation $\deg(A) \leq \deg(B)$ iff $A \leq_{\mathrm{T}} B$. We write $\deg(A) < \deg(B)$ if $A <_{\mathrm{T}} B$, i.e., if $A \leq_{\mathrm{T}} B$ and $B \nleq_{\mathrm{T}} A$.

(vi) A degree $\mathbf{a}$ is *recursively enumerable* if it contains an r.e. set. Let $\mathbf{R}$ denote the class of r.e. degrees with the same ordering as for $\mathbf{D}$.

(vii) A degree $\mathbf{a}$ is *recursively enumerable in* $\mathbf{b}$ if $\mathbf{a}$ contains some set A r.e. in some set $B \in \mathbf{b}$.

Two sets of the same degree should be thought of as coding the same information and hence as equally difficult to compute, while $\mathbf{a} < \mathbf{b}$ asserts that sets of degree $\mathbf{b}$ are more difficult to compute than those of degree $\mathbf{a}$. Furthermore, $\deg(A \oplus B)$ is clearly the least upper bound for $\deg(A)$ and $\deg(B)$ in this partial ordering (see Exercise 2.6). Thus, the degrees form an upper semi-lattice, $(\mathbf{D}, \leq, \cup)$, as defined in Definition II.1.14. Unfortunately, the greatest lower bound of two degrees need not always exist (Corollary VI.4.4) and similarly even for the r.e. degrees (Corollary IX.3.3 or Exercise IX.3.12). Hence neither $\mathbf{D}$ nor $\mathbf{R}$ forms a lattice.

In Chapter I, diagonalization produced a nonrecursive r.e. set K. Relativizing this procedure to A we get a set K^A r.e. in A such that $A <_{\mathrm{T}} K^A$.

2.2 Definition. (i) Let $K^A = \{ x : \Phi_x^A(x) \downarrow \} = \{ x : x \in W_x^A \}$. K^A is called the *jump* of A and is denoted by A' (read as "A prime").

(ii) $A^{(n)}$, the nth jump of A, is obtained by iterating the jump n times; i.e., $A^{(0)} = A$, $A^{(n+1)} = (A^{(n)})'$.

It follows from the Relativized *s-m-n* Theorem 1.5, the proof of Theorem II.4.2, and Myhill's Isomorphism Theorem I.5.4, that $K^A \equiv K_0^A \equiv K_1^A$ where $K_0^A = \{ \langle x, y \rangle : \Phi_x^A(y) \text{ converges} \}$ and $K_1^A = \{ x : W_x^A \neq \emptyset \}$. These alternate characterizations of the jump are useful.

The crucial properties of the jump operator can be summarized as follows.

2.3 Jump Theorem. (i) A' *is r.e. in* A.

(ii) $A' \nleq_{\mathrm{T}} A$.

(iii) B *is r.e. in* A *iff* $B \leq_1 A'$.

(iv) *If* A *is r.e. in* B *and* $B \leq_{\mathrm{T}} C$ *then* A *is r.e. in* C.

(v) $B \leq_{\mathrm{T}} A$ *iff* $B' \leq_1 A'$.

(vi) *If* $B \equiv_{\mathrm{T}} A$ *then* $B' \equiv_1 A'$ *(and therefore* $B' \equiv_{\mathrm{T}} A'$*)*.

(vii) A *is r.e. in* B *iff* A *is r.e. in* $\overline{B}$.

Proof. Parts (i)–(iii) follow by relativizing the proofs in Chapters I and II, although (iii) ($\implies$) uses $K^A \equiv K_0^A$.

(iv) If $A \neq \emptyset$, then A is the range of some B-recursive function, and hence of some C-recursive function, since $B \leq_{\mathrm{T}} C$.

(v) ($\Longrightarrow$). If $B \leq_T A$ then B' is r.e. in A by (iv), because B' is r.e. in B by (i). Hence, $B' \leq_1 A'$ by (iii).

(v) ($\Longleftarrow$). If $B' \leq_1 A'$, then both B and $\overline{B}$ are r.e. in A by (iii) (since $B, \overline{B} \leq_1 B'$). Hence, $B \leq_T A$ by the Complementation Theorem 1.12.

(vi) follows immediately from (iv).

(vii) follows immediately from (iv).   ▯

Let $a' = \deg(A')$ for $A \in \mathbf{a}$. Note that $\mathbf{a}' > \mathbf{a}$ and $\mathbf{a}'$ is r.e. in $\mathbf{a}$. By Theorem 2.3(v) the jump is well-defined on degrees. Let $\mathbf{0}^{(n)} = \deg(\emptyset^{(n)})$. Thus, we have an infinite hierarchy of degrees, $\mathbf{0} < \mathbf{0}' < \mathbf{0}'' < \cdots < \mathbf{0}^{(n)} < \cdots$. The first few degrees in this hierarchy are of special importance and will be shown to be the degrees of certain unsolvable problems considered in Chapter I.

$$\mathbf{0} = \deg(\emptyset) = \{\, B : B \text{ is recursive} \,\};$$
$$\mathbf{0}' = \deg(\emptyset'), \text{ where } \emptyset' =_{\mathrm{dfn}} K^{\emptyset} \equiv K \equiv K_0 \equiv K_1;$$
$$\mathbf{0}'' = \deg(\emptyset'') = \deg(\mathrm{Fin}) = \deg(\mathrm{Tot});$$
$$\mathbf{0}''' = \deg(\emptyset''') = \deg(\mathrm{Cof}) = \deg(\mathrm{Rec}) = \deg(\mathrm{Ext}).$$

By definition, $\mathbf{0}$ is the least degree and consists precisely of the recursive sets. Degree $\mathbf{0}'$ (read "zero prime") is the degree of the halting problem and also the degree of the problem $\{\, x : W_x \neq \emptyset \,\}$ since $K \equiv K_1$. The above characterizations for $\mathbf{0}''$ and $\mathbf{0}'''$ will be shown in Chapter IV, although using current methods we can show $\emptyset'' \equiv \mathrm{Fin}$ (see Exercise 2.9).

2.4 Definition. Fix $A \subseteq \omega$ and let $\mathbf{a} = \deg(A)$. Relativizing Definition II.2.5 to A we define a sequence of sets $\{\, V_n \,\}_{n \in \omega}$ to be *uniformly recursive in A* (*uniformly of degree $\leq \mathbf{a}$*) if there is an A-recursive function $g(x, n)$ such that $\lambda x[g(x, n)]$ is the characteristic function of V_n, for all n.

2.5–2.9 Exercises

2.5. Using the Axiom of Choice prove that there are at most $2^{\aleph_0}$ degrees. (Using the methods in Chapter VI we can show that there are *at least* $2^{\aleph_0}$ degrees by constructing a 1:1 map from the power set of ω into the degrees (see Exercise VI.1.8), and hence there are *exactly* $2^{\aleph_0}$ degrees.) *Hint.* For any A there are only $\aleph_0$ sets B such that $B \leq_T A$. Now by the Axiom of Choice $\aleph_0 \cdot \kappa = \kappa$ for $\kappa \geq \aleph_0$.

2.6. Show that $\deg(A \oplus B)$ is the least upper bound for $\deg(A)$ and $\deg(B)$ in $(\mathbf{D}, \leq, \cup)$.

2.7. Let $\{\, A_y \,\}_{y \in \omega}$ be any countable sequence of sets. Define the *infinite join*

$$\oplus\{\, A_y \,\}_{y \in \omega} =_{\mathrm{dfn}} \{\, \langle x, y \rangle : x \in A_y \ \& \ y \in \omega \,\}.$$

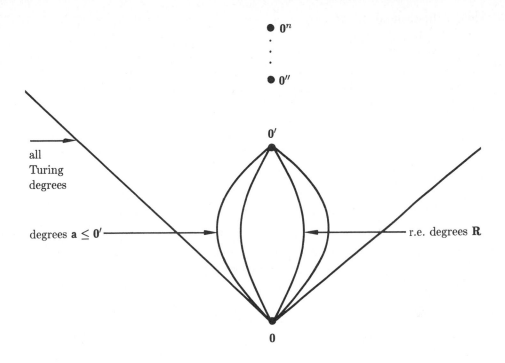

Diagram 2.1. Turing Degrees $(\mathbf{D}, \leq)$

Let $\oplus_y A_y$ denote $\oplus\{\, A_y \,\}_{y\in\omega}$. Prove that $\deg(\oplus_y A_y)$ is the *uniform* least upper bound for $\{\, \deg(A_y) : y \in \omega \,\}$ in the sense that if there exist a set C and a recursive function f such that $A_y = \{\, f(y) \,\}^C$ for all y, then $\oplus_y A_y \leq_{\mathrm{T}} C$.

2.8. Prove that $\emptyset' \equiv K$. (Recall that $\{\, e \,\}^\emptyset$ is an acceptable numbering by Exercise 1.14.)

2.9.$^\diamond$ Prove that Fin $\equiv_1 K'$ (and hence Fin $\equiv K'$). *Hint.* For Fin $\leq_1 K'$, note that $x \in$ Fin iff $(\exists y)\, [W_x \cap [y,\infty) = \emptyset]$, so Fin is Σ_1 in K_1. To show $K' \leq_1$ Fin, let $K_s = W_{i,s}$ for some i such that $W_i = K$. Note that

$$e \in A' \iff \{\, e \,\}^A(e)\!\downarrow \iff (\exists\sigma)\,(\exists s)\, [\sigma \subset A \;\&\; \{\, e \,\}^\sigma_s(e)\!\downarrow\,].$$

Define the recursive function

$$h(e, s) = \begin{cases} lh(\sigma) & \text{where } \sigma = (\mu\tau)\,[\tau \subseteq K_s \;\&\; \{e\}_s^\tau(e)\downarrow\,], \text{ if such } \tau \text{ exists;} \\ s & \text{if no such } \tau \text{ exists.} \end{cases}$$

Let $W_{f(e)} = \{x : (\exists s)\,[x \leq h(e, s)]\}$. Show that $e \in K'$ iff $f(e) \in \mathrm{Fin}$.

3. The Modulus Lemma and Limit Lemma

Although we are interested in the degrees in general, we are especially interested in those degrees $\leq \mathbf{0}'$, and among these particularly the r.e. degrees. We now show that $\deg(A) \leq \mathbf{0}'$ iff there is a recursive function $\hat{f}(x, s)$ such that $\chi_A(x) = \lim_s \hat{f}(x, s)$, i.e., roughly, iff A can be recursively approximated. Furthermore, $\deg(A)$ is actually r.e. iff $\hat{f}$ may be chosen to have modulus of convergence $m \leq_T A$. These characterizations, particularly the former, will be very useful.

3.1 Definition. (i) A sequence of (total) functions $\{f_s(x)\}_{s\in\omega}$ *converges* (*pointwise*) to $f(x)$, written $f = \lim_s f_s$, if for all x, $f_s(x) = f(x)$ for a.e. s (all but finitely many s).

(ii) A *modulus* (*of convergence*) for $\{f_s\}_{s\in\omega}$ is a function $m(x)$ such that for all s, if $s \geq m(x)$ then $f_s(x) = f(x)$. (Hence, $f_{m(x)}(x) = f(x)$.) The *least modulus* is the function $m(x) = (\mu s)\,(\forall t \geq s)\,[f_t(x) = f(x)]$.

(iii) The sequence $\{f_s(x)\}_{s\in\omega}$ is *recursive* if there is a recursive function $\hat{f}(x, s)$ such that $f_s(x) = \hat{f}(x, s)$ for all x, s.

Let $\{f_s(x)\}_{s\in\omega}$ be a recursive sequence. Note that the least modulus is recursive in any modulus. If $f = \lim_s f_s$ and m is any modulus then

$$(3.1) \qquad\qquad\qquad f \leq_T m$$

because $f_{m(x)}(x) = f(x)$. However, $m \leq_T f$ usually fails even for the least modulus. Remarkably, if f has r.e. degree, then $m \leq_T f$ holds for some modulus m of a particular recursive sequence, as we prove in the following lemma.

3.2 Modulus Lemma. *If A is r.e. and $f \leq_T A$ then there is a recursive sequence $\{f_s\}_{s\in\omega}$ such that $\lim_s f_s = f$, and a modulus m of $\{f_s\}_{s\in\omega}$ which is recursive in A.*

Proof. Let A be r.e. and $f = \{e\}^A$. Let $A_s = W_{i,s}$ for some $W_i = A$. Define the functions

$$f_s(x) = \begin{cases} \{e\}_s^{A_s}(x) & \text{if } \{e\}_s^{A_s}(x)\downarrow, \\ 0 & \text{otherwise;} \end{cases}$$

$$m(x) = (\mu s) \, (\exists z \le s) \, [\{e\}_s^{A_s \upharpoonright z}(x)\!\downarrow \ \& \ A_s \upharpoonright z = A \upharpoonright z].$$

Now $\{f_s\}_{s \in \omega}$ is a recursive sequence. Furthermore, m is A-recursive because the quantifier on z is bounded by s, the first clause is recursive, and the second clause is A-recursive because A is r.e. Furthermore, m is a modulus because, by the Use Principle 1.9, for $s \ge m(x)$,

$$\{e\}_s^{A_s \upharpoonright z}(x) = \{e\}_s^{A \upharpoonright z}(x) = \{e\}^A(x) = f(x). \quad \blacksquare$$

3.3 Limit Lemma. *For any function f, $f \le_T A'$ iff there is an A-recursive sequence $\{f_s\}_{s \in \omega}$ (i.e., an A-recursive function $\hat{f}(x,s) = f_s(x)$) such that $f = \lim_s f_s$.*

Proof. ($\Longrightarrow$). Let $f \le_T A'$. Now A' is r.e. in A. Hence, the A-recursive sequence $\{f_s\}_{s \in \omega}$ exists by the Modulus Lemma relativized to A.
 ($\Longleftarrow$). Let $f = \lim_s f_s$. Define

$$A_x = \{s : (\exists t) \, [s \le t \ \& \ f_t(x) \ne f_{t+1}(x)]\}.$$

Now A_x is finite, and $B = \oplus_x A_x = \{\langle s, x\rangle : s \in A_x\}$ is Σ_1^A and hence A-r.e., so $B \le_T A'$. Thus, given x we can B-recursively (and therefore A'-recursively) compute the least modulus $m(x) = (\mu s) \, [s \notin A_x]$. Hence, $f \le_T m \oplus A \le_T B \oplus A \le_T A'$. $\quad \blacksquare$

 In particular, $f \le_T \emptyset'$ iff $f = \lim_s f_s$ for some recursive sequence $\{f_s\}_{s \in \omega}$. This will be the most useful characterization of the degrees below $\mathbf{0}'$. Since not all the degrees below $\mathbf{0}'$ are r.e., the following corollary selects those that are.

3.4 Corollary. *A function f has r.e. degree iff f is the limit of a recursive sequence $\{f_s\}_{s \in \omega}$ which has a modulus $m \le_T f$.*

Proof. ($\Longrightarrow$). Let $f \equiv_T A$ with A r.e. Apply the Modulus Lemma to obtain $m \le_T A \equiv_T f$.
 ($\Longleftarrow$). Let $f = \lim_s f_s$ with modulus $m \le_T f$. Define the r.e. set B as in the proof of the Limit Lemma, and recall that $f \le_T B$. Now $B \le_T m \le_T f$. Hence, $f \equiv_T B$ and f has r.e. degree. $\quad \blacksquare$

3.5–3.13 Exercises

3.5. Show that the m defined in the Modulus Lemma is not necessarily the least modulus. Explicitly define the least modulus as an A-recursive function.

3.6. A set D is the *difference of r.e. sets (d.r.e.)* if $D = A - B$ where A and B are r.e. sets. Show that the d.r.e. sets are closed under intersection.

3.7. Define $C_n = \{\, e : |W_e| = n \,\}$.

   (a) For $n \geq 0$, show that C_n is d.r.e.

   (b) For $n \geq 0$, show that C_n is not r.e. *Hint.* Show that $\overline{K} \leq_1 C_n$.

3.8. A set $A \leq_T K$ is *n-r.e.* if $A = \lim_s A_s$ for some recursive sequence $\{\, A_s \,\}_{s \in \omega}$ such that for all x, $A_0(x) = 0$ and $\operatorname{card}\{\, s : A_s(x) \neq A_{s+1}(x) \,\} \leq n$. For example, the only 0-r.e. set is $\emptyset$, the 1-r.e. sets are the usual r.e. sets, and the 2-r.e. sets are the d.r.e. sets.

   (a) Show that for each n there is an $(n+1)$-r.e. set which is not n-r.e. Hence, the d.r.e. sets are not closed under union. *Hint.* Note that the approximation $\{\, A_s \,\}_{s \in \omega}$ may be chosen so that the predicate $n \in A_s$ is primitive recursive, and hence all such approximations are uniformly recursive.

   (b) Show that a set A is $(2n+2)$-r.e. iff A is the union of $n+1$ d.r.e. sets.

   (c) Show that A is $(2n+1)$-r.e. iff A is the union of n d.r.e. sets and an r.e. set.

3.9. Let $\mathcal{B}$ be the least Boolean algebra containing $\mathcal{E}$ (i.e., the Boolean algebra obtained by closing the r.e. sets under complementation, union and intersection).

   (a) Prove that $A \in \mathcal{B}$ iff A is a finite union of d.r.e. sets.

   (b) Show that there exists a set $A \leq_T K$ such that $A \notin \mathcal{B}$.

3.10 (Jockusch and others). Show by induction on n that if A is n-r.e. then either A or $\overline{A}$ contains an infinite r.e. set.

3.11. Suppose that A and B are r.e. sets and A is nonrecursive. Prove that $A \times \overline{A} \not\leq_m B \oplus \overline{B}$. *Hint* (Jockusch). Suppose for a contradiction that f is a recursive function such that for all x and y

$$x \in A \ \& \ y \in \overline{A} \iff f(x, y) \in B \oplus \overline{B}.$$

Define $C = \{\, y : (\exists x)\, [f(x, y) \in B \oplus \emptyset] \,\}$. Show that C is r.e., $C \subseteq \overline{A}$, and that there exists $a \in \overline{A} - C$. Show that for all x,

$$x \in A \iff f(x, a) \in \emptyset \oplus \overline{B},$$

so A is co-r.e.

3.12. (a) Show that $K \times \overline{K}$ is d.r.e.

   (b) Show that if D is d.r.e. then $D \leq_1 K \times \overline{K}$. (Hence, $K \times \overline{K}$ is 1-complete with respect to d.r.e. sets.)

   (c) Show that $K \oplus \overline{K}$ is d.r.e.

   (d) Using Exercise 3.11 show that it is false that $K \times \overline{K} \leq_m K \oplus \overline{K}$. (Hence, $K \oplus \overline{K}$ is neither 1-complete nor m-complete with respect to d.r.e. sets.) (See Exercise VII.2.4 for a proof that there are d.r.e. degrees which are not r.e.)

3.13. Give an alternative proof that $K \times \overline{K} \not\leq_m K \oplus \overline{K}$ by showing that $K \times \overline{K} \equiv_m A \equiv_{\text{dfn}} \{\, x : W_x = \{\, 0\,\}\,\}$, and that $A \not\equiv_m \overline{A}$ by the Recursion Theorem.

Chapter IV

The Arithmetical Hierarchy

In the previous chapter the jump operator gave us a hierarchy of degrees $\mathbf{0} < \mathbf{0'} < \cdots < \mathbf{0}^{(n)} < \cdots$. In this chapter we give a different hierarchy of sets A according to the quantifier complexity in the syntactic definition of A. The two hierarchies are related in §2 by Post's Theorem. These tools enable us to exactly classify in both hierarchies the unsolvable problems (nonrecursive index sets) mentioned in Chapter I. Finally, we consider the high and low r.e. degrees which will play an important role in classifying the degree and algebraic structure of r.e. sets.

1. Computing Levels in the Arithmetical Hierarchy

We define the Σ_n and Π_n sets. (The classes Σ_n and Π_n sometimes appear in the literature as Σ_n^0 or Π_n^0 to distinguish them from the classes Σ_n^1 or Π_n^1 of the analytical hierarchy.)

1.1 Definition. (i) A set B is in Σ_0 (Π_0) iff B is recursive.

(ii) For $n \geq 1$, B is *in* Σ_n (written $B \in \Sigma_n$) if there is a recursive relation $R(x, y_1, y_2, \ldots, y_n)$ such that

$$x \in B \iff (\exists y_1)\,(\forall y_2)\,(\exists y_3)\,\cdots\,(Q y_n)\, R(x, y_1, y_2, \ldots, y_n),$$

where Q is $\exists$ if n is odd, and $\forall$ if n is even. Likewise, B is *in* Π_n ($B \in \Pi_n$) if

$$x \in B \iff (\forall y_1)\,(\exists y_2)\,(\forall y_3)\,\cdots\,(Q y_n)\, R(x, y_1, y_2, \ldots, y_n),$$

where Q is $\exists$ or $\forall$ according as n is even or odd.

(iii) B is *in* Δ_n if $B \in \Sigma_n \cap \Pi_n$.

(iv) B is *arithmetical* if $B \in \bigcup_n (\Sigma_n \cup \Pi_n)$.

Note that B is arithmetical iff B can be obtained from a recursive relation by finitely many applications of projection and complementation. (See Exercise 1.9.)

1.2 Definition. Fix a set A. In Definition 1.1 if we replace everywhere "recursive" by "A-recursive" then we have the definition of B being Σ_n in A (written $B \in \Sigma_n^A$), B being Π_n in A ($B \in \Pi_n^A$), $B \in \Delta_n^A$, and B being *arithmetical in A*.

We say that a formula is Σ_n (Π_n) if it is Σ_n (Π_n) as a relation of its free variables. We assume familiarity with the usual rules of quantifier manipulation from elementary logic for converting a formula to an equivalent one in prenex normal form consisting of a string of quantifiers (*prefix*) followed by a formula with no quantifiers (*matrix*), which will in our case be a recursive relation. Using these rules we can show the following facts which will be frequently used to prove that a particular set is in Σ_n or Π_n. The only nontrivial fact is (vi) concerning bounded quantifiers. A *bounded quantifier* is one of the form $(Qx \le y)F$, which abbreviates $(\forall x)\,[x \le y \implies F]$ if Q is $\forall$, and $(\exists x)\,[x \le y \ \& \ F]$, if Q is $\exists$. Part (vi) asserts that bounded quantifiers may be moved to the right past ordinary quantifiers and thus may be ignored in counting quantifier complexity.

1.3 Theorem. (i) $A \in \Sigma_n \iff \overline{A} \in \Pi_n$;

(ii) $A \in \Sigma_n(\Pi_n) \implies (\forall m > n)\,[A \in \Sigma_m \cap \Pi_m]$;

(iii) $A, B \in \Sigma_n(\Pi_n) \implies A \cup B, \ A \cap B \in \Sigma_n(\Pi_n)$;

(iv) $[R \in \Sigma_n \ \& \ n > 0 \ \& \ A = \{\, x : (\exists y)\, R(x,y) \,\}] \implies A \in \Sigma_n$;

(v) $[B \le_m A \ \& \ A \in \Sigma_n] \implies B \in \Sigma_n$;

(vi) *If* $R \in \Sigma_n(\Pi_n)$, *and A and B are defined by*

$$\langle x, y \rangle \in A \iff (\forall z < y)\, R(x,y,z),$$

and

$$\langle x, y \rangle \in B \iff (\exists z < y)\, R(x,y,z),$$

then $A, B \in \Sigma_n(\Pi_n)$.

Proof. (i) If $A = \{\, x : (\exists y_1)\, (\forall y_2) \ \cdots \ R(x, \overrightarrow{y}) \,\}$, then

$$\overline{A} = \{\, x : (\forall y_1)\, (\exists y_2) \ \cdots \ \neg R(x, \overrightarrow{y}) \,\}.$$

(ii) For example, if $A = \{\, x : (\exists y_1)\, (\forall y_2)\, R(x, y_1, y_2) \,\}$, then

$$A = \{\, x : (\exists y_1)\, (\forall y_2)\, (\exists y_3)\, [R(x, y_1, y_2) \ \& \ y_3 = y_3] \,\}.$$

(iii) Let $A = \{\, x : (\exists y_1)\, (\forall y_2) \ \cdots \ R(x, \overrightarrow{y}) \,\}$, and

$$B = \{\, x : (\exists z_1)\, (\forall z_2) \ \cdots \ S(x, \overrightarrow{z}) \,\}.$$

2. Post's Theorem and the Hierarchy Theorem

2.1 Definition. A set A is Σ_n-*complete* (Π_n-*complete*) if $A \in \Sigma_n(\Pi_n)$ and $B \leq_1 A$ for every $B \in \Sigma_n(\Pi_n)$. (By Exercises 2.6 and 2.7 it makes no difference whether we use "$B \leq_m A$" or "$B \leq_1 A$" in the definition of Σ_n-complete and Π_n-complete.)

Note that A is Σ_1-complete iff A is 1-complete as defined in Definition II.4.1. Hence, K is Σ_1-complete and $\overline{K}$ is Π_1-complete. The following fundamental theorem relates the jump hierarchy of degrees from Chapter III to the arithmetical hierarchy.

2.2 Post's Theorem. *For every $n \geq 0$,*

(i) $B \in \Sigma_{n+1} \iff B$ *is r.e. in some Π_n set* $\iff B$ *is r.e. in some Σ_n set, by Theorem III.2.3(vii);*
(ii) $\emptyset^{(n)}$ *is Σ_n-complete for $n > 0$;*
(iii) $B \in \Sigma_{n+1} \iff B$ *is r.e. in $\emptyset^{(n)}$;*
(iv) $B \in \Delta_{n+1} \iff B \leq_T \emptyset^{(n)}$.

Proof. (i) ($\Longrightarrow$). Let $B \in \Sigma_{n+1}$. Then $x \in B \iff (\exists y)\, R(x, y)$ for some $R \in \Pi_n$. Hence B is Σ_1 in R and therefore r.e. in R by Theorem III.1.13.

(i) ($\Longleftarrow$). Suppose B is r.e. in some Π_n set C. Then for some e,

$$x \in B \iff x \in W_e^C$$
$$\iff (\exists s)\, (\exists \sigma)\, [\sigma \subset C \ \& \ x \in W_{e,s}^\sigma].$$

Clearly $x \in W_{e,s}^\sigma$ is recursive by Theorem III.1.8. Hence by Theorem 1.3(iv) it suffices to show that $\sigma \subset C$ is Σ_{n+1}. Now

$$\sigma \subset C \iff (\forall y < lh(\sigma))\, [\sigma(y) = C(y)],$$
$$\iff (\forall y < lh(\sigma))\, [[\sigma(y) = 1 \ \& \ y \in C] \vee [\sigma(y) = 0 \text{ and } y \notin C]],$$
$$\iff (\forall y < lh(\sigma))\, [\Pi_n \vee \Sigma_n],$$

since $C \in \Pi_n$. Hence $\sigma \subset C$ is Σ_{n+1} by Theorems 1.3(ii), 1.3(iii), and 1.3(vi).

(ii) is proved by induction on n and is clear for $n = 1$. Fix $n \geq 1$ and assume $\emptyset^{(n)}$ is Σ_n-complete. Hence $\overline{\emptyset^{(n)}}$ is Π_n-complete. Now

$$B \in \Sigma_{n+1} \iff B \text{ is r.e. in some } \Sigma_n \text{ set, by (i)},$$
$$\iff B \text{ is r.e. in } \emptyset^{(n)}, \text{ by inductive hypothesis,}$$
$$\iff B \leq_1 \emptyset^{(n+1)}, \text{ by Theorem III.2.3(vii).}$$

Hence $\emptyset^{(n+1)}$ is Σ_{n+1}-complete.

(iii) By (i) and (ii) since $\overline{\emptyset^{(n)}}$ is Π_n-complete.

(iv)
$$B \in \Delta_{n+1} \iff B, \overline{B} \in \Sigma_{n+1},$$
$$\iff B, \overline{B} \text{ are r.e. in } \emptyset^{(n)}, \text{ by (iii)},$$
$$\iff B \leq_T \emptyset^{(n)}. \quad \square$$

2.3 Corollary (Hierarchy Theorem). $(\forall n > 0) [\Delta_n \subset \Sigma_n \text{ and } \Delta_n \subset \Pi_n]$.

Proof. $\emptyset^{(n)} \in \Sigma_n - \Pi_n$ (by Theorems 2.2(ii), 2.2(iv), and III.2.3(ii)), and likewise $\overline{\emptyset^{(n)}} \in \Pi_n - \Sigma_n$. $\quad \square$

2.4–2.7 Exercises

2.4. Recall that e is a Σ_1-index for A if $A = W_e$. In this case we define e to be a Π_1-index for $\overline{A}$. Define e to be a Σ_{n+1}-index for A and a Π_{n+1}-index for $\overline{A}$ if $A = \{ x : (\exists y) \, B(x,y) \}$ where B is the set with Π_n-index e. Prove that A has Π_2-index e iff $A = \{ x : (\forall y_1) (\exists y_2) \, T(e, \langle x, y_1 \rangle, y_2) \}$.

2.5. By Theorem 2.2(iv) and Lemma III.3.3 we know that the following are equivalent: (a) $A \in \Delta_2$; (b) $A \leq_T K$; (c) $A = \lim_s f_s$ for some recursive sequence $\{ f_s \}_{s \in \omega}$. Prove that we can pass effectively from an index for one characterization to an index for any other. (A Δ_2-index for A is a number $\langle e, i \rangle$ such that e is a Σ_2 index for A and i is a Π_2 index for A; an index for (b) is an e such that $A = \{ e \}^K$, and an index for (c) is an e such that $\varphi_e(s, x) = f_s(x)$.)

2.6. Prove that if $B \leq_m A$ and $A = \emptyset^{(n)}$ for $n \geq 1$ then $B \leq_1 A$. *Hint.* Use the Padding Lemma I.3.2. An alternative proof is to show that $B \in \Sigma_n$ and hence B is r.e. in $\emptyset^{(n-1)}$ by Theorem 2.2(iii) so we can apply Theorem III.2.3(iii).

2.7. Prove that if $B = \emptyset^{(n)}$ for $n \geq 1$, and $B \leq_m A$ then $B \leq_1 A$. (Hence, to prove that B is Σ_n-complete it suffices to prove that $\emptyset^{(n)} \leq_m A$ rather than proving $\emptyset^{(n)} \leq_1 A$, although our proofs usually establish the latter because the function s in the s-m-n Theorem I.3.5 is 1:1.) *Hint.* Use Corollary II.4.7.

3. Σ_n-Complete Sets

We have shown that $\emptyset^{(n)}$ is Σ_n-complete for all n. However, there are other Σ_n-complete sets with natural definitions which will be useful in later applications. For example, we know that K, K_0 and K_1 are all Σ_1-complete and we shall now show that Fin is Σ_2-complete and Cof and Rec are Σ_3-complete. Once we have classified a set A as being in Σ_n by the method of §1, we attempt to show that the classification is best possible by proving that $B \leq_1 A$ for some known Σ_n-complete set B, thus showing that A is Σ_n-complete. Recall from Definition II.4.9 that $(A, B) \leq_m (C, D)$ via f if $f(A) \subseteq C$, $f(B) \subseteq D$ and $f(\overline{A \cup B}) \subseteq \overline{C \cup D}$. We write "$\leq_1$" if f is 1:1.

3.1 Definition. For $n \geq 1$ we define $(\Sigma_n, \Pi_n) \leq_m (C, D)$ if $(A, \overline{A}) \leq_m (C, D)$ for some Σ_n-complete set A, and similarly for $\leq_1$ in place of $\leq_m$. In this case we also write $\Sigma_n \leq_m C$ and $\Pi_n \leq_m D$. (By the same remark as in Definition 2.1 it makes no difference whether we write "$\leq_m$" or "$\leq_1$" here.)

(Although this notation seems strange at first because Σ_n and Π_n are classes rather than sets, it is justified by the fact that if $(\Sigma_n, \Pi_n) \leq_m (C, D)$ then $(A, \overline{A}) \leq_m (C, D)$ and $(\overline{B}, B) \leq_m (C, D)$ for any Σ_n set A and Π_n set B. The notation will prove convenient in Chapter XII.)

3.2 Theorem. $(\Sigma_2, \Pi_2) \leq_1$ (Fin, Tot). *Thus* Fin *is* Σ_2-complete, Inf *and* Tot *are* Π_2-complete, *and* Inf $\equiv$ Tot.

Proof. By Proposition 1.4 and Corollary 1.7, Fin $\in \Sigma_2$ (so Inf $\in \Pi_2$) and Tot $\in \Pi_2$. Fix $A \in \Sigma_2$, so $\overline{A} \in \Pi_2$, and hence there is a recursive relation R such that

$$x \in \overline{A} \iff (\forall y)\,(\exists z)\, R(x, y, z).$$

Using the *s-m-n* Theorem I.3.5, define a 1:1 recursive function f by

$$\varphi_{f(x)}(u) = \begin{cases} 0 & \text{if } (\forall y \leq u)\,(\exists z)\, R(x, y, z), \\ \uparrow & \text{otherwise.} \end{cases}$$

Now

$$x \in \overline{A} \implies W_{f(x)} = \omega \implies f(x) \in \text{Tot},$$

but

$$x \in A \implies W_{f(x)} \text{ is finite} \implies f(x) \in \text{Fin}. \quad \blacksquare$$

3.3 Definition. Comp $= \{\, x : W_x \equiv_T K \,\}$, i.e., the set of indices of (*Turing*) *complete* r.e. sets.

3.4 Theorem. $(\Sigma_3, \Pi_3) \leq_1$ (Cof, Comp) $\leq_1$ (Rec, Comp).

3.5 Corollary. Cof *is* Σ_3-complete.

Proof. By Proposition 1.5 and Theorem 3.4. $\quad \blacksquare$

3.6 Corollary (Rogers). Rec *is* Σ_3-complete.

Proof. By Corollary 1.8 and Theorem 3.4 because Cof $\subseteq$ Rec and because Rec $\cap$ Comp $= \emptyset$. $\quad \blacksquare$

In Corollary XII.1.7 we shall show that Comp is Σ_4-complete.

Proof of Theorem 3.4. Fix a recursive enumeration $\{K_s\}_{s\in\omega}$ of K. Fix $A \in \Sigma_3$. Now for some relation $R \in \Pi_2$, $x \in A$ iff $(\exists y) R(x,y)$. Since $R \in \Pi_2$ there is a recursive function g by Theorem 3.2 such that $R(x,y)$ iff $W_{g(x,y)}$ is infinite. Hence,

$$(3.1) \qquad\qquad x \in A \iff (\exists y)\, [W_{g(x,y)} \text{ is infinite}].$$

We shall define an r.e. set $W_{f(x)} = \bigcup_s W^s_{f(x)}$ by stages, where $W^s_{f(x)}$ consists of the elements enumerated in $W_{f(x)}$ by the end of stage s, and such that $x \in A$ iff $W_{f(x)}$ is cofinite.

Stage $s = 0$. Set $W^0_{f(x)} = \emptyset$.

Stage $s + 1$. Let $\overline{W}^s_{f(x)} = \{\, b^s_{x,0} < \cdots < b^s_{x,y} < \cdots \,\}$. For each $y \leq s$ such that $W_{g(x,y),s} \neq W_{g(x,y),s+1}$ or $y \in K_{s+1} - K_s$ enumerate $b^s_{x,y}$ in $W^{s+1}_{f(x)}$. If no such y exists set $W^{s+1}_{f(x)} = W^s_{f(x)}$.

Now if $x \in A$ then for some y, $W_{g(x,y)}$ is infinite, $\lim_s b^s_{x,y} = \infty$, so $|\overline{W}_{f(x)}| \leq y$ and hence $\overline{W}_{f(x)}$ is finite. If $x \notin A$, then for all y, $W_{g(x,y)}$ is finite, so that $b_{x,y} =_{\text{dfn}} \lim_s b^s_{x,y} < \infty$, $\overline{W}_{f(x)} = \{\, b_{x,y} \,\}_{y\in\omega}$, and $\overline{W}_{f(x)}$ is infinite. Furthermore, if $x \notin A$ then $K \leq_T W_{f(x)}$ because for each y we can $W_{f(x)}$-recursively compute a stage $s(y)$ such that $b^{s(y)}_{x,y} = b_{x,y}$, and we note that by the construction $y \in K$ iff $y \in K_{s(y)}$. ▯

This method of construction should be visualized as follows. We fix x and enumerate the r.e. set $W_{f(x)}$ by focusing on its complement and using a set of "movable markers" $\{\,\Gamma_{x,y}\,\}_{y\in\omega}$. At stage 0 the marker $\Gamma_{x,y}$ is placed on integer y. The markers move in order, always to larger elements, and any element not covered by a marker is enumerated in $W_{f(x)}$. Hence, at the end of stage s, marker $\Gamma_{x,y}$ rests on the $(y+1)$st member of $\overline{W}^s_{f(x)}$. (In the above proof let $b^s_{x,y}$ denote the position of $\Gamma_{x,y}$ at the end of stage s.) At stage $s + 1$ enumerate in $W_{f(x)}$ the current position of $\Gamma_{x,y}$ for each $y \leq s$ such that $W_{g(x,y)}$ has just increased in cardinality, or y has just entered K, and move all markers greater than or equal to the displaced marker positions to elements not yet in $W_{f(x)}$ maintaining the order of the markers. Thus, we view $W_{g(x,y)}$ as a "kicking" set which forces the yth marker $\Gamma_{x,y}$ to move at all stages $s > y$ at which $W_{g(x,y)}$ increases in cardinality. If $x \in A$, $W_{g(x,y)}$ is infinite for some y, so marker $\Gamma_{x,y}$ is moved ("kicked") infinitely often. Hence $\overline{W}_{f(x)}$ is finite, and indeed $|\overline{W}_{f(x)}| = y_0$ for the least such y_0. If $x \notin A$, each $W_{g(x,y)}$ is finite, so each $\Gamma_{x,y}$ is moved finitely often and $\overline{W}_{f(x)}$ is infinite. The movable marker method is one of the most useful tools for visualizing more complicated constructions of r.e. sets, and will be used many times.

In Theorem XII.1.7 we improve the classification in Theorem 3.4 of Comp by proving that Comp is Σ_4-complete. In Chapter XII numerous other index sets of r.e. sets are classified at various higher levels of the arithmetical hierarchy.

The following is probably the most useful characterization for approximating a Σ_3 set A, i.e., for "guessing" whether $x \in A$. It should be viewed as a strengthening of (3.1).

3.7 Corollary. *If $A \in \Sigma_3$ then there is a recursive function g such that*

$$(3.2) \qquad\qquad x \in A \iff (\text{a.e. } y) \ [W_{g(x,y)} = \omega];$$

and

$$(3.3) \qquad\qquad x \in \overline{A} \iff (\forall y) \ [W_{g(x,y)} \text{ is finite}].$$

Proof. Let $A \leq_1 \text{Cof}$ via f. Define g by

$$z \in W_{g(x,y)} \iff (\forall u) \ [y \leq u \leq z \implies u \in W_{f(x)}].$$

Hence,

$$x \in A \implies W_{f(x)} \text{ cofinite} \implies (\exists y) \ (\forall z \geq y) \ [z \in W_{f(x)}]$$
$$\implies (\exists y) \ (\forall z \geq y) \ [W_{g(x,z)} = \omega];$$

and

$$x \in \overline{A} \implies W_{f(x)} \text{ coinfinite} \implies (\forall y) \ (\exists z \geq y) \ [z \notin W_{f(x)}]$$
$$\implies (\forall y) \ [W_{g(x,y)} \text{ finite}]. \quad \blacksquare$$

To "guess" about membership in a Σ_2 set A we have a recursive function f such that $x \in A$ iff $W_{f(x)}$ is finite. For a Σ_3 set A, Corollary 3.7 is the two-dimensional analogue where $W_{g(x,y)}$ is viewed as the yth row of a matrix. If $x \in A$ then almost all rows are ω, and otherwise all are finite.

3.8–3.13 Exercises

3.8. Prove that $(\Sigma_3, \Pi_3) \leq_1 (\text{Cof}, \overline{\text{Ext}})$ and hence that Ext is Σ_3-complete. *Hint.* Use the notation and method of Theorem 3.4 to construct $\varphi_{f(x)}$ such that if $x \in A$ then $f(x) \in A \subseteq \text{Ext}$, and if $x \notin A$ then φ_y total implies $\varphi_{f(x)}(z) \neq \varphi_y(z)$ for some $z < b_{x,y}$, where $b_{x,y}$ is the final position of some movable marker $\Gamma_{x,y}$ as described after the proof of Theorem 3.4. Modify the motion of marker $\Gamma_{x,y}$ from the earlier proof as follows. Let $b_{x,y}^s$ be the element upon which $\Gamma_{x,y}$ is resting at the end of stage s. As before if $W_{g(x,y)}$ increases in cardinality at stage $s+1$, put $b_{x,y}^s$ into $W_{f(x)}$ by defining $\varphi_{f(x)}(b_{x,y}^s)$, and moving $\Gamma_{x,y}$ to a fresh element. In addition if $\varphi_{y,s}(b_{x,y}^s) \downarrow = v$ and

$$\neg(\exists z < b_{x,y}^s) \ [\varphi_{f(x)}^s(z) \downarrow \neq \varphi_{y,s}(z) \downarrow],$$

then define $\varphi_{f(x)}(b_{x,y}^s) = 1 \div v$, and move $\Gamma_{x,y}$ to a fresh element. (Exercise 3.8 will be used in Theorem XII.3.4.)

3.9. Prove that $\{\langle x, y \rangle : W_x \text{ and } W_y \text{ are recursively separable}\}$ is Σ_3-complete. *Hint.* Make $\varphi_{f(x)}$ of Exercise 3.8 take values $\subseteq \{0, 1\}$.

3.10. Prove that $\{\langle x, y \rangle : W_x \subseteq^* W_y\}$ and $\{\langle x, y \rangle : W_x =^* W_y\}$ are each Σ_3-complete.

3.11. Prove that $\{x : \varphi_x \text{ total } \& \neg[\lim_y \varphi_x(y) = \infty]\}$ is Σ_3-complete. *Hint.* For Σ_3-completeness use Corollary 3.7 and define

$$\varphi_{f(x)}(s) = (\mu y) \, [W_{g(x,y),s+1} \neq W_{g(x,y),s}].$$

3.12. Define the *weak jump* $H_A = \{x : W_x \cap A \neq \emptyset\}$. Prove that if A is Π_n-complete, then H_A is Σ_{n+1}-complete. (This gives an interesting sequence of sets complete at each level.) Conclude that

$$H_{\mathrm{Inf}} =_{\mathrm{dfn}} \{x : (\exists y) \, [y \in W_x \; \& \; W_y \text{ is infinite}]\}$$

is Σ_3-complete.

3.13. Show that if A is an r.e. set then $G_m(A) \in \Sigma_3$ where

$$G_m(A) =_{\mathrm{dfn}} \{x : W_x \equiv_m A\}.$$

4. *The Relativized Arithmetical Hierarchy and High and Low Degrees*

The definition of Σ_n^A (Π_n^A) is the same as Definition 1.1 for Σ_n (Π_n) except that the matrix R is A-recursive instead of recursive. If $\mathbf{a} = \deg(A)$, we use the notation $\Sigma_n^{\mathbf{a}}$ in place of Σ_n^A since the class Σ_n^A is independent of the particular representative $A \in \mathbf{a}$.

Everything in this chapter can be relativized to an arbitrary set A with virtually the same proofs, and with Σ_n^A, Π_n^A and $A^{(n)}$ in place of Σ_n, Π_n and $\emptyset^{(n)}$, respectively. For example:

4.1 Relativized Post's Theorem. *For every $n \geq 0$,*

    (i) $A^{(n)}$ *is* Σ_n^A-*complete if* $n > 0$;

    (ii) $B \in \Sigma_{n+1}^A \iff B$ *is r.e. in* $A^{(n)}$; *and*

    (iii) $B \leq_T A^{(n)} \iff B \in \Delta_{n+1}^A =_{\mathrm{dfn}} \Sigma_{n+1}^A \cap \Pi_{n+1}^A$.

Define Fin^A, Tot^A, and Cof^A as before but with W_e^A in place of W_e. The proofs in §3 establish that Fin^A is Σ_2^A-complete, Tot^A is Π_2^A-complete, Cof^A

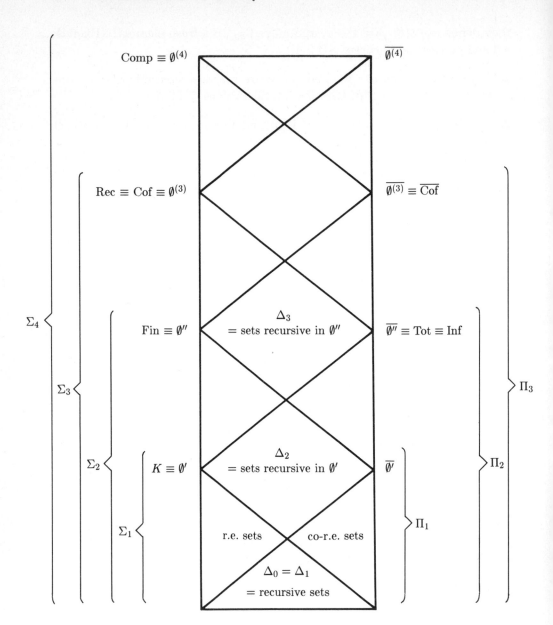

Diagram 3.1. Arithmetical Hierarchy of Sets of Integers

and Rec^A are Σ_3^A-complete, etc. Hence, if $\mathbf{a} = \deg(A)$, then $\mathbf{a}' = \deg(A')$, $\mathbf{a}'' = \deg(\mathrm{Fin}^A)$, and $\mathbf{a}''' = \deg(\mathrm{Cof}^A)$.

It will be useful to classify a degree in terms of its jump. By Theorem III.2.3, if $\mathbf{a} \leq \mathbf{b}$, then $\mathbf{a}' \leq \mathbf{b}'$. Hence if $\mathbf{0} \leq \mathbf{a} \leq \mathbf{0}'$ then $\mathbf{0}' \leq \mathbf{a}' \leq \mathbf{0}''$.

4.2 Definition. (i) A degree $\mathbf{a} \leq \mathbf{0}'$ is *low* if $\mathbf{a}' = \mathbf{0}'$ (i.e., if the jump $\mathbf{a}'$ has the lowest degree possible), and *high* if $\mathbf{a}' = \mathbf{0}''$ (the highest possible value).

(ii) A set $A \leq_T \emptyset'$ is *low* (*high*) if $\deg(A)$ is low (high). (See also Exercise 4.5 and Definition VII.1.3.)

Intuitively, the high sets are those with high information content ("close" to the complete set), and the low sets have low information content (like the recursive sets). Nontrivial high and low r.e. sets will be exhibited in Chapters VII and VIII and will play an important role both in classifying the r.e. degrees and particularly in relating the algebraic structure of an r.e. set to its degree. For these later applications it is useful to have the following characterizations of low and high sets.

4.3 Theorem. *For $A \leq_T \emptyset'$, the following are equivalent:*

   (i) *A is low;*
   (ii) $\Sigma_1^A \subseteq \Pi_2$;
   (iii) $A' \leq_1 \overline{\emptyset^{(2)}} \equiv \mathrm{Tot}$.

Proof.

$$A \text{ is low} \iff A' \leq_T \emptyset',$$
$$\iff A' \in \Delta_2, \text{ by Post's Theorem 2.2,}$$
$$\iff \Sigma_1^A \subseteq \Delta_2, \text{ since } A' \text{ is } \Sigma_1^A\text{-complete,}$$
$$\iff \Sigma_1^A \subseteq \Pi_2, \text{ since } \Sigma_1^A \subseteq \Sigma_1^{\emptyset'} = \Sigma_2,$$
$$\iff A' \leq_1 \overline{\emptyset^{(2)}}, \text{ since } \overline{\emptyset^{(2)}} \text{ is } \Pi_2\text{-complete.} \quad \square$$

4.4 Theorem. *For $A \leq_T \emptyset'$, the following are equivalent:*

   (i) *A is high;*
   (ii) $\Sigma_2 \subseteq \Pi_2^A$;
   (iii) $\emptyset^{(2)} \leq_1 A^{(2)}$ *(i.e., $\mathrm{Fin} \leq_1 \mathrm{Tot}^A$).*

Proof.

$$A \text{ is high} \iff \emptyset'' \leq_T A',$$
$$\iff \emptyset'' \in \Delta_2^A, \text{ by Theorem 4.1,}$$
$$\iff \Sigma_2 \subseteq \Delta_2^A, \text{ since } \emptyset'' \text{ is } \Sigma_2\text{-complete,}$$
$$\iff \Sigma_2 \subseteq \Pi_2^A, \text{ since } \Sigma_2 \subseteq \Sigma_2^A \text{ trivially,}$$
$$\iff \emptyset^{(2)} \leq_1 \overline{A^{(2)}}, \text{ since } \overline{A^{(2)}} \text{ is } \Pi_2^A\text{-complete.} \quad \square$$

4.5–4.11 Exercises

4.5. Note that if $a \leq 0'$ then $0^{(n)} \leq a^{(n)} \leq 0^{(n+1)}$. For each $n > 0$, define a degree $a \leq 0'$ to be low_n ($high_n$) if $a^{(n)} = 0^{(n)}$ ($a^{(n)} = 0^{(n+1)}$), and a set $A \leq_T 0'$ to be low_n ($high_n$) if $\deg(A)$ is low_n ($high_n$). We let $\mathbf{L}_n$ ($\mathbf{H}_n$) consist of the r.e. degrees which are low_n ($high_n$). For convenience, we say that a set A is low_0 ($high_0$) if $A \equiv_T \emptyset$ ($A \equiv_T \emptyset'$) so that $\mathbf{L}_0 = \{\mathbf{0}\}$ and $\mathbf{H}_0 = \{\mathbf{0}'\}$. We abbreviate low_1 ($high_1$) by low (high) as in Definition 4.2. (For more on low_n and $high_n$ r.e. degrees see Definition VII.1.3, Exercise VII.1.6, Corollary VIII.3.4, and Corollary VIII.3.5.) For a set $A \leq_T 0'$, prove that for all $n > 0$,

(a) A is low_n $\iff$ $\Sigma_n^A \subseteq \Pi_{n+1}$ $\iff$ $A^{(n)} \leq_1 \overline{\emptyset^{(n+1)}}$;

(b) A is $high_n$ $\iff$ $\Sigma_{n+1} \subseteq \Pi_{n+1}^A$ $\iff$ $\emptyset^{(n+1)} \leq_1 \overline{A^{(n+1)}}$;

(c) A is low_2 $\iff$ $\mathrm{Fin}^A \leq_1 \overline{\mathrm{Cof}}$;

(d) A is $high_2$ $\iff$ $\mathrm{Cof} \leq_1 \mathrm{Cof}^A$.

(See the application of (c) in the proof of Theorem XI.5.1.)

4.6. (a) For any set $A \subseteq \omega$ define the *weak jump* $H_A = \{e : W_e \cap A \neq \emptyset\}$. Show that $H_A \leq_T K^A$.

(b) We say A is *semi-low* if $H_A \leq_T \emptyset'$. (By (a) if A is low then A is semi-low.) Prove that if $H_A \leq_T \emptyset'$ then $A \leq_T \emptyset'$. (See also Exercises XI.3.5 and XI.3.6 and Notes XIII.1.15 for more about coinfinite r.e. sets A such that $\overline{A}$ is semi-low.)

4.7. Let A be an r.e. set. Prove that the following are equivalent:

(a) $\overline{A}$ is semi-low;

(b) there is a recursive enumeration $\{A_s\}_{s \in \omega}$ of A such that for all e,

$$(\exists^\infty s)\, [W_{e,s} - A_s \neq \emptyset] \implies W_e - A \neq \emptyset.$$

(c) there is a recursive function f such that for all e

(1) $W_e \cap \overline{A} = W_{f(e)} \cap \overline{A}$, and

(2) $W_e \cap \overline{A} = \emptyset \iff W_{f(e)}$ is finite.

Hint. To show (a) implies (b) use the Limit Lemma III.3.3 to find a recursive function $g(e, s)$ such that $\hat{g}(e) = \lim_s g(e, s)$ is the characteristic function of $H_{\overline{A}}$. Choose i such that $W_i = A$. For each $e \leq s$, if $x \in W_{e,s+1} - W_{e,s}$ define

$$t_e = (\mu t \geq s)\, [x \in W_{i,t} \text{ or } g(e, t) = 0],$$

and let $A_{s+1} = W_{i,t}$, where $t = \max\{t_e : e \leq s\}$. (See also Exercise XI.3.6.)

4.8. Prove that in Exercise 4.7(c)(2) the hypothesis $W_e \cap \overline{A} = \emptyset$ may be replaced by $W_e \cap \overline{A} =^* \emptyset$.

4.9.   Using the method of Exercise 4.7, prove that if A is r.e. then A is low iff there is a recursive function f such that for all j,

(1) $(\forall x)\,[D_x \subseteq \overline{A} \implies [x \in W_j \iff x \in W_{f(j)}]]$, and
(2) $\{\, x : D_x \subseteq \overline{A} \ \& \ x \in W_j \,\} = \emptyset \iff W_{f(j)}$ is finite.

State and prove an equivalent enumeration condition analogous to Exercise 4.7(b).

4.10.   Prove that for every r.e. set A there exists an r.e. set $B \equiv_T A$ such that $H_{\overline{B}} \equiv_T A'$. (Hence if A is not low then $\overline{B}$ is not semi-low.) (Exercise 4.10 is used in Exercise XII.4.22.) *Hint.* Let $B = \{\, n : D_n \cap A \neq \emptyset \,\}$.

4.11.   Prove that for every r.e. set A there exists an r.e. set $B \equiv_T A$ such that $\overline{B}$ is semi-low. *Hint.* Let $\{\, A_s \,\}_{s \in \omega}$ be a recursive enumeration of A. Let $\{\, D_{f(n)} \,\}_{n \in \omega}$ be a canonical sequence of disjoint finite sets such that $|D_{f(n)}| = n + 1$ and $\bigcup_n D_{f(n)} = \omega$. Let $\Gamma_e^s = \mu x[x \in W_{e,s} - B_s]$ if x exists and $= 0$ otherwise. If $n \in A_{s+1} - A_s$, let

$$x_n = \mu x[x \in D_{f(n)} - \bigcup\{\,\Gamma_e^s : e < n \,\}],$$

and enumerate x_n in B_{s+1}. Show that $n \in A$ iff $D_{f(n)} \cap B \neq \emptyset$, so $A \equiv_T B$.

Part B

Post's Problem, Oracle Constructions, and the Finite Injury Priority Method

Chapter V
Simple Sets and Post's Problem

Part A included the basic results and methods developed during the period 1931 to 1943. Part B does the same for the period roughly from 1944 to 1960. In each case results are included which were proved later, but which use only the methods developed during that period. In his fundamental paper [1944], Post stripped away the formalism associated with the development of recursive functions in the 1930's and revealed in a clear informal style the essential properties of r.e. sets and their application to Gödel's Incompleteness Theorem. He went on to attempt to classify the r.e. sets and their degrees. He raised the question (which became known as Post's problem) of whether there are more than two r.e. degrees. In his attempt to solve this problem he introduced various classes of simple sets which we study in Chapter V. This chapter also includes a very pleasing extension of the Recursion Theorem and a completeness criterion for r.e. sets which was proved much later by Martin, Lachlan and Arslanov. Attempts to give a positive answer to Post's problem by producing a simplicity-like property which guaranteed incompleteness were unsuccessful. Meanwhile, Kleene and Post [1954] studied the upper semilattice of the Turing degrees in general (rather than just the r.e. degrees) and used diagonal arguments with oracle constructions (i.e., nonrecursive constructions) to produce incomparable degrees below $\mathbf{0}'$. These methods are presented in Chapter VI along with the Friedberg Completeness Criterion and the Spector minimal degree construction. Finally, Friedberg [1957] and independently Muchnik [1956] solved Post's problem by inventing the finite injury priority method which is a more subtle and effective version of the Kleene-Post method where the construction must now be recursive in order to yield r.e. degrees. These results are included in Chapter VII along with a splitting theorem of Sacks whose method will be very useful later. For a solution to Post's problem in the style of Post see the remark following Corollary 2.6. For a recent injury free solution see Remark VII.1.9.

1. Immune Sets, Simple Sets, and Post's Construction

The only r.e. degrees constructed so far are $\mathbf{0}$ and $\mathbf{0}'$. *Post's Problem* was to construct other r.e. degrees, i.e., to construct an r.e. set A such that

$\emptyset <_T A <_T \emptyset'$. *Post's Program* for constructing such a set A was to find some easily definable property on $\overline{A}$ (compatible with A being nonrecursive) which guarantees incompleteness ($K \not\leq_T A$).

Recall (from Theorem II.4.5) that K (and hence any creative set) has an infinite r.e. set (and therefore infinitely many infinite r.e. sets) contained in its complement. Post's idea for constructing A incomplete was to make $\overline{A}$ sufficiently "thin" with respect to containment of infinite r.e. sets so that $\overline{K}$ could not be Turing reduced to $\overline{A}$.

1.1 Definition. (i) A set is *immune* if it is infinite but contains no infinite r.e. set.

(ii) A set A is *simple* if A is r.e. and $\overline{A}$ is immune.

1.2 Proposition.   *If A is simple then:*

   (i) *A is not recursive;*
   (ii) *A is not creative;*
   (iii) *A is not m-complete (i.e., $K \not\leq_m A$).*

Proof. (i) $\overline{A}$ cannot be r.e., since otherwise it contains an infinite r.e. set (namely $\overline{A}$).

(ii) The complement of a creative set does contain an infinite r.e. set (by Theorem II.4.5).

(iii) If $K \leq_m A$, then A is creative (by Corollary II.4.8).   ▯

1.3 Theorem (Post [1944]).   *There exists a simple set S.*

Proof. Let $A = \{ (e,x) : x \in W_e$ & $x > 2e \}$. Now A is Σ_1 and hence r.e. Let ψ be any p.r. selector function for A (in the sense of Theorem II.1.6) and let $S = \operatorname{rng} \psi$. (Intuitively, enumerate W_e until the first element $\psi(e) > 2e$ appears in W_e and then put $\psi(e)$ into S.) The following facts give the simplicity of S.

(1) S is r.e. (S is the range of a p.r. function.)

(2) $\overline{S}$ is infinite. To see this, note that S contains at most e elements out of $\{ 0,1,2,\dots,2e \}$, namely $\psi(0), \psi(1), \dots, \psi(e-1)$. Hence $e+1 \leq \operatorname{card}(\overline{S} \restriction (2e+1))$, so $\overline{S}$ is infinite.

(3) If W_e is infinite, then $W_e \cap S \neq \emptyset$, because $(e,x) \in A$ for some $x > 2e$ and so $\psi(e)$ is defined and $\psi(e) \in S \cap W_e$.   ▯

For another construction of a simple set which will be useful see Exercise 1.6.

1.4 Definition. A simple set A is *effectively simple* if there is a recursive function f such that

(1.1) $$(\forall e) \, [W_e \subseteq \overline{A} \implies |W_e| \leq f(e)].$$

Note that Post's simple set is effectively simple via $f(e) = 2e+1$, because if $W_e \subseteq \overline{S}$ then $W_e \subseteq \{\, 0, 1, \ldots, 2e \,\}$.

1.5–1.10 Exercises

1.5. Let S be the class of simple sets and C the class of cofinite sets. Prove that $S \cup C$ is a filter in $\mathcal{E}$. (In Exercise 2.8 we shall see this for the h-simple sets, and in Exercise X.2.17 for the hh-simple sets.)

1.6. Give a direct "movable marker" style construction of a simple set A, by satisfying the following requirements for $e \in \omega$:

$$P_e : W_e \text{ infinite} \implies W_e \cap A \neq \emptyset,$$
$$N_e : |\overline{A}| \geq e.$$

Hint. Let $A = \bigcup_s A_s$, where A_s consists of the elements enumerated in A by the end of stage s in the following recursive construction. (Thus, the sequence $\{\, A_s \,\}_{s \in \omega}$ is recursive, so A is r.e. because $x \in A$ iff $(\exists s) \, [x \in A_s]$.) Let $A_0 = \emptyset$. Given A_s, let $\overline{A}_s = \{\, a_0^s < a_1^s \ldots \,\}$. (Think of a_n^s as the position at the end of stage s of a movable marker Γ_n.) At stage $s+1$ find the least $e \leq s$ such that: (1) $W_{e,s} \cap A_s = \emptyset$; and (2) $a_n^s \in W_{e,s}$ for some $n \geq e$. Enumerate a_n^s in A. If e fails to exist pass to the next stage. Note that clauses (1) and (2) guarantee that $\lim_s a_e^s < \infty$ so requirement N_e is met.

1.7. Show that the set A constructed using the hint in Exercise 1.6 is effectively simple via the function $f(e) = e$.

1.8. Show that $\{\, e : W_e \text{ is simple} \,\}$ is Π_3-complete. (See also Corollary XII.4.8(ii).) *Hint.* Combine the methods of Exercise 1.6 and Corollary IV.3.7 as follows. Let A be Σ_3-complete and g be as in Corollary IV.3.7. Construct a recursive function f such that if $x \in A$ then $W_{f(x)}$ is cofinite, and if $x \notin A$ then $W_{f(x)}$ is simple. Construct $W_{f(x)}$ using markers $\{\, \Gamma_n^x \,\}_{n \in \omega}$ as in Exercise 1.6 but also move Γ_n^x whenever a new element is enumerated in $W_{g(x,n)}$.

1.9. Prove that if $A \leq_m S$ and A is part of a recursively inseparable pair (as defined in Exercise I.4.22) of r.e. sets then S is not simple. Note that $K \equiv \{\, x : \varphi_x(x) = 0 \,\}$, and hence by Exercise II.4.13, K is part of an effectively inseparable (and therefore recursively inseparable) pair of r.e. sets. (Hence, a simple set is not m-complete.) *Hint.* Let (A, B) be recursively inseparable, $A \leq_m S$ via f, and consider $f(B)$.

1.10. Prove that the set S of minimal indices of p.r. functions is r.e. where

$$S =_{\text{dfn}} \{\, x : \neg(\exists y < x)\, [\varphi_x = \varphi_y]\,\}.$$

Hint. Use the Recursion Theorem.

2. Hypersimple Sets and Majorizing Functions

Although Post proved that simple sets are necessarily m-incomplete (see Exercise 1.9) and even incomplete for a certain intermediate reducibility called bounded truth-table denoted by btt and defined in Exercise 2.13, Post realized that simple sets could be Turing complete. Indeed in §4 we shall show that *every* effectively simple set is T-complete. Thus, Post continued by defining coinfinite r.e. sets with even thinner complements called hypersimple (h-simple) and hyperhypersimple (hh-simple) sets. Although these sets also failed to solve Post's problem, they were later shown to have very interesting characterizations which gave considerable information about the structure of nonrecursive r.e. sets and about the relationship between an r.e. set and its degree. (Recall that D_y is the finite set with canonical index y as defined in Definition II.2.4.)

2.1 Definition. (i) A sequence $\{\, F_n \,\}_{n \in \omega}$ of *finite* sets is a *strong (weak) array* if there is a recursive function f such that $F_n = D_{f(n)}$ $(F_n = W_{f(n)})$. (See also Definition II.2.4.)

(ii) An array is *disjoint* if its members are pairwise disjoint.

(iii) An infinite set B is *hyperimmune*, abbreviated *h-immune*, (*hyperhyperimmune*, abbreviated *hh-immune*) if there is no disjoint strong (weak) array $\{\, F_n \,\}_{n \in \omega}$ such that $F_n \cap B \neq \emptyset$ for all n.

(iv) An r.e. set A is *hypersimple*, abbreviated *h-simple* (*hyperhypersimple*, abbreviated *hh-simple*) if $\overline{A}$ is h-immune (hh-immune).

The intention behind h-simplicity and hh-simplicity is that instead of specifying an r.e. set $\{\, a_n \,\}_{n \in \omega} \subseteq \overline{A}$ we specify a disjoint r.e. array of finite sets $\{\, F_n \,\}_{n \in \omega}$ such that each F_n contains some $x \in \overline{A}$ but we cannot tell which $x \in F_n$ has this property. In a strong array we can explicitly compute $\max(F_n)$ and all its members, but in a weak array we can merely enumerate F_n. It can be easily shown (Exercise 2.18) that hh-simple implies h-simple and h-simple implies simple. We shall prove later that these implications are not reversible.

2.2 Definition. (i) A function f *majorizes* a function g if $f(x) \geq g(x)$ for all x, and f *dominates* g if $f(x) \geq g(x)$ for a.e. x. (See also Definition I.5.5.)

(ii) If $A = \{ a_0 < a_1 < a_2 < \cdots \}$ is an infinite set, the *principal function* of A is p_A, where $p_A(n) = a_n$.

(iii) A function f *majorizes* (*dominates*) an infinite set A if f majorizes (dominates) p_A. Similarly, A *majorizes* (*dominates*) f if p_A does.

2.3 Theorem (Kuznecov, Medvedev, Uspenskii). *An infinite set A is h-immune iff no recursive function f majorizes A.*

Proof. ($\Longleftarrow$). Let $\{ D_{g(x)} \}_{x \in \omega}$ be a disjoint strong array such that $D_{g(x)} \cap A \neq \emptyset$ for all x. Set $f(x) = \max(\bigcup \{ D_{g(y)} \}_{y \leq x})$. Then $f(x) \geq p_A(x)$ for all x.

($\Longrightarrow$). Assume f is a recursive function such that $f(x) \geq p_A(x)$ for all x. Set $D_{g(0)} = [0, f(0)]$ where $[n, m] = \{ n, n+1, \ldots, m \}$. Suppose we are given $g(0), g(1), \ldots, g(n)$. Let $k = 1 + \max(\bigcup \{ D_{g(i)} \}_{i \leq n})$ and define $D_{g(n+1)} = [k, f(k)]$. Now $k \leq p_A(k) \leq f(k)$ so $p_A(k) \in D_{g(n+1)} \cap A$. ▯

2.4 Corollary. *A coinfinite r.e. set A is h-simple iff no recursive function majorizes $\overline{A}$.* ▯

Note that Post's simple set S of Theorem 1.3 is not hypersimple since $\overline{S}$ is majorized by $f(x) = 2x$. If A is r.e. and $p_{\overline{A}}$ dominates *every* recursive function then A is called *dense simple* and must have high degree. (See Exercise XI.1.11.) Contrast this with the following theorem which asserts that hypersimple sets exist in *every* nonzero r.e. degree.

2.5 Theorem (Dekker [1954]). *For every nonrecursive r.e. set A there is an h-simple set $B \equiv_T A$.*

Proof. Let $A = \operatorname{rng} f$, for f a 1:1 recursive function, and let $a_s = f(s)$, and $A_s = \{ f(0), \ldots, f(s) \}$. Define $B = \{ s : (\exists t > s) [a_t < a_s] \}$, the *deficiency set* of A for the enumeration f. Clearly B is Σ_1 and hence r.e. and clearly $\overline{B}$ is infinite. Note that

$$(2.1) \qquad\qquad x \in A \text{ iff } x \in \{ a_0, a_1, \ldots, a_{p_{\overline{B}}(x)} \}$$

by the definition of B and the fact that $x \leq a_{p_{\overline{B}}(x)}$. Hence, $A \leq_T B$. Next, $\overline{B}$ is h-immune, for if some recursive function g majorizes $p_{\overline{B}}$ then by (2.1) $x \in A$ iff $x \in \{ a_0, a_1, \ldots, a_{g(x)} \}$, which would imply that A is recursive. Finally, $B \leq_T A$ since to test whether $s \in B$, we A-recursively compute the least t such that $A_t \upharpoonright a_s = A \upharpoonright a_s$. Now $s \in B$ iff $t > s$, i.e., iff $s \in B_t$. Indeed $B \leq_{tt} A$ as shown in Exercise 2.12. ▯

The crucial point about the deficiency set B is that any *nondeficiency* stage $s \in \overline{B}$ is a "true" stage in the enumeration $\{ A_s \}_{s \in \omega}$ of A in the sense that $A_s \upharpoonright f(s) = A \upharpoonright f(s)$. These stages will prove very useful in our study of the infinite injury priority method in Chapter VIII.

2.6 Corollary. *Every nonrecursive r.e. degree contains a simple (indeed h-simple) set.*

It follows from Corollary 2.6 that h-simplicity does not guarantee incompleteness. Although h-simple sets can be easily constructed directly (see Exercise 2.9 below) it is not obvious how to construct an hh-simple set, and indeed the first and most natural example is a maximal set which we shall construct in Chapter X. Maximal sets will be seen to have the thinnest possible complement that a coinfinite r.e. set can have. However, Yates [1965] constructed a complete maximal set, thereby refuting Post's idea that thinness of complement of a coinfinite r.e. set could guarantee incompleteness. Later Marchenkov [1976] found a combination of properties in the style of Post guaranteeing incompleteness, but the proof of the existence of a set, with these properties requires the same kind of finite injury priority argument introduced originally by Friedberg and Muchnik to solve Post's problem, which we shall present in Chapter VII. For an exposition of the Marchenkov solution in the style of Post see Odifreddi [1981].

2.7–2.18 Exercises

2.7. A set S is *introreducible* if $S \leq_T T$ for every infinite set $T \subseteq S$. For the deficiency set B of Theorem 2.5 prove that $\overline{B}$ is introreducible.

2.8. Show that the h-simple sets together with the cofinite sets form a filter in $\mathcal{E}$. (In Exercise 1.5 we saw this for the simple sets.)

2.9. Give a direct "movable marker" type construction of an h-simple set. *Hint.* Modify the construction of Exercise 1.6 with P_e replaced by the requirement

$$\hat{P}_e : \{ D_{\varphi_e(x)} \}_{x \in \omega} \text{ a disjoint strong array} \implies (\exists x) [D_{\varphi_e(x)} \subseteq A].$$

At stage $s + 1$ the strategy for meeting $\hat{P}_e$ is to find an x such that:

$$(\forall y \leq x) [\varphi_{e,s}(y) \downarrow];$$

$$(\forall y, z \leq x) [D_{\varphi_{e,s}(y)} \cap D_{\varphi_{e,s}(z)} = \emptyset];$$

and

$$D_{\varphi_{e,s}(x)} \subseteq [a_{e+1}^s, \infty).$$

Then enumerate in A all members of $D_{\varphi_{e,s}(x)}$.

2.10. Show that $\{ x : W_x \text{ is h-simple} \}$ is Π_3-complete. (See the hint for Exercise 1.8.)

2.11. Show that any coinfinite r.e. set B has an h-simple superset A. *Hint.* If B is not already h-simple, then by Theorem 2.3 there is an increasing recursive function f such that for all n, $[f(n), f(n + 1)) \cap \overline{B} \neq \emptyset$. Let $F_n = [f(n), f(n+1))$. Play the strategy of Exercise 2.9 but with the markers Γ_e each associated with some *set F_n* instead of an *integer n*. When x is enumerated in B, it is also enumerated in A. In addition, when F_n has no marker then all its members are put into A. Begin by associating Γ_e with F_e. To meet $\hat{P}_e$ of Exercise 2.9 wait for some x such that $\varphi_e(x) \downarrow$ and $D_{\varphi_e(x)} \cap F_n = \emptyset$ for each F_n associated currently with a marker Γ_i, $i \leq e$. For each $y \in D_{\varphi_e(x)}$ find n such that $y \in F_n$, enumerate *all* members of F_n into A and move the marker (if any) associated with F_n to some F_m, $m \geq n$, F_m not yet a subset of A. Prove that each Γ_e moves finitely often and comes to rest on some F_n such that $F_n \cap \overline{A} \neq \emptyset$. Hence, $\overline{A}$ is infinite. (Note that A is not obtained *uniformly* from B.)

2.12. A set A is *truth-table reducible* to a set B, abbreviated *tt-reducible* ($A \leq_{tt} B$), if there are recursive functions f and g such that $x \in A$ if and only if $B \upharpoonright f(x) = D_y$ for some $y \in D_{g(x)}$. (In the truth table imagine a row corresponding to each of the $2^{f(x)}$ possible choices for the characteristic function of $B \upharpoonright f(x)$. The set $D_{g(x)}$ contains code numbers y for those rows where "true" is the correct answer on that row to the question "Is $x \in A$?".) Prove for A and the deficiency set B as in Theorem 2.5 that $B \leq_{tt} A$.

2.13. A set A is *bounded truth-table reducible* to a set B (written $A \leq_{btt} B$) if there is some n (called the *norm* of the reduction) and recursive functions f and g such that for all x: (1) $|D_{f(x)}| \leq n$ and (2) $x \in A$ iff $B \cap D_{f(x)} = D_y$ for some y in $D_{g(x)}$. (For example, for any set A, $A \leq_{btt} \overline{A}$.) Let $\mathcal{B}$ be the Boolean algebra generated by the r.e. sets as in Exercise III.3.9. Prove that $A \in \mathcal{B}$ iff $A \leq_{btt} K$. (Note that $A \leq_m B \implies A \leq_{btt} B \implies A \leq_{tt} B$, and that $\leq_{btt}$ and $\leq_{tt}$ are reflexive and transitive and therefore induce equivalence relations $\equiv_{btt}$ and $\equiv_{tt}$.)

2.14. Prove that if A and B are r.e. and $\neq \omega$, $\neq \emptyset$, and $A \leq_{btt} B$ with norm 1, then $A \leq_m B$.

2.15$^{\diamond\diamond}$ (Kobzev). Prove by induction on the norm n that if $A \leq_{btt} B$ and A is part of a recursively inseparable pair of r.e. sets then B is not simple. For $n = 1$ use Exercises 2.14 and 1.9. (For a sketch see Odifreddi [1981, Theorem 8.2].)

2.16 (Friedberg and Rogers [1959]). A set A is *weak truth-table reducible* to a set B (written $A \leq_{wtt} B$) if $A = \{e\}^B$ for some e, and there is a recursive function $f(x)$ such that $f(x) \geq u(B; e, x)$, the use function defined in Definition III.1.7, for the computation $\{e\}^B(x)$. (This resembles $A \leq_{tt} B$

except that the function g of Exercise 2.12 is missing.) It is also called *bounded Turing reducibility* and is sometimes written $A \leq_{bT} B$. We also define $g \leq_{wtt} h$ for functions g and h by using their graphs in the above definition as for $g \leq_T h$ in Definition III.1.2(ii). Show that we cannot have $K \leq_{wtt} H$ for H h-simple. *Hint* (Owings). Assume $K = \{e\}^H$ with f as above. It suffices to show that for every interval $[0, y]$ which intersects $\overline{H}$ we can find uniformly in y some $z > y$ such that $[y + 1, z] \cap \overline{H} \neq \emptyset$. Suppose $F = \overline{H} \cap [0, y]$. Let W_n consist of all x such that

$$(\exists \sigma) \, [\{e\}^\sigma(x) = 0 \ \& \ (\forall y < lh(\sigma) \, [[\sigma(y) = 0 \implies y \in F]$$
$$\& \ [\sigma(y) = 1 \implies y \in H]]].$$

Apply to W_n the productive function for $\overline{K}$ and then apply f to obtain z^F. Let $z = \max\{z^F : F \subseteq [0, y]\}$.

2.17 (Miller and Martin [1968]). A degree **a** is *hyperimmune* if **a** contains a hyperimmune set, and **a** is *hyperimmune-free* otherwise.

(a) Using Theorem 2.3 prove that if **a** is hyperimmune and **a** < **b** then **b** is hyperimmune.

(b) Show that **a** is hyperimmune iff some function of degree $\leq$ **a** is majorized by no recursive function.

(c) Show that if $(\exists \mathbf{a}) \, [\mathbf{a} < \mathbf{b} \leq \mathbf{a}']$, then **b** is hyperimmune. (Use (b) and the Limit Lemma.)

(d) Show that every nonzero degree comparable with $\mathbf{0}'$ is hyperimmune.

2.18. (a) Prove that every h-simple set is simple.

(b) Prove that every hh-simple set is h-simple. (The converse is false since by Theorem XI.1.7 hh-simple sets must have high degree but by Theorem 2.5 h-simple sets lie in every nonzero r.e. degree.)

3. The Permitting Method

Friedberg [1957b] and Yates [1965] introduced an easy method for constructing an r.e. set A which is recursive in a given nonrecursive r.e. set B by enumerating an element x in A at some stage s only when B *permits* x in the sense that some element $y < x$ appears in B at stage s. By combining this permitting with coding of B into A we can often construct an r.e. set A of the same degree as B such that A has some specified property. For example, here we construct a simple nonhypersimple set of arbitrary nonzero r.e. degree. (In Exercise 5.6 we shall see that step 2 is really unnecessary and step 1 alone suffices.)

3.1 Proposition. *If* $\{A_s\}_{s\in\omega}$ *and* $\{B_s\}_{s\in\omega}$ *are recursive enumerations of r.e. sets* A *and* B *such that* $x \in A_{s+1} - A_s$ *implies* $(\exists y < x)\,[y \in B - B_s]$ *then* $A \leq_T B$.

Proof. To B-recursively compute whether $x \in A$ find a stage s such that $B_s \restriction x = B \restriction x$. Now $x \in A$ iff $x \in A_s$. ▯

3.2 Theorem. *For any nonrecursive r.e. set* B *there is a simple nonhypersimple set* $A \equiv_T B$.

Proof. Let f be a 1:1 recursive function with range B and let

$$B_s = \{\, f(0), \ldots, f(s) \,\}.$$

Let A_s be the set of elements enumerated in A by the end of stage s. At stage $s = 0$ do nothing. Suppose $\overline{A}_s = \{\, a_0^s < a_1^s < \cdots \}$.

> *Stage $s + 1$.*
>
>   *Step 1.* (To make A simple and $A \leq_T B$.) For all e (necessarily $e \leq s$) such that $W_{e,s} \cap A_s = \emptyset$ and

$$(3.1) \qquad\qquad (\exists x)\,[x > 3e \ \& \ x \in W_{e,s} \ \& \ f(s+1) < x],$$

enumerate in A the least such x corresponding to e. If no such e exists do nothing. In either case go to Step 2.

>   *Step 2.* (To code B into A.) Enumerate $a^s_{3f(s+1)+1}$ in A.

Note that $|\overline{A} \cap [0, 3e]| \geq e$ since at most e elements $y \leq 3e$ can be enumerated in A under Step 1 (one for each W_i, $i < e$), and at most e elements under Step 2 (one for each $i \in B$, $i < e$, because $y \leq a_y^s$ so if $y \in A_{s+1} - A_s$ under Step 2 then $y = a^s_{3x+1}$ for $x \in B_{s+1} - B_s$ so $3e \geq y \geq 3x+1$, and hence $e > x$). Hence $\overline{A}$ is majorized by the function $f(x) = 3x$, so $\overline{A}$ is infinite and A is not h-simple.

Note that $A \leq_T B$ by the third clause of (3.1) and the fact that $y \leq a_y^s$, so $B_s \restriction x = B \restriction x$ implies $A_s \restriction x = A \restriction x$ as in Proposition 3.1. Now $B \leq_T A$ since given x, we can A-recursively compute s such that $a^s_{3x+1} = \lim_t a^t_{3x+1}$. Now $x \in B$ iff $x \in B_s$.

To see that A is simple assume for a contradiction that W_e is infinite and $W_e \cap A = \emptyset$ with e minimal. Then by Step 1 the set

$$U_e = \{\, x : (\exists s)\,[x \in W_{e,s} \ \& \ f(s+1) < x]\,\}$$

is finite, contrary to the hypothesis that B is nonrecursive. ▯

The permitting of Proposition 3.1 is the simplest kind of permitting and in fact achieves the stronger reducibility $A \leq_{wtt} B$ as defined in Exercise 2.16. A more general and useful permitting is to permit x to enter A at a stage $s+1$ only if some $y < \Gamma_x^s$ enters B at stage $s+1$ where Γ_x^s is the position at the end of stage s of a marker Γ_x which comes to a limit and which moves only when a smaller element enters B. For example in Chapter XI we wish to have the principal function of B permit so we define $\Gamma_x^s = b_x^s$ where $\overline{B}_s = \{ b_0^s < b_1^s < \cdots \}$. In the following proposition think of $f(x, s)$ as Γ_x^s.

Proposition 3.3. *Let* $\{ A_s \}_{s \in \omega}$ *and* $\{ B_s \}_{s \in \omega}$ *be recursive enumerations of r.e. sets A and B. Let $f(x, s)$ be a recursive function such that for every x and s,*

> (i) $f(x, s) \neq f(x, s+1) \implies (\exists y < f(x, s)) [y \in B_{s+1} - B_s];$
> (ii) $\hat{f}(x) =_{\text{dfn}} \lim_s f(x, s) < \infty;$ *and*
> (iii) $x \in A_{s+1} - A_s \implies (\exists y < f(x, s)) [y \in B_{s+1} - B_s];$

then $A \leq_{\mathrm{T}} B$.

Proof. For each x we B-recursively compute t such that $B \upharpoonright f(x, t) = B_t \upharpoonright f(x, t)$. Now $f(x, t) = \lim_s f(x, s)$ and $x \in A$ iff $x \in A_t$. (Note that (i) and (ii) alone imply that $\hat{f} \leq_{\mathrm{T}} B$.)   ☐

We should think of this permitting as an attempt to build a Turing reducibility Ψ such that $A = \Psi(B)$. From time to time we issue a value $\Psi(B_s; x) = y$ and we assign "use" function $\psi(x, s)$ for argument x. Namely, we enumerate into a certain r.e. set V_Ψ, which defines Ψ, the "axiom" $\langle \sigma, x, y \rangle$, where $\sigma = B_s \upharpoonright \psi(x, s)$. This indicates that $\Psi_s^\sigma(x) = y$ with use $\psi(x, s)$. At a later stage $t+1$ we can change our mind about the value y or redefine the use function only if $B_{t+1} \upharpoonright \psi(x, s) \neq B_t \upharpoonright \psi(x, s)$, so that $\sigma \not\subseteq B_{t+1}$.

It is useful to contrast this Yates permitting method with the Martin permitting method of Chapter XI §2 where an r.e. set A satisfying certain *infinitary* positive requirements is constructed below a *high* r.e. set. See Remark XI.2.6.

3.4–3.6 Exercises

3.4. Prove that the permitting of Proposition 3.3 is the most general kind in the sense that given nonrecursive r.e. sets A and B with $A \leq_{\mathrm{T}} B$ we can find recursive enumerations and a recursive function f satisfying (i), (ii) and (iii) of Proposition 3.3. *Hint.* Let $A = \{ e \}^B$. Fix recursive enumerations $\{ \hat{A}_t \}_{t \in \omega}$ and $\{ \hat{B}_t \}_{t \in \omega}$ of A and B. Given A_s, B_s, $b_s \in B$, and $\{ f(x, s) \}_{x \in \omega}$ find the least t such that $A_s \subset \hat{A}_t$, $\{ b_s \} \cup B_s \subseteq \hat{B}_t$ and $\hat{A}_t(y) = \{ e \}_t^{\hat{B}_t}(y)$ for all $y \leq s$. Set $A_{s+1} = \hat{A}_t$ and $B_{s+1} = \hat{B}_t$. For each $y \leq s$ such that $\hat{B}_t \upharpoonright f(y, s) \neq B_s \upharpoonright f(y, s)$ define $f(y, s+1) = u(\hat{B}_t; e, y, t)$, where the latter was defined in Definition III.1.7. Let b_{s+1} be some element in $B - \hat{B}_t$. For $y > s$ define $f(y, s+1) = 1 + b_{s+1}$.

3.5 (Yates).   An r.e. set A is *semicreative* if there is a recursive function f such that if $W_x \subseteq \overline{A}$ then $W_{f(x)} \subseteq \overline{A}$ and $W_{f(x)} \not\subseteq W_x$. Show that for every nonrecursive r.e. set B there is a semicreative set $A \equiv_T B$.   *Hint.* To get $A \leq_T B$ enumerate $2^x 3^y$ in A at stage $s + 1$ if $2^x 3^y \in W_{x,s}$ and $B_{s+1} \upharpoonright y \neq B_s \upharpoonright y$. Let $W_{f(x)} = \{\, 2^x 3^y : y \in \omega \,\}$. (See also Exercises VII.2.7 and XI.2.10 and Notes XI.2.19.)

3.6.   Use the permitting method to prove that there is a hypersimple set in every nonrecursive r.e. degree.

4. Effectively Simple Sets Are Complete

Although Post constructed a complete simple set, he did not realize that his original simple set and indeed all effectively simple sets are complete.

4.1 Theorem (Martin [1966a]).   *If A is effectively simple, then A is complete.*

Proof. Let $\{\, K_s \,\}_{s \in \omega}$ be a recursive enumeration of K and $\theta(x)$ a p.r. function such that $\theta(x) = (\mu s)\, [x \in K_s]$ if $x \in K$ and $\theta(x)$ diverges otherwise. Let A be effectively simple via f, $\{\, A_s \,\}_{s \in \omega}$ an enumeration of A, and $\overline{A}_s = \{\, a_0^s < a_1^s < a_2^s < \cdots \,\}$. By the Recursion Theorem with parameters, define the recursive function h by

$$(4.1) \qquad W_{h(x)} = \begin{cases} \{\, a_0^{\theta(x)}, a_1^{\theta(x)}, \ldots, a_{fh(x)}^{\theta(x)} \,\} & \text{if } x \in K; \\ \emptyset & \text{otherwise.} \end{cases}$$

Set $r(x) = (\mu s)\, [a_{fh(x)}^s = a_{fh(x)}]$. Then $r \leq_T A$ since f and h are recursive. Now if $x \in K$ and $r(x) \leq \theta(x)$, then $W_{h(x)} \subseteq \overline{A}$ and $|W_{h(x)}| = fh(x) + 1$, contrary to the hypothesis on f. Hence, for all $x \in K$, $r(x) > \theta(x)$. Thus for all x, $x \in K$ iff $x \in K_{r(x)}$. Therefore $K \leq_T A$.   ▯

4.2–4.4 Exercises

4.2.   A set A is *effectively immune* if A is infinite and there is a recursive function f such that if $W_x \subseteq A$ then $|W_x| < f(x)$. Prove that if A is effectively immune, B is r.e., and $A \leq_T B$, then $B \equiv_T \emptyset'$.   *Hint.* Use the Modulus Lemma III.3.2 and modify the proof of Theorem 4.1.

4.3 (Smullyan-Martin).   (a) Prove that if A is effectively simple and $A \subseteq B \subset_\infty \omega$ and B is r.e. then B is effectively simple. (Recall that $X \subset_\infty Y$ denotes that $X \subset Y$ and $Y - X$ is infinite.)

(b) Show that there is an effectively simple hypersimple set. (Use Exercise 2.11. Note that Post's simple set S of Theorem 1.3 is effectively simple and *not* hypersimple.)

4.4 (Shoenfield [1957]).   *A* is *quasi-creative* if *A* is r.e. and

$$(\exists \text{ recursive } f)\,(\forall x)\,[W_x \subseteq \overline{A} \implies [D_{f(x)} \subset \overline{A} \ \& \ D_{f(x)} \not\subseteq W_x]].$$

Prove that *A* quasi-creative implies *A* complete. (Use the Recursion Theorem in the manner of Theorem 4.1 or Theorem II.4.6.)

5. A Completeness Criterion for R.E. Sets

Martin [1966a] actually proved considerably more than Theorem 4.1 by giving a sufficient condition for an r.e. set to be complete. His condition applied to numerous other sets which are "effectively" nonrecursive r.e. sets including creative and quasi-creative sets and truth-table complete sets. Lachlan [1968a] modified Martin's condition so that it became both necessary and sufficient. Arslanov [1977] then converted Lachlan's condition to the following form which can be viewed as a generalization of the Recursion Theorem. It asserts that not only recursive functions but all those of r.e. degree less than $\mathbf{0}'$ have fixed points. (The following proof of Theorem 5.1 is a simplification due to Soare and appears in Arslanov [1981]. Arslanov's original proof used Lachlan's completeness criterion (Lachlan [1968e]) and can be found in Arslanov, Nadirov, and Solov'ev [1977].)

5.1 Theorem (Arslanov Completeness Criterion) (Arslanov [1981]).   *An r.e. set A is complete iff there is a function $f \leq_T A$ such that $W_{f(x)} \neq W_x$ for all x.*

Proof. ($\implies$). Recall that $\{\, x : W_x = \emptyset \,\} \equiv_T \emptyset'$ by Theorem II.4.2. Define $f \leq_T \emptyset'$ by

$$W_{f(x)} = \begin{cases} \emptyset & \text{if } W_x \neq \emptyset, \\ \{0\} & \text{otherwise.} \end{cases}$$

($\impliedby$). Assume $(\forall x)\,[W_{f(x)} \neq W_x]$ where $f \leq_T A$. By the Modulus Lemma there is a recursive function $\hat{f}(x, s)$ such that $f(x) = \lim_s \hat{f}(x, s)$ for every x, and the sequence $\{\, \lambda x \hat{f}(x, s)\,\}_{s \in \omega}$ has a modulus $m \leq_T A$. Let $\{\, K_s \,\}_{s \in \omega}$ be a recursive enumeration of K. Let $\theta(x) = (\mu s)\,[x \in K_s]$ if $x \in K$, and let $\theta(x)$ diverge otherwise. By the Recursion Theorem with Parameters define the recursive function h by

$$W_{h(x)} = \begin{cases} W_{\hat{f}(h(x),\theta(x))} & \text{if } x \in K; \\ \emptyset & \text{otherwise.} \end{cases}$$

Now if $x \in K$ and $\theta(x) \geq m(h(x))$, then $\hat{f}(h(x), \theta(x)) = f(h(x))$ and $W_{f(h(x))} = W_{h(x)}$ contrary to the hypothesis on f. Hence, for all x,

$$x \in K \iff x \in K_{m(h(x))}$$

so $K \leq_T A$.   ◻

5.2 Corollary (Arslanov [1981]).   *Given an r.e. degree* $\mathbf{a}$, $\mathbf{a} < \mathbf{0}'$ *iff for every function* $f \in \mathbf{a}$ *there exists* n *such that* $W_n = W_{f(n)}$.   ◻

Note that the hypothesis of "A is r.e." in Theorem 5.1 is necessary by Exercise 5.9. Secondly, if f does not have a fixed point n such that $W_n = W_{f(n)}$, we might at least hope for a *-*fixed point* (*almost fixed point*), namely n such that $W_n =^* W_{f(n)}$. Exercise 5.5(a) is analogous to the Recursion Theorem since it shows that every function $f \leq_T \emptyset'$ has a *-fixed point, while Exercise 5.10 gives a $\emptyset''$-completeness criterion which is the analogue for *-fixed points of Theorem 5.1 for fixed points. Finally, if f has no almost fixed points we might hope that f has a *Turing fixed point*, namely n such that $W_n \equiv_T W_{f(n)}$. In Corollary XII.5.2 we prove that this is true if $f \leq_T \emptyset^{(2)}$. In Chapter XII §6 we shall extend the Fixed Point Theorem and the Completeness Criterion to analogous statements for all levels of the arithmetical hierarchy. For a strengthening of Theorem 5.1 see Corollary XII.6.5.

5.3–5.10 Exercises

5.3.   For each of the following properties define f appropriately and apply Theorem 5.1 to show that A is complete.
   (a) A creative.
   (b) A quasi-creative (as defined in Exercise 4.4).
   (c) A effectively simple.
   (d) B of Exercise 4.2.

5.4 (Morris-Gill).   An r.e. set A is *subcreative* (M. Blum) iff there is a recursive function g such that for every x if $W_x \cap A$ is finite (possibly empty), then $A \subset W_{g(x)} \subseteq A \cup \overline{W}_x$. Apply Theorem 5.1 to prove that every subcreative set is complete. *Hint.* Find the least s such that either $(\exists y) [y \in W_{g(x),s} - A]$ or $W_{x,s} \neq \emptyset$. In the former case set $W_{f(x)} = \{y\}$, and otherwise set $W_{f(x)} = \emptyset$. Show that $f \leq_T A$, and f satisfies Theorem 5.1.

5.5 (Arslanov, Nadirov, and Solov'ev [1977]).   An integer n is an *almost fixed point* for a function f if $W_{f(n)} =^* W_n$.
   (a) Show that for any function ψ partial recursive in $\emptyset'$ there is a recursive function f such that

$$(\forall x \in \mathrm{dom}\ \psi)\ [W_{\psi(x)} =^* W_{f(x)}].$$

(For a generalization see Theorem XII.6.2.) *Hint.* Choose a total recursive function $\hat{f}(x, s)$ such that $\psi(x) = \lim_s \hat{f}(x, s)$ for every $x \in \mathrm{dom}\ \psi$. Define

$$W_{f(x)} = \bigcup \{W_{\hat{f}(x,s),s} : s \in \omega\}.$$

(b) Show that any function $f \leq_T \emptyset'$ has an almost fixed point. (See also Theorem XII.6.3.)

(c) Use (b) to show that for no coinfinite r.e. set A does there exist a function $f \leq_T \emptyset'$ such that for all x with $A \subseteq W_x$: (1) $W_x \cap W_{f(x)} =^* A$; and (2) $W_x \cup W_{f(x)} =^* \omega$. (The point is that in Corollary X.2.7 we show that A is hh-simple iff $\mathcal{L}^*(A)$ is a Boolean algebra, namely there is a function f satisfying (1) and (2). By this exercise we cannot have $f \leq_T \emptyset'$.)

5.6 (Jockusch-Soare [1972a]).   In Theorem 3.2 suppose A is constructed using Step 1 only (with "$x > 2e$" in place of "$x > 3e$") and without Step 2 to code B into A. Use the method of Theorem 4.1 to prove that $B \leq_T A$ automatically. (This is an example of a certain "maximum degree principle" which asserts roughly that an r.e. set A being constructed has the highest degree not explicitly ruled out.) (See also Exercise XI.2.9 and Notes XI.2.19.) *Hint.* Define h and r as in (4.1) with $\hat{f}(x) = 2x$ in place of $f(x)$ in (4.1), and B in place of K. Let $\hat{B} = \{ x : x \in B - B_{r(x)} \}$. Note that $\hat{B}$ is r.e. in A. Consider separately the cases $\hat{B}$ finite and $\hat{B}$ infinite. If $\hat{B}$ is finite then clearly $\hat{B} \leq_T A$. If $\hat{B}$ is infinite, prove $\hat{B} \leq_T A$ as follows. Clearly $\hat{B}$ is r.e. in A. Let y be a given number. Recursively in A find $x \in \hat{B}$ such that $2h(x) > y$. Let s be the first stage such that $W_{h(x),s}$ contains a number $z > 2h(x)$. Show first that such a z exists, and then that $y \in B$ iff $y \in B_s$. (Note that if y later enters B then z later enters A but $W_{h(x)} = \{ a_0^{\theta(x)}, \ldots, a_{fh(x)}^{\theta(x)} \} \subseteq \overline{A}$ since $x \in B - B_{r(x)}$ so $|W_{h(x)}| = 2h(x) + 1 > \hat{f}(x)$.) (See also Exercise XI.2.9 and Notes XI.2.19.)

5.7 (Lachlan [1968a]).   (a) An r.e. set A is *weakly creative* if there is a function $f \leq_T A$ such that for all x, $W_x \subseteq \overline{A}$ implies $f(x) \notin W_x \cup A$. Prove that an r.e. set A is weakly creative iff A is complete.

(b) (Martin). An r.e. set S is *weakly effectively simple* if $\overline{S}$ is infinite and there is a function $f \leq_T S$ such that for all x, $W_x \subseteq \overline{S}$ implies that $f(x) > |W_x|$. Show that every weakly effectively simple set is complete.

(c) Show that any complete simple set is weakly effectively simple.

5.8 (Jockusch).   Prove that the following are equivalent for an arbitrary set A:

(a) $(\exists f \leq_T A) (\forall e) [W_e \neq W_{f(e)}]$,
(b) $(\exists g \leq_T A) (\forall e) [\varphi_e \neq \varphi_{g(e)}]$,
(c) $(\exists h \leq_T A) (\forall e) [h(e) \neq \varphi_e(e)]$.

Hint. For (b) implies (c) let d be the recursive function defined in the proof of the Recursion Theorem II.3.1, and let $h(e) = gd(e)$. For (c) implies (a) fix a p.r. function ψ such that if $W_e \neq \emptyset$ then $\psi(e) \in W_e$. Choose a recursive function q such that for every e, $\varphi_{q(e)} = \lambda y[\psi(e)]$. Let

$W_{f(e)} = \{ h(q(e)) \}$, for all e. (Note that the implication (a) implies (c) is a generalization of the Recursion Theorem.)

5.9 (Arslanov [1979] and [1981]).   Show that there exists a set $A \leq_T \emptyset'$, $A' \equiv_T \emptyset'$, and $f \leq_T A$ such that $(\forall e)\, [W_e \neq W_{f(e)}]$. *Hint* (Jockusch). Show that the class $\mathcal{C}$ of $\{0,1\}$-valued functions h such that $(\forall e)\, [h(e) \neq \varphi_e(e)]$ is a Π_1^0 class (as defined in VI.5.11). Apply the Low Basis Theorem VI.5.13 to get $h \in \mathcal{C}$, h of low degree, and apply Exercise 5.8.

5.10 $\emptyset''$-Completeness Criterion (Arslanov [1981]).   Suppose $A \in \Sigma_2$ and $\emptyset' \leq_T A$. Prove that

$$A \equiv_T \emptyset'' \iff (\exists f \leq_T A)\, (\forall e)\, [W_e \neq^* W_{f(e)}].$$

(Hence, this theorem is to Exercise 5.5(a) as Theorem 5.1 is to the Recursion Theorem.) See Theorem XII.6.5 for a generalization. *Hint* (Jockusch). First show by the method of Exercise 5.5(a) that if ψ is a function partial recursive in $\emptyset'$ then there exists a recursive function $g(e)$ such that

$$(\forall e \in \text{ dom } \psi)\, [W_{\psi(e)} =^* W_{g(e)}].$$

Replace the construction of Theorem 5.1 by a $\emptyset'$-construction. Let $B = \text{Fin}$, which is r.c. in $\emptyset'$. Let $\{A_s\}_{s \in \omega}$ and $\{B_s\}_{s \in \omega}$ be $\emptyset'$-recursive enumerations of A and B. Let $\theta(x) = \mu s[x \in B_s]$ if $x \in B$ and $\theta(x)$ diverge otherwise. From $f = \{e\}^A$ define $\hat{f}$ as in Theorem 5.1 except that now $\hat{f} \leq_T \emptyset'$. Hence $\hat{f}(y, \theta(x))$ is a function partial recursive in $\emptyset'$ so there is a recursive function $g(y, x)$ such that

$$(5.1) \qquad \begin{aligned} (\forall x)\, (\forall y)\, [[\hat{f}(y, \theta(x)) {\downarrow} &\implies W_{\hat{f}(y,\theta(x))} =^* W_{g(y,x)}] \\ \& \; [\hat{f}(y, \theta(x)) {\uparrow} &\implies W_{g(y,x)} = \emptyset]]. \end{aligned}$$

Apply the Recursion Theorem with parameter x to the function g to obtain a recursive function $h(x)$ such that for all x, $W_{h(x)} = W_{g(h(x),x)}$, and hence

$$(5.2) \qquad (\forall x \in B)\, [W_{h(x)} = W_{g(h(x),x)} =^* W_{\hat{f}(h(x),\theta(x))}].$$

Now complete the proof as in Theorem 5.1.

Chapter VI

Oracle Constructions of Non-R.E. Degrees

A more basic question than Post's problem is whether there exist *any* degrees (not necessarily r.e.) between $\mathbf{0}$ and $\mathbf{0}'$. Kleene and Post [1954] gave a positive answer by constructing Turing incomparable sets $A, B \leq_T \emptyset'$. Their method was to replace a complicated condition like $A \not\leq_T B$ by an infinite sequence $\{ R_e \}_{e \in \omega}$ of simpler conditions called *requirements*, where R_e asserts that $A \neq \{e\}^B$. The characteristic functions of A and B are constructed in a sequence of stages, $\chi_A = \bigcup_s f_s$, $\chi_B = \bigcup_s g_s$, where at stage s we define partial functions f_s and g_s in such a way as to meet a single requirement. The requirements are assigned priority according to their order in the above listing, and a requirement once acted upon and satisfied at some stage s remains satisfied forever. (Namely, it is never later *injured*.) These arguments are often called "diagonal", "wait and see", "initial segment", or "Kleene-Post" arguments to distinguish them from the finite injury priority arguments of Chapter VII, where action taken at some stage s to satisfy a requirement R_e may later be injured by action to satisfy a higher priority requirement R_i, $i < e$, so that we must then begin all over the attempt to satisfy R_e.

In Chapter IV and V we built r.e. sets A as the union of finite sets $\{ A_s \}_{s \in \omega}$ where the construction of A_s at stage s was recursive. In this chapter the construction of f_s at stage s will be nonrecursive, requiring some oracle C, for example $C = \emptyset'$. This is called a *C-recursive oracle construction* (or simply a *C-recursive construction*). We shall arrange that $f_s \subseteq f_{s+1}$. If dom f_s is finite for all s, we call this a *finite extension* construction (as in §1, §2, and §3 below), and otherwise an *infinite extension* construction (as in §4).

In Chapter VII and later chapters we shall want to construct *recursively enumerable* sets with similar properties. To do this we must abandon the oracle constructions in favor of more complicated recursive constructions which approximate them, and where requirements may be injured and therefore must receive attention many times. However, since the fundamental strategy for meeting each requirement is the same as here, it is best understood first in the simpler setting of an oracle construction.

For this chapter only, the subscripted symbols f_s, g_s, ... will be used to represent strings and *partial* functions, although unsubscripted f and g will still denote only *total* functions. Lower-case Greek letters ρ, σ, τ, and ν denote strings in either $\omega^{<\omega}$ or $2^{<\omega}$. We let $\sigma^\frown\tau$ denote the string which is the concatenation of string σ followed by τ. Recall that we use $\emptyset$ to denote the empty string; $\sigma \subseteq \tau$ to denote that string σ is an initial segment of τ; $\sigma \subset \tau$ to denote that $\sigma \subseteq \tau$ and $\sigma \neq \tau$; and we use $\sigma \subseteq \rho, \tau$ ($a \leq b, c$) to abbreviate $\sigma \subseteq \rho$ and $\sigma \subseteq \tau$ ($a \leq b$ and $a \leq c$). The context will remove any ambiguity.

1. A Pair of Incomparable Degrees Below $\mathbf{0}'$

1.1 Proposition. *If f is total and $f = \bigcup_s \psi_s$, where $\{\psi_s\}_{s \in \omega}$ is an A-recursive sequence of partial functions (i.e., $\psi_s(x) = \theta(x, s)$ for some partial A-recursive function θ) then f is A-recursive.*

Proof. To compute $f(x)$, A-recursively find an s such that $\psi_s(x)$ is defined. Then $f(x) = \psi_s(x)$. ∎

1.2 Theorem (Kleene-Post). *There exist degrees $\mathbf{a}, \mathbf{b} \leq \mathbf{0}'$ such that $\mathbf{a}$ is incomparable with $\mathbf{b}$, written $\mathbf{a} \mid \mathbf{b}$ (i.e., $\mathbf{a} \nleq \mathbf{b}$ and $\mathbf{b} \nleq \mathbf{a}$); hence $\mathbf{0} < \mathbf{a}, \mathbf{b}$ and $\mathbf{a}, \mathbf{b} < \mathbf{0}'$.*

Proof. We shall construct sets $A, B \leq_T \emptyset'$ in stages by a finite extension $\emptyset'$-oracle construction so that $\chi_A = \bigcup_s f_s$ and $\chi_B = \bigcup_s g_s$ where f_s and g_s are finite strings in $2^{<\omega}$ of length $\geq s$ viewed as initial segments of χ_A and χ_B. Since the construction of f_s and g_s at stage s is recursive in $\emptyset'$, $\{f_s\}_{s \in \omega}$ and $\{g_s\}_{s \in \omega}$ are $\emptyset'$-recursive sequences, and so $A, B \leq_T \emptyset'$ by Proposition 1.1. It suffices to meet, for each e, the following *requirements*:

$$R_e : A \neq \{e\}^B;$$

and

$$S_e : B \neq \{e\}^A;$$

to ensure that $A \nleq_T B$ and $B \nleq_T A$; hence $\mathbf{a} = \deg(A)$ and $\mathbf{b} = \deg(B)$ are incomparable. (There is a hidden requirement that $\bigcup_s f_s$ and $\bigcup_s g_s$ are total functions, which is accomplished by guaranteeing that $f_s \subset f_{s+1}$ and $g_s \subset g_{s+1}$.)

  Stage $s = 0$. Define $f_0 = g_0 = \emptyset$.

  Stage $s + 1 = 2e + 1$. (We satisfy R_e.) We are given $f_s, g_s \in 2^{<\omega}$ of length $\geq s$. Let $n = lh(f_s) = (\mu x)\, [x \notin \text{dom } f_s]$. Using a $\emptyset'$-oracle we test whether

(1.1) $(\exists t)\, (\exists \sigma)\, [\sigma \supset g_s \ \& \ \{e\}_t^\sigma(n)\!\downarrow].$

Note that $\sigma \supset g_s$ is recursive as a relation of strings σ and g_s, and the second clause of (1.1) is recursive by Theorem III.1.8 so (1.1) is a Σ_1 statement and hence can be decided recursively in $\emptyset'$.

Case 1. Suppose (1.1) is satisfied. Enumerate the recursive set

$$\{ \langle \sigma, t \rangle : \{e\}_t^\sigma(n)\!\downarrow \},$$

and choose the least $\langle \sigma', t' \rangle$ satisfying the matrix of (1.1). Set $g_{s+1} = \sigma'$ and $f_{s+1}(n) = 1 \dot- \{e\}_{t'}^{\sigma'}(n)$.

Case 2.   Suppose (1.1) fails.   Then define $f_{s+1}(lh(f_s)) = 0$, and $g_{s+1}(lh(g_s)) = 0$. (Namely, define $f_{s+1} = f_s\,\hat{}\,0$ and $g_{s+1} = g_s\,\hat{}\,0$.)

In either case $lh(f_{s+1}), lh(g_{s+1}) \geq s+1$, so $\chi_A = \bigcup_s f_s$ and $\chi_B = \bigcup_s g_s$ are defined on all arguments. In either case if $f \supset f_{s+1}$ and $g \supset g_{s+1}$ then $f(n) \neq \{e\}^g(n)$ by the Use Principle III.1.9.

Stage $s+1 = 2e+2$. (We satisfy S_e.) Proceed exactly as above but with the roles of f and g interchanged.   ▯

1.3 Theorem (Relativized Version).   *For any degree* **c**, *there are degrees* **a**, **b** *such that* $\mathbf{c} \leq \mathbf{a}, \mathbf{b}$ *and* $\mathbf{a}, \mathbf{b} \leq \mathbf{c}'$ *and* $\mathbf{a} \mid \mathbf{b}$.

Proof. Fix a set $C \in \mathbf{c}$. Relativize the above proof to C, using a C'-recursive construction to build sets A and B such that $A \oplus C \mid_T B \oplus C$ and $A, B \leq_T C'$. In place of (1.1) we use a C'-oracle to test whether

$$(1.2) \qquad (\exists t)\, (\exists \tau_1)\, (\exists \tau_2)\, [\tau_1 \supset g_s \ \& \ \tau_2 \subset C \ \& \ \{e\}_t^{\tau_1 \oplus \tau_2}(n)\!\downarrow],$$

where $\tau_1 \oplus \tau_2$ is defined to be the shortest string $\rho \in 2^{<\omega}$ such that $\rho(2x) = \tau_1(x)$ and $\rho(2x+1) = \tau_2(x)$. The obvious modification of cases 1 and 2 ensures that $A \neq \{e\}^{B \oplus C}$. At stage $2e+2$ we ensure that $B \neq \{e\}^{A \oplus C}$. Finally let $\mathbf{a} = \deg(A \oplus C)$ and $\mathbf{b} = \deg(B \oplus C)$.   ▯

1.4 Definition.   A countable sequence of sets $\{A_i\}_{i \in \omega}$ is *recursively indepen-dent* if for each i, $A_i \not\leq_T \oplus\{A_j : j \neq i\}$, where the latter is defined as in Exercise III.2.7.

1.5–1.8 Exercises

1.5.   Modify the proof of Theorem 1.2 to build a recursively independent sequence $\{A_j\}_{j \in \omega}$ of sets each recursive in $\emptyset'$ (indeed $\oplus_j A_j \leq_T \emptyset'$). *Hint.* Use a finite extension $\emptyset'$-recursive construction to build at stage s strings $\{f_j^s\}_{s \in \omega}$ such that if $A_j = \bigcup_s f_j^s$ then we meet for each e and i the requirement

$$R_{\langle e,i \rangle} : A_i \neq \{e\}^{\oplus\{A_j : j \neq i\}}.$$

At stage $s = 0$, set $f_j^0 = \emptyset$ for all j. At stage $s+1 = \langle e, i \rangle + 1$ we meet requirement $R_{\langle e,i \rangle}$ as follows. Given $\{f_j^s\}_{j \in \omega}$, only finitely many of which are nonempty, let $n = lh(f_i^s)$, and use a $\emptyset'$-oracle to test whether there exists m and (a code number for) a finite sequence of strings $\sigma_1, \sigma_2, \ldots, \sigma_k$ such that

(1.3) $\qquad \{e\}^{\oplus\{\sigma_j : j \neq i\}}(n) \downarrow = m$ & $(\forall j \leq k)$ $[j \neq i \implies f_j^s \subset \sigma_j]$.

Now according to whether (1.3) holds proceed as in Theorem 1.2 case 1 (letting $f_i^{s+1}(n) = 1 \dotminus m$, and $f_j^{s+1} = \sigma_j$ for $j \neq i$), or as in case 2 otherwise. (Be sure to make each f_i total.)

1.6. A partially ordered set $\mathcal{P} = (P, \leq_P)$ is *countably universal* if every countable partially ordered set is order isomorphic to a subordering of $\mathcal{P}$. Prove that there is a recursive partial ordering $\leq_R$ of ω which is countably universal. *Hint.* This can be done either by considering a recursively presented atomless Boolean algebra, or by a direct construction where at stage $s + 1$ given a finite set P_s of elements in $\leq_R$, one obtains P_{s+1} by adding a new point for each possible order type over P_s. A Boolean algebra $\mathcal{B} = (\{b_i\}_{i \in \omega}; \leq, \vee, \wedge, ')$ is *recursively presented* if there exist a recursive relation $P(i, j)$ and recursive functions f, g and h such that $P(i, j)$ holds iff $b_i \leq b_j$, and such that $b_{f(i,j)} = b_i \vee b_j$, $b_{g(i,j)} = b_i \wedge b_j$, and $b_{h(i)} = b_i'$.

1.7. Prove that for any countable partially ordered set $\mathcal{P} = \langle P, \leq_P \rangle$ there is a 1:1 order preserving map from P into $\mathbf{D}(< \mathbf{0}')$, the degrees $< \mathbf{0}'$. *Hint.* By Exercise 1.6 we may assume $P = \omega$ and $\leq_P$ is a recursive relation. Let $\{A_i\}_{i \in \omega}$ be as in Exercise 1.5. Define $f : \omega \to \mathbf{D}(\leq \mathbf{0}')$ by $f(i) = \mathbf{a}_i = \deg(\oplus A_j : j \leq_P i)$. Show that if $i \leq_P j$ then $\mathbf{a}_i \leq \mathbf{a}_j$ (by definition and the fact that $\leq_P$ is recursive), and if $i \not\leq_P j$ then $\mathbf{a}_i \not\leq \mathbf{a}_j$ (by the recursive independence of $\{A_i\}_{i \in \omega}$).

1.8. Prove that there are $2^{\aleph_0}$ mutually incomparable degrees. *Hint.* Using the notation of Definition 5.11 and Exercise 5.12 construct a tree $T \subseteq 2^{<\omega}$ such that $f \mid_T g$ for every pair $f, g \in [T]$, $f \neq g$. Let $T = \bigcup_e T_e$ where tree $T_{e+1} \supset T_e$ and T_{e+1} consists of 2^{e+1} incomparable strings such that for all $\sigma, \tau \in T_{e+1}$ if $\sigma \neq \tau$ then

(1.4) $\qquad (\forall f \supset \sigma)\, (\forall g \supset \tau)\, [\{e\}^f \neq g$ & $\{e\}^g \neq f]$.

Let $T_0 = \{\emptyset\}$. Given T_e define L_e to be the *leaves* of tree T_e, namely

$$L_e = \{\sigma : \sigma \in T_e \ \& \ (\forall \tau \supset \sigma)\, [\tau \notin T_e]\}.$$

Next define

$$S = \{\sigma \hat{\ } 0 : \sigma \in L_e\} \cup \{\sigma \hat{\ } 1 : \sigma \in L_e\}.$$

Suppose $S = \{\rho_i : i \leq 2^{e+1}\}$. Fix $i, j \leq 2^{e+1}$, $i \neq j$. Use the method of Theorem 1.2 to replace ρ_i and ρ_j by strings $\sigma \supset \rho_i$, $\tau \supset \rho_j$ satisfying (1.4). Repeat this procedure for all $i, j \leq 2^{e+1}$, $i \neq j$.

2. Avoiding Cones of Degrees

So far the degrees we have constructed (such as $\mathbf{0}^{(n)}$ or degrees below $\mathbf{0}'$) are comparable to $\mathbf{0}'$. In this section we show how to construct a degree $\mathbf{a}$ incomparable with a given degree $\mathbf{b} > \mathbf{0}$. To achieve this, $\mathbf{a}$ must avoid the *lower cone* of degrees $\{\mathbf{d} : \mathbf{d} \leq \mathbf{b}\}$ and the *upper cone* $\{\mathbf{d} : \mathbf{d} \geq \mathbf{b}\}$. The strategy for accomplishing the latter (which we play on the even stages) will be used in §4 and §5 of this chapter, and will be refined and often used in constructions of r.e. degrees, such as the Sacks Splitting Theorem VII.3.2.

2.1 Theorem. *For every degree $\mathbf{b} > \mathbf{0}$ there exists a degree $\mathbf{a} < \mathbf{b}'$ such that $\mathbf{a} \mid \mathbf{b}$.*

Proof. Fix $B \in \mathbf{b}$. Construct f, the characteristic function of A, by a B'-recursive finite extension construction, $f = \bigcup_s f_s$, to meet the requirements R_e and S_e of Theorem 1.2.

    Stage $s = 0$. Set $f_0 = \emptyset$.

    Stage $s + 1 = 2e + 1$. (Satisfy $R_e : A \neq \{e\}^B$.) Let $n = lh(f_s)$. Now B'-recursively compute whether $\{e\}^B(n)$ converges (i.e., whether $\langle n, e\rangle \in K_0^B \equiv K^B = B'$). If so, set $f_{s+1}(n) = 1 \,\dot{-}\, \{e\}^B(n)$. If not, set $f_{s+1}(n) = 0$.

    Stage $s+1 = 2e+2$. (Satisfy $S_e : B \neq \{e\}^A$.) Given f_s, first $\emptyset'$-recursively test whether

$$(2.1) \qquad (\exists\sigma)\,(\exists\tau)\,(\exists x)\,(\exists y)\,(\exists z)\,(\exists t)\,[f_s \subset \sigma, \tau \ \& \ \{e\}_t^\sigma(x)\downarrow \\ = y \neq z = \{e\}_t^\tau(x)\downarrow].$$

    If so, one of the values y, z must differ from $B(x)$; we let f_{s+1} be the first $\sigma \supset f_s$ such that $\{e\}^\sigma(x)\downarrow \neq B(x)$ for some x. (This is B'-recursive since $B \oplus \emptyset' <_T B'$.) If (2.1) fails, we let $f_{s+1} = f_s\widehat{}0$. In this case, we claim that for any $f \supset f_s$ if $\{e\}^f = g$ is total, then g is recursive (and hence $\{e\}^f \neq B$ since $\emptyset <_T B$). To recursively compute $g(x)$, recursively enumerate L of Theorem III.1.8 until the first $\sigma \supset f_s$ is found such that $\{e\}^\sigma(x)$ converges. Now $g(x) = \{e\}^f(x) = \{e\}^\sigma(x)$, because otherwise for some $\tau, f_s \subseteq \tau \subset f$, $\{e\}^\tau(x)\downarrow \neq \{e\}^\sigma(x)$, thereby satisfying (2.1).   ▯

2.2–2.3 Exercises

2.2. Let $\{\mathbf{b}_n\}_{n\in\omega}$ be any countable sequence of nonrecursive degrees. Prove that there exist pairwise incomparable degrees $\{\mathbf{a}_n\}_{n\in\omega}$ such that $\mathbf{a}_n \mid \mathbf{b}_m$ for all $m, n \in \omega$. *Hint.* Combine the constructions of Theorems 1.2 and 2.1 to meet for all e, m, and n requirements of the form $R_{\langle e,n,m\rangle} : A_n \neq \{e\}^{B_m}$, $S_{\langle e,n,m\rangle} : B_m \neq \{e\}^{A_n}$, and also $T_{\langle e,n,m\rangle} : A_n \neq \{e\}^{A_m}$, for $n \neq m$.

2.3. Given $\{B_n\}_{n\in\omega}$ a $\emptyset'$-recursive sequence of nonrecursive sets uniformly recursive in $\emptyset'$, show that there exists a set A, $\emptyset <_T A <_T \emptyset'$ such that

$(\forall n)\,[B_n \not\leq_T A]$. *Hint.* The strategy on the even stages in Theorem 2.1 uses only a $\emptyset' \oplus B$-oracle.

3. Inverting the Jump

Note that for any degree $\mathbf{a}$, $\mathbf{0} \leq \mathbf{a}$ and hence $\mathbf{0}' \leq \mathbf{a}'$, i.e., any jump is above $\mathbf{0}'$. Hence the jump, viewed as a map on degrees, has range contained in $\{\,\mathbf{b} : \mathbf{b} \geq \mathbf{0}'\,\}$. The next theorem asserts that this map is onto this set $\{\,\mathbf{b} : \mathbf{b} \geq \mathbf{0}'\,\}$. A degree $\mathbf{a}$ is called *complete* if $\mathbf{a} \geq \mathbf{0}'$, so the result also gives a criterion for $\mathbf{a}$ being complete.

3.1 Theorem (Friedberg Completeness Criterion). *For every degree* $\mathbf{b} \geq \mathbf{0}'$ *there is a degree* $\mathbf{a}$ *such that* $\mathbf{a}' = \mathbf{a} \cup \mathbf{0}' = \mathbf{b}$.

Proof. Fix $B \in \mathbf{b}$. We shall construct f, the characteristic function of A, by finite initial segments $\{\,f_s\,\}_{s \in \omega}$ using a B-recursive finite extension construction.

    Stage $s = 0$. Set $f_0 = \emptyset$.

    Stage $s + 1 = 2e + 1$. (We decide whether $e \in A'$.) We meet the requirement

(3.1) $$R_e : (\exists \sigma \subset A)\,[\{\,e\,\}^\sigma(e)\!\downarrow \lor (\forall \tau \supseteq \sigma)\,[\{\,e\,\}^\tau(e)\!\uparrow]].$$

(If A meets R_e we say that A *forces the jump* on argument e.) Given f_s, use a $\emptyset'$-oracle to test whether

(3.2) $$(\exists \sigma)\,(\exists t)\,[f_s \subset \sigma \;\&\; \{\,e\,\}^\sigma_t(e)\!\downarrow].$$

(Note that the matrix is recursive, so (3.2) is a Σ_1 condition and thus is recursive $\mathbf{0}'$.) If (3.2) is satisfied, let f_{s+1} be the first such σ (say, in the standard enumeration of L of Theorem III.1.8). If not, set $f_{s+1} = f_s$.

    Stage $s + 1 = 2e + 2$. (We code $B(e)$ into A.) Let $n = lh(f_s)$. Define $f_{s+1}(n) = B(e)$. (This completes the construction.)

    Now $f = \bigcup_s f_s$ is total since $lh(f_{2e}) \geq e$. Let $A = \{\,x : f(x) = 1\,\}$, and $\mathbf{a} = \deg(A)$. The construction is B-recursive since at odd stages we use a $\emptyset'$-oracle, at even stages we use a B-oracle, and $\emptyset' \leq_T B$. Since $A \oplus \emptyset' \leq_T A'$ for any A, to prove $A' \equiv_T B \equiv_T A \oplus \emptyset'$ it suffices to prove the following two lemmas.

Lemma 1. $A' \leq_T B$.

Lemma 2. $B \leq_T A \oplus \emptyset'$.

Proof of Lemma 1. Since the construction is B-recursive, the sequence $\{\,f_s\,\}_{s \in \omega}$ is also B-recursive. To decide whether $e \in A'$, we B-recursively compute using $\emptyset' \leq_T B$ whether (3.2) holds for f_s, $s = 2e$. If so, $e \in A'$ and otherwise $e \notin A'$ because no $\sigma \supseteq f_s$ has $\{e\}^\sigma(e)$ defined.

Proof of Lemma 2. It suffices to show that $\{f_s\}_{s\in\omega}$ is an $(A \oplus \emptyset')$-recursive sequence since $B(e)$ is the last value of f_{2e+2}. The proof is by induction on s. Given $\{f_s : s \leq 2e\}$, use a $\emptyset'$-oracle to compute f_{2e+1}. If $n = lh(f_{2e+1})$ then $f_{2e+2} = f_{2e+1}\hat{\ }A(n)$, so f_{2e+2} can be computed from f_{2e+1} using an A-oracle. ▯

3.2 Theorem (Relativized Friedberg Completeness Criterion). *For every degree* $\mathbf{c}$,

$$F_1(\mathbf{c}) : (\forall \mathbf{b}) \; [\mathbf{b} \geq \mathbf{c}' \implies (\exists \mathbf{a}) \; [\mathbf{a} \geq \mathbf{c} \; \& \; \mathbf{a}' = \mathbf{c}' \cup \mathbf{a} = \mathbf{b}]].$$

Proof. Do the proof of Theorem 3.1 with $\mathbf{c}$ and $\mathbf{c}'$ in place of $\mathbf{0}$ and $\mathbf{0}'$. ▯

3.3 Corollary. *For every* $n \geq 1$, *and every degree* $\mathbf{c}$,

$$F_n(\mathbf{c}) : (\forall \mathbf{b}) \; [\mathbf{b} \geq \mathbf{c}^{(n)} \implies (\exists \mathbf{a}) \; [\mathbf{a} \geq \mathbf{c} \; \& \; \mathbf{a}^{(n)} = \mathbf{a} \cup \mathbf{c}^{(n)} = \mathbf{b}]].$$

Proof. To prove $(\forall \mathbf{c})F_n(\mathbf{c})$ holds for all $n \geq 1$, use induction on n and the fact that $F_{n+1}(\mathbf{c})$ follows from $F_n(\mathbf{c})$ and $F_1(\mathbf{c}^{(n)})$. ▯

Although Theorem 3.1 demonstrates a pleasant property of the jump operator, it also demonstrates an unpleasant property, namely that the jump map is not 1:1. To see this, apply Theorem 3.1 with $\mathbf{b} = \mathbf{0}''$ to obtain $\mathbf{a}$ such that $\mathbf{a}' = \mathbf{a} \cup \mathbf{0}' = \mathbf{0}''$. Clearly $\mathbf{a} \mid \mathbf{0}'$ yet they have the same jump. We shall see in Exercise 3.5 that it is also possible to have $\mathbf{a} < \mathbf{b}$ and $\mathbf{a}' = \mathbf{b}'$.

3.4–3.10 Exercises

3.4. Prove

$$(\forall \mathbf{b} \geq \mathbf{0}') \; (\exists \mathbf{a}_0) \; (\exists \mathbf{a}_1) \; [\mathbf{a}_0 \mid \mathbf{a}_1 \; \& \; \mathbf{a}_0' = \mathbf{a}_0 \cup \mathbf{0}' = \mathbf{b} = \mathbf{a}_1 \cup \mathbf{0}' = \mathbf{a}_1'].$$

Hint. Combine the constructions of Theorems 1.2 and 3.1 to handle four types of requirements (the two types from Theorem 1.2 and the two from Theorem 3.1).

3.5. Use Exercise 3.4 to prove $(\exists \mathbf{a}) \; [\mathbf{0} < \mathbf{a} < \mathbf{0}' \; \& \; \mathbf{a}' = \mathbf{0}']$, i.e., there is a nonrecursive low degree. (In Theorem VII.1.1 we shall see that $\mathbf{a}$ can also be made r.e.)

3.6 (Jockusch-Posner). We say that a set A is *1-generic* if it satisfies for each e the requirement R_e of Theorem 3.1. Show that if A is 1-generic then $A \oplus \emptyset' \equiv_T A'$.

3.7 (Jockusch-Posner).   Prove that A is 1-generic if and only if for any r.e. set W_e of strings we meet requirement

$$(3.3) \qquad S_e : (\exists \sigma \subset A) \, [\sigma \in W_e \vee (\forall \tau \supseteq \sigma) \, [\tau \notin W_e]].$$

Hint. Given e, we define a recursive function $f(e)$ as follows:

$$\{f(e)\}^\sigma (x) = \begin{cases} 1 & \text{if } (\exists s \leq |\sigma|) \, (\exists \tau \subseteq \sigma) \, [\tau \in W_{e,s}], \\ \text{undefined} & \text{otherwise.} \end{cases}$$

Show that if A satisfies $R_{f(e)}$ of (3.1) then A satisfies S_e.

3.8 (Jockusch-Posner).   Assume A is 1-generic.
  (a) Prove that A is immune. *Hint.* Let V be an r.e. subset of A. Define $W_e = \{\sigma : (\exists x \in V) \, [\sigma(x) = 0]\}$ and use S_e of (3.3) to prove that V is finite.
  (b) Prove that A is hyperimmune.
  (c) Prove that there is no nonrecursive r.e. set $V \leq_T A$. *Hint.* Assume $V = \{i\}^A$, define

$$W_e = \{\sigma : (\exists x) \, [\{i\}^\sigma (x) = 0 \ \& \ x \in V]\},$$

and apply S_e of (3.3) to show that $\overline{V}$ is r.e.
  (d) Prove that A_0 and A_1 are Turing incomparable where $A_0(x) = A(2x)$ and $A_1(x) = A(2x+1)$. *Hint.* To see that $A_0 \neq \{e\}^{A_1}$ consider the r.e. set of strings

$$V_e = \{\sigma : (\exists x) \, [\{e\}^{\sigma_1}(x) \downarrow \neq \sigma_0(x)]\}$$

where $\sigma_0(x) = \sigma(2x)$ and $\sigma_1(x) = \sigma(2x+1)$.
  (e) Prove that there are sets $B_i \leq_T A$, $i \in \omega$, such that for every i, $B_i \not\leq_T \oplus \{B_j : j \neq i\}$.

3.9.$^\diamond$ Prove that if V is a nonrecursive r.e. set then there is a 1-generic set $A \leq_T V$. *Hint.* (Shore) Define the *computation function* for V to be

$$c(n) = (\mu s) \, [V_s \upharpoonright n = V \upharpoonright n].$$

Note that $c \equiv_T V$. To satisfy R_e of Theorem 3.1 replace the $\emptyset'$-construction at stage $s+1$ of Theorem 3.1 by the following V-recursive construction. Choose the least $e \leq s$ such that

$$\{e\}^{f_s}_s(e) \uparrow \ \text{ and } (\exists \sigma)_{|\sigma| \leq c(s)} \, [f_s \subset \sigma \text{ and } \{e\}^\sigma_{c(s)}(e) \downarrow].$$

If e exists, let f_{s+1} be the first corresponding string σ. If R_e is not satisfied show that c is a recursive function by recursively computing at each stage $s+1$: (1) a string $\sigma_s \supset f_s$ such that $\{e\}^{\sigma_s}(e) \downarrow$; (2) $c(s)$; (3) f_{s+1}. (Note that this argument is similar to the permitting argument in Theorem V.3.2 for building a simple set $S \leq_T V$. In each case if V or c fails to "permit" soon enough to carry out a standard construction, then we argue that V is recursive.)

3.10 (Jockusch-Shore [1983]).   Prove that for any $i \in \omega$ and any B such that $\emptyset' \leq_T B$ there exists A such that

$$A \oplus W_i^A \equiv_T A \oplus \emptyset' \equiv_T B.$$

(Note that Theorem 3.1 is a special case of this setting where i is defined by $W_i^X = X'$.) *Hint.* Do the proof of Theorem 3.1 but in (3.1) replace $\{e\}^\rho(e)\downarrow$ by $e \in W_i^\rho$ for $\rho = \sigma$ or τ. (Note that this construction is uniform in B and in any j such that $\{j\}^B = \emptyset'$.)

4. Upper and Lower Bounds for Degrees

By Chapter III §2, every finite set of degrees has a least upper bound (lub). In this section we show that this is false for greatest lower bounds (glb's). Hence, the degrees do not form a lattice, but merely an upper semilattice.

4.1 Definition.   For any set A define the *ω-jump* of A,

$$A^{(\omega)} = \{ \langle x, n \rangle : x \in A^{(n)} \},$$

and let $\mathbf{a}^{(\omega)} = \deg(A^{(\omega)})$ for $A \in \mathbf{a}$.

An infinite sequence of degrees $\{\mathbf{a}_n\}_{n\in\omega}$ is *ascending* if $\mathbf{a}_n < \mathbf{a}_{n+1}$ for all n. For example, $\mathbf{0}, \mathbf{0}^{(1)}, \mathbf{0}^{(2)}, \ldots$ is ascending, and $\mathbf{0}^{(\omega)}$ is a natural upper bound for the sequence, although by the next theorem the sequence has no lub.

4.2 Theorem (Kleene-Post-Spector).   *For any ascending sequence $\{\mathbf{a}_n\}_{n\in\omega}$ of degrees there exist upper bounds $\mathbf{b}$, $\mathbf{c}$ (called an* exact pair *for the sequence) such that*

$$(4.1) \qquad\qquad (\forall \mathbf{d})\,[[\mathbf{d} \leq \mathbf{b}\ \&\ \mathbf{d} \leq \mathbf{c}] \implies (\exists n)\,[\mathbf{d} \leq \mathbf{a}_n]].$$

4.3 Corollary.   *No infinite ascending sequence has a least upper bound.*   ▯

4.4 Corollary.   *There are degrees $\mathbf{b}$ and $\mathbf{c}$ with no greatest lower bound.*   ▯

Before proving Theorem 4.2 we make some definitions and introduce some new notation.

4.5 Notation.   For any set $A \subseteq \omega$ define the *y-section* of A,

$$A^{[y]} = \{ \langle x, z \rangle : \langle x, z \rangle \in A \ \& \ z = y \},$$

and define

$$A^{[<y]} = \bigcup \{ A^{[z]} : z < y \}.$$

(Using the pairing function we can identify A with a subset of $\omega \times \omega$ and view $A^{[y]}$ as the yth row of A. We use the square bracket notation $A^{[y]}$ to distinguish from the yth jump $A^{(y)}$.)

4.6 Definition.   A subset $B \subseteq A$ is a *thick subset* of A if we meet for every y the thickness requirement

$$T_y : B^{[y]} =^* A^{[y]},$$

where $X =^* Y$ denotes that the symmetric difference $X \triangle Y =_{\mathrm{dfn}} (X - Y) \cup (Y - X)$ is finite.

Thick subsets will be very useful here and in later constructions of r.e. sets and degrees in Chapter VIII, particularly for the Thickness Lemma VIII.1.1.

4.7 Definition.   Partial functions θ, ψ are *compatible* (written $compat(\theta, \psi)$) if they have a common extension, i.e., if there is no x for which $\theta(x)$ and $\psi(x)$ are defined and unequal.

Proof of Theorem 4.2. Choose $A_y \in \mathbf{a}_y$ for each y and define $A = \{ \langle x, y \rangle : x \in A_y \}$, so that $\langle x, y \rangle \in A^{[y]}$ iff $x \in A_y$. We shall construct characteristic functions f and g of sets B and C which are thick in A (so that $A_y \equiv_T B^{[y]} \leq_T B$, and likewise for C). This ensures that $\mathbf{b} = \deg(B)$ and $\mathbf{c} = \deg(C)$ are upper bounds for $\{ \mathbf{a}_n \}_{n \in \omega}$. In addition to the thickness requirements

$$T_y^B : B^{[y]} =^* A^{[y]},$$

and

$$T_y^C : C^{[y]} =^* A^{[y]},$$

we must meet, for all e and i, the requirements

$$R_{\langle e, i \rangle} : \{ e \}^B = \{ i \}^C = h \ \text{total} \ \implies (\exists y) \, [h \leq_T A_y],$$

by looking for "e-splittings" as we did in proving Theorem 2.1.

Let f_s, g_s, B_s, and C_s be the portions of f, g, B, and C defined by the end of stage s of the following construction.

Stage $s = 0$. Set $f_0 = g_0 = \emptyset$.

Stage $s + 1$. Assume that f_s and g_s are defined on $\omega^{[<s]}$ and that

(4.2) $(\forall y < s) \, [B_s^{[y]} =^* C_s^{[y]} =^* A^{[y]}],$

and

(4.3) $$(\text{dom } f_s - \omega^{[<s]}) =^* \emptyset =^* (\text{dom } g_s - \omega^{[<s]}).$$

Step 1. (Satisfy $R_{\langle e,i \rangle}$ for $s = \langle e,i \rangle$.) If

(4.4) $(\exists \sigma)\,(\exists \tau)\,(\exists x)\,(\exists t)\,[\text{compat}(\sigma, f_s)\ \&\ \text{compat}(\tau, g_s)$
$$\&\ \{e\}_t^\sigma(x){\downarrow} \neq \{i\}_t^\tau(x){\downarrow}],$$

then let σ and τ be the first such strings and extend f_s to $\hat{f} = f_s \cup \sigma$ and g_s to $\hat{g} = g_s \cup \tau$. Otherwise, let $\hat{f} = f_s$, and $\hat{g} = g_s$. Note that $f_s \equiv_T A^{[<s]} \equiv_T g_s$ by (4.2) and (4.3). Hence, compat(σ, f_s) is an $A^{[<s]}$-recursive relation on σ. (Note that for $s > 0$ Step 1 requires a $A'_{s-1} \equiv_T (A^{[<s]})'$ oracle.)

Step 2. (Satisfy T_s^B and T_s^C.)

Let $f_{s+1} = \hat{f}$ on dom $\hat{f}$, and on all $x \in \omega^{[s]} - \text{dom } \hat{f}$ define $f_{s+1}(x) = A(x)$. Similarly let $g_{s+1} = \hat{g}$ on dom $\hat{g}$ and $g_{s+1}(x) = A(x)$ for all

$$x \in \omega^{[s]} - \text{dom } \hat{g}.$$

By (4.3), f_s (and hence $\hat{f}$) is already defined on at most finitely many elements of $\omega^{[s]}$, and similarly for g_s, so $B_{s+1}^{[s]} =^* A^{[s]} =^* C_{s+1}^{[s]}$, and f and g are now defined on $\omega^{[\leq s]}$. This ends the construction.

If (4.4) holds, then $\{e\}^B \neq \{i\}^C$. If (4.4) fails and $\{e\}^B = \{i\}^C = h$ is total, then for $s = \langle e,i \rangle$ we shall show that $h \leq_T A^{[<s]}$. Notice that $A^{[<s]} \leq_T A_s$ because $A^{[<s]} \equiv_T A^{[0]} \oplus \cdots \oplus A^{[s-1]} \equiv_T A_0 \oplus \cdots \oplus A_{s-1} \leq_T A_s$. To $A^{[<s]}$-recursively compute $h(x)$, find the first string σ in some enumeration of $\{\sigma : \{e\}^\sigma(x){\downarrow}\}$ such that compat(σ, f_s) and set $h(x) = \{e\}^\sigma(x)$. Now $h(x) = \{e\}^f(x)$, or else for some $\sigma' \subset f$, compat(σ', f_s) holds and $\{e\}^{\sigma'}(x){\downarrow} = y \neq \{e\}^\sigma(x)$, so (4.4) holds for either σ or σ' and for any $\tau \subset C$ such that $\{i\}^\tau(x)$ converges. ▯

4.8–4.11 Exercises

4.8. Show that the proof of Theorem 4.2 automatically produces sets B and C recursive in $\oplus\{A'_y : y \in \omega\}$.

4.9. Show that in the proof of Theorem 4.2 we can set $B = A$ and appropriately modify Steps 1 and 2 to construct C such that B and C satisfy the same requirements as before.

4.10. Let **I** be a countable ideal contained in the Turing degrees **D**. Prove that there exist degrees **b**, **c** such that for all $\mathbf{a} \in \mathbf{D}$,

$$\mathbf{a} \in \mathbf{I} \iff [\mathbf{a} \leq \mathbf{b}\ \&\ \mathbf{a} \leq \mathbf{c}].$$

(We call **b** and **c** an *exact pair* for **I**.)

4.11° (Shoenfield Jump Inversion Theorem). Fix S such that $\emptyset' \leq_T S$ and S is r.e. in $\emptyset'$. Construct $A \leq_T \emptyset'$ such that $A' \equiv_T S$. (In Theorem VIII.3.1 we strengthen this by making A r.e. using a different proof.) *Hint.* Define a $\emptyset'$-sequence $\{f_s\}_{s \in \omega}$ of $\{0,1\}$-valued partial functions such that $f_s \subseteq f_{s+1}$ and $\lim_s f_s = \chi_A$. We ensure that $S \leq_T A'$ by arranging that for all y, $\lim_x A(\langle x,y \rangle) = \chi_S(y)$. We ensure $A' \leq_T S$ by forcing the jump $\{e\}^A(e)$. Fix a $\emptyset'$-recursive enumeration $\{S_s\}_{s \in \omega}$ of S such that $|S_{s+1} - S_s| = 1$. Let $f_0 = \emptyset$. The following is a $\emptyset'$-construction.

Stage $s+1$. Assume that if $y \in S_s$ then $f_s(\langle x,y \rangle) = 1$ for almost every x, and otherwise $f_s(\langle x,y \rangle) \downarrow = 0$ for at most finitely many x such that $\langle x,y \rangle < s$, and $f_s(\langle x,y \rangle) \uparrow$ for all other x.

Step 1. Now f_{s+1} has a recursive domain and is recursive on its domain. Hence, we can $\emptyset'$-recursively test for each $e \leq s$ which has not yet been forced in A' whether:

(4.5) $(\exists t)\,(\exists \sigma)\,[\text{compat}(\sigma, f_s)\ \&\ \{e\}^\sigma_t(e) \downarrow$
$$\&\ (\forall y < e)\,(\forall x)\,[\langle x,y \rangle \notin \text{dom } f_s \implies \sigma(\langle x,y \rangle) = 0].$$

If so choose the least e and the least corresponding string σ. Define $g_{s+1} = f_s \cup \sigma$ and say that e is *forced in A'*. Otherwise, define $g_{s+1} = f_s$.

Step 2. Enumerate the next element $z \in S_{s+1} - S_s$. Define

$$f_{s+1}(\langle x,y \rangle) = \begin{cases} g_{s+1}(\langle x,y \rangle) & \text{if } \langle x,y \rangle \in \text{dom } g_{s+1}, \\ 1 & \text{if } y = z \text{ and } \langle x,y \rangle \notin \text{dom } g_{s+1}, \\ 0 & \text{if } y \notin S_{s+1},\ \langle x,y \rangle \leq s, \\ & \text{and } \langle x,y \rangle \notin \text{dom } g_{s+1}. \end{cases}$$

The last clause of (4.5) is to ensure that if $y \notin S$ then $\lim_x A(\langle x,y \rangle) = 0$. To see that $A' \leq_T S$ fix e, assume that membership of $i \in A'$ has been decided for all $i < e$, and find s such that $S_s \upharpoonright e = S \upharpoonright e$. Show that if e has not been forced in A' by stage s it will never be.

5.* Minimal Degrees

This section contains one of the few fundamental methods of recursion theory on ω which does not arise in connection with r.e. sets and degrees. It is included only for the sake of completeness of the former methods and can be omitted without loss in reading any later material of this book. Variations on this method have produced many important results in degrees and their initial segments. These may be found in Lerman [1983], which is a book in the same series as the present book, complementing it in areas of recursion theory not dealing with r.e. sets and degrees. Also many references about theorems using the minimal degree construction can be found in the bibliography at the end of Epstein [1979].

5.1 Definition. A degree **a** is *minimal* if **a > 0** and there is no **b** such that **0 < b < a**.

Spector [1956] proved the existence of minimal degrees and Sacks [1963a] proved their existence below $\mathbf{0'}$. Our method is essentially that of Shoenfield [1966] which is a simplification of Sacks's method. We begin with some terminology and lemmas which will be useful in both proofs.

Let $\alpha, \beta, \gamma \in 2^{<\omega}$ be strings. We say β and γ *split* α if $\alpha \subset \beta$, $\alpha \subset \gamma$, and β and γ are incompatible in the sense of Definition 4.7. An *f-tree* is a partial recursive function $T : 2^{<\omega} \to 2^{<\omega}$ such that if one of $T(\alpha\hat{~}0), T(\alpha\hat{~}1)$ is defined, then all of $T(\alpha), T(\alpha\hat{~}0), T(\alpha\hat{~}1)$ are defined and $T(\alpha\hat{~}0)$ and $T(\alpha\hat{~}1)$ split $T(\alpha)$. (For example, the identity function $\text{Id}(\sigma) = \sigma$ is an f-tree.)

We use the term "f-tree" (abbreviating "function tree") to distinguish this notion from the standard notion of a *tree* as a set of nodes $T \subseteq \omega^{<\omega}$ which is closed under initial segments as defined in Definition 5.11. The relationship between these notions is that if T is a total recursive f-tree then T_1 is a recursive tree where T_1 is the closure of ran T under initial segments. The notion of f-tree will be used in this section only, but trees and recursive trees will frequently occur. (See Exercises 5.12–5.14, Chapter X and Chapter XIV.)

We say that a string σ is *on* T if $\sigma \in$ rng T. A set A is *on* T if $\sigma \subseteq A$ for infinitely many σ on T. (As usual we identify a set A with its characteristic function χ_A in 2^ω.) An f-tree T_1 is a *sub-f-tree* of T (written $T_1 \subseteq T$) if σ on T_1 implies σ on T (and hence A on T_1 implies A on T). Strings ρ and τ *e-split* σ *on* T, and we say σ *e-splits on* T, if ρ and τ are on T, $\sigma \subset \rho$, $\sigma \subset \tau$, and $\{e\}^\rho$ and $\{e\}^\tau$ are incompatible, namely

$$(5.1) \qquad (\exists x)\,(\exists y)\,(\exists z)\,(\exists t)\,[\{e\}_t^\rho(x)\!\downarrow = y \ \& \ \{e\}_t^\tau(x)\!\downarrow = z \ \& \ y \neq z].$$

(If (5.1) holds then necessarily ρ and τ split σ.)

5.2 Lemma. *If A is on T, $\sigma \subset A$, σ does not e-split on T, and $\{e\}^A$ is total, then $\{e\}^A$ is recursive.*

Proof. Suppose $\{e\}^A = g$ is total. To recursively compute $g(x)$ find any τ on T, $\tau \supseteq \sigma$, such that $\{e\}^\tau(x)$ converges, and set $g(x) = \{e\}^\tau(x)$. Such a τ exists because $\{e\}^A(x)$ converges, so $\{e\}^\rho(x)$ converges for some ρ on T, $\sigma \subseteq \rho \subset A$. Furthermore, $\{e\}^\rho(x) = \{e\}^\tau(x)$ since σ has no e-splittings on T. Hence, $\{e\}^\tau(x) = \{e\}^A(x)$. Finally, g is recursive because T is a p.r. function, so rng T is r.e., and can be enumerated until τ is found. ▯

An f-tree T is *e-splitting* if whenever $T(\alpha{}^\frown 0)$, $T(\alpha{}^\frown 1)$ are defined, they e-split $T(\alpha)$.

5.3 Lemma. *If an f-tree T is e-splitting, A is on T, and $\{e\}^A = g$ is total then $A \leq_T g$.*

Proof. Fix g as an oracle. We shall g-recursively compute a sequence of strings $\{\sigma_s\}_{s \in \omega}$ on T such that $A = \bigcup_s \sigma_s$.

Stage $s = 0$. Set $\sigma_0 = T(\emptyset)$.

Stage $s + 1$. Suppose we are given $\sigma_s = T(\alpha)$ for some α of length s such that $\sigma_s \subset A$. Compute $\rho = T(\alpha{}^\frown 0)$ and $\tau = T(\alpha{}^\frown 1)$. (Both exist since σ_s is extendible.) Now ρ and τ e-split σ_s, so (5.1) holds for ρ and τ. Exactly one value y or z of (5.1) agrees with $g(x)$ because $\{e\}^A(x) = g(x)$ and $\{e\}^A(x) = y$ or $\{e\}^A(x) = z$. Enumerate the quadruples $\langle x, y, z, t \rangle$ until the first is found satisfying (5.1). Set $\sigma_{s+1} = \rho$ if $g(x) = y$ and $\sigma_{s+1} = \tau$ if $g(x) = z$. Now $\sigma_{s+1} \subseteq A$ because A extends exactly one of ρ and τ, and because $\{e\}^A(x) = g(x)$. $\Box$

5.4 Definition. Given an f-tree T, a string σ on T, and $e \in \omega$, define $T_1 = \mathcal{T}(T, \sigma, e)$, the *e-splitting sub-f-tree of T above σ* by induction on $lh(\alpha)$ as follows. Set $T_1(\emptyset) = \sigma$. If $T_1(\alpha) = \nu$ is defined, enumerate all tuples $\langle \rho, \tau, t, x, y, z \rangle$ such that $\nu \subseteq \rho, \tau$ and ρ and τ are on T until the first such tuple is found satisfying the matrix of (5.1). Define $T_1(\alpha{}^\frown 0) = \rho$ and $T_1(\alpha{}^\frown 1) = \tau$. If they do not exist $T(\alpha{}^\frown i)$ is undefined.

5.5 Lemma. *The definition of T_1 obviously satisfies the following:*

(i) T_1 *is an f-tree;*
(ii) $T_1 \subseteq T$ *(namely, any τ on T_1 is also on T);*
(iii) σ *is on T_1;*
(iv) τ *is on T_1 implies $\sigma \subseteq \tau$;*
(v) T_1 *is e-splitting;*
(vi) *If τ is on T_1 and e-splits on T, then τ e-splits on T_1.* $\Box$

5.6 Theorem (Spector [1956]). *There exists a minimal degree $\mathbf{a}$. (Indeed $\mathbf{a} \leq \mathbf{0}''$.)*

Proof. By Lemmas 5.2 and 5.3 it suffices to construct $A \subseteq \omega$ to meet for all e the requirements:

(minimality) R_e : A lies on an f-tree T which is either e-splitting or else there is some string $\sigma \subset A$ such that σ does not e-split on T;

(nonrecursiveness) S_e : $A \neq \{e\}$.

At stage $s + 1$ we shall construct a *total* f-tree T_{s+1} and a string σ_{s+1} on T_{s+1} such that $A = \bigcup_s \sigma_s$ and:

(5.2) $$\sigma_s \subset \sigma_{s+1};$$

(5.3) $$T_s \supseteq T_{s+1};$$

and

(5.4)   either T_{s+1} is s-splitting or else σ_{s+1} does not s-split on T_{s+1}.

Note that A is on T_s for each s by (5.2) and (5.3). Hence, by (5.4), A satisfies R_e for all e.

Stage $s = 0$. Set $\sigma_0 = \emptyset$, and $T_0 = \text{Id}$.

Stage $s + 1$. We are given σ_s on T_s.

Step 1. (We meet S_s.) Choose σ' on T_s such that $\sigma' \supset \sigma_s$ and σ' is incompatible with $\{s\}$ if $\{s\}$ is total, and $\sigma' = \sigma$ otherwise. This is possible since T_s is total, so T_s has a splitting above σ_s. (Note that Step 1 requires a $\emptyset''$-oracle to test whether $\{s\}$ is total.)

Step 2. (We meet R_s.) Test whether

(5.5) $(\exists \tau) [\tau \text{ on } T_s \ \& \ \sigma' \subseteq \tau \ \& \ \tau \text{ has no } s\text{-splittings on } T_s]$.

Note that (5.5) is a Σ_2 statement and hence can be tested recursively in $\emptyset''$, since T_s is total.

If (5.5) holds, let σ_{s+1} be the first such τ and set $T_{s+1} = T_s$. Finding τ from σ' is r.e. in $\emptyset'$ and hence recursive in $\emptyset''$. If (5.5) fails then set $\sigma_{s+1} = \sigma'$ and $T_{s+1} = T(T_s, \sigma', s)$. In this case every τ on T_s s-splits on T_s, so T_{s+1} is total and s-splitting by Lemma 5.5(vi).

Clearly $A = \bigcup_s \sigma_s$ has minimal degree and $A \leq_T \emptyset''$ since the constructions at Steps 1 and 2 are each recursive in $\emptyset''$.   ▯

5.7 Theorem (Sacks).   *There is a minimal degree $\mathbf{a} < \mathbf{0}'$.*

Proof. We shall use a $\emptyset'$-construction to specify at each stage s a string σ_s and f-trees T_e^s, $e \leq s$, so that $T_{e+1} = \lim_s T_{e+1}^s$ exists and ensures that A satisfies R_e where $A = \bigcup_s \sigma_s$. In this construction the f-trees T_e^s may be partial. Define $T_0^s = \text{Id}$ for all $s \in \omega$. Note that $T_{e+1}^s \subseteq T_e^s$.

Stage $s = 0$. Set $\sigma_0 = \emptyset$.

Stage $s + 1$.

Step 1. (To meet S_s.) Let n_s be the largest integer $\leq s$ such that σ_s has a splitting ρ, τ on $T_{n_s}^s$. (Such n_s exists because $T_0^s = \text{Id}$. Furthermore, n_s can be found using a $\emptyset'$-oracle because each T_e^s, $e \leq s$, is a partial recursive f-tree.) Now using a $\emptyset'$-oracle let σ_{s+1} be ρ if $\{s\}$ and ρ are incompatible or if $\{s\}(x)\uparrow$ for some $x < lh(\rho)$, and let $\sigma_{s+1} = \tau$ otherwise.

Step 2. (To meet R_s.) Set $T_e^{s+1} = T_e^s$ for all $e \leq n_s$. If $n_s = s$, set $T_{s+1}^{s+1} = \mathcal{T}(T_s^{s+1}, \sigma_{s+1}, s)$. If $n_s < s$, choose α such that $T_{n_s}^s(\alpha) = \sigma_{s+1}$. Then set $T_{n_s+1}^{s+1}(\gamma) = T_{n_s}^{s+1}(\alpha^\frown\gamma)$ for all γ, and set $T_{e+1}^{s+1} = \mathcal{T}(T_e^{s+1}, \sigma_{s+1}, e)$ for $n_s < e \leq s$.

This ends the construction. Let $A = \bigcup_s \sigma_s$. Note that $A \leq_T \emptyset'$ since Step 1 requires only a $\emptyset'$-oracle, and Step 2 no oracle.

(The intuition behind Step 2 is the following. The construction begins by letting T_1^s be a 0-splitting sub-f-tree of T_0^s. If at a later stage $s+1$, the $\emptyset'$-oracle tells us that there are no further 0-splittings above σ_s on T_1^s, then we set $n_s = 0$, we abandon the old 0-splitting f-tree T_1^s, and we define T_1^{s+1} to be the restriction of T_0^{s+1} to nodes extending σ_{s+1}, namely we define $T_1^{s+1}(\gamma) = T_0^{s+1}(\alpha^\frown\gamma)$ for all γ. Now T_1^{s+1} presents no further obstacles in finding splittings, because T_0^s presents none. Note that by this change in T_1^{s+1} we have switched our strategy for meeting requirement R_0 from that of Lemma 5.3 to that of Lemma 5.2. Naturally, all the old f-trees T_e^s, $e > 1$, must be replaced by new f-trees T_e^{s+1} lying nested within the new f-tree T_1^{s+1}.)

Finally, we prove by induction on e that for all e,

$$(5.6) \qquad\qquad T_e = \lim_s T_e^s \text{ exists,}$$

and

$$(5.7) \qquad\qquad (\text{a.e. } s) \, [n_s \geq e].$$

These are clearly true for $e = 0$. Fix $e \geq 0$ and by induction choose s_e minimal such that

$$(5.8) \qquad\qquad (\forall s \geq s_e) \, [T_e^{s+1} = T_e^{s_e} \text{ and } n_s \geq e].$$

Now (5.8) also holds for $e + 1$ unless $n_t = e$ for some $t \geq s_e$. In this case choose α such that $T_e^t(\alpha) = \sigma_{t+1}$. Then $T_{e+1}^{t+1}(\gamma) = T_e^t(\alpha^\frown\gamma)$ for all γ. It follows by induction on $s \geq t+1$ that for all $s \geq t+1$, $T_{e+1}^s = T_{e+1}^{t+1}$ and that any splittings of σ_s on T_e^s exist on T_{e+1}^s as well. Thus, (5.8) holds for $e+1$ with t in place of s_e. Therefore, T_{e+1} is either e-splitting (if $s_{e+1} = s_e$) or contains no nodes which e-split (if $s_{e+1} > s_e$). ▯

(If one analyzes this construction carefully, it may be viewed as a finite injury construction with oracle $\emptyset'$, but we prefer to defer such analysis until Chapter VII since the above proof can be understood without explicit mention of the priority method.)

5.8–5.15 Exercises

5.8 (Posner-Epstein). Show that in the proof of Theorem 5.6 if we meet all the minimality requirements R_e, $e \in \omega$, then we have automatically met all the requirements S_e, $e \in \omega$. *Hint.* Assume that A is recursive. Define the recursive functional $\{i\}^X$ by

$$\{i\}^\sigma(x) = \begin{cases} \sigma(x) & \text{if } \sigma \text{ is incompatible with } A; \\ \text{undefined} & \text{if } \sigma \subset A. \end{cases}$$

Show that requirement R_i is not met.

5.9 (Sacks). Show that any countable ascending sequence of degrees $\mathbf{b}_0 < \mathbf{b}_1 < \cdots$ has a minimal upper bound $\mathbf{a}$. (Namely, $\mathbf{a}$ is an upper bound but there is no upper bound $\mathbf{d} < \mathbf{a}$.) *Hint.* Choose a set $B_n \in \mathbf{b}_n$ for each n. Use Theorem 5.6 to find a *total* recursive f-tree T_1 satisfying R_0. Define a total B_0-recursive f-tree $\widetilde{T}_1 \subseteq T_1$ which still satisfies R_0 and such that $B_0 \leq_T A$ for every A on $\widetilde{T}_1$. Define $\widetilde{T}_s(\alpha)$ by induction on $|\alpha|$ as follows. $\widetilde{T}_1(\emptyset) = T_1(\emptyset)$. Assume $\widetilde{T}_1(\alpha)$ has been defined and equals $T_1(\rho)$ for some ρ. If $|\alpha| = n$, define $\widetilde{T}_1(\alpha^\frown i) = T_1(\rho^\frown B_0(n)^\frown i)$, for $i = 0, 1$. Choose an appropriate σ_1 on $\widetilde{T}_1$, $|\sigma_1| \geq n$. In general apply the above procedure to $\widetilde{T}_n$ to obtain a B_n-recursive total f-tree $\widetilde{T}_{n+1} \subseteq \widetilde{T}_n$ such that $\widetilde{T}_{n+1}$ satisfies R_n and $B_n \leq_T A$ for every A on $\widetilde{T}_{n+1}$. Next choose $\sigma_{n+1} \in \widetilde{T}_{n+1}$ such that $\sigma_n \subset \sigma_{n+1}$. Let $A = \bigcup_n \sigma_n$.

5.10.° Show that there exist $2^{\aleph_0}$ different minimal degrees. *Hint.* Combine the methods of Exercise 1.8 and Theorem 5.6 using a tree of f-trees.

The following exercises use a slightly different notion of tree from that above and require finite and infinite extension arguments.

5.11 Definition. (i) A set $T \subseteq \omega^{<\omega}$ is a *tree* if for all σ and τ

$$[\sigma \in T \ \& \ \tau \subseteq \sigma] \implies \tau \in T.$$

(ii) If T is a tree let $[T] = \{f : (\forall x) \overline{f}(x) \in T\}$, the infinite paths through T, where $\overline{f}(x) = p_0^{1+f(0)} p_1^{1+f(1)} \cdots p_{x-1}^{1+f(x-1)}$, and where we identify nodes $f \upharpoonright x \in T$ with their code numbers $\overline{f}(x)$.

(iii) A tree T is *recursive* if the relation $\sigma \in T$ is recursive, where we identify σ with its sequence number $\overline{\sigma}(lh(\sigma))$, and where

$$\overline{\sigma}(x) = p_0^{1+\sigma(0)} p_1^{1+\sigma(1)} \cdots p_{x-1}^{1+\sigma(x-1)}.$$

(iv) If R is a recursive predicate of one free variable then the class of all functions $f \in \omega^\omega$ satisfying $(\forall x) R(\overline{f}(x))$ is called a Π_1^0 *class.*

5.12. (a) Prove that for any Π_1^0 class $\mathcal{C} \subseteq \omega^\omega$ there is a recursive tree T such that $\mathcal{C} = [T]$. (The converse is obvious.)

(b) Let $T \subseteq 2^{<\omega}$ be an infinite recursive tree. Prove that if $[T] = \{f\}$ then f is recursive.

(c) Let $T \subseteq 2^{<\omega}$ be an infinite recursive tree. Prove that $[T]$ has a member of r.e. degree. *Hint.* Use Corollary III.3.4.

5.13 Low Basis Theorem (Jockusch-Soare [1972b]). Prove that any non-empty Π_1^0 class $C \subseteq 2^\omega$ has a member f of low degree. *Hint.* Given C apply Exercise 5.11 to obtain a recursive tree $T \subseteq 2^{<\omega}$ such that $C = [T]$. Define a sequence of infinite recursive trees $\{ T_e \}_{e \in \omega}$ and choose $f \in \bigcap\{ [T_e] \}_{e \in \omega}$. Tree T_{e+1} will have the property that $\{ e \}^g(e)$ is defined for all or no $g \in [T_{e+1}]$. Let $T_0 = T$. Assume tree T_e has been defined. Let

$$U_e = \{ \sigma : \{ e \}_{lh(\sigma)}^\sigma (e) \text{ is undefined} \},$$

and note that U_e forms a recursive tree. Let $T_{e+1} = T_e$ if $T_e \cap U_e$ is finite, and $T_{e+1} = T_e \cap U_e$ otherwise. Choose $f \in \bigcap\{ [T_e] \}_{e \in \omega}$ and show that $\deg(f') = \mathbf{0}'$ since the entire construction is recursive in $\mathbf{0}'$.

5.14. Prove that there is an effectively immune set of low degree and hence a function f of low degree with no fixed points. (Thus the hypothesis of recursive enumerability in Exercise V.4.2 and Theorem V.5.1 was necessary.) *Hint.* Let S be Post's simple set of Theorem V.1.3 which is effectively simple but not hypersimple. Let $\{ D_{f(x)} \}_{x \in \omega}$ be a disjoint strong array of finite sets each of which intersects $\overline{S}$. Define

$$C = \{ A : A \subseteq \overline{S} \text{ and } (\forall x) \, [D_{f(x)} \cap A \neq \emptyset] \}.$$

Show that C is a nonempty Π_1^0 class, and apply the Low Basis Theorem.

5.15 (Jockusch-Soare). If T is an infinite recursive tree prove that $[T]$ has a member of hyperimmune-free degree (as defined in Exercise V.2.17). *Hint.* Define a sequence of infinite recursive trees $\{ T_e \}_{e \in \omega}$ and choose $f \in \bigcap\{ [T_e] \}_{e \in \omega}$. Given T_e define $U_e^x = \{ \sigma : \{ e \}_{lh(\sigma)}^\sigma (x) \text{ is undefined} \}$. If $T_e \cap U_e^x$ is finite for all x, let $T_{e+1} = T_e$. Otherwise, let $T_{e+1} = T_e \cap U_e^{x'}$ where x' is minimal such that the tree is infinite. Use Exercise V.2.17(b) to see that f is of hyperimmune-free degree.

Chapter VII

The Finite Injury Priority Method

A positive solution to Post's problem was finally achieved by Friedberg [1957] and independently by Muchnik [1956]. In their method the desired r.e. set A is constructed by stages to meet certain requirements $\{ R_n \}_{n \in \omega}$ like those of Chapter VI. However, now the construction must be recursive to ensure that A is r.e. If $n < m$, then requirement R_n is given *priority* over R_m, and we say that R_n has *stronger* priority (also called *higher* priority) than R_m. Action taken for R_m at some stage s may at a later stage $t > s$ be undone when action is taken for R_n, thereby *injuring* R_m at stage t. The proofs in this chapter have the property that each requirement is injured at most finitely often. Later we shall consider more complicated methods allowing infinite injury.

In these three sections we give three different results, each of which yields a solution to Post's problem, and each of which uses a different kind of finite injury priority argument. Each of the three methods is fundamental and will be essential for results in later chapters. The three methods are in ascending order with respect to technical complexity because if $f(e)$ is the number of times that requirement R_e is injured then $f(e) \le e$ in §1, $f(e) \le 2^e - 1$ in §2, and $f(e)$ is not even recursively bounded in §3.

1. Low Simple Sets

Probably the easiest solution to Post's problem is the construction of a low simple set A. Simplicity guarantees that A is r.e. and nonrecursive while the lowness of A (namely $A' \equiv_T \emptyset'$) guarantees incompleteness, namely $A <_T \emptyset'$. Furthermore, in this construction the requirements may be separated into the *positive* requirements P_e, which attempt to put elements *into* A, and the *negative* requirements N_e, which attempt to keep elements *out of* A. Low sets and degrees have several pleasant structural properties as we shall see in Chapter XI.

1.1 Theorem.   *There is a simple set A which is low ($A' \equiv_T \emptyset'$).*

1.2 Corollary (Friedberg-Muchnik).   *There is a nonrecursive incomplete r.e. degree $\mathbf{a}$ (i.e., $\mathbf{0} < \mathbf{a} < \mathbf{0}'$).*

Proof of Theorem 1.1. It suffices to construct a coinfinite r.e. set A to meet for all e the requirements:

(simplicity) $$P_e : W_e \text{ infinite} \implies W_e \cap A \neq \emptyset,$$

(lowness) $$N_e : (\exists^\infty s) \, [\{e\}_s^{A_s}(e)\downarrow] \implies \{e\}^A(e)\downarrow,$$

where $(\exists^\infty s) \, Q(s)$ denotes "there exist infinitely many s such that $Q(s)$," A_s consists of the elements enumerated in A by the end of stage s, and $A = \bigcup_s A_s$. The priority ranking of the requirements is $N_0, P_0, N_1, P_1, \dots$. (Since the construction will be recursive the sequence of finite sets $\{A_s\}_{s \in \omega}$ will be recursive, namely there is a recursive function f such that $f(s)$ is the canonical index of A_s.)

Note that the requirements $\{N_e\}_{e \in \omega}$ guarantee $A' \leq_T \emptyset'$ as follows. Define the recursive function g by

$$g(e, s) = \begin{cases} 1 & \text{if } \{e\}_s^{A_s}(e)\downarrow; \\ 0 & \text{otherwise.} \end{cases}$$

If requirement N_e is satisfied for all e, then $\hat{g}(e) = \lim_s g(e, s)$ exists for all e. But $\hat{g} \leq_T \emptyset'$ by the Limit Lemma III.3.3, and $\hat{g}$ is the characteristic function of A' so $A' \leq_T \emptyset'$.

Recall from Definition III.1.7 the use function $u(A_s; e, x, s)$ which is 1 plus the maximum element used in the computation $\{e\}_s^{A_s}(x)$ if the latter converges and 0 otherwise. To aid in meeting N_e, given A_s define for all e

(restraint function) $$r(e, s) = u(A_s; e, e, s).$$

Now $r(e, s)$ will be a recursive function because $\{A_s\}_{s \in \omega}$ is a recursive sequence. To meet N_e we attempt to restrain with priority N_e any elements $x \leq r(e, s)$ from entering A_{s+1}. (The point is that if $\{e\}_s^{A_s}(e)\downarrow$, $r = u(A_s; e, e, s)$, and N_e succeeds in preventing any $x \leq r$ from later entering A, then $A \upharpoonright r = A_s \upharpoonright r$, so $\{e\}^A(e)\downarrow$.) Thus, such elements can only enter A for the sake of some P_i of stronger priority (namely $i < e$). The strategy for meeting P_i is the same as for Post's simple set, Theorem V.1.3.

Construction of A.
    Stage $s = 0$. Let $A_0 = \emptyset$.
    Stage $s + 1$. Given A_s we have $r(e, s)$ for all e. Choose the least $i \leq s$ such that

(1.1) $$W_{i,s} \cap A_s = \emptyset;$$

and

$$(1.2) \qquad (\exists x) \, [x \in W_{i,s} \;\&\; x > 2i \;\&\; (\forall e \leq i) \, [r(e, s) < x]].$$

If i exists, choose the least x satisfying (1.2). Enumerate x in A_{s+1}, and say that requirement P_i *receives attention*. Hence $W_{i,s} \cap A_{s+1} \neq \emptyset$ so P_i is satisfied, (1.1) fails for stages $> s + 1$, and hence P_i never again receives attention. If i does not exist, do nothing, so $A_{s+1} = A_s$.

Let $A = \bigcup_s A_s$. This ends the construction.

We say that x *injures* N_e at stage $s + 1$ if $x \in A_{s+1} - A_s$ and $x \leq r(e, s)$. Define the *injury set for* N_e,

$$I_e = \{\, x : (\exists s) \, [x \in A_{s+1} - A_s \;\&\; x \leq r(e, s)] \,\}$$
$$= \{\, x : x \text{ injures } N_e \text{ at some stage } s + 1 \,\}.$$

(The positive requirements, of course, are never injured.)

Lemma 1. $(\forall e) \, [I_e \text{ is finite}]$. *(Namely, N_e is injured at most finitely often.)*

Proof. Each positive requirement P_i contributes at most one element to A by (1.1). But by the last clause of (1.2), N_e can be injured by P_i only if $i < e$. Hence $|I_e| \leq e$.

Lemma 2. *For every e, requirement N_e is met and $r(e) = \lim_s r(e, s)$ exists.*

Proof. Fix e. By Lemma 1, choose stage s_e such that N_e is not injured at any stage $s > s_e$. However, if $\{e\}_s^{A_s}(e)$ converges for $s > s_e$, then by induction on $t \geq s$, $r(e, t) = r(e, s)$ and $\{e\}_t^{A_t}(e) = \{e\}_s^{A_s}(e)$ for all $t \geq s$, so $A_s \restriction r = A \restriction r$ for $r = r(e, s)$, and hence $\{e\}^A(e)$ is defined by the Use Principle III.1.9.

Lemma 3. *For every i, requirement P_i is met.*

Proof. Fix i such that W_i is infinite. By Lemma 2, choose s such that

$$(\forall t \geq s) \, (\forall e \leq i) \, [r(e, t) = r(e)].$$

Choose $s' \geq s$ such that no P_j, $j < i$, receives attention after stage s'. Choose $t > s'$ such that

$$(\exists x) \, [x \in W_{i,t} \;\&\; x > 2i \;\&\; (\forall e \leq i) \, [r(e) < x]].$$

Now either $W_{i,t} \cap A_t \neq \emptyset$ or else P_i receives attention at stage $t + 1$. In either case $W_{i,t} \cap A_{t+1} \neq \emptyset$ so P_i is met by the end of stage $t + 1$.

Note that $\overline{A}$ is infinite by the second clause of (1.2). Hence A is simple and low. □

In this and later constructions one should think of $r(e, s)$ as a "wall" imposed by N_e, extending from 0 to $r(e, s)$. For fixed e the wall $r(e, s)$ is not monotonically increasing in s since after a wall $r(e, s)$ is erected, penetration of that wall by some x injuring N_e may cause $\{e\}_{s+1}^{A_{s+1}}(e)\uparrow$ so that the wall drops to 0. If P_i wishes to contribute some element x to A, then x must lie beyond all walls $r(e, s)$, $e \leq i$. The crucial feature of all finite injury constructions is that each wall is penetrated (injured) finitely often and so comes to a limit. But then each positive requirement is satisfied because it merely chooses a witness beyond all walls of stronger priority.

In Definition IV.4.2 we defined the high (low) r.e. degrees as those whose first jump has the highest (lowest) possible value. In Exercise IV.4.5 we extended this hierarchy by replacing the first jump by the nth jump as follows.

1.3 Definition. For every $n \geq 0$, define the following subclasses of $\mathbf{R}$, the r.e. degrees,

$$\mathbf{H}_n = \{\mathbf{d} : \mathbf{d} \in \mathbf{R} \text{ and } \mathbf{d}^{(n)} = \mathbf{0}^{(n+1)}\},$$

and

$$\mathbf{L}_n = \{\mathbf{d} : \mathbf{d} \in \mathbf{R} \text{ and } \mathbf{d}^{(n)} = \mathbf{0}^{(n)}\},$$

where $\mathbf{d}^{(0)} = \mathbf{d}$; and let $\overline{\mathbf{L}}_n = \mathbf{R} - \mathbf{L}_n$. Note that $\mathbf{H}_0 = \{\mathbf{0}'\}$, $\mathbf{L}_0 = \{\mathbf{0}\}$, and that $\mathbf{H}_1$ ($\mathbf{L}_1$) are simply the high (low) r.e. degrees of Definition IV.4.2. For $n > 1$ we refer to the degrees in $\mathbf{H}_n$ ($\mathbf{L}_n$) as $high_n$ (low_n). We also write $\mathbf{H}_n^B$ and $\mathbf{L}_n^B$ for these classes relativized to an arbitrary oracle set B.

Clearly, $\mathbf{L}_n \subseteq \mathbf{L}_{n+1}$ and $\mathbf{H}_n \subseteq \mathbf{H}_{n+1}$ for every n. Theorem 1.1 demonstrates that $\mathbf{L}_0 \neq \mathbf{L}_1$, and we show in Exercise 1.6(c) that $\mathbf{H}_0 \neq \mathbf{H}_1$, namely that there is an incomplete high r.e. degree. In Corollary VIII.1.2 we prove this by the original Sacks method using an infinite injury argument. The proof here is a combination of two finite injury constructions. By extending these methods we can show that all the inclusions $\mathbf{L}_n \subset \mathbf{L}_{n+1}$ and $\mathbf{H}_n \subset \mathbf{H}_{n+1}$ are strict. (See Exercise 1.6(d) or Corollary VIII.3.4.) Furthermore, there is an r.e. degree $\mathbf{a}$ such that

$$(1.3) \qquad\qquad (\forall n)\, [\mathbf{0}^{(n)} < \mathbf{a}^{(n)} < \mathbf{0}^{(n+1)}],$$

i.e., $\mathbf{a} \in \mathbf{R} - \bigcup_n (\mathbf{H}_n \cup \mathbf{L}_n)$. (See Exercise 1.6(e) or Corollary VIII.3.5.) Such a degree is called *intermediate*. See also Exercise VIII.3.9.

1.4–1.7 Exercises

1.4. In the proof of Theorem 1.1 replace the requirements P_e by the requirements of Exercise V.2.9 to give a direct construction of a low hypersimple set.

1.5.   Combine the method of Theorem 1.1 with the permitting method of Chapter V §3 to show that for any nonrecursive r.e. set B there is a low simple set $A \leq_T B$.

1.6 (Jockusch-Shore [1983]).   For $e \in \omega$ and $Y \subseteq \omega$ define $J_e(Y) = Y \oplus W_e^Y$.

(a) Prove that for every e there is a nonrecursive r.e. set A such that $J_e(A) \equiv_T \emptyset'$. (This result is related to Theorem 1.1 as Exercise VI.3.10 is related to the Friedberg Completeness Criterion VI.3.1.) We say that $A \equiv_T B$ *via* $\langle k_0, k_1 \rangle$ if $A = \{k_0\}^B$ and $B = \{k_1\}^A$. Show that the proof is uniform in e and can be relativized to any oracle $Y \subseteq \omega$, namely that there exist recursive functions f and g such that for all $e \in \omega$ and $Y \subseteq \omega$,

$$(1.4) \qquad\qquad J_e(J_{f(e)}(Y)) \equiv_T Y' \quad \text{via } g(e).$$

Hint.   Use the method of Theorem 1.1 to construct A and a function $h \leq_T A \oplus W_e^A$ to meet for all x the requirements:

$$N_x : (\exists^\infty s)\, [x \in W_{e,s}^{A_s}] \implies x \in W_e^A;$$
$$P_{2x} : x \in K \iff (\exists y \leq h(2x))\, [y \in \omega^{[2x]} \text{ and } y \in A];$$

and

$$P_{2x+1} : A \neq \overline{W}_x.$$

At stage $s+1$ given A_s define the recursive functions

$$r(x, s) = \begin{cases} u(A_s; e, x, s) & \text{if } \{e\}_s^{A_s}(x)\!\downarrow, \\ 0 & \text{otherwise}; \end{cases}$$

and

$$h(x, s) = (\mu y)\, [y \in \omega^{[x]} \ \& \ h(x-1, s) < y$$
$$\& \ h(x, s-1) \leq y \ \& \ (\forall j \leq x)\, [r(j, s) < y]].$$

For each $x \in K_{s+1} - K_s$ enumerate $h(2x, s)$ into A_{s+1}. Furthermore, for each $x < s$, if $W_{x,s} \cap A_s = \emptyset$ and $h(2x+1, s) \in W_{x,s}$, enumerate $h(2x+1, s)$ into A_{s+1}, and otherwise do nothing. This defines A.

Show that $\lim_s h(x, s)$, $\lim_s r(x, s)$ and $K(x)$ can be computed recursively in $A \oplus W_e^A$.

(b) Use the uniformity of the proof of (a) to show that there is a recursive function h such that for all Y and e, $J_e(J_{h(e)}(Y)) \equiv_T Y'$ and $Y <_T J_{h(e)}(Y)$.

(c) (Sacks). Show that there exists an incomplete high r.e. set A. *Hint* (Jockusch-Shore). Note that Theorem 1.1 can be relativized to an arbitrary set Y to produce an index e_1 such that $Y <_T W_{e_1}^Y$ and $Y' \equiv_T (W_{e_1}^Y)'$, namely such that $W_{e_1}^Y$ is *low over* Y. Apply part (a) to e_1 to obtain A. (Note that if $\emptyset'$ is low over A then A is high.)

(d) (Sacks [1963c]). Prove that for every n there exist r.e. degrees in $\mathbf{H}_{n+1} - \mathbf{H}_n$ and $\mathbf{L}_{n+1} - \mathbf{L}_n$. *Hint.* Note that part (c) produces an index e_2

such that for any set Y, $W_{e_2}^Y$ is high over Y, namely $Y <_T W_{e_2}^Y <_T Y'$ and $(W_{e_2}^Y)' \equiv_T Y''$. Apply (a) to e_2 to obtain a low$_2$ r.e. set A, and continue this pattern using the fact that Y' is low$_n$ over C if and only if C is high$_n$ over Y.

(e) (Martin, Lachlan, Sacks). Prove that there is an r.e. degree $\mathbf{a}$ which is not in $\mathbf{H}_n$ or $\mathbf{L}_n$ for any n. *Hint* (Jockusch-Shore [1983]). Apply the Relativized Recursion Theorem III.1.6 to the recursive function h of part (b) to obtain a number $\hat{e}$ such that $W_{h(\hat{e})}^Y = W_{\hat{e}}^Y$ for all Y. (Since $Y <_T J_{\hat{e}}(Y)$ and $J_{\hat{e}}(J_{\hat{e}}(Y)) \equiv_T Y'$ we can think of $J_{\hat{e}}$ as a kind of "*half-jump*" although we have no reason to believe that it is degree-invariant.) Show that $J_{\hat{e}}^{2n}(Y) \equiv_T Y^{(n)}$, so $J_{\hat{e}}^{2n}(Y) <_T J_{\hat{e}}^{2n+1}(Y) <_T J_{\hat{e}}^{2n+2}(Y)$ and hence $J_{\hat{e}}(Y) \notin \mathbf{L}_n^Y \cup \mathbf{H}_n^Y$.

1.7° (Jockusch-Shore).   Given $e, i \in \omega$ find an r.e. set A such that: (i) $A \oplus W_e^A$ is of r.e. degree; and (ii) $A \oplus W_e^A \oplus W_i^A \equiv_T K$. *Hint*. In the proof of Exercise 1.6(a) view $h(x, s)$ as the position of a "movable marker" Γ_x which moves so that its position exceeds the use function for $x \in W_{e,s}^{A_s}$ and the position of Γ_{x-1}. In addition to these "e-markers" $\{\Gamma_x\}_{x \in \omega}$ we have a list $\{\Lambda_x\}_{x \in \omega}$ of i-markers which behave analogously for computations $x \in W_{i,s}^{A_s}$ except that the i-markers must move on the set of the e-marker positions. If x enters K we enumerate the current position of Λ_x into A. To show (ii), note that given $A \oplus W_i^A \oplus W_e^A$ we can compute the use functions for $x \in W_e^A$ or $x \in W_i^A$, the final position of all markers, and thus whether $x \in K$. To see that $A \oplus W_e^A$ has r.e. degree, note that the motion of the Γ-markers is as in Exercise 1.6(a) and hence A satisfies each negative requirement N_x there. By our construction there is a function $m \leq_T A \oplus W_e^A$ such that for all $s \geq m(x)$, $x \in W_e^A$ if and only if $x \in W_{e,s}^{A_s}$. Hence, $A \oplus W_e^A$ has r.e. degree by the Modulus Lemma III.3.2.

1.8 Remark.   Exercise 1.6 can be used to give a short and finite injury style proof of the Sacks Jump Theorem VIII.3.1 which in the notation of Exercise 1.6 asserts that

$$(\forall i)\, (\exists p)\, [(J_p(\emptyset))' \equiv_T J_i(\emptyset')].$$

Proof (M. Simpson [1985]). Fix i. Let f and g satisfy (1.4). Then setting $Y = \emptyset$ in (1.4) we have for all e,

(1.5) $$J_e(J_{f(e)}(\emptyset)) \equiv_T \emptyset' \quad \text{via } g(e).$$

Clearly, there is a recursive function h such that

(1.6) $$B \equiv_T \emptyset' \text{ via } k \implies J_{h(k)}(B) \equiv_T J_i(\emptyset').$$

Now by (1.5) and (1.6) with $B = J_e(J_{f(e)}(\emptyset))$,

(1.7) $$J_{hg(e)}(J_e(J_{f(e)}(\emptyset))) \equiv_T J_i(\emptyset').$$

We wish to get the left hand side of (1.7) as the jump of an r.e. set using (1.4). To do this we first replace e by $f(e)$ in (1.7) to get

$$(1.8) \qquad J_{hgf(e)}(J_{f(e)}(J_{ff(e)}(\emptyset))) \equiv_T J_i(\emptyset').$$

Next apply the Recursion Theorem to hgf to obtain e_0 such that for all $Y \subseteq \omega$,

$$(1.9) \qquad W^Y_{hgf(e_0)} = W^Y_{e_0}, \text{ and hence } J_{e_0}(Y) = J_{hgf(e_0)}(Y).$$

Substituting e_0 for e in (1.8) we get by (1.8) and (1.9) that

$$(1.10) \qquad J_{e_0}(J_{f(e_0)}(J_{ff(e_0)}(\emptyset))) \equiv_T J_i(\emptyset').$$

Hence, by (1.10) and (1.4) with $Y = J_{ff(e_0)}(\emptyset)$,

$$(1.11) \qquad (J_{ff(e_0)}(\emptyset))' \equiv_T J_i(\emptyset').$$

Define $p = ff(e_0)$. (Note also that p can be found uniformly in i because in (1.6) we can replace $h(k)$ by a recursive function $h(i,k)$, and the Recursion Theorem with Parameters gives e_0 uniformly in i. Note also that Remark 1.8 holds in the relativized version with Y in place of $\emptyset$.   ⬚

To better understand Exercise 1.6 and Remark 1.8 think of $J_e(Y)$ as the "hop" of Y since it behaves as a kind of half-jump. By Exercise 1.6 for every choice of a second hop there is a first hop such that the two hops are (Turing) equivalent to one jump. The Sacks Jump Theorem says that a jump followed by any hop is equivalent to some hop followed by a jump. The above shows that both are equivalent to a sequence of three appropriately chosen hops.

1.9 Remark.   A. Kučera has recently given the following solution to Post's problem which does not use the priority method and therefore has no injury to requirements. We say that a function f is *fixed point free* if $W_{f(x)} \neq W_x$ for all x. (See Theorem V.5.1.)

1.10 Theorem (Kučera [ta]).   *If $f \leq_T \emptyset'$ and f is fixed point free then there is a simple set $A \leq_T f$.*

1.11 Corollary.   *There is a low simple set A.*

Proof. By Exercise V.5.9 there is a fixed point free function f of low degree. Apply Theorem 1.10 to produce $A \leq_T f$. (Note that Exercises V.5.8, V.5.9, and VI.5.13 do not use the priority method but only use oracle constructions. Hence, there is no injury to requirements.)   ⬚

Proof of Theorem 1.10. Since $f \leq_T \emptyset'$ we can choose a recursive function $\hat{f}(x, s)$ such that $f(x) = \lim_s \hat{f}(x, s)$. It suffices to construct a coinfinite r.e. set $A \leq_T f$ such that for all e, A meets requirement P_e of Theorem 1.1. To accomplish the latter we shall use the Recursion Theorem to define a recursive function $h(e)$ which plays the same role as $h(e)$ did in Theorem V.5.1.

Construction of A.

Stage $s = 0$. Let $A_0 = \emptyset$.

Stage $s + 1$. Find the least e (if it exists) such that $W_{e,s} \cap A_s = \emptyset$, and

(1.12)
$$(\exists y) \, [y \in W_{e,s} \, \& \, y > 2e \, \& \, y > h(e)$$
$$\& \, (\forall t)_{y \leq t \leq s}[\hat{f}(h(e), t) = \hat{f}(h(e), s)]].$$

Enumerate in A_{s+1} the least y satisfying (1.12). Using the Recursion Theorem define $h(e)$ by $W_{h(e)} = W_{\hat{f}(h(e),s)}$. (If P_e never receives attention then $W_{h(e)} = \emptyset$.)

Lemma 1. *If W_e is infinite then $W_e \cap A \neq \emptyset$.*

Proof. Choose $y \in W_e$ such that $y > 2e$, $y > h(e)$, and $y > m(h(e))$, where $m(x)$ is the modulus function for $\{ \hat{f}(x, s) \}_{s \in \omega}$ (as defined in Definition III.3.1). Eventually y (or another element of W_e) is enumerated in A.

Lemma 2. $A \leq_T f$.

Proof. Define the "weak" modulus function for $\{ \hat{f}(x, s) \}_{s \in \omega}$ as follows,

$$c(y) = (\mu s > y) \, (\forall x \leq y) \, [\hat{f}(x, s) = f(x)].$$

Clearly, $c \leq_T f$. We claim that for all y, $y \in A$ iff $y \in A_{c(y)}$, so that $A \leq_T f$. Assume to the contrary that $y \in A - A_{c(y)}$. Suppose that $y \in A_{s+1} - A_s$ and y enters A for the sake of P_e. Then $\hat{f}(h(e), t) = \hat{f}(h(e), s)$ for all t, $y \leq t \leq s$. But $y < c(y) < s$ (by definition of $c(y)$ and our hypothesis on y respectively). Hence,

(1.13)
$$W_{h(e)} = W_{\hat{f}(h(e),s)} = W_{\hat{f}(h(e),c(y))} = W_{f(h(e))},$$

contrary to f being fixed point free. (The first equality of (1.13) follows by the definition of h, and the last uses the definition of $c(y)$ and the fact that $h(e) \leq y$.) ▯

1.12 Remark. Notice that if f has *recursively enumerable* degree then $m \leq_T f$ as in Theorem V.5.1 and we can prove that $K \leq_T f$. Here since f is only Δ_2^0 we do not have $m \leq_T f$ but only that the *weak* modulus satisfies $c \leq_T f$ so if W_e is infinite we eventually get *some* y satisfying (1.12) which will enter A, but we cannot choose a priori a *specific* such y and thus cannot achieve $K \leq_T f$ as before.

Using a more delicate argument in the same spirit, Kučera has given a priority-free proof of Theorem 2.1, the existence of two incomparable r.e. degrees.

2. The Original Friedberg-Muchnik Theorem

Friedberg and Muchnik originally solved Post's problem by constructing r.e. sets A and B of incomparable degree, namely satisfying exactly the same requirements as in Theorem VI.1.2. Unlike the requirements in §1, these requirements, for example $A \neq \{e\}^B$, have both a positive and negative component since they try to put elements into A and simultaneously keep elements out of B. We shall describe first the strategy for meeting a single such requirement and then how the strategies may be combined.

2.1 Theorem (Friedberg [1957a]–Muchnik [1956a]).   *There exist r.e. sets A and B such that $A \mid_T B$ (i.e., $A \not\leq_T B$ and $B \not\leq_T A$), and hence $\emptyset <_T A, B$, and $A, B <_T \emptyset'$.*

Proof. It suffices to recursively enumerate A and B to meet for all e the requirements:

$$R_{2e} : A \neq \{e\}^B,$$
$$R_{2e+1} : B \neq \{e\}^A.$$

The strategy for meeting a *single* such requirement R_{2e} is to attach to R_{2e} a potential "witness" x not yet enumerated in A and to look for a stage $s+1$ such that

(2.1) $$\{e\}_s^{B_s}(x)\downarrow = 0.$$

(If no such stage exists we do nothing and R_{2e} is automatically satisfied by the witness x because $A(x) = 0$ and either $\{e\}^B(x)\uparrow$ or $\{e\}^B(x)\downarrow \neq 0$.) If $s+1$ exists, we say R_{2e} *requires attention* at stage $s+1$. Now R_{2e} *receives attention* and we: (1) enumerate x in A_{s+1}; (2) define the *restraint function* (i.e., the "wall") $r(2e, s+1) = u(B_s; e, x, s)$ and attempt (with priority R_{2e}) to restrain any numbers $y \leq r = r(2e, s+1)$ from later entering B. If we achieve the latter objective then $B \upharpoonright r = B_s \upharpoonright r$, so by the Use Principle,

$$\{e\}^B(x) = \{e\}^{B\upharpoonright r}(x) = \{e\}^{B_s\upharpoonright r}(x) = 0.$$

However, $A(x) = 1$ so requirement R_{2e} is satisfied. (The strategy for R_{2e+1} is the same but with the roles of A and B reversed.)

To accommodate all requirements simultaneously, we must occasionally, but only finitely often, change the witness x_e for requirement R_e. Let x_e^s be its approximation at the end of stage s and $x_e = \lim_s x_e^s$. (One may think of x_e^s as the position at the end of stage s of a movable marker Γ_e which comes to rest on x_e.) To keep witnesses for different requirements distinct we choose all witnesses for R_e from $\omega^{[e]} = \{ \langle x, y \rangle : y = e \}$, so that $x_e^s \notin A$ unless it was put into A for the sake of R_e.

If R_e receives attention at stage $s + 1$ we redefine all witnesses x_i^s, for all $i > e$, choosing $x_i^{s+1} > r = r(e, s+1)$. Thus, only requirements R_i of stronger priority (namely $i < e$) can later *injure* R_e by contributing some $x \leq r$ to A or B. After every R_i, $i < e$, has ceased to receive attention, R_e will receive attention at most once, at which time it will become satisfied and remain satisfied forever.

Construction of A and B.

   Stage $s = 0$. Set $A_0 = B_0 = \emptyset$, $x_e^0 = \langle 0, e \rangle$ and $r(e, 0) = -1$ for all e.

   Stage $s + 1$. Requirement R_{2e} *requires attention* if

$$(2.2) \qquad \{e\}_s^{B_s}(x_{2e}^s)\downarrow = 0 \ \& \ r(2e, s) = -1,$$

and R_{2e+1} *requires attention* if (2.2) holds with A_s and $2e + 1$ in place of B_s and $2e$. Choose the least $i \leq s$ such that R_i requires attention. We say R_i *receives attention*. Suppose that $i = 2e$. Enumerate x_{2e}^s into A_{s+1}; set $x_{2e}^{s+1} = x_{2e}^s$ and reset the wall $r(2e, s + 1) = u(B_s; e, x_{2e}^s, s)$; this action satisfies R_{2e}. For $j < 2e$ set $r(j, s + 1) = r(j, s)$ and $x_j^{s+1} = x_j^s$, which leaves untouched any previous wall or action for stronger priority requirements. For $j > 2e$ set $r(j, s + 1) = -1$ and define x_j^{s+1} to be the least $y \in \omega^{[j]}$ such that $y \notin A_{s+1} \cup B_{s+1}$, $y > \max\{ r(k, s + 1) : k \leq 2e \}$, and $y > x_j^s$. (Thus, action previously taken for weaker priority requirements R_j, $j > 2e$, is undone, the walls $r(j, s + 1)$ are reset to -1 to indicate that R_j may again require attention in the future, and the marker positions x_j^{s+1} are reset to new elements greater than the current R_{2e} wall so that their possible later entry into B will not disturb the present computation for R_{2e}.) If $i = 2e + 1$ do likewise with A and B interchanged. If i fails to exist do nothing (i.e., set $A_{s+1} = A_s$, $B_{s+1} = B_s$, $r(e, s + 1) = r(e, s)$, and $x_e^{s+1} = x_e^s$ for all e).

Lemma. *For every i, requirement R_i receives attention at most finitely often and is eventually satisfied.*

Proof. Fix i and assume by induction that the Lemma holds for all $j < i$. Choose s minimal so that no R_j, $j < i$, receives attention at a stage $t \geq s$. Hence, for all $t \geq s$, $x_i^t = x_i^s = x_i$, and $x_i \notin A_s \cup B_s$. Suppose $i = 2e$.

(The case of i odd is similar.) If R_{2e} never receives attention after stage s then $A(x_{2e}) = 0$ and it is not the case that $\{e\}^B(x_{2e})\downarrow = 0$. If R_{2e} receives attention at some stage $t+1 \geq s$ then $\{e\}_t^{B_t}(x_{2e})\downarrow = 0$, $x_{2e} \in A_{t+1} - A_t$, and $B_t \upharpoonright r = B \upharpoonright r$ for $r = u(B_t; e, x_{2e}, t)$ so $\{e\}^B(x_{2e})\downarrow = 0 \neq 1 = A(x_{2e})$. In the latter case R_{2e} never requires or receives attention at any stage $v > t+1$. □

Note that R_i can be injured at most $2^i - 1$ times.

2.2–2.7 Exercises

2.2. (a) Show that there is an r.e. sequence of r.e. sets $\{A_i\}_{i \in \omega}$ such that for every i, $A_i \nleq_T \oplus\{A_j : j \neq i\}$. *Hint.* Modify the construction of Theorem 2.1 to meet the requirements $R_{\langle e,i\rangle}$ of Exercise VI.1.5.

(b) Show that any countable partially ordered set can be embedded in the r.e. degrees $(\mathbf{R}, \leq)$ by an order-preserving map. (See the hint for Exercise VI.1.7.)

2.3 (Soare [1972]). Prove that the r.e. sets A and B constructed in Theorem 2.1 automatically are low and satisfy $K \leq_T A \oplus B$. *Hint.* To show that $K \leq_T A \oplus B$ prove that $\lambda e[x_e]$ is a function recursive in $A \oplus B$ (because $x_e^s \neq x_e^{s+1}$ only if some $y < x_e^s$ enters A or B at stage $s+1$), and that there is a recursive function f such that for all e, $e \in K$ if and only if $R_{2f(e)}$ causes $x_{2f(e)}$ to enter A, where $\{f(e)\}^X(y) = 0$ for all y and X if $e \in K$, and is undefined otherwise. To show that A and B are low, first define a recursive function h such that for all X and e,

$$(2.3) \qquad \{e\}^X(e)\downarrow \implies (\forall y)\,[\{h(e)\}^X(y) = 0],$$

$$(2.4) \qquad \{e\}^X(e)\uparrow \implies (\forall y)\,[\{h(e)\}^X(y)\uparrow].$$

Next define the recursive function $g(e, s)$ by

$$g(e, s) = \begin{cases} 1 & \text{if } \{h(e)\}_s^{A_s}(x_{2h(e)+1}^s)\downarrow = 0, \\ 0 & \text{otherwise.} \end{cases}$$

Show that the restraint functions in the construction of Theorem 2.1 guarantee that $\hat{g}(e) =_{\mathrm{dfn}} \lim_s g(e, s)$ exists. Now by (2.3) and (2.4) $\hat{g}$ is the characteristic function of A'.

2.4 (Cooper-Epstein-Lachlan). Construct a pair of r.e. sets A and B such that the d.r.e. set $D = A - B$ does not have r.e. degree. *Hint.* For each e, i, j meet the requirement

$$R_{\langle e,i,j\rangle} : D \neq \{i\}^{W_e} \text{ or } W_e \neq \{j\}^D.$$

To meet a single $R_{\langle e,i,j\rangle}$ choose x not yet in A or B and wait for a stage s such that

$$0 = \{i\}_s^{W_{e,s}\upharpoonright u}(x) \quad \& \quad W_{e,s}\upharpoonright u = \{j\}_s^{D_s\upharpoonright v}\upharpoonright u$$

for some u and v, where $D_s = A_s - B_s$. Now enumerate x in A_{s+1} and restrain D from changing on any other elements $\leq v$. We win $R_{\langle e,i,j\rangle}$ by the first clause unless there is some stage $t \geq s+1$ and $y < u$ such that $y \in W_{e,t} - W_{e,s}$. In this case we enumerate x in B_t so that $D_t \upharpoonright v = D_s \upharpoonright v$ and $\{j\}_t^{D_t\upharpoonright v}(y) = W_{e,s}(y) \neq W_{e,t}(y)$.

2.5 (Trakhtenbrot).   A set A is *autoreducible* if there is an e such that for all x, $A(x) = \{e\}^{A-\{x\}}$. (The idea is that $\{e\}$ determines whether $x \in A$ by using oracle questions "$y \in A$?" only for $x \neq y$.)

(a) Construct an r.e. set A which is not autoreducible. *Hint.* For each e choose a witness x, and attempt to meet $A(x) \neq \{e\}^{A-\{x\}}(x)$ by waiting for $\{e\}_s^{A_s-\{x\}}(x)\downarrow = y$ for some s, then: (1) enumerating x in A iff $y = 0$; and (2) attempting to restrain elements $z \leq u(A_s; e, x, s)$ from entering A.

(b) For any nonzero r.e. set B construct an r.e. set $A \leq_T B$ such that A is not autoreducible. (Ladner [1973a] has shown that one cannot achieve $A \equiv_T B$ because he proved that A is autoreducible iff A is *mitotic*, namely A is the disjoint union of r.e. sets A_0 and A_1 such that $A \equiv_T A_0 \equiv_T A_1$, and in [1973b] that there is a nonzero r.e. degree containing only mitotic r.e. sets.)

2.6.   Recall the definitions of $\leq_{tt}$ and $\leq_{wtt}$ from Exercises V.2.12 and V.2.16. Note that if A_0 and A_1 are disjoint r.e. sets then $A_i \leq_{wtt} A_0 \cup A_1$, $i = 0, 1$. Use a priority argument to construct disjoint r.e. sets A_0 and A_1 such that $A_i \not\leq_{tt} A_0 \cup A_1$, $i = 0, 1$. *Hint.* Pick a witness x to meet the requirement that $A_i \not\leq_{tt} A_0 \cup A_1$ via the eth tt-reduction; wait until the latter converges on $A_{0,s} \cup A_{1,s} \cup \{x\}$; and put x into A_0 or A_1 to achieve a disagreement.

2.7.   Combine the permitting method of Theorem V.3.2 with the Friedberg-Muchnik construction of Theorem 2.1 to show that for any nonrecursive r.e. set C there exist Turing incomparable r.e. sets $A, B \leq_T C$. (In Theorem 3.2 we even achieve $A \cup B = C$ and $A \cap B = \emptyset$, so that $\deg(A) \cup \deg(B) = \deg(C)$.)

3. Splitting Theorems

Shortly after proving Theorem 2.1, Friedberg [1958a] proved (see Theorem X.2.1) that any nonrecursive r.e. set B could be split as the disjoint union of nonrecursive r.e. sets A_0 and A_1. Sacks [1963b] then generalized these two Friedberg theorems simultaneously by showing that A_0 and A_1 could be chosen not merely nonrecursive but even Turing incomparable. If C is a given nonrecursive r.e. set, Sacks's method for constructing an r.e. set A to meet a requirement of the form $C \neq \{e\}^A$ is to preserve *agreement* between C_s and $\{e\}_s^{A_s}$ rather than disagreement as in the Friedberg strategy. Sufficient preservation of agreement will guarantee that if $C = \{e\}^A$ then C is recursive, contrary to hypothesis. This method is very powerful and will be essential in the next chapter on infinite injury. To isolate those key features which we shall extend in Chapter VIII we first prove a weaker result which illustrates the method, and then we prove the full Sacks Splitting Theorem.

3.1 Theorem (Sacks).   *For every nonrecursive r.e. set C there is a simple set A such that $C \not\leq_T A$ (and hence $\emptyset <_T A <_T \emptyset'$).*

Proof. It clearly suffices to construct A to be coinfinite and to satisfy, for all e, the requirements:

$$N_e : C \neq \{e\}^A;$$

and

$$P_e : W_e \text{ infinite} \implies W_e \cap A \neq \emptyset.$$

Let $\{C_s\}_{s \in \omega}$ be a recursive enumeration of C.

Construction of A.
    Stage $s = 0$. Set $A_0 = \emptyset$.
    Stage $s + 1$. Since A_s has already been defined, we can define for all e,

(length function) $l(e, s) = \max\{ x : (\forall y < x) [\{e\}_s^{A_s}(y){\downarrow} = C_s(y)] \};$

(restraint function) $r(e, s) = \max\{ u(A_s; e, x, s) : x \leq l(e, s) \}.$

We say P_i *requires attention* at stage $s + 1$ if $i \leq s$, $W_{i,s} \cap A_s = \emptyset$, and

(3.1) $(\exists x) [x \in W_{i,s} \ \& \ x > 2i \ \& \ (\forall e \leq i) [r(e, s) < x]].$

We enumerate in A_{s+1} the least such x for each P_i which requires attention at $s + 1$, and we say that P_i *receives attention* at stage $s + 1$.

Remark. Fix e. Assume that no P_j, $j < e$, receives attention at stage $s + 1$.
    (i) For $y < l(e, s)$, if $z \in A_{s+1} - A_s$ then $z > u(A_s; e, y, s)$ so

$$\{e\}_{s+1}^{A_{s+1}}(y){\downarrow} = \{e\}_s^{A_{s+1}}(y){\downarrow} = \{e\}_s^{A_s}(y) = C_s(y),$$

and $u(A_{s+1}; e, y, s + 1) = u(A_s; e, y, s)$.
    (ii) For $y = l(e, s)$, if $z \in A_{s+1} - A_s$ then $z > u(A_s; e, y, s)$, by the use of "$x \leq l(e, s)$" in the definition of $r(e, s)$, so if $\{e\}_s^{A_s}(y){\downarrow}$ then

$$\{e\}_{s+1}^{A_{s+1}}(y){\downarrow} = \{e\}_s^{A_{s+1}}(y){\downarrow} = \{e\}_s^{A_s}(y) \neq C_s(y),$$

and $u(A_{s+1}; e, y, s + 1) = u(A_s; e, y, s)$.

The remark is immediate from the defining clauses. The negative require-ment N_e is *injured at stage* $s+1$ *by element* x if $x \leq r(e, s)$ and $x \in A_{s+1} - A_s$. These elements form an r.e. set:

(injury set) $I_e = \{ x : (\exists s) [x \in A_{s+1} - A_s \ \& \ x \leq r(e, s)] \}$.

As in §1 note that each I_e is finite (indeed $|I_e| \leq e$) because N_e is injured at most once by each P_i, $i < e$, whereupon P_i is satisfied forever as in Theorem 1.1.

Lemma 1. $(\forall e) [C \neq \{ e \}^A]$.

Proof. Assume for a contradiction that $C = \{ e \}^A$. Then $\lim_s l(e, s) = \infty$. Choose s' such that N_e is never injured after stage s'. We shall recursively compute C contrary to hypothesis. To compute $C(p)$ for some $p \in \omega$ find the least $s > s'$ such that $l(e, s) > p$. It follows by induction on $t \geq s$ that

(3.2) $(\forall t \geq s) [l(e, t) > p \ \& \ r(e, t) \geq \max\{ u(A_s; e, x, s) : x \leq p \}]$,

and hence that $\{ e \}_s^{A_s}(p) = \{ e \}^{A_s}(p) = \{ e \}^A(p) = C(p)$, whence C is recur-sive. To prove (3.2) assume it holds for t. Then by the Remark, $r(e, t)$ and $s > s'$ ensure that $A_{t+1} \upharpoonright z = A_t \upharpoonright z$ for all numbers z used in a computation $\{ e \}_t^{A_t}(x) = y$, for any $x \leq p$. Thus, $\{ e \}_{t+1}^{A_{t+1}}(x) = y$ so $l(e, t + 1) > p$ unless $C_{t+1}(x) \neq C_t(x)$ for some $x < l(e, t)$. But if $C_t(x) \neq C_s(x)$ for some $t \geq s$ and $x \leq p$, where x is minimal, then our use of "$\leq l(e, t)$" rather than "$< l(e, t)$" in the definition of $r(e, t)$ ensures that the *disagreement* $C_t(x) \neq \{ e \}_t^{A_t}(x)$ is preserved forever, contrary to our hypothesis that $C = \{ e \}^A$. (Note that even though the Sacks strategy is always described as one which preserves agreements, it is crucial that we preserve at least one *disagreement* as well whenever possible. The reason that the disagreement is preserved is that since C is r.e. $C_s(x)$ can only change from 0 to 1 and never back again.) (See Exercise 3.8 to extend to the case $C \leq_T \emptyset'$ in place of C r.e.)

Lemma 2. $(\forall e) [\lim_s r(e, s)$ *exists and is finite*].

Proof. By Lemma 1, choose $p = (\mu x) [C(x) \neq \{ e \}^A(x)]$. Choose s' suffi-ciently large such that for all $s \geq s'$,

   (1) $(\forall x < p) [\{ e \}_s^{A_s}(x) \downarrow = \{ e \}^A(x)]$;
   (2) $(\forall x \leq p) [C_s(x) = C(x)]$; and
   (3) N_e is not injured at stage s.

   Case 1. $(\forall s \geq s') [\{ e \}_s^{A_s}(p) \uparrow]$. Then $r(e, s) = r(e, s')$ for all $s \geq s'$.
   Case 2. $\{ e \}_t^{A_t}(p) \downarrow$ for some $t \geq s'$. Then $\{ e \}_s^{A_s}(p) = \{ e \}_t^{A_t}(p)$ for all $s \geq t$ because $l(e, s) \geq p$, and so, by the definition of $r(e, s)$, and the fact

that N_e is not injured after stage t, the computation $\{e\}_t^{A_t}(p)$ is preserved forever. Thus, $\{e\}^A(p) = \{e\}_s^{A_s}(p)$. But $C(p) \neq \{e\}^A(p)$. Hence,

$$(\forall s \geq t)\, [C_s(p) \neq \{e\}_s^{A_s}(p) \ \& \ l(e,s) = p \ \& \ r(e,s) = r(e,t)].$$

Therefore, $r(e,t) = \lim_s r(e,s)$.

Lemma 3. $(\forall e)\, [W_e \text{ infinite} \implies W_e \cap A \neq \emptyset]$.

Proof. By Lemma 2, let $r(i) = \lim_s r(i,s)$ for $i \leq e$, and $R(e) = \max\{r(i) : i \leq e\}$. Now if

$$(\exists x)\, [x \in W_e \ \& \ x > 2e \ \& \ x > R(e)]$$

then $W_e \cap A \neq \emptyset$.

Note that $\overline{A}$ is infinite by the clause "$x > 2e$" in (3.1), and hence A is simple.   ▯

3.2 Theorem (Sacks Splitting Theorem) (Sacks [1963b]).   *Let B and C be r.e. sets such that C is nonrecursive. Then there exist low r.e. sets A_0 and A_1 such that:*

(i) $A_0 \cup A_1 = B$ and $A_0 \cap A_1 = \emptyset$, and
(ii) $C \not\leq_T A_i$, for $i = 0, 1$.

3.3 Proposition.   *If the r.e. set B is the disjoint union of r.e. sets A_0 and A_1 then $B \equiv_T A_0 \oplus A_1$.*

Proof. First note that $A_i \leq_T B$ for $i = 0, 1$, and hence $A_0 \oplus A_1 \leq_T B$, because to decide whether $x \in A_i$, we ask whether $x \in B$, and if so enumerate A_0 and A_1 until x appears in one of them. Clearly $B \leq_T A_0 \oplus A_1$.   ▯

3.4 Corollary.   *If $\mathbf{b}$ is any nonzero r.e. degree there are incomparable low r.e. degrees $\mathbf{a_0}$ and $\mathbf{a_1}$ such that $\mathbf{b} = \mathbf{a_0} \cup \mathbf{a_1}$.*

Proof. Choose an r.e. set $B \in \mathbf{b}$. Apply Theorem 3.2 with $C = B$ to obtain A_0 and A_1. Hence, $A_0, A_1 <_T B$. By Proposition 3.3, $B \equiv_T A_0 \oplus A_1$. Now A_0 and A_1 cannot have comparable degree because if, say, $A_0 \leq_T A_1$ then $A_1 \equiv_T B$. Let $\mathbf{a_0} = \deg(A_0)$ and $\mathbf{a_1} = \deg(A_1)$.   ▯

3.5 Corollary.   *The low r.e. degrees generate the r.e. degrees when closed under join.*

3.6 Corollary.   *No r.e. degree is minimal.*

3.7 Corollary.   *For any r.e. degree $\mathbf{c}$, $\mathbf{0} < \mathbf{c} < \mathbf{0'}$, there is an r.e. degree incomparable with $\mathbf{c}$.*

Proof. Let $B = K$, let $C \in \mathbf{c}$ be r.e. and apply Theorem 3.2. One of A_0, A_1 must have degree incomparable with $\mathbf{c}$, else $A_0 \leq_T C$ and $A_1 \leq_T C$, so $K \leq_T C$, contrary to $C <_T \emptyset'$.   ▯

Note that in Corollary 3.7 we do not produce A incomparable with C *uniformly* in C but merely give a pair A_0, A_1 *one* of which succeeds. In Theorem VIII.4.3 we shall be able to find A uniformly from C.

Proof of Theorem 3.2. (We may assume that B is infinite, else the result is trivial. This, however, does not affect the uniformity of A_0, A_1 from B and C.) Let $\{ B_s \}_{s \in \omega}$ and $\{ C_s \}_{s \in \omega}$ be recursive enumerations of B and C such that $B_0 = \emptyset$ and $|B_{s+1} - B_s| = 1$ for all s. We shall give recursive enumerations $\{ A_{i,s} \}_{s \in \omega}$, $i = 0, 1$, satisfying the single positive requirement

$$P : x \in B_{s+1} - B_s \implies [x \in A_{0,s+1} \lor x \in A_{1,s+1}],$$

and the negative requirements for $i = 0, 1$, and all e,

$$N_{\langle e, i \rangle} : C \neq \{ e \}^{A_i}.$$

(The lowness of A_i will follow automatically from the $N_{\langle e, i \rangle}$ requirements as explained below.)

Stage $s = 0$. Define $A_{i,0} = \emptyset$, $i = 0, 1$.

Stage $s+1$. Given $A_{i,s}$ define the recursive functions $l^i(e, s)$ and $r^i(e, s)$ as in the proof of Theorem 3.1 but with $A_{i,s}$ in place of A_s. Let $x \in B_{s+1} - B_s$. Choose $\langle e', i' \rangle$ to be the least $\langle e, i \rangle$ such that $x \leq r^i(e, s)$, and enumerate x in $A_{1-i',s+1}$. (Namely, choose the highest priority requirement $N_{\langle e', i' \rangle}$ which would be injured by enumerating x in $A_{i'}$, and enumerate x on the "other side" $A_{1-i'}$.) If $\langle e', i' \rangle$ fails to exist, enumerate x into A_0.

To see that the construction succeeds, define the injury set I_e^i as in the proof of Theorem 3.1 but with A_i in place of A. It follows by induction on $\langle e, i \rangle$ that for $i = 0, 1$ and all e,

(1) I_e^i is finite;

(2) $C \neq \{ e \}^{A_i}$; and

(3) $r^i(e) = \lim_s r^i(e, s)$ exists and is finite.

Namely, fix $\langle e, i \rangle$ and assume (1), (2), and (3) hold for all $\langle k, j \rangle < \langle e, i \rangle$. By (3) choose t such that $r^j(k, s) = r^j(k)$ for all $\langle k, j \rangle < \langle e, i \rangle$ and all $s \geq t$.

Choose r greater than all such $r^j(k)$. Choose $v > t$ such that $B \restriction r = B_v \restriction r$. Now $N_{\langle e, i \rangle}$ is never injured after stage v so (1) holds for $\langle e, i \rangle$. Now (2) and (3) hold for $\langle e, i \rangle$ exactly as in Lemmas 1 and 2 of Theorem 3.1.

To see that each A_i is low define the recursive function g as follows:

$$\{ g(e) \}^X (y) = \begin{cases} C(0) & \text{if } y = 0 \text{ and } \{ e \}^X(e) \downarrow, \\ \text{undefined} & \text{otherwise.} \end{cases}$$

Note that for $i = 0, 1$,

$$e \in A_i' \iff \{ e \}^{A_i}(e) \downarrow \iff \{ g(e) \}^{A_i}(0) = C(0) \iff \lim_s l^i(g(e), s) > 0,$$

so $A'_i \in \Delta^0_2$ because l^i and g are recursive functions, and $\lim_s l^i(g(e), s)$ exists, and hence $A'_i \leq_T \emptyset'$. ▯

Note that the injury set I^i_e, although finite, has no obvious bounding function as in §1 and §2. Indeed, in general there is no recursive function f such that $|I^i_e| \leq f(i, e)$.

3.8–3.11 Exercises

3.8. Prove that Theorems 3.1 and 3.2 hold with the hypothesis "$C \leq_T \emptyset'$ and nonrecursive" in place of "C r.e. and nonrecursive." *Hint.* Using the Limit Lemma III.3.3 choose a recursive canonical sequence $\{C_s\}_{s\in\omega}$ of finite sets such that $C = \lim_s C_s$. Given A_s and $l(e, s)$ as in Theorem 3.1 define

(maximum length function) $m(e, s) = \max\{l(e, t) : t \leq s\}$.

Use $m(e, s)$ in place of $l(e, s)$ in the definition of $r(e, s)$, and show that the former proof succeeds. (The key point is to use $m(e, s)$ to prove (3.2), since $l(e, s)$ would not have sufficed.)

3.9. Prove that if $\{C_j\}_{j\in\omega}$ is a recursive sequence of nonrecursive Δ^0_2 sets and B is an r.e. set then there exist disjoint r.e. sets A_0 and A_1 such that $B = A_0 \cup A_1$ and $C_j \not\leq_T A_i$ for $j \in \omega$, $i \in \{0, 1\}$. Conclude that there is no recursive enumeration $\{A_i\}_{i\in\omega}$ of r.e. (Δ^0_2) sets such that $\{\deg(A_i)\}_{i\in\omega}$ consists precisely of the nonzero r.e. (respectively Δ^0_2) degrees. *Hint.* Replace the negative requirements in Theorem 3.2 by $N_{\langle e,i,j\rangle} : C_j \neq \{e\}^{A_i}$ and use the same method.

3.10. Given the sequence $\{C_j\}_{j\in\omega}$ as in Exercise 3.9, find an infinite sequence $\{A_i\}_{i\in\omega}$ of pairwise Turing incomparable r.e. sets meeting all the requirements $N_{\langle e,i,j\rangle}$ of Exercise 3.9.

3.11. Prove that there exist low r.e. degrees $\mathbf{a}_0$ and $\mathbf{a}_1$ such that for any r.e. degree $\mathbf{c}$ there exist r.e. degrees $\mathbf{b}_0 \leq \mathbf{a}_0$ and $\mathbf{b}_1 \leq \mathbf{a}_1$ such that $\mathbf{c} = \mathbf{b}_0 \cup \mathbf{b}_1$. *Hint.* Fix $W_e \in \mathbf{c}$ and apply Theorem 3.2 to $K_0 = \{\langle x, e\rangle : x \in W_e\}$ to obtain A_0 and A_1.

Part C

Infinitary Methods for Constructing R.E. Sets and Degrees

Chapter VIII
The Infinite Injury Priority Method

The infinite injury priority method was invented by Shoenfield [1961] and independently by Sacks [1963a, 1963c, and 1964a]. Shoenfield proved a slightly weaker form of the Thickness Lemma of §1 (namely with $C = \emptyset'$) and used it to construct an incomplete r.e. theory in which every recursive function is representable. Sacks used a different approach and combined it with his preservation method of Theorem VII.3.1 to obtain many powerful new results on r.e. degrees, the most striking of which is the Density Theorem 4.1, which asserts that for any r.e. degrees $\mathbf{d} < \mathbf{c}$ there is an r.e. degree $\mathbf{a}$ such that $\mathbf{d} < \mathbf{a} < \mathbf{c}$. Another version of the infinite injury method was introduced by Yates [1966b] to study index sets as we shall discuss in Chapter XII.

In a typical infinite injury argument, a negative requirement such as $N_e : C \neq \{e\}^A$ of Theorem VII.3.1 may be injured infinitely often (namely, the injury set I_e may be infinite), thereby causing the associated restraint function $r(e, s)$ to have $\limsup_s r(e, s) = \infty$ even though we may arrange that $\liminf_s r(e, s) < \infty$. Thus a positive requirement P_i may have considerable difficulty enumerating an element $x \in A_{s+1} - A_s$ while still respecting restraint functions of stronger priority (i.e., while arranging that $x > r(e, s)$ for all $e \leq i$).

There are two solutions to this difficulty. First, one may have the element x associated with P_i (called a *follower* of P_i) pass in turn each stronger priority N_e (viewed as a "gate") waiting for the corresponding "wall" $r(e, s)$ to drop below x before x is allowed to pass to the next gate. This is closer to the original method and is recast in §5 in the "pinball machine" model, which is used in many other arguments as well. A second somewhat neater approach is to produce an infinite sequence of "true stages" in the enumeration of A, during each of which *all* the restraint functions $r(e, s)$, $e \leq i$, drop back *simultaneously*, thereby allowing the follower x of P_i to pass all the gates at once and enter A_{s+1} immediately. We follow the latter approach which was first suggested by Lachlan [1973] and was refined in Soare [1976].

We begin in §1 with the Thickness Lemma which is the easiest example of an infinite injury argument. In §2 a stronger form is proved and the key ingredients of the method, namely, the Injury and Window Lemmas, are isolated and proved. These are applied in §3 and §4 to prove the elegant and useful Jump Theorem and the Density Theorem of Sacks.

1. The Obstacles in Infinite Injury and the Thickness Lemma

For this chapter it will be convenient to use the notation $\Phi_{e,s}(A_s; x)$ in place of $\{e\}_s^{A_s}(x)$. Let C be a nonrecursive r.e. set. In the easiest applications of the infinite injury priority method we wish to construct an r.e. set A where the negative requirements are the same as in Theorem VII.3.1, namely, $N_e : C \neq \Phi_e(A)$, and the positive requirements are of the form

$$P_e : W_{p(e)} \subseteq^* A$$

(where $X \subseteq^* Y$ denotes that $X - Y$ is finite and where p is a recursive function) so that a single positive requirement P_e may contribute infinitely many elements to A. In the simplest cases, the r.e. sets $\{W_{p(e)}\}_{e \in \omega}$ will each be recursive. For each N_e we would like a restraint function $\hat{r}(e, s)$ like $r(e, s)$ of Theorem VII.3.1 so that we can enumerate x in A_{s+1} for the sake of P_e just if $x \in W_{p(e),s+1}$ and $x > \hat{r}(i, s)$ for all $i \leq e$. The negative requirement N_e can now be injured infinitely often by those P_i, $i < e$, but the recursiveness of $W_{p(i)}$, $i < e$, will enable us to meet N_e as in Lemma 1 of Theorem VII.3.1. The main difficulty will be that some P_e remains unsatisfied forever because of the restraint functions $\hat{r}(i, s)$, $i \leq e$, which may now be unbounded in s (i.e., $\limsup_s \hat{r}(i, s) = \infty$). To satisfy P_e it clearly suffices to define $\hat{r}(e, s)$ to satisfy

$$(1.1) \qquad\qquad\qquad \liminf_s \hat{R}(e, s) < \infty,$$

where $\hat{R}(e, s) = \max\{\hat{r}(i, s) : i \leq e\}$, because then P_e has a "window" through the negative restraints at least infinitely often.

The *first* obstacle to achieving (1.1) is that if we let $\hat{r}(e, s)$ be $r(e, s)$ as defined in Theorem VII.3.1, then we may have $\lim_s r(e, s) = \infty$ for some e. (For example, suppose $\Phi_{1,s}(X; 0) = C(0)$ just if $n \notin X$ for some even $n < s$, $n > \liminf_s R(0, s)$ but P_0 eventually forces every such number into A so that $\Phi_1(A; 0)$ is undefined. Then N_1 is satisfied by divergence, but it could happen that $C(0) = \Phi_{1,s}(A_s; 0)$ for almost every s, so $\lim_s u(A_s; 1, 0, s) = \infty$, $\lim_s r(1, s) = \infty$, and P_1 is not satisfied.) This difficulty arises only if there are infinitely many stages s such that $A_s \upharpoonright u \neq A_{s+1} \upharpoonright u$ where $u = u(A_s; 1, 0, s)$. We shall find that we can easily remove the first obstacle by replacing $\Phi_{e,s}$ everywhere by $\widehat{\Phi}_{e,s}$ as defined below and by letting $\hat{r}(e, s)$ be the resulting restraint function. We can then show that if $C \neq \Phi_e(A)$ then we have $\liminf_s \hat{r}(e, s) < \infty$.

The *second* obstacle to (1.1) is that $\lim_s \hat{R}(e, s)$ may be infinite even though $\liminf_s \hat{r}(i, s) < \infty$ for each $i \leq e$. (For example, N_1 and N_2 may together permanently restrain all elements because their restraint functions do not drop back simultaneously.) Surprisingly, the $\widehat{\Phi}_{e,s}$ solution to the first obstacle automatically removes the second, as Lachlan [1973] first observed.

Suppose we wish to give a recursive enumeration $\{A_s\}_{s\in\omega}$ of an r.e. set A. Given $\{A_t\}_{t\leq s}$, define

$$a_s = \begin{cases} \mu x[x \in A_s - A_{s\dot-1}] & \text{if } A_s - A_{s\dot-1} \neq \emptyset, \\ \max(A_s \cup \{s\}) & \text{otherwise;} \end{cases}$$

$$\widehat{\Phi}_{e,s}(A_s; x) = \begin{cases} \Phi_{e,s}(A_s; x) & \text{if defined and } u(A_s; e, x, s) < a_s, \\ \text{undefined} & \text{otherwise;} \end{cases}$$

$$\hat{u}(A_s; e, x, s) = \begin{cases} u(A_s; e, x, s) & \text{if } \widehat{\Phi}_{e,s}(A_s; x) \text{ is defined,} \\ 0 & \text{otherwise;} \end{cases}$$

and

$$T = \{s : A_s \restriction a_s = A \restriction a_s\}.$$

If $\{A_s\}_{s\in\omega}$ is any recursive enumeration of an infinite r.e. set A we refer to T as the set of *true* (*nondeficiency*) stages of this enumeration. Note that T is infinite and $T \equiv_T A$ uniformly in (the enumeration of) A. If $\Phi_e(A; x) = y$ then clearly $\lim_s \widehat{\Phi}_{e,s}(A_s; x) = y$ as before. The crucial point about $\widehat{\Phi}_{e,s}$ is that for any true stage t any *apparent* computation $\widehat{\Phi}_{e,t}(A_t; x) = y$ is a *true* computation $\Phi_e(A; x) = y$. Namely, we have

(1.2) $$(\forall t \in T) \, [\widehat{\Phi}_{e,t}(A_t; x) = y \implies$$

$$(\forall s \geq t) \, [\widehat{\Phi}_{e,s}(A_s; x) = \Phi_e(A; x) = y \ \& \ \hat{u}(A_s; e, x, s) = u(A_t; e, x, t)]],$$

because if $\widehat{\Phi}_{e,t}(A_t; x)$ is defined then $u(A_t; e, x, t) \leq a_t$ and $A_t \restriction a_t = A \restriction a_t$.

The simplest application of this method is the Thickness Lemma. For any $A \subseteq \omega$ and $y \in \omega$ recall the Definitions VI.4.5 and VI.4.6 of the y-section, $A^{[y]} = \{\langle x, z\rangle : \langle x, z\rangle \in A \ \& \ z = y\}$, $A^{[<y]} = \bigcup\{A^{[z]} : z < y\}$, and that a subset $A \subseteq B$ is a *thick* subset if $A^{[y]} =^* B^{[y]}$ for all y. We say that B is *piecewise recursive* if $B^{[y]}$ is recursive for all y. (The following lemma was proved by Shoenfield [1961] for the special case $C = \emptyset'$, but for arbitrary nonrecursive r.e. sets C requires the Sacks preservation method of VII §3 combined with the infinite injury method as first developed in Sacks [1963a and 1963c].)

1.1 Thickness Lemma (Shoenfield). *Given a nonrecursive r.e. set C and a piecewise recursive r.e. set B there is a thick r.e. subset A of B such that $C \not\leq_T A$.*

Proof. Fix recursive enumerations $\{B_s\}_{s\in\omega}$ and $\{C_s\}_{s\in\omega}$ of B and C. Let $A_0 = \emptyset$. Given $\{A_t\}_{t\leq s}$ define $\widehat{\Phi}_{e,s}(A_s)$ as above. Define the remaining functions as in Theorem VII.3.1 with $\widehat{\Phi}_{e,s}$ in place of $\Phi_{e,s}$, namely,

(length function) $\hat{l}(e, s) = \max\{x : (\forall y < x) \, [C_s(y) = \widehat{\Phi}_{e,s}(A_s; y)\downarrow]\},$

(restraint function) $\hat{r}(e,s) = \max\{\,\hat{u}(A_s;e,x,s) : x \leq \hat{l}(e,s)\,\}$,

(injury set) $\hat{I}_e = \bigcup_s \hat{I}_{e,s}$,

where
$$\hat{I}_{e,s} = \{\,x : (\exists v \leq s)\,[x \leq \hat{r}(e,v)\ \&\ x \in A_{v+1} - A_v]\,\}.$$

To meet the requirements
$$P_e : B^{[e]} =^* A^{[e]} \qquad \text{and} \qquad N_e : C \neq \Phi_e(A),$$

we enumerate x in $A_{s+1}^{[e]}$ just if $x \in B_{s+1}^{[e]}$ and $x > \hat{r}(i,s)$ for all $i \leq e$. Let $A = \bigcup_s A_s$.

Note that $\hat{I}_e \subseteq A^{[<e]}$ because N_e is injured by P_i only if $i < e$. Thus we have $\hat{I}_e \leq_T A^{[<e]}$ because if $x \in A^{[<e]}$, say $x \in A_s^{[<e]}$, then $x \in \hat{I}_e$ just if $x \in \hat{I}_{e,s}$.

Fix e and assume by induction that $C \neq \Phi_i(A)$ and $A^{[i]} =^* B^{[i]}$ for all $i < e$. Then $A^{[<e]} =^* B^{[<e]}$ is recursive, and hence $\hat{I}_e$ is recursive because $\hat{I}_e \leq_T A^{[<e]}$. The following lemmas are similar to those of Theorem VII.3.1 but with the true stages T in place of all stages.

Lemma 1. $C \neq \Phi_e(A)$.

Proof. Assume for a contradiction that $C = \Phi_e(A)$. Then $\lim_s \hat{l}(e,s) = \infty$. We shall show that $C \leq_T \hat{I}_e$. But $\hat{I}_e$ is recursive so C is recursive contrary to hypothesis. To $\hat{I}_e$-recursively compute $C(p)$ for $p \in \omega$ find some s such that $\hat{l}(e,s) > p$ and

$$(\forall x \leq p)\,(\forall z)\,[z \leq u(A_s;e,x,s) \implies [z \notin \hat{I}_e \text{ or } z \in A_s]].$$

Such s exists since $C = \Phi_e(A)$. By the same remarks as in Lemma 1 of Theorem VII.3.1, it follows by induction on $t \geq s$ that

(1.3) $(\forall t \geq s)\,[\hat{l}(e,t) > p\ \&\ \hat{r}(e,t) \geq \max\{\,\hat{u}(A_s;e,x,s) : x \leq p\,\}]$,

and hence that
$$\Phi_{e,s}(A_s;p) = \Phi_e(A_s;p) = \Phi_e(A;p) = C(p).$$

Lemma 2.   *Let T be the set of true stages in the enumeration $\{A_s\}_{s\in\omega}$ of A. Then $\lim_{t\in T} \hat{r}(e,t) < \infty$. (Hence, if $C \neq \Phi_i(A)$, for all $i \leq e$, then $\lim_{t\in T} \hat{R}(e,t) < \infty$, where $\hat{R}(e,s) = \max\{\,\hat{r}(i,s) : i \leq e\,\}$ thereby satisfying (1.1).)*

Proof. By Lemma 1 we know $C \neq \Phi_e(A)$. Define $p = \mu x[C(x) \neq \Phi_e(A;x)]$. Choose $s' \in T$ sufficiently large such that, for all $s \geq s'$,

$$(\forall x < p)\,[\hat{\Phi}_{e,s}(A_s;x) = \Phi_e(A;x)]\ \&\ (\forall x \leq p)\,[C_s(x) = C(x)].$$

Case 1. $(\forall t \geq s')$ $[t \in T \implies \widehat{\Phi}_{e,t}(A_t; p) \uparrow]$. Then for any $t \geq s'$, such that $t \in T$, we have $\hat{l}(e,t) = p$ and $\hat{r}(e,t) = \max\{ u(A_{s'}; e, x, s') : x < p \}$.

Case 2. $(\exists t \geq s')$ $[t \in T \ \& \ \widehat{\Phi}_{e,t}(A_t; p) \downarrow]$. Then $\Phi_e(A; p) = \widehat{\Phi}_{e,s}(A_s; p)$ for all $s \geq t$ by (1.2). But $C(p) \neq \Phi_e(A; p)$. Hence, we have

(1.4)
$$(\forall s \geq t) \ [\hat{l}(e, s) = p \ \& \ \hat{r}(e, s) = \hat{r}(e, t) = \max\{ u(A_s; e, x, s) : x \leq p \}].$$

Lemma 3. $(\forall e) \ [A^{[e]} =^* B^{[e]}]$.

Proof. Any $x \in B^{[e]}$ such that $x > \hat{R}(e) =_{\mathrm{dfn}} \lim_{t \in T} \hat{R}(e, t)$ will be enumerated in $A^{[e]}$. ▯

The following is the dual of Theorem VII.1.1 which asserted the existence of a nonrecursive low r.e. set.

1.2 Corollary (Sacks [1963c]). *For any nonrecursive r.e. set C there exists an r.e. set A such that $A' \equiv_T \emptyset''$ (namely A is high) and $C \not\leq_T A$ (and hence A is incomplete.)*

Proof. By Post's Theorem IV.2.2, any set S which is r.e. in $\emptyset'$ (such as $\emptyset''$) is in Σ_2 form, and hence by Theorem IV.3.2 there is a recursive function $h(y)$ such that for all y, $y \in S$ implies $W_{h(y)}$ is finite, and $y \notin S$ implies $W_{h(y)} = \omega$. For each such S define the r.e. set B_S by $B_S = \{ \langle x, y \rangle : x \in W_{h(y)} \}$. If A is a thick subset of B_S then

$$y \in S \implies B_S^{[y]} \text{ finite} \implies A^{[y]} \text{ finite} \implies \lim_x A(\langle x, y \rangle) = 0,$$

and

$$y \notin S \implies B_S^{[y]} = \omega^{[y]} \implies A^{[y]} =^* \omega^{[y]} \implies \lim_x A(\langle x, y \rangle) = 1.$$

The function $f(y) = \lim_x A(\langle x, y \rangle)$ is recursive in A' by the Limit Lemma, so $S \leq_T A'$. Now choose $S = \emptyset''$ and apply Theorem 1.1 to B_S to obtain A. ▯

(Note that a priority argument with finitary positive requirements can in general be combined with the lowness requirements of Chapter VII §1 to make the constructed r.e. set low. Thus it is not surprising that to construct an incomplete high r.e. set we need either the infinitary positive requirements of Theorem 1.2, or else one finite injury construction over another as in Exercise VII.1.6(c), or as in Exercise VII.1.8 combined with Theorem VII.1.1 relativized to $\emptyset'$. These constructions, when carefully examined, amount to infinitary positive requirements.)

2. The Injury and Window Lemmas and the Strong Thickness Lemma

To prove the Jump Theorem and the Density Theorem we shall need not just the Thickness Lemma, but slightly more general versions of Lemmas 1 and 2 of §1 which we now isolate and prove. Given $\emptyset <_T C \leq_T \emptyset'$ fix a recursive sequence of (canonically indexed) finite sets $\{C_s\}_{s\in\omega}$ such that $C(x) = \lim_s C_s(x)$. Suppose $A = \bigcup_s A_s$ where $\{A_s\}_{s\in\omega}$ is a recursive sequence of finite sets, either given ahead of time or defined by us during a recursive construction. Given $\{A_v\}_{v\leq s}$ define:

(length function) $\hat{l}(e,s) = \max\{x : (\forall y < x)\, [C_s(y) = \hat{\Phi}_{e,s}(A_s;y)]\}$;

(modified length function) $\hat{m}(e,s) = \max\{x : (\exists v \leq s)\, [x \leq \hat{l}(e,v)$

$\&\ (\forall y \leq x)\, [A_s \restriction u = A_v \restriction u]]\}$,   where $u = \hat{u}(A_v;e,y,v)$;

(restraint functions) $\hat{r}(e,s) = \max\{\hat{u}(A_s;e,x,s) : x \leq \hat{m}(e,s)\}$;

$$\hat{R}(e,s) = \max\{\hat{r}(i,s) : i \leq e\};$$

(injury set) $\hat{I}_e = \bigcup_s \hat{I}_{e,s}$, where

$$\hat{I}_{e,s} = \{x : (\exists v < s)\, [x \leq \hat{r}(e,v)\ \&\ x \in A_{v+1} - A_v]\}.$$

The point of $\hat{m}$ is to record a length of agreement established at some stage v so long as the "$\Phi_e(A)$ side" is unchanged even though a change in the "C side" at some stage $s > v$ may cause $\hat{l}(e,s) < \hat{l}(e,v)$. The use of $\hat{m}$ is necessary in order to prove Lemma 1.1 with the weaker hypothesis $C \leq_T \emptyset'$ in place of "C r.e.", and it immediately yields by (1.2) certain convenient properties such as

(2.1) $(\forall t \in T)\,(\forall s \geq t)\, [\hat{m}(e,t) \leq \hat{m}(e,s)\ \&\ \hat{R}(e,t) \leq \hat{R}(e,s)]$.

(From now on $\hat{r}$, $\hat{R}$ and $\hat{I}_e$ will refer to the above definitions rather than to those in §1.) The infinite injury method depends upon the following two lemmas, whose proofs are essentially the same as those of §1 Lemmas 1 and 2.

2.1 Injury Lemma. *If $C \not\leq_T \hat{I}_e$ then $C \neq \Phi_e(A)$.*

2.2 Window Lemma. *Let T be the set of true stages in the enumeration $\{A_s\}_{s\in\omega}$ of the r.e. set A. If $C \neq \Phi_e(A)$ then $\lim_{t\in T} \hat{r}(e,t) < \infty$. (Hence, if $C \neq \Phi_i(A)$ for all $i \leq e$, then $\lim_{t\in T} \hat{R}(e,t) < \infty$.)*

Proof of Injury Lemma 2.1. Apply the proof of §1 Lemma 1 with the new definitions of $\hat{r}(e,s)$, $\hat{R}(e,s)$, $\hat{I}_e$, and the function $\hat{m}$ in place of $\hat{l}$ in (1.3).   □

Proof of Window Lemma 2.2. Apply the proof of §1 Lemma 2, with the new definitions of $\hat{r}(e,s)$, $\hat{R}(e,s)$, and $\hat{I}_e$. Now Case 2 of §1 Lemma 2 is proved as in §1, and Case 1 of §1 Lemma 2 is also proved as before because of our use of $\widehat{\Phi}_{e,s}$. Thus, in Case 2 by (2.1) we have in addition to (1.4),

$$(2.2) \qquad (\forall s \geq t)\,[\hat{m}(e,t) = \hat{m}(e,s) \ \& \ \hat{R}(e,t) = \hat{R}(e,s)]. \quad \square$$

We now obtain a stronger form of the Thickness Lemma. Similar versions have been obtained implicitly or explicitly by Robinson, Shoenfield, Sacks, and Yates.

2.3 Thickness Lemma—Strong Form. *Given $\emptyset <_T C \leq_T \emptyset'$ and an r.e. set B there is an r.e. set $A \subseteq B$ such that $A \leq_T B$,*

(i) *$(\forall e)\,[C \not\leq_T B^{[<e]}] \implies [C \not\leq_T A \ \& \ A$ is a thick subset of $B]$, and indeed*

(ii) *$(\forall e)\,[C \not\leq_T B^{[<e]} \implies (\forall j \leq e)\,[C \neq \Phi_j(A) \ \& \ A^{[j]} =^* B^{[j]}]].$*

Furthermore, an index for A can be computed uniformly in indices for B and C.

Proof. It suffices to prove (ii) since (ii) implies (i). Fix recursive sequences of finite sets $\{B_s\}_{s\in\omega}$ and $\{C_s\}_{s\in\omega}$ such that $B = \bigcup_s B_s$ and $C = \lim_s C_s$. Define A exactly as in Theorem 1.1 but with the new definitions of $\hat{r}(e,s)$, $\hat{R}(e,s)$, and $\hat{I}_e$. Note that as before we have

$$(2.3) \qquad\qquad \hat{I}_e \leq_T A^{[<e]}.$$

Fix e, assume that

$$(2.4) \qquad\qquad [C \not\leq_T B^{[<e]}],$$

and assume by induction that $C \neq \Phi_k(A)$ and $A^{[k]} =^* B^{[k]}$ for all $k < e$. Then $A^{[<e]} =^* B^{[<e]}$, so $\hat{I}_e \leq_T B^{[<e]}$ by (2.3). By this, (2.4), and the Injury Lemma, $C \neq \Phi_e(A)$. Thus, by the Window Lemma $\lim_{t\in T} \hat{R}(e,t) < \infty$; and hence $A^{[e]} =^* B^{[e]}$ by construction. This establishes the conclusion of (ii). The fact that $A \leq_T B$ will follow from the following lemma which will be useful later. $\square$

2.4 Lemma. *Let $\{C_s\}_{s\in\omega}$ be as above and let $\{A_s\}_{s\in\omega}$ be a recursive enumeration of an r.e. set A. Recall that $A_s^{[<e]} = \bigcup\{A_s^{[i]} : i < e\}$. Define*

$$T^e = \{\, t : A^{[<e]} \restriction a_t = A_t^{[<e]} \restriction a_t \,\},$$

where a_t is defined as in §1 with $A_t^{[<e]}$ in place of A_t. Then

$$(2.5) \qquad\qquad T^e \leq_T A^{[<e]} \quad \text{uniformly in } e,$$

and if

(2.6) $$\hat{I}_e \subseteq A^{[<e]},$$

then

(2.7) $(\forall t \in T^e)\,(\forall x \leq \hat{m}(e,t))\,(\forall y)\,[\hat{\Phi}_{e,t}(A_t; x) = y \implies \Phi_e(A; x) = y],$

and

(2.8) $(\forall t \in T^e)\,(\forall s \geq t)\,[\hat{m}(e,t) \leq \hat{m}(e,s)\ \&\ \hat{R}(e,t) \leq \hat{R}(e,s)].$

Proof. Clearly (2.5) follows as did $T \leq_T A$ in §1. Now (2.7) follows as did (1.2) now using (2.6) and the fact that $\hat{m}$ (not $\hat{l}$) is used in the definition of $\hat{r}(e,s)$ and hence in the definition of $\hat{I}_e$ also. Finally, (2.8) follows for T^e as did (2.1) for T.   ▯

2.5 Lemma. *For A, B, and C, satisfying the hypotheses of Lemmas 2.3 and 2.4 including (2.6), we have automatically achieved $A \leq_T B$.*

Proof. Fix e and assume that for all $i < e$ we have B-recursively computed $g(i)$ such that $A^{[i]} = \Phi_{g(i)}(B)$. Then B-recursively compute $A^{[<e]}$ and T^e. Now if $x \in B^{[e]}$, say $x \in B_s^{[e]}$, let $t' = \mu t[t \geq s\ \&\ t \in T^e]$. Now $\hat{R}(e,t') = \min\{\,\hat{R}(e,v) : t' \leq v\,\}$ by (2.8). (Namely, the restraint $\hat{R}(e,t')$ is as small now as it will ever be again.) Hence, $x \in A^{[e]}$ just if $x \in A_{t'+1}^{[e]}$ by the construction of $A_{t'+1}$.   ▯

Many results on r.e. degrees follow easily from the Strong Thickness Lemma, such as the following theorem.

2.6 Theorem (Sacks [1963c]).   *Let $\mathbf{d}_0 < \mathbf{d}_1 < \mathbf{d}_2 < \cdots$ be an infinite ascending sequence of uniformly r.e. degrees. Then there is an r.e. degree $\mathbf{a}$ such that $\mathbf{d}_0 < \mathbf{d}_1 < \cdots < \mathbf{a} < \mathbf{0}'$. (Hence, $\mathbf{0}'$ is not a minimal upper bound for the sequence.)*

Proof. Fix a recursive function h such that $\deg(W_{h(x)}) = \mathbf{d}_x$ for all x. Define the r.e. set B by $B^{[y]} = \{\,\langle x, y \rangle : x \in W_{h(y)}\,\}$. Let $C = \emptyset'$ and apply Lemma 2.3(i) to obtain a thick r.e. subset $A \subseteq B$ such that $C \not\leq_T A$. By thickness $W_{h(x)} \equiv_T B^{[x]} =^* A^{[x]}$ so that $\mathbf{d}_i < \deg(A)$ for all i.   ▯

In contrast to Theorem 2.6, Cooper [1972a] has shown that there does exist an r.e. degree $\mathbf{a}$ and an ascending r.e. sequence of r.e. degrees $\{\,\mathbf{d}_i\,\}_{i \in \omega}$ such that $\mathbf{a}$ is a minimal upper bound for the sequence. Another result which follows easily from the Strong Thickness Lemma 2.3 is the Yates Index Set Theorem XII.1.5. From this it is possible to give a short and elegant proof of the Sacks Density Theorem 4.1 as we show in Theorem XII.1.8.

2.7 Exercise

2.7 (Interpolation Theorem) (Sacks).   Prove that if C and D are r.e. sets such that $D <_T C$ then there is an r.e. set A such that $D \leq_T A <_T C$ and $A' \equiv_T C'$. *Hint.* Fix recursive enumerations $\{C_s\}_{s \in \omega}$ and $\{D_s\}_{s \in \omega}$ of C and D and define the recursive function

$$h(x, s) = \begin{cases} x & \text{if } \widehat{\Phi}_{x,s}(C_s; x)\downarrow, \\ s & \text{otherwise.} \end{cases}$$

This is the canonical representation for the jump C' of an r.e. set C because since h is recursive we have $\lim_s h(x, s) < \infty$ iff $x \in C'$. Define $B^{[0]} = \{\langle n, 0 \rangle : n \in D\}$ and $B^{[y+1]} = \{\langle n, y+1 \rangle : (\exists s) [n \leq h(y, s)]\}$. Apply Lemma 2.3 to obtain A from B and C and prove that

$$y \in C' \iff \lim_x A(\langle x, y+1 \rangle) = 0.$$

One needs to prove that $B \leq_T C$ in order to get that $A \leq_T C$ and $A' \leq_T C'$. (For a stronger version see the Jump Interpolation Theorem 4.4.)

3. *The Jump Theorem*

If A is r.e. and $S = A'$, then clearly S is r.e. in $\emptyset'$ and $\emptyset' \leq_T S$. The Sacks Jump Theorem asserts that these conditions on S are also *sufficient* so that $S \equiv_T A'$ for some r.e. set A, and in addition that we can guarantee $C \not\leq_T A$ where $\emptyset <_T C \leq_T \emptyset'$ is given. Historically, this was Sacks's first application of the infinite injury priority method.

3.1 Sacks Jump Theorem [1963c].   *Given sets S and C such that S is r.e. in $\emptyset'$, $\emptyset' \leq_T S$, and $\emptyset <_T C \leq_T \emptyset'$ there exists a nonrecursive r.e. set A such that $A' \equiv_T S$ and $C \not\leq_T A$. Furthermore, an index of A can be found uniformly from indices for S and C.*

Proof. Any set S r.e. in $\emptyset'$ is Σ_2, so by Theorem IV.3.2 there is a recursive function h such that

$$y \in S \implies W_{h(y)} \text{ is finite, and } y \notin S \implies W_{h(y)} = \omega.$$

Define the r.e. set $B = \{\langle x, y \rangle : x \in W_{h(y)}\}$ so that for all y,

(3.1) $y \in S \implies B^{[y]}$ is finite, and $y \notin S \implies B^{[y]} = \omega^{[y]}$.

Note that since $\emptyset' \leq_T S$ there is an S-recursive function f such that $\varphi_{f(y)}$ is the characteristic function for $B^{[y]}$.

We shall build a thick subset A of B such that $C \not\leq_T A$ by meeting for every e the requirements

$$N_e : C \neq \{e\}^A,$$

and

$$P_e : A^{[e]} =^* B^{[e]}.$$

The thickness requirements guarantee that $S \leq_T A'$ because

$$F(y) =_{\mathrm{dfn}} \lim_x A(\langle x, y \rangle)$$

exists for all y and is the characteristic function of $\overline{S}$ by (3.1) and because $F \leq_T A'$ by the Limit Lemma.

To keep the jump *down* and ensure that $A' \leq_T S$ we have for each $e \in \omega$, a negative "pseudo-requirement"

$$Q_e : (\exists^\infty s) \, [\widehat{\Phi}_{e,s}(A_s; e)\!\downarrow] \implies \{e\}^A(e)\!\downarrow,$$

and its corresponding restraint function $\hat{q}(e, s) =_{\mathrm{dfn}} \hat{u}(A_s; e, e, s)$. If $\emptyset' <_T S$ we cannot hope to meet Q_e for every e, or otherwise $A' \equiv_T \emptyset'$ as in Theorem VII.1.1. Nevertheless, we *attempt* to meet Q_e in exactly the same way as in Theorem VII.1.1 by allowing P_i to injure Q_e (namely, contribute an element $x \leq \hat{q}(e, s)$ to A_{s+1}) only if $i < e$. We then prove that the injuries to Q_e (although possibly infinite in number) are sufficiently well-behaved so that we can verify $A' \leq_T S$ using the hypothesis $\emptyset' \leq_T S$. This is accomplished by defining an S-recursive function g such that for every e, $\varphi_{g(e)}$ is the characteristic function of $A^{[e]}$.

Construction of A. Fix recursive sequences $\{B_s\}_{s\in\omega}$ and $\{C_s\}_{s\in\omega}$ such that $B = \bigcup_s B_s$ and $C = \lim_s C_s$. Given $\{A_t\}_{t\leq s}$ define $\hat{r}(e, s)$ and $\hat{u}(A_s; e, x, s)$ as in §2 and $\hat{q}(e, s)$ as above. Let $A_0 = \emptyset$. Enumerate x in $A^{[e]}_{s+1}$ iff $x \in B^{[e]}_{s+1}$ and $x > \max\{\hat{r}(i, s), \hat{q}(i, s)\}$ for all $i \leq e$. Let $A = \bigcup_s A_s$.

Lemma 1.   For all e, $A^{[e]} =^* B^{[e]}$ and $C \neq \Phi_e(A)$. *(Hence, $C \not\leq_T A$ and $S \leq_T A'$.)*

Proof. Let T be the set of true stages in the enumeration $\{A_s\}_{s\in\omega}$. Fix e and assume that P_i and N_i hold for all $i < e$. The $A^{[<e]} =^* B^{[<e]}$ is recursive (because B is piecewise recursive), N_e is satisfied by the Injury Lemma 2.1, and $\lim_{t\in T} \hat{R}(e, t) < \infty$ by the Window Lemma 2.2. Furthermore, $\lim_{t\in T} \hat{q}(i, t) < \infty$ for all $i \leq e$, by (1.2). Hence, $A^{[e]} =^* B^{[e]}$ by construction.

Lemma 2. $A' \leq_T S$.

Proof. Let T^e be the set of true stages of the enumeration $\{A^{[<e]}_s\}_{s\in\omega}$ (as in Lemma 2.4, which is very similar). Now by construction, the requirement N_e

or Q_e can be injured by an element x only if $x \in A^{[<e]}$. Hence, (2.6) holds. Thus, as in (2.7) and (2.8) we now have

$$(3.2) \qquad (\forall t \in T^e) \, [\hat{\Phi}_{e,t}(A_t; e) \downarrow \implies \Phi_e(A; e) \downarrow],$$

and

$$(3.3) \qquad (\forall t \in T^e) \, (\forall s \geq t) \, [\hat{Q}(e, t) \leq \hat{Q}(e, s) \ \& \ \hat{R}(e, t) \leq \hat{R}(e, s)],$$

where $\hat{Q}(e, s) = \max\{ \hat{q}(i, s) : i \leq e \}$.

Fix e and assume that we have S-recursively computed, for all $i < e$, an index $g(i)$ such that $A^{[i]}$ has a (recursive) characteristic function $\varphi_{g(i)}$. From this we can S-recursively compute indices for the (recursive) characteristic function of $A^{[<e]}$ and hence of T^e by (2.5). To compute $g(e)$ we first S-recursively compute an index $f(e)$ for the recursive characteristic function of $B^{[e]}$ as above. Now if $x \in B^{[e]}$, say $x \in B_s^{[e]}$, define $t' = (\mu t > s) \, [t \in T^e]$. Then $x \in A^{[e]}$ iff $x \in A_{t'+1}$ by (3.3). This completes the definition of g.

To establish that $A' \leq_T S$ note that since $\emptyset' \leq_T S$ and S can compute uniformly in e an index for the (recursive) characteristic function of T^e, using g, we can now S-recursively decide whether $e \in A'$ because by (3.2),

$$(3.4) \qquad e \in A' \iff \Phi_e(A; e) \downarrow \iff (\exists t) \, [t \in T^e \ \& \ \hat{\Phi}_{e,t}(A_t; e) \downarrow].$$

Note that given a recursive index for T^e we have that (3.4) is in Σ_1 form and is hence recursive in $\emptyset'$ (and not merely recursive in S). *This is the crucial point of the entire proof* and should be thoroughly understood before proceeding. (If the matrix of the last clause of (3.4) were merely S-recursive rather than recursive then we would require an S'-oracle to decide whether (3.4) holds.)

To verify that A is nonrecursive note that we may assume $\emptyset' <_T S$ (else Theorem 3.1 is an immediate corollary of the Sacks Splitting Theorem VII.3.2). Hence, A is nonrecursive since $A' \equiv_T S$. We can also give a uniform construction of a nonrecursive r.e. set A satisfying Theorem 3.1 by adding to the above construction additional finitary positive requirements of the form $R_e : A \neq \{ e \}$ and using some witness $x \in \omega^{[e]}$ to meet R_e in the manner of Chapter VII without disturbing the strategies for meeting requirements P_i and N_i above. ▯

For future reference note that using $\emptyset' \leq_T S$ and (3.3) we have

$$(3.5) \qquad \hat{R}(e) = \liminf_s \hat{R}(e, s) \text{ is an } S\text{-recursive function.}$$

(Since we know that $\lim_{t \in T^e} \hat{R}(e, t) < \infty$, we ask for each $t \in T^e$ whether

$$(\exists t' > t) \, [t' \in T^e \ \& \ \hat{R}(e, t) < \hat{R}(e, t')]$$

until the answer is negative. Each such question is Σ_1 because T^e is recursive.)

3.2 Remark (Sacks).   *Given S and C as in Theorem 3.1 if D is any r.e. set such that $D' \leq_T S$ and $C \not\leq_T D$, then we can add to the conclusion of Theorem 3.1 that $D \leq_T A$.*

Proof. Replace B by the r.e. set $\tilde{B}$ where $\tilde{B}^{[0]} = \{ \langle x,0 \rangle : x \in D \}$ and $\langle x, y + 1 \rangle \in \tilde{B}$ iff $\langle x, y \rangle \in B$. The construction and proof are now the same as above except that A and $\tilde{B}$ are now "piecewise recursive in D" instead of "piecewise recursive" because $A^{[0]} =^* \tilde{B}^{[0]} \equiv_T D$. The Injury Lemma still applies in Lemma 1 because $\hat{I}_e \leq_T A^{[<e]} \leq_T D$ and $C \not\leq_T D$. Now $T^e \leq_T D$ so that in (3.4) we must use a D'-oracle instead of a $\emptyset'$-oracle. Since $D' \leq_T S$, the proof of Lemma 2 now is simply relativized to D.   ▯

3.3 Remark.   *Given S and C as in Theorem 3.1 we can obtain A directly by applying the Thickness Lemma 2.3 to B and C without introducing the second restraint function $\hat{q}(e, s)$.*

Proof. Given $C(0)$ define the recursive function h by

$$\Phi_{h(e)}(X; y) = \begin{cases} C(0) & \text{if } y = 0 \text{ and } \Phi_e(X; e)\!\downarrow, \\ \text{undefined} & \text{otherwise.} \end{cases}$$

Hence,

$$\Phi_e(A; e)\!\downarrow \iff \Phi_{h(e)}(A; 0) = C(0) \iff \liminf_s \hat{l}(h(e), s) > 0,$$
$$\iff (\exists t)\, [t \in T^{h(e)} \And \hat{\Phi}_{h(e),t}(A_t; 0) = C(0)],$$

by (3.4).   ▯

In Definition IV.4.2 we defined the high (low) r.e. degrees by considering those whose first jump had highest (lowest) possible value. This hierarchy was extended in Definition VII.1.3 (and Exercise IV.4.5) by replacing the first jump by the nth jump to obtain the classes $\mathbf{H}_n$ and $\mathbf{L}_n$ for $n \geq 0$. The next results establish that these classes are all distinct and yet do not exhaust the r.e. degrees $\mathbf{R}$. (For alternative proofs see Exercises VII.1.6(d) and VII.1.6(e).)

3.4 Corollary.   *For all $n \in \omega$, $\mathbf{L}_n \neq \mathbf{L}_{n+1}$ and $\mathbf{H}_n \neq \mathbf{H}_{n+1}$.*

Proof. By relativizing Corollary 1.2 and Theorem VII.1.1 to $\mathbf{0}^{(n)}$ and using the Jump Theorem 3.1, it follows that for all n, $\mathbf{H}_n \neq \mathbf{H}_{n+1}$ and $\mathbf{L}_n \neq \mathbf{L}_{n+1}$.   ▯

Martin [1966d], Lachlan [1965b], and Sacks [1967] each proved that the union of these classes does not exhaust **R**. We use the proof of Sacks.

3.5 Corollary (Martin, Lachlan, Sacks). *There is an r.e. degree* **a** *such that for all* n, $[\mathbf{0}^{(n)} < \mathbf{a}^{(n)} < \mathbf{0}^{(n+1)}]$, *i.e.,* $\mathbf{a} \in \mathbf{R} - \bigcup_n (\mathbf{H}_n \cup \mathbf{L}_n)$. *(Such an* **a** *is called an* intermediate *degree.)*

Proof (Sacks [1967]). The uniformity of the Jump Theorem 3.1 combined with Corollary 3.4 both relativized to any set $Y \subseteq \omega$ produces a recursive function $q(x)$ such that for all $x \in \omega$ and $Y \subseteq \omega$,

$$(3.6) \qquad Y <_T W^Y_{q(x)} <_T Y' \quad \text{and} \quad (W^Y_{q(x)})' \equiv_T (W^{Y'}_x) \oplus Y'.$$

Now apply the Relativized Recursion Theorem III.1.6 to obtain a fixed point n such that $W^Y_{q(n)} = W^Y_n$ for all $Y \subseteq \omega$. Define $\mathbf{a} = \deg(W^{\emptyset}_n)$. (Note that $q(x)$ is 1:1, recursive, and not merely Y-recursive by the Relativized Recursion Theorem III.1.6(ii).) ◻

The following strengthening of (3.6) will be useful in Chapter XII. We prove a certain proposition $(\forall n)\,(\forall Y)\,P_n(Y)$ by induction on n. We are primarily interested in the unrelativized version $P_n(\emptyset)$, but we need the relativized inductive hypothesis $(\forall Y)\,P_n(Y)$ in order to prove the induction step. This form resembles the relativized inductive proofs in Corollary VI.3.3 and Exercise VII.1.6. We shall use Corollary 3.6 in proving Theorem XII.4.4.

3.6 Corollary. *Let $q(x)$ be as in Corollary 3.5. Then for all $x \in \omega$, $Y \subseteq \omega$, and $n \geq 0$,*

$$(3.7) \qquad Y <_T W^Y_{q^n(x)} <_T Y' \;\&\; (W^Y_{q^n(x)})^{(n)} \equiv_T W^{Y^{(n)}}_x \oplus Y^{(n)}.$$

(Recall that $q^n(x)$ denotes the function $q(x)$ composed with itself n times.)

Proof. Let $P_n(Y)$ denote line (3.7) preceded by "$(\forall x)$". We prove $(\forall Y)\,P_n(Y)$ by induction on n. The case $n = 1$ is simply (3.6). Note that $P_{n+1}(Y)$ follows from $P_n(Y')$ and $P_1(Y)$. ◻

3.7 Proposition. *There is a recursive function $f(i,x)$ such that for every $Y \subseteq \omega$ and $i, x \in \omega$,*

$$W^Y_{f(i,x)} = W^Y_{\varphi^Y_i(x)},$$

where $W^Y_{f(i,x)}$ is defined to be $\emptyset$ if $\varphi^Y_i(x)$ diverges.

Proof. The oracle Turing program $\hat{P}_{f(i,x)}$ of Chapter III §1 first uses its oracle tape (containing the characteristic function of Y) to attempt to compute $\varphi^Y_i(x) = z$ and then enumerates W^Y_z. While $\varphi^Y_i(x)$ is undefined let $W^Y_{f(i,x)} = \emptyset$. ◻

3.8–3.9 Exercises

3.8. Strengthen (3.7) by proving that there is a recursive function $q(n, i, x)$ such that for every $Y \subseteq \omega$, $n \geq 1$, and $i, x \in \omega$,

$$(3.8) \qquad Y <_T W^Y_{q(n,i,x)} <_T Y' \; \& \; (W^Y_{q(n,i,x)})^{(n)} \equiv_T W^{Y^{(n)}}_{\varphi_i^{Y^{(n)}}(x)} \oplus Y^{(n)}.$$

Hint. Combine Corollary 3.6 with Proposition 3.7.

3.9 (Martin, Lachlan). Show that there are r.e. degrees $\mathbf{a}$ and $\mathbf{b}$ such that for all n, $\mathbf{a}^{(n)} \mid \mathbf{b}^{(n)}$. *Hint.* Modify the proof of Theorem 3.1 and line (3.6) to get recursive functions $p(x, y)$ and $q(x, y)$ such that for all $x, y \in \omega$ and $Y \subseteq \omega$,

$$(W^Y_{p(x,y)})' \equiv_T (W^{Y'}_x) \oplus Y';$$

$$(W^Y_{q(x,y)})' \equiv_T (W^{Y'}_y) \oplus Y';$$

and

$$W^Y_{p(x,y)} \mid_T W^Y_{q(x,y)}.$$

Now use the Double Recursion Theorem II.3.15.

4. The Density Theorem and the Sacks Coding Strategy

Sacks [1964a] proved the Density Theorem by inventing a new coding strategy for meeting the positive requirements and combining it with the infinite injury method above. This coding strategy is the major new idea in the following proof and has numerous other applications. (A different proof by Yates using index sets will be given in Theorem XII.1.8.)

4.1 Density Theorem (Sacks [1964a]). *Given r.e. sets $D <_T C$ there exists an r.e. set A such that $D <_T A <_T C$.*

Proof. Fix recursive enumerations $\{C_s\}_{s \in \omega}$ and $\{D_s\}_{s \in \omega}$ of C and D such that $D_0 = \emptyset$. Define $A_s^{[0]} = \{\langle x, 0 \rangle : x \in D_s\}$ so that $D \leq_T A$. We shall arrange that $A \leq_T C$ as in Lemma 2.5 by finding a C-recursive function g such that $A^{[e]} = \Phi_{g(e)}(C)$ for all e. To make both inequalities strict it suffices to meet, for all $e > 0$, the requirements

$$N_e : C \neq \Phi_e(A), \quad \text{and} \quad P_e : A \neq \Phi_e(D).$$

(We need to consider only $e > 0$, rather than $e \geq 0$, since every Turing reduction has infinitely many indices.) To meet P_e we attempt to code C into $A^{[e]}$ so that if $A = \Phi_e(D)$ then $C \leq_T D$ contrary to hypothesis.

Let $A_0 = \emptyset$. Given $\{A_t\}_{t \leq s}$, for $e > 0$ define $\hat{r}(e, s)$ as in §2, $\hat{R}(e, s) = \max\{\hat{r}(i, s) : 0 < i \leq e\}$, and define

$$\hat{l}^D(e, s) = \max\{x : (\forall y < x)\,[A_s(y) = \widehat{\Phi}_{e,s}(D_s; y)]\}.$$

(Recall by Notation I.3.6 that $\langle x, s, e\rangle$ denotes $\langle\langle x, s\rangle, e\rangle$.) For $t \leq s$ and e, such that $0 < e \leq s$, enumerate $\langle x, t, e\rangle$ in A_{s+1} iff $\langle x, t, e\rangle > \hat{R}(e, s)$, $x \in C_{s+1}$ and $x < \hat{l}^D(e, v)$ for all v, $t \leq v \leq s$. Let $A = \bigcup_s A_s$. (This completes the construction of A.)

(We visualize this coding strategy as follows. Fixing e, the elements $\{\langle x, y, e\rangle : x, y \in \omega\}$ are arranged in a plane. A (nonmovable) "coding marker" is assigned to $\langle x, t, e\rangle$ at stage t if $\hat{l}^D(e, t) > x$. This marker is later removed (forever) at some stage $s > t$ if $\hat{l}^D(e, s) \leq x$. If $x \in C_{s+1}$ then all elements $\langle x, t, e\rangle$, $t \leq s$, still possessing markers at stage s and not restrained with higher priority are enumerated in A_{s+1}.)

To see that A succeeds we shall verify the requirements by induction on e and simultaneously C-recursively define $g(e)$ such that $A^{[e]} = \Phi_{g(e)}(C)$. Clearly $g(0)$ exists because $A^{[0]} \equiv_T D <_T C$. Fix $e > 0$ and assume for all i, $0 < i < e$, that

(4.1) $$C \neq \Phi_i(A),$$

(4.2) $$A \neq \Phi_i(D),$$

(4.3) $$A^{[i]} \text{ is recursive,}$$

and

(4.4) $\quad A^{[i]} = \Phi_{g(i)}(C)$, where $g(i)$ has been C-recursively computed.

Lemma 1. $C \neq \Phi_e(A)$.

Proof. Now $A^{[<e]} \leq_T D$ because $A^{[i]}$ is recursive for every i, $0 < i < e$. Let $\hat{I}_e$ be the injury set for N_e as defined in §2. Then $\hat{I}_e \subseteq A^{[<e]}$ by construction, and hence $\hat{I}_e \leq_T A^{[<e]}$ as in (2.3). Thus, $\hat{I}_e \leq_T A^{[<e]} \leq_T D <_T C$. Hence, $C \neq \Phi_e(A)$ by the Injury Lemma 2.1.

Lemma 2. $A \neq \Phi_e(D)$.

Proof. By Lemma 1 and the Window Lemma 2.2, $\hat{R}(e) =_{\text{dfn}} \liminf_s \hat{R}(e, s) < \infty$. If $A = \Phi_e(D)$ then $\lim_s \hat{l}^D(e, s) = \infty$. Define the modulus function

$$M(x) = (\mu s)\,[x < \hat{l}^D(e, s) \ \& \ D_s \upharpoonright u = D \upharpoonright u],$$

where $u = \max\{\, u(D_s; e, y, s) : y \leq x \,\}$. Now $M \leq_T D$ by the Modulus Lemma III.3.2 since D is r.e. However, we also have

$$M(x) = (\mu s)\,(\forall t \geq s)\,[x < \hat{l}^D(e, t)],$$

because for all $s \geq M(x)$ and for all $y \leq x$, $\hat{\Phi}_{e,s}(D_s; y) = \Phi_e(D; y)$ so if $A(y) \neq A_s(y)$ then $\Phi_e(D) \neq A$ contrary to hypothesis. For each x D-recursively choose the least $y_x > \max\{\, M(x), \hat{R}(e) \,\}$. Now $x \in C$ iff $\langle x, y_x, e \rangle \in A$. Hence, $C \leq_T A \oplus D \leq_T D$ contrary to hypothesis.

Lemma 3. $A^{[e]}$ *is recursive.*

Proof. By Lemma 2, $A \neq \Phi_e(D)$. Let $p_e = (\mu x)\,[A(x) \neq \Phi_e(D; x)]$. Then $\liminf_s \hat{l}^D(e, s) = p_e$. For $x \geq p_e$ given x and t find $s \geq t$ such that $\hat{l}^D(e, s) = p_e$. Then $\langle x, t, e \rangle \in A$ iff $\langle x, t, e \rangle \in A_{s+1}$. For $x < p_e$ fix s' such that $C_{s'} \restriction p_e = C \restriction p_e$, $p_e \leq \hat{l}^D(e, t)$ and $\hat{R}(e) \leq \hat{R}(e, t)$ for all $t \geq s'$. Given t, define

$$v_t = (\mu v)\,[v \geq s' \ \& \ v \geq t \ \& \ \hat{R}(e, v) = \hat{R}(e)].$$

Then for $t \geq s'$, $\langle x, t, e \rangle \in A$ iff $\langle x, t, e \rangle \in A_{v_t+1}$.

(Note that to decide whether $\langle x, t, e \rangle \in A$, the bound exists on s for different reasons: if $x \geq p_e$, it exists because $\hat{l}^D(e, s)$ drops back; if $x < p_e$ (and t is large) it exists because $\hat{l}^D(e, s)$ is eventually $> x$ and $\liminf_s \hat{R}(e, s) = \hat{R}(e)$. Hence, the decision procedure is *not* uniform in e.)

Lemma 4. *We can C-recursively compute $g(e)$ such that* $A^{[e]} = \Phi_{g(e)}(C)$, *uniformly in e.*

Proof. Define T^e for $\{\, A_s^{[<e]} \,\}_{s \in \omega}$ as in Lemma 2.4. From $\{\, g(i) : i < e \,\}$ we C-recursively compute $A^{[<e]}$ and hence T^e. Fix $\langle x, t, e \rangle$. If $x \notin C$ then $\langle x, t, e \rangle \notin A$. If $x \in C$ choose s such that $C_s \restriction x + 1 = C \restriction x + 1$ and define

$$v' = (\mu v)\,[v \geq t \ \& \ v \geq s \ \& \ v \in T^e].$$

Then $\langle x, t, e \rangle \in A$ iff $\langle x, t, e \rangle \in A_{v'+1}$ because $\hat{R}(e, v') \leq \hat{R}(e, v)$ for all $v \leq v'$. ☐

4.2 Remark. Notice that unlike in Lemma 2 of Theorem 3.1 we do not claim here that (4.3) and (4.4) can be combined to produce a C-recursive function g such that $\varphi_{g(e)}$ is the characteristic function of $A^{[e]}$ for all $e > 0$, but merely that $A^{[e]} = \Phi_{g(e)}(C)$. The point is that even though $A^{[e]}$ is recursive for all $e > 0$, the proof of Lemma 3 above depends upon parameters p_e and $\hat{R}(e)$ which cannot be C-recursively computed uniformly in e, and the proof of

Lemma 4 clearly uses a C-oracle for each x. This subtle distinction is crucial in Theorem 4.4. This explains the difference between Exercises 4.5 and 4.6.

The above coding strategy for avoiding lower cones has many other applications such as the following uniform version of Corollary VII.3.7, which was nonuniform.

4.3 Theorem (Yates [1966b]).   *Given any r.e. set C such that $\emptyset <_T C <_T \emptyset'$, there exists an r.e. set A such that $A \mid_T C$. Furthermore, an index for A can be found uniformly from one for C.*

Proof. It suffices to meet for all $e > 0$ the requirements

$$N_e : C \neq \Phi_e(A) \quad \text{and} \quad P_e : A \neq \Phi_e(C).$$

Let $K = \{ e : e \in W_e \} \equiv_T \emptyset'$. In place of the hypothesis $D <_T C$ of Theorem 4.1 we use the hypothesis $C <_T K$ so that C and K play the former roles of D and C, respectively. Let $A_0 = \emptyset$. Given enumerations $\{ C_s \}_{s \in \omega}$, $\{ K_s \}_{s \in \omega}$, and $\{ A_t \}_{t \leq s}$ as usual define

$$\hat{l}^C(e, s) = \max\{ x : (\forall y < x) \, [A_s(y) = \hat{\Phi}_{e,s}(C_s; y)] \}.$$

For $t \leq s$ enumerate $\langle x, t, e \rangle$ in A_{s+1} just if $\langle x, t, e \rangle > \hat{R}(e, s)$, $x \in K_{s+1}$, and $x < \hat{l}^C(e, v)$, for all v, $t \leq v \leq s$. Let $A = \bigcup_s A_s$.

Fix e and assume by induction that, for all $i < e$, $C \neq \Phi_i(A)$, $A \neq \Phi_i(C)$, and $A^{[i]}$ is recursive. The proofs above establish (with C and K in place of D and C, respectively, in Lemmas 2 and 3): $C \neq \Phi_e(A)$; $A \neq \Phi_e(C)$; and $A^{[e]}$ is recursive.   ▯

It is natural to ask to what extent the previous theorem can be combined with the Jump Theorem 3.1 so that $A' \equiv_T S$ for some given S. Yates [1969] proved that this could be done for $S \equiv_T \emptyset''$ while still preserving the uniformity of Theorem 4.3. Later Robinson [1971b, Corollary 3] extended the result to any S r.e. in $\emptyset'$ such that $\emptyset' \leq_T S$, but without the Yates uniformity. These results are given in Exercises 4.5 and 4.6.

We can also use the coding strategy to simultaneously combine the results of the Jump Theorem, the Density Theorem, and the Interpolation Theorem 2.7. Given r.e. sets $D <_T C$, suppose that $D <_T A <_T C$ and $S \equiv_T A'$. What can be said of S? Clearly it is necessary that $D' \leq_T S$ and that S be r.e. in C. The following theorem asserts that these conditions are also sufficient. It is one of the most important and widely used theorems in recursion theory.

4.4 Jump Interpolation Theorem (Robinson [1971b]). *Given r.e. sets* $D <_T C$ *and a set* S *r.e. in* C *such that* $D' \leq_T S$, *there exists an r.e. set* A *such that* $D <_T A <_T C$ *and* $A' \equiv_T S$.

Proof. Since S is r.e. in C (and hence $S \leq_1 C'$) we can use the method of the Interpolation Theorem 2.7 to find an r.e. set $B \leq_T C$ such that $B^{[0]} \equiv_T D$ and, for all $x > 0$, if $x \in S$ then $B^{[x]}$ is finite and if $x \notin S$ then $B^{[x]} = \omega^{[x]}$. The strategy for enumerating elements in $A^{[e]}$ is the same as that for the thickness requirement P_i of Exercise 2.7 (with B as above) if $e = 2i$ and for P_i of the Density Theorem 4.1 if $e = 2i + 1$.

Since $B \leq_T C$ we can compute (as in Lemma 4 of Theorem 4.1) a C-recursive function g such that, for all $e \geq 0$, $A^{[e]} = \Phi_{g(e)}(C)$. Since $D' \leq_T S$ we can also compute as in Theorem 3.1, Lemma 2, and Exercise 4.5 an S-recursive function $h(e)$ such that $\varphi_{h(e)}$ is the characteristic function of $A^{[e]}$. The rest follows as in the previous proofs.   ▯

4.5–4.10 Exercises

4.5 (Yates [1969]).   Given any r.e. set C such that $\emptyset <_T C <_T \emptyset'$ and any set S r.e. in $\emptyset'$ such that $C' \leq_T S$ there exists an r.e. set A such that $A' \equiv_T S$ and $A \mid_T C$. Furthermore, an index for A can be found uniformly in indices for C and S. *Hint.* The strategy for enumerating elements in $A^{[e]}$ is the same as that for P_i of Theorem 4.3 if $e = 2i$ and as that for P_i of Theorem 3.1 if $e = 2i + 1$ subject to the usual restraint function $\hat{R}(e, s)$. For the former proofs to suffice it remains only to show that there is an S-recursive function g such that $\varphi_{g(e)}$ is the characteristic function of $A^{[e]}$ for all $e \geq 0$. Refer to Theorem 3.1, Lemma 2, if e is odd, and to Theorem 4.3, Lemma 3 if e is even. (Note that the parameters $\hat{R}(e)$ and p_e in Lemma 3 are S-recursive as functions of e using (3.5) and $C' \leq_T S$.)

4.6 (Robinson [1971b]).   Given any r.e. set C such that $\emptyset <_T C <_T \emptyset'$ and a set S r.e. in $\emptyset'$ such that $\emptyset' \leq_T S$ there is an r.e. set A such that $A \mid_T C$ and $A' \equiv_T S$. *Hint.* Use the Sacks Splitting Theorem VII.3.2 to find disjoint r.e. sets D_0, D_1 such that (1) $K = D_0 \cup D_1$, (2) $C \nleq_T D_i$ for $i = 0, 1$, and (3) $D_i' \equiv_T \emptyset'$ for $i = 0, 1$. Show that one of the sets, say D_0, is incomparable with C and apply Remark 3.2 with D_0 in place of D.

4.7 (Robinson [1971a, Corollary 3]).   Suppose D and C are r.e., $D <_T C$, $\{f_i\}_{i \in \omega} \leq_T \emptyset'$, $\{g_i\}_{i \in \omega} \leq_T C$, and for all i, $f_i \nleq_T D$ and $C \nleq_T g_i$. Prove there is an r.e. set A such that $D <_T A <_T C$ and for all i, $f_i \nleq_T A$ and $A \nleq_T g_i$. (The notation $\{g_i\}_{i \in \omega} \leq_T C$ means that there is a function $h \leq_T C$ such that $\lambda x[h(i, x)] = g_i$.) *Hint.* First consider a single f and g.

4.8 (Robinson [1971a]).   Assume $D <_T C$. A finite sequence $\{A_1, \ldots, A_n\}$ is *independent relative to* D *and* C if $D \oplus (\oplus\{A_i\}_{i \leq n}) <_T C$ and

$A_i \not\leq_T D \oplus (\oplus\{\, A_j : j \neq i \,\})$. Use Exercise 4.7 and the method of avoiding upper cones to show that if $D <_T C$ are r.e. sets and $\{\, A_i \,\}_{i \leq n}$ is a finite sequence of r.e. sets which is independent relative to D and C then there is an r.e. set A_{n+1} such that $\{\, A_1, \ldots, A_n, A_{n+1} \,\}$ is independent relative to D and C. Note that we need more than Exercise 4.7 to show for example that $A_i \neq \{\, e \,\}^{\oplus\{\, A_j \,\}_{j \neq i}}$.

4.9 (Robinson [1971a]).   Given r.e. sets $D <_T C$ show that there is an infinite r.e. sequence of r.e. sets $\{\, A_n \,\}_{n \in \omega}$ such that for all n, $D <_T A_n <_T C$ and $A_n \not\leq_T \oplus\{\, A_m \,\}_{m \neq n}$. *Hint.* First repeatedly apply Exercise 4.8 to obtain an r.e. sequence $\{\, B_n \,\}_{n \in \omega}$ of r.e. sets uniformly recursive in C and such that for each n, $\{\, B_i \,\}_{i \leq n}$ is independent relative to D and C. Let $\widetilde{B} = \oplus_n B_n$. Use the method of Lemma 2.3 to construct an r.e. set $\widetilde{A} \subseteq \widetilde{B}$, $\widetilde{A} \leq_T \widetilde{B}$, meeting the requirements $P_e : \widetilde{A}^{[e]} =^* \widetilde{B}^{[e]}$, and $N_{\langle e,i \rangle} : \widetilde{B}^{[i]} \neq \Phi_e(\oplus_{j \neq i} \widetilde{A}^{[j]})$. Let $A_n = \widetilde{A}^{[n]}$. Consider the simultaneous enumeration of *all* the r.e. sets $\{\, A_i \,\}_{i \in \omega}$, $\{\, B_i \,\}_{i \in \omega}$ and consider the true stages for this combined enumeration.

4.10 (Robinson [1971a, Corollary 7]).   Prove that if **d** and **c** are r.e. degrees and $\mathbf{d} < \mathbf{c}$ then any countable partially ordered set can be embedded in the r.e. degrees between **d** and **c**, with joins preserved. *Hint.* Use Exercise 4.9 and the method of Exercises VI.1.7 and VII.2.2. (Robinson [1971a] proved that if **d** is also low then the embedding preserves greatest and least elements whenever they exist (see Notes XI.3.7).)

5.* The Pinball Machine Model for Infinite Injury

The pinball machine model is an alternative method for dealing with the second obstacle (as described in §1) arising in infinite injury constructions. It is slightly more cumbersome but more versatile than the "true stages" method above. It can be used to give the results on minimal pairs in Chapter IX and indeed is essential in understanding certain more complicated constructions of r.e. degrees. Furthermore, this method is necessary in proofs using certain constructions involving more than one r.e. set where the "true stages" method of §1 to §4 will not suffice. See Exercise XII.2.13.

   The crucial point of the pinball machine model is that we allow followers of positive requirements to pass by the negative requirements of stronger priority one by one instead of waiting for the restraint functions to drop back simultaneously as in §1. (A similar approach was used in the original proofs by Sacks [1963a], Yates [1966] and others, but the illuminating pinball machine model was first suggested by Lerman [1973].) We illustrate this method by giving an alternative proof of the Thickness Lemma 1.1. Let

B, C, $\{B_s\}_{s\in\omega}$, $\{C_s\}_{s\in\omega}$, and $\hat{r}(e,s)$ be as in §1. As before we have the requirements $P_e : A^{[e]} =^* B^{[e]}$, and $N_e : C \neq \Phi_e(A)$.

The pinball machine is pictured as having a *hole* H_e, corresponding to each positive requirement P_e, and a *gate* G_e, corresponding to each negative requirement N_e. These are arranged alternately in order of priority, $G_0, H_0, G_1, H_1, \ldots$, as shown in Diagram 5.1. The gate G_{-1}, corresponding to the pocket at the bottom of the machine, is added for the sake of convenience.

Informally, if x is enumerated in $B^{[e]}$ at stage $s+1$ then x drops from hole H_e to the *surface of the machine* (called the *board*), x moves in the direction of the arrows (called *down*) and x stops at the first gate G_i, $i \leq e$, such that $x \leq \hat{r}(i,s)$. If at a later stage $t+1$ we have $\hat{r}(i,t) < x$ then x is *released by gate* G_i and x passes to the next gate G_j, $j < i$, such that $x \leq \hat{r}(j,t)$. Element x continues in this fashion until x is either permanently held at some gate or x reaches the final gate G_{-1}, at which point x is enumerated in A. Let $G_{e,s}$ denote the residents of gate G_e at the end of stage s and $G_{e,\infty}$ the permanent residents of gate G_e. Let $G_s^{\leq e} = \bigcup_{i<e} G_{i,s}$ and $G_\infty^{\leq e} = \bigcup_{i<e} G_{i,\infty}$.

Just as in §1, the use of $\hat{r}(e,s)$ (instead of $r(e,s)$) guarantees that if $C \neq \Phi_e(A)$ then $\liminf_s \hat{r}(e,s) < \infty$; hence $|G_{e,\infty}| < \infty$ and P_e is satisfied. Furthermore, N_e is met because the P_i, $i < e$, cause only a recursive set of injuries to N_e, and because $|G_\infty^{\leq e}| < \infty$. A formal proof follows.

Alternative proof of the Thickness Lemma 1.1. Fix recursive sequences $\{B_s\}_{s\in\omega}$ and $\{C_s\}_{s\in\omega}$ such that $B = \bigcup_s B_s$ and $C = \bigcup_s C_s$. Let $A_0 = \emptyset$. Given $\{A_t\}_{t\leq s}$ we define $\hat{l}(e,s)$, and $\hat{r}(e,s)$ as in §1.

Stage $s+1$.

Step 1. For each e and $x \in B_{s+1}^{[e]} - B_s^{[e]}$, release x from hole H_e and place x at gate G_e.

Step 2. For each x now at gate G_e (including those x newly placed there) if $\hat{r}(e,s) < x$ then G_e *releases* x. Find the largest $i < e$ such that $x \leq \hat{r}(i,s)$ and place x at G_i. We arbitrarily define $r(-1,s) = \infty$ so that i exists. If x is at G_{-1} enumerate x in A_{s+1}.

Lemma. *For all e,*

 (i) $C \neq \Phi_e(A)$,
 (ii) $\liminf_s \hat{r}(e,s) < \infty$, *and hence* $|G_{e,\infty}| < \infty$,
 (iii) $A^{[e]} =^* B^{[e]}$.

Proof. Fix e and assume the lemma for all $j < e$. Then by (ii), $|G_\infty^{\leq e}| < \infty$, and by (iii) $A^{[<e]} =^* B^{[<e]}$, which is recursive since B is piecewise recursive.

To prove part (i) assume for a contradiction that $C = \Phi_e(A)$. Then $\lim_s \hat{l}(e,s) = \infty$. To recursively compute $C(p)$ find the least stage s such that: (1) $\hat{l}(e,s) > p$; and (2) for all $z < u =_{\text{dfn}} \max\{\hat{u}(A_s; e, x, s) : x \leq p\}$

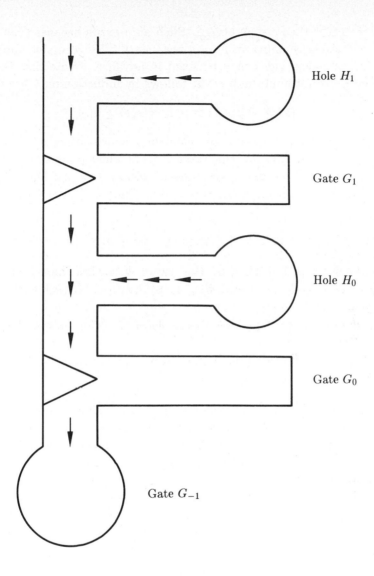

Diagram 5.1. Pinball Machine

if $z \in A^{[<e]} \cup G_s^{\le e}$ then $z \in A_s \cup G_\infty^{\le e}$. Such an s exists because $C = \Phi_e(A)$. We can recursively compute such an s because $A^{[<e]}$ is recursive and $G_\infty^{\le e}$ is finite so its canonical index may be fixed in advance. (Note that therefore this procedure is not uniform in e.) It follows by induction on $t \ge s$ that

$$(\forall t \ge s)\, [\hat{l}(e,t) > p \;\&\; \hat{r}(e,t) \ge u].$$

Then $C(p) = \Phi_{e,s}(A_s; p)$ because any element x which enters A after stage s must first be released by G_e at some stage $t \ge s$ at which point $x > \hat{r}(e,t) \ge u$. Thus, C is recursive contrary to hypothesis. Hence, $C \ne \Phi_e(A)$.

To prove (ii) let $p = \mu x[\Phi_e(A; x) \ne C(x)]$. Choose s' sufficiently large so that, for all $s \ge s'$,

$$(\forall x < p)\, [\widehat{\Phi}_{e,s}(A_s; x) = \Phi_e(A; x)] \;\&\; (\forall x \le p)\, [C_s(x) = C(x)].$$

Case 1. $\Phi_e(A; p) \uparrow$. Then by the use of $\widehat{\Phi}_e$ rather than Φ_e there are infinitely many stages s such that $\widehat{\Phi}_{e,s}(A_s; p)$ diverges. At each such stage s, $\hat{l}(e,s) = p$ and $\hat{r}(e,s) = \hat{r}(e,s')$.

Case 2. $\Phi_e(A; p) \downarrow$. Then for almost every s, $\hat{l}(e,s) = p$ and $\hat{r}(e,s) = \hat{r}(e,s')$.

Finally part (iii) follows immediately from (ii) since each $x \in B^{[e]}$ is either in $A^{[e]}$ or in $G_{i,\infty}$ for some $i \le e$. ▯

Notice that requirements of stronger priority are arranged *below* those of weaker priority in the pinball machine model. Thus, if $i < j$ we have used the terminology that requirement R_i has *stronger priority* than R_j (rather than "higher priority" as is customary in the literature) to avoid confusion in this and later arguments.

Chapter IX
The Minimal Pair Method and Embedding Lattices into the R.E. Degrees

Consider the language $L(\leq, \cup, 0, 1)$ for describing upper semi-lattices such as the r.e. degrees $\mathbf{R}$, where $\leq$ and $\cup$ are interpreted in the usual way, and 0 and 1 are constant symbols to be interpreted as least and greatest element, respectively. After seeing the Sacks Density Theorem VIII.4.1, Shoenfield [1965] tried to describe $\mathbf{R}$ algebraically by formulating a conjecture which implies that $\mathbf{R}$ is a dense structure as an upper semi-lattice analogous to the rationals being a dense structure as a linearly ordered set. Shoenfield's conjecture asserts that if $\vec{\mathbf{a}} \in \mathbf{R}$ satisfies the diagram $D(\vec{x})$, and if $D_1(\vec{x}, y)$ is any consistent diagram in $L(\leq, \cup, \mathbf{0}, \mathbf{0}')$ extending D, then there exists $\mathbf{b} \in \mathbf{R}$ such that $D_1(\vec{\mathbf{a}}, \mathbf{b})$. In addition to the Sacks Density Theorem, Shoenfield listed two consequences of his conjecture:

(0.1)   If $\mathbf{a}, \mathbf{b} \in \mathbf{R}$ are incomparable then they have no infimum (greatest lower bound) in $\mathbf{R}$;

(0.2)   Given r.e. degrees $\mathbf{0} < \mathbf{b} < \mathbf{a}$ there exists an r.e. degree $\mathbf{c} < \mathbf{a}$ such that $\mathbf{a} = \mathbf{b} \cup \mathbf{c}$, namely, that every such $\mathbf{b}$ can be nontrivially *cupped* to $\mathbf{a}$. (See Definition XIII.4.1.)

Both of these consequences are false (and hence Shoenfield's conjecture fails), but as he anticipated they have led to the development of two important new areas and methods of proof. In this chapter we study the first area which led to the embedding of lattices in $\mathbf{R}$. In Notes XIII.4.4 we discuss the second which led to the cupping and anti-cupping theorems and ultimately to the undecidability of the elementary theory of $\mathbf{R}$.

If P is any countable partially ordered set (poset), then the finite injury priority method enables us to embed P (by an order preserving map) into the r.e. degrees $\mathbf{R}$ (see Exercise VII.2.2), and the infinite injury priority method allows P even to be embedded in any interval $[\mathbf{a}, \mathbf{b}]$, for $\mathbf{a}, \mathbf{b} \in \mathbf{R}$ and $\mathbf{a} < \mathbf{b}$ (see Exercise VIII.4.9). If P happens also to be a lattice, then these embeddings naturally preserve supremums (sups) but not necessarily infimums (infs).

We now introduce a new method for embedding certain lattices into $\mathbf{R}$, preserving both infs and sups. Consider the four element Boolean algebra $\diamond$ which we call the *diamond*. In §1 we prove that the diamond can be embedded into **R** preserving not only sups and infs but the least element as well. In §2 we replace the diamond by an arbitrary countable distributive lattice. On the other hand, in §3 we prove the surprising fact that not even the diamond can be embedded as a lattice in **R** by a map preserving both least and greatest elements. A corollary is that the r.e. degrees are not closed under infs and thus fail to form a lattice. In §4 and §5 we consider nonbranching and noncappable degrees, whose properties are exactly the opposite of those of the degrees constructed in §1 and §2.

The embedding technique of §1 (called the *minimal pair method* because of Definition 1.1) is also one where some requirements are infinitary, but it is quite different from the method of Chapter VIII. The version we present here (due essentially to Lachlan [1973]) uses special stages analogous to the nondeficiency stages (i.e., the true stages) of Chapter VIII, so that the negative restraints drop back simultaneously, and the proof closely resembles a finite injury argument. An alternative proof in the style of Chapter VIII §5 will be given in Exercise 1.7.

1. Minimal Pairs and Embedding the Diamond Lattice

1.1 Definition.   Recursively enumerable degrees **a** and **b** form a *minimal pair* if **a**, **b** > **0** and
$$(\forall \text{ r.e. } \mathbf{c}) \, [\mathbf{c} \le \mathbf{a} \,\, \& \,\, \mathbf{c} \le \mathbf{b} \implies \mathbf{c} = \mathbf{0}].$$

1.2 Theorem (Lachlan [1966b]—Yates [1966a]).   *There exists a minimal pair of r.e. degrees* **a** *and* **b**.

1.3 Corollary.   *The diamond lattice can be embedded in* **R** *preserving sups, infs, and least element.*

Proof. Let $\mathbf{c} = \mathbf{a} \cup \mathbf{b}$ in Theorem 1.2. Then $\{\mathbf{0}, \mathbf{a}, \mathbf{b}, \mathbf{c}\}$ embeds the diamond lattice.   ▯

Proof of Theorem 1.2. If suffices to construct r.e. sets A and B satisfying for all e, i, j the requirements
$$P_{2e} : A \ne \{e\},$$
$$P_{2e+1} : B \ne \{e\},$$
and
$$N_{\langle i,j \rangle} : \{i\}^A = \{j\}^B = f \text{ total} \implies f \text{ is recursive.}$$

The following remark allows us to simplify the form of the negative requirements.

1.4 Remark (Posner). *To satisfy all $N_{\langle i,j \rangle}$, $i,j \in \omega$, it suffices to satisfy for all e the requirement*

$$N'_e : \{e\}^A = \{e\}^B = f \text{ total} \implies f \text{ is recursive.}$$

Proof. The requirements $\{N'_e\}_{e \in \omega}$ and $\{P_e\}_{e \in \omega}$ clearly imply that $A \neq B$, say, $n_0 \in A - B$. For each i and j there is an index e such that

$$\{e\}^X(x) = \begin{cases} \{i\}^X(x) & \text{if } n_0 \in X, \\ \{j\}^X(x) & \text{otherwise.} \end{cases}$$

The remark follows immediately. ▯

From now on we shall replace all occurrences of negative requirements similar to $N_{\langle i,j \rangle}$ by equivalent requirements N'_e, and we shall write the latter as N_e.

Given $\{A_t\}_{t \leq s}$ and $\{B_t\}_{t \leq s}$ we define as usual the functions

(length function) $l(e,s) = \max\{x : (\forall y < x) [\{e\}_s^{A_s}(y){\downarrow} = \{e\}_s^{B_s}(y){\downarrow}]\}$,

(maximum length function) $m(e,s) = \max\{l(e,t) : t \leq s\}$.

A stage s is called 0-*expansionary* if $s = 0$ or if $l(0,s) > m(0, s-1)$. Define the *restraint function*

$$r(0,s) = \begin{cases} 0 & \text{if } s \text{ is 0-expansionary,} \\ \text{the greatest 0-expansionary stage } t < s, & \text{otherwise.} \end{cases}$$

(Notice that we can define the restraint function in terms of a stage s rather than an element z used in a computation at stage s since $z < u(A_s; e, y, s) < s$ by our convention in line III (1.1) following Definition III.1.7.)

The strategy σ_0 for meeting a *single* negative requirement N_0 is to allow x to enter A or B at stage $s+1$ only if $x \geq r(0,s)$, and even then at most *one* of the sets A, B receives an element x at any stage. Thus, if x destroys one of the computations $\{0\}_s^{A_s}(p) = q$ or $\{0\}_s^{B_s}(p) = q$ for some $p < l(0,s)$, say $\{0\}_s^{A_s}(p)$, then the other computation, $\{0\}_s^{B_s}(p) = q$, will be preserved until the A-computation is restored and both sides output q again. In this way, if $\{0\}^A = \{0\}^B = f$ is a total function, then f is recursive. (To compute $f(p)$ we find the least s such that $p < l(0,s)$ and set $f(p) = \{0\}_s^{A_s}(p)$.) Furthermore, $\liminf_s r(0,s) < \infty$, since $\liminf_s r(0,s) = 0$ unless there is a largest 0-expansionary stage t in which case $r(0,s) = t$ for all $s \geq t$.

This fundamental strategy of having one side or the other hold the computation at all times is applied to the other negative requirements N_e, $e > 0$, but with some crucial modifications to force the negative restraints to drop back simultaneously, thus creating "windows" through the restraints, as in Chapter VIII.

For example, to drop back simultaneously with N_0, N_1 must guess the value of $k = \liminf_s r(0, s)$. Thus, N_1 must simultaneously play infinitely many strategies σ_1^k, $k \in \omega$, one for each possible value of k. (See the explanation in Chapter XIV §3.1 using trees to define $r(e, s)$.) Each strategy σ_1^k is played like σ_0 but with $S^k = \{\, s : r(0, s) = k \,\}$ in place of ω as the set of stages during which it is active and on which its length functions l and m are defined. This allows σ_1^k to open its window more often since its length functions ignore the stages in $\omega - S^k$. Strategy σ_1^k still succeeds, providing any restraint it imposes is maintained during intermediate stages $s \notin S^k$ while σ_1^k is dormant. Thus, at stage s if $k = r(0, s)$, we play σ_1^k, maintain the restraints previously imposed by the dormant σ_1^i, $i < k$, and discard any restraints imposed by σ_1^j, $j > k$. Thus, if $k = \liminf_s r(0, s)$, then: (1) strategy σ_1^k succeeds in meeting N_1; (2) the strategies σ_1^i, $i < k$, impose finitely much restraint over the whole construction; and (3) the strategies σ_1^j, $j > k$, drop all restraint at each stage $s \in S^k$. Thus, the entire restraint $r(1, s)$ imposed by N_0 and N_1 *together* has $\liminf_s r(1, s) < \infty$.

Construction of A and B.
   Stage $s = 0$. Do nothing.
   Stage $s + 1$. Given A_s and B_s, define the restraint function $r(e, s)$ for N_e by induction on e as follows. Define $r(0, s)$ as above. A stage s is $(e + 1)$-*expansionary* if $s = 0$ or

$$(\forall t < s) \, [r(e, t) = r(e, s) \implies l(e + 1, t) < l(e + 1, s)].$$

Let $r(e + 1, s)$ be the maximum of

   (i) $r(e, s)$,
   (ii) those $t < s$ such that $r(e, t) < r(e, s)$, and
   (iii) those $t < s$ such that $r(e, t) = r(e, s)$ and t is $(e + 1)$-expansionary, if s is not $(e + 1)$-expansionary.

   (In the spirit of Chapter XIV §3 we could view this construction as being done on a tree. Part (i) is for those requirements N_i, $i \le e$; part (ii) for those nodes k' to the left of $k = r(e, s)$, namely those strategies which assume that $r(e) = k' < k$; and part (iii) is for the non-$(e+1)$-expansionary stages for the sake of node k, namely for the sake of the strategy which assumes $r(e) = k$.)
   Requirement P_{2e} *requires attention* if

(1.1) $\neg(\exists y) \, [y \in A_s \ \& \ \{e\}_s(y) \!\downarrow\, = 0]$,

and

(1.2) $(\exists x) \, [\{e\}_s(x) \!\downarrow\, = 0 \ \& \ r(2e, s) < x \ \& \ x \in \omega^{[2e]}]$,

and likewise for P_{2e+1} with B in place of A and $2e + 1$ in place of $2e$. Choose the highest priority requirement P_e which requires attention and the least x

satisfying (1.2) for that e. Enumerate x in A if e is even (in B if e is odd). (Note that $A \cap B = \emptyset$, since any witness x for P_e lies in $\omega^{[e]}$.)

Lemma 1. $(\forall e) [\liminf_s r(e, s) < \infty]$.

Proof. The proof is by induction on e. First let $e = 0$. If there are infinitely many 0-expansionary stages then by definition there are infinitely many s such that $r(0, s) = 0$, and hence $\liminf_s r(0, s) = 0$. If there are only finitely many 0-expansionary stages, and the largest is, say v, then $\lim_s r(0, s) = v$. For the inductive step, fix e and assume $k = \liminf_s r(e, s)$. Then there are only finitely many stages s such that $r(e, s) < k$. Let t be the largest such. Let $S = \{ s : r(e, s) = k \}$. Either there are infinitely many $(e + 1)$-expansionary stages in S, in which case $\liminf_s r(e + 1, s) = \max\{ t, k \}$, or else there is a largest $(e + 1)$-expansionary stage $v \in S$, in which case $\liminf_s r(e + 1, s) = \max\{ t, k, v \}$.

Lemma 2. *Every positive requirement P_e is satisfied and acts at most once.*

Proof. Consider requirement P_{2e} (since P_{2e+1} is similar). If $A = \{ e \}$, then P_{2e} is never met so $A \cap \omega^{[2e]} = \emptyset$. Hence, there exists $x \in \overline{A} \cap \omega^{[2e]}$ such that $\{ e \}(x) \downarrow = 0$, and $x > \liminf_s r(2e, s)$. Some such x is eventually enumerated into A satisfying P_{2e}. Clearly, P_{2e} acts at most once by (1.1).

Lemma 3. $(\forall e)$ [Requirement N_e is met].

Proof. Fix e and let $k = \liminf_s r(e - 1, s)$, and $S = \{ s : r(e - 1, s) = k \}$. (If $e = 0$ let $S = \omega$ and $k = 0$.) Choose s' such that no P_i, $i < e$, acts after stage s' and $r(e - 1, s) \geq k$ for all $s \geq s'$. Now assume that $\{ e \}^A = \{ e \}^B = f$ is a total function. To recursively compute $f(p)$, $p \in \omega$, find an e-expansionary stage $s'' \in S$, $s'' > s'$, such that $l(e, s'') > p$. There are infinitely many e-expansionary stages in S because $\{ e \}^A = \{ e \}^B$. Let $q = \{ e \}_{s''}^{A_{s''}}(p) = \{ e \}_{s''}^{B_{s''}}(p)$. We shall prove by induction on t that for all $t \geq s''$ either

(1) $\{ e \}_t^{A_t}(p) = q$, or
(2) $\{ e \}_t^{B_t}(p) = q$,

and hence that $f(p) = q$. Let $s_1 < s_2 < \cdots$ be the e-expansionary stages in S which are greater than or equal to s''. Both (1) and (2) hold for $t = s_1$ since $s_1 = s''$. Fix n and assume by induction that both (1) and (2) hold for $t = s_n$ (and that either (1) or (2) holds at every t, $s_1 \leq t \leq s_n$). Now at stage $s_n + 1$ at most one element enters A or B so at most one of the computations (1), (2) for $t = s_n$ is destroyed, and the other, say (1), holds for $t = s_n + 1$. Now by choice of s'',

$$(\forall t) [[s_n < t < s_{n+1} \ \& \ t \notin S] \implies r(e - 1, t) > k],$$

and hence
$$(\forall t)\,[[s_n < t < s_{n+1}] \implies r(e,t) \geq s_n],$$

by clauses (ii) and (iii) in the definition of $r(e,t)$, namely by (ii) for $t \notin S$, and by (iii) for $t \in S$. (If $e = 0$, only the latter clause applies.) Hence, for all t, $s_n \leq t \leq s_{n+1}$,

$$A_t \restriction s_n = A_{s_n} \restriction s_n \ \& \ \{e\}_t^{A_t}(p) = \{e\}_{s_n}^{A_{s_n}}(p) = q,$$

so (1) holds for $t = s_{n+1}$. But s_{n+1} is e-expansionary, so the A and B computations must converge and agree. Thus, (2) also holds for $t = s_{n+1}$. □

See Chapter XIV §3 for a tree version of this proof which sheds light on the interplay between various strategies working on the same requirement and clarifies the definition above of $r(e,s)$.

1.5–1.7 Exercises

1.5. Construct a minimal pair of r.e. degrees $\mathbf{a}$, $\mathbf{b}$ such that $\mathbf{a}$ and $\mathbf{b}$ and even $\mathbf{a} \cup \mathbf{b}$ are low. *Hint.* Mix the minimal pair construction with the negative lowness requirements N_e of Theorem VII.1.1.)

1.6. Construct an r.e. sequence of r.e. degrees $\{\mathbf{a}_i\}_{i \in \omega}$, such that $\mathbf{a}_i$, $\mathbf{a}_j$ is a minimal pair for each i and j such that $i \neq j$.

1.7 (Lerman [1973]). Give a proof of Theorem 1.2, using the pinball machine model of VIII §5. To meet the requirements P_e and N_e of §1, let P_e correspond to hole H_e and N_e to gate G_e. Define $r(0,s)$ as above and define $r(e,s)$ exactly like $r(0,s)$ but with e in place of 0. Note that $r(e) = \liminf_s r(e,s) < \infty$.

Stage $s = 0$. For each e place all elements $x \in \omega^{[e]}$ above hole H_e. They are now *unrealized*.

Stage $s + 1$. A follower x *requires attention* if

(1) x lies above some hole H_e, x is now unrealized, $x > y$ for every previously realized follower y, and $\{e\}_s(x)\!\downarrow = 0$; or

(2) x lies at some gate G_e and $r(e,s) < x$.

Case 1. If some follower x requires attention, choose the least such x, and cancel (forever) all realized followers $y > x$. If (1) holds then that x becomes *realized*. In either case, choose the greatest i, $i \leq e$, such that $x \leq r(i,s)$ and move x to gate G_i. (We arbitrarily define $r(-1,s) = \infty$ so that i exists.) If $i = -1$, enumerate x in A if x follows P_j for some even j, and in B otherwise, thereby satisfying P_j.

Case 2. If no follower requires attention, then choose the least j such that P_j is unsatisfied and has no unrealized follower, and appoint as (unrealized) follower of P_j the least $y \in \omega^{[j]}$ greater than any previously appointed follower. Cancel all followers of any requirement P_k, $k > j$.

This ends the construction of A and B.

Now as in VIII §5 prove that: $G_{e,\infty}$, the set of permanent residents of gate G_e, is finite; that each P_j is satisfied and emits finitely many followers; and that each N_e is satisfied. At any given time we think of the realized but uncancelled followers as *active* (i.e., on the *surface of the machine*). Notice that if at a given time followers x and y are active, x follows P_i, and y follows P_j, then $x < y$ iff $i < j$ by the procedures for realizing and cancelling followers. Hence, we can prove by induction on i that for every i at most finitely often does some follower of P_i receive attention.

2.* Embedding Distributive Lattices

Using a fairly easy modification of the preceding method, we shall now replace the diamond lattice of Theorem 1.2 by *any* countable distributive lattice. Since any countable distributive lattice can be embedded in the countable atomless Boolean algebra, it suffices to prove the following.

2.1 Theorem (Lachlan-Lerman-Thomason).   *There is an embedding of the countable atomless Boolean algebra $\mathcal{B}$ into the r.e. degrees $\mathbf{R}$ which preserves sups, infs, and least element.*

Proof. Let $\{\alpha_i\}_{i\in\omega}$ be any uniformly recursive sequence of recursive sets (i.e., $\{\langle x,i\rangle : x \in \alpha_i\}$ is a recursive relation) which forms an atomless Boolean algebra $\mathcal{B}$ under $\cup$, $\cap$, and complementation, contains ω, and has $\emptyset$ as its only finite member. We shall construct r.e. sets A_i, $i \in \omega$, and define $A_\alpha = \{\langle i,x\rangle : x \in A_i \ \& \ i \in \alpha\}$ for $\alpha \in \mathcal{B}$. Notice that we immediately have

$$(2.1) \qquad \deg(A_{\alpha\cup\beta}) = \deg(A_\alpha) \cup \deg(A_\beta),$$

$$(2.2) \qquad \alpha \subseteq \beta \implies \deg(A_\alpha) \leq \deg(A_\beta),$$

and

$$(2.3) \qquad \deg(A_{\alpha\cap\beta}) \leq \deg(A_\alpha), \deg(A_\beta).$$

We shall meet for all i, j, α, β the requirements:

$$P_{\langle i,j\rangle} : A_i \neq \overline{W}_j,$$
$$N_{(\alpha,\beta,j)} : \{j\}^{A_\alpha} = \{j\}^{A_\beta} = f \text{ total} \implies f \leq_{\mathrm{T}} A_{\alpha\cap\beta}.$$

These requirements ensure that

$$(2.4) \qquad \deg(A_{\alpha\cap\beta}) = \deg(A_\alpha) \cap \deg(A_\beta),$$

and

(2.5) $$\deg(A_\alpha) \leq \deg(A_\beta) \implies \alpha \subseteq \beta.$$

Note that (2.1)–(2.5) guarantee that the map $\alpha \to \deg(A_\alpha)$ is the desired embedding, and (2.5) guarantees that the map is 1:1. (To see that the negative requirements ensure (2.5) suppose: (1) $\deg(A_\alpha) \leq \deg(A_\beta)$; but (2) $\alpha \nsubseteq \beta$, say $i \in \alpha - \beta$. Then $\deg(A_i) \leq \deg(A_\alpha) \leq \deg(A_\beta)$ by the definition of A_α and (1), but $\deg(A_i) \leq \deg(A_{\bar\beta})$ by the definition of A_α and (2). Hence, $\deg(A_i) \leq \deg(A_{\beta \cap \bar\beta}) = \deg(A_\emptyset) = \mathbf{0}$ by (2.4), contradicting that A_i is nonrecursive.)

The strategy for meeting the negative requirements $N_{(\alpha,\beta,j)}$ begins as before. Denoting $N_{(\alpha,\beta,j)}$ by N_e, where $\alpha = \alpha_{i_1}$, $\beta = \alpha_{i_2}$, and $e = \langle i_1, i_2, j \rangle$, we define the restraint function $r(e,s)$ and e-expansionary stages by induction on e exactly as in §1 but with A_α, B_α, and j in place of A, B and e used in §1. However, new difficulties in proving Lemma 3 (that N_e is satisfied) require greater care in enumerating elements for the positive requirements. To meet requirement $P_{(i,j)}$ we shall appoint *followers* $x \in \omega$ so called because the eventual enumeration of x in A_i will satisfy $P_{(i,j)}$ (although x may be cancelled before this happens). If x is a follower of P_i and y a follower of P_j then we say x has *stronger priority* than y (written $x \prec y$) if $i < j$, or if $i = j$ and x was appointed before y. We shall arrange for all uncancelled followers x and y existing at stage s that $x \prec y$ iff $x < y$.

Construction of A_i, $i \in \omega$:

   Stage $s = 0$. Do nothing.

   Stage $s+1$. Requirement $P_{(i,j)}$ is *satisfied* if $A_{i,s} \cap W_{j,s} \neq \emptyset$. Requirement $P_{(i,j)}$ *requires attention* if $P_{(i,j)}$ is not satisfied and either:

(2.6) $x \in W_{j,s}$ and $x > r(\langle i, j \rangle, s)$ for some uncancelled follower x of $P_{(i,j)}$;

or

(2.7) $x \in W_{j,s}$ for every uncancelled follower x of $P_{(i,j)}$.

   Let $P_{(i,j)}$ be the strongest priority requirement which requires attention. We say $P_{(i,j)}$ *receives attention* (*acts*) at stage $s+1$. If (2.6) holds for some x, enumerate the least such x in A_i. If (2.6) fails and (2.7) holds, then appoint $x = s+1$ as a follower of $P_{(i,j)}$. In either case cancel all followers y of weaker priority than x (i.e., $x \prec y$). (If no $P_{(i,j)}$ requires attention, then do nothing.)

Lemma 1. $(\forall e)\, [\liminf_s r(e,s) < \infty]$.

Proof. Exactly as in §1, Lemma 1.

Lemma 2. $(\forall e)$ [*Requirement P_e receives attention at most finitely often and is satisfied*].

Proof. Fix e and choose s_0 such that for no $e' < e$ does $P_{e'}$ receive attention after stage s_0. Let $k = \liminf_s r(e, s)$ by Lemma 1. Let $e = \langle i, j \rangle$. Now if $P_{\langle i,j \rangle}$ receives attention infinitely often then some follower $x > k$ is appointed to follow $P_{\langle i,j \rangle}$ and x is never cancelled. Furthermore, $x \in W_j$ by (2.7). Hence, there is a stage $t + 1 > s_0$ such that $r(e, t) < x$ and $x \in W_{j,t}$. Now x or some smaller follower of P_e is enumerated in A at stage $t + 1$, P_e is met, and P_e never again requires attention. Therefore, P_e receives attention at most finitely often.

Finally, requirement $P_{\langle i,j \rangle}$ is satisfied because otherwise $\overline{A}_i = W_j$, $x \in W_j$ for every uncancelled follower x of $P_{\langle i,j \rangle}$, and $P_{\langle i,j \rangle}$ receives attention infinitely often under (2.7).

Lemma 3. $(\forall \alpha) (\forall \beta) (\forall j)$ [*Requirement $N_{(\alpha,\beta,j)}$ is met*].

Proof. Fix $N_e = N_{(\alpha,\beta,j)}$. Choose k, S and stage s' as in Lemma 3 of §1. Assume that $\{j\}^{A_\alpha} = \{j\}^{A_\beta} = f$ is a total function. A computation $\{j\}_s^{A_{\gamma,s}}(x)$ is A_δ-*correct* if $A_{\delta,s} \upharpoonright u = A_\delta \upharpoonright u$ where

$$u = u(A_{\gamma,s}; j, x, s).$$

To $A_{\alpha \cap \beta}$-recursively compute $f(p)$ find an e-expansionary stage $s \in S$, $s > s'$, such that $l(e, s) > p$ and both computations $\{j\}_s^{A_{\alpha,s}}(p) = q$ and $\{j\}_s^{A_{\beta,s}}(p) = q$ are $A_{\alpha \cap \beta}$-correct. We shall show by induction on t that for all $t \geq s$ we have either

(2.8) $\{j\}_t^{A_{\alpha,t}}(p) = q,$

or

(2.9) $\{j\}_t^{A_{\beta,t}}(p) = q,$

via an $A_{\alpha \cap \beta}$-correct computation. Now if x destroys either computation (2.8) or (2.9) by entering $A_\alpha \cup A_\beta$ at stage $t + 1$, then t must have been e-expansionary and $t \in S$ (as in Lemma 3 of §1), so *both* computations existed at the end of stage t. By inductive hypothesis, at least one computation, say (2.8), is $A_{\alpha \cap \beta}$-correct. Suppose x is enumerated in A_α at stage $t+1$ destroying this computation. Then x cancels at stage $t+1$ all followers y such that $x < y$ (since these are exactly those followers y such that $x \prec y$). Furthermore, $z > t + 1 \geq u_\beta =_{\text{dfn}} u(A_{\beta,t}; j, t, p)$ for any follower z later appointed. But $x < u_\alpha =_{\text{dfn}} u(A_{\alpha,t}; j, t, p)$ since the A_α-computation is destroyed by x. Also $A_{\alpha \cap \beta,t} \upharpoonright u_\alpha = A_{\alpha \cap \beta} \upharpoonright u_\alpha$ since the A_α-computation (2.8) was $A_{\alpha \cap \beta}$-correct. Hence $A_{\alpha \cap \beta,t} \upharpoonright (t+1) = A_{\alpha \cap \beta} \upharpoonright (t+1)$, and $u_\beta \leq t+1$ so the A_β-computation

(2.9) now becomes assured of remaining $A_{\alpha \cap \beta}$-correct. This is the crucial new element of this proof.

2.2–2.6 Exercises

2.2.   Prove that the embedding of Theorem 2.1 actually is an embedding of $\mathcal{B}$ into the m-degrees. (This can be made into an embedding into the 1-degrees by replacing $A_\emptyset$ by an infinite coinfinite recursive set.)

2.3.   An r.e. degree $\mathbf{c}$ is *branching* if there are r.e. degrees $\mathbf{a}, \mathbf{b} > \mathbf{c}$ such $\mathbf{c} = \mathbf{a} \cap \mathbf{b}$ and $\mathbf{c}$ is *nonbranching* otherwise. For example, Theorem 1.2 asserts that $\mathbf{0}$ is branching. Apply Theorem 2.1 to show that there is an r.e. branching degrees $\mathbf{c} > \mathbf{0}$. (For nonbranching degrees see §4.)

2.4.   (a) (Lachlan). Give a direct construction of a nonzero branching degree $\mathbf{c}$ with branches $\mathbf{a} = \deg(A \oplus C)$, $\mathbf{b} = \deg(B \oplus C)$ by constructing sets A, B, C to meet the requirements: $\{e\}^C \neq A$, $\{e\}^C \neq B$, $\{e\} \neq C$, and

$$\{e\}^{A \oplus C} = \{e\}^{B \oplus C} = f \text{ total} \implies f \leq_T C.$$

(b) (Fejer). Combine this with the permitting method of Chapter V §3 to show that for every r.e. degree $\mathbf{d} > \mathbf{0}$ there is a nonzero r.e. branching degree $\mathbf{c} < \mathbf{d}$.

2.5 (P. F. Cohen [1975]).   In the definition of branching replace Turing reducibility by *wtt*-reducibility of Exercise V.2.16 to get the concept of a *wtt*-branching degree. Prove that every r.e. *wtt*-degree $\mathbf{c} < \mathbf{0'}$ is *wtt*-branching. *Hint.* Given an r.e. set $C \in \mathbf{c}$ construct r.e. sets A, B whose degrees form the branches. Ensure that $\{e\}^C \neq A$ using the Sacks coding strategy of Theorem VIII.4.1.

2.6 (Lachlan [1972a]).   (a) Prove that if A, B, and C are r.e. sets such that $A \leq_{wtt} B \oplus C$, then A is the disjoint union of r.e. sets A_0, A_1 such that $A_0 \leq_{wtt} B$ and $A_1 \leq_{wtt} C$.

(b) Prove that the following nondistributive lattices M_5 and N_5 of Diagram 2.1 cannot be embedded in the r.e. *wtt*-degrees by maps preserving supremums and infimums. Any nondistributive lattice contains a copy of M_5 or N_5 (see Birkhoff [1967, pp. 13, 39]). Hence, no nondistributive lattice can be so embedded in the r.e. *wtt*-degrees. *Hint.* Use (a).

(c) Show that the proof of Theorem 2.1 establishes that every countable distributive lattice can be embedded in the *wtt*-degrees. Hence, (b) and (c) give a complete characterization of which lattices can be so embedded.

2.7 Notes. Thomason [1971] and independently Lerman (unpublished) proved that all finite distributive lattices could be embedded in **R**. Lachlan and independently Lerman extended the result to countable distributive lattices. The nondistributive case, however, was much more difficult. Lachlan [1972a] showed that the nondistributive lattices M_5 and N_5 of Diagram 2.1 *can* be so embedded. Robinson [1971a, p. 313] and Shoenfield [1975, p. 976] conjectured that *every* finite lattice could be so embedded. Lachlan and Soare [1980] refuted this by proving that the lattice S_8 of Diagram 2.1 cannot be so embedded. Ambos-Spies and Lerman [1986] obtained a further nonembeddability result. Other partial results are known but a general characterization of the embeddable lattices has not been found.

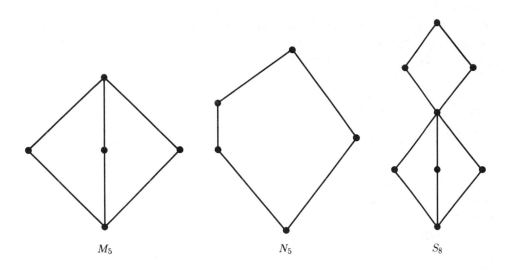

M_5 N_5 S_8

Diagram 2.1. Certain Nondistributive Finite Lattices

3. The Non-Diamond Theorem

One might expect to extend Theorem 2.1 by constructing lattice embeddings which preserve both greatest and least elements. The following surprising theorem shows that this is impossible even for the diamond lattice. However, this fact is true if we replace r.e. degrees by Δ_2^0 degrees (see Cooper [1972b] or Posner and Robinson [1981]).

3.1 Non-Diamond Theorem (Lachlan [1966b]). *If* **a** *and* **b** *are nonrecursive r.e. degrees such that* $\mathbf{a} \cup \mathbf{b} = \mathbf{0}'$, *then there is a nonrecursive r.e. degree* **c** *such that* $\mathbf{c} \leq \mathbf{a}$ *and* $\mathbf{c} \leq \mathbf{b}$.

Proof. Let A and B be r.e. sets in degrees **a** and **b**, respectively such that A is contained in the evens and B in the odds so that $A \cup B \equiv A \oplus B \equiv_T \emptyset'$. Let $\{a_s\}_{s\in\omega}$ and $\{b_s\}_{s\in\omega}$ be recursive enumerations of A and B. Let $A_s = \{a_t : t \leq s\}$ and $B_s = \{b_t : t \leq s\}$. We shall construct coinfinite r.e. sets $E, F_0, F_1, F_2, \ldots$ such that one of these sets has the desired degree **c**. We attempt to meet for each i and j the requirement

$$P_{i,j} : E \cap W_i \neq \emptyset \quad \text{or} \quad F_i \cap W_j \neq \emptyset.$$

As an aid in the construction, we shall also construct an r.e. set D. Since $\deg(A \cup B) = \mathbf{0}'$, we may assume (using the Recursion Theorem) that we can fix at the beginning an index e such that $D = \{e\}^{A\cup B}$. (Note that it would be sufficient to make D Δ_2^0 instead of r.e., and we do this in Exercise 3.8.) Let $u(k,s)$ denote the use function $u(A_s \cup B_s; e, k, s)$. (For $X = D$, E, or F_i, let X_s denote the elements in X by the end of stage s.)

In order to ensure that $E \leq_T A$ and $E \leq_T B$, we shall enumerate a number x in E at stage s only if numbers $\leq x$ are enumerated in *both* A and B. For example, to satisfy the requirement $R_i : W_i \cap E \neq \emptyset$, suppose $x \in W_{i,s}$, $a_s \leq x$, and $u < x$ for some $k \notin D_s$ such that $\{e\}_s^{A_s\cup B_s}(k) = 0$, and $u = u(k,s)$. We can now attack this requirement by enumerating the *attacker* k into D. Since $D = \{e\}^{A\cup B}$ this forces a number $z < u$ to be enumerated in either A or B. If z is enumerated in B, then the attack is successful and we enumerate x in E, satisfying R_i forever.

If there are infinitely many unsuccessful attacks on R_i, we must ensure that $F_i \cap W_j \neq \emptyset$ for all infinite W_j. Thus, we do not attack as above, unless there is $t < s$ such that $y \in W_{j,t}$, $b_t < y$, and $u < y$. Now after the attack, if z is enumerated in A, then we can enumerate y in F_i and argue that $F_i \leq_T A$ and $F_i \leq_T B$. Thus, we do not attack R_i directly but rather requirement $P_{i,j}$ via the attacker $k = \langle i, j \rangle$. An attack on $P_{i,j}$ succeeds in satisfying $P_{i,j}$ forever so that $P_{i,j}$ is never attacked again.

Construction.

Stage $s = 0$. Do nothing.

Stage $s + 1$. Requirement $P_{i,j}$ *requires attention* if $P_{i,j}$ is not yet satisfied (i.e., $E_s \cap W_{i,s} = \emptyset$ and $F_{i,s} \cap W_{j,s} = \emptyset$), and there is an ordered pair (x, y) satisfying:

(3.1) $$\{e\}_s^{A_s\cup B_s}(\langle i,j \rangle) = 0,$$

(3.2) $$x > 2i, \quad a_s, \quad u(\langle i,j \rangle, s),$$

(3.3) $$x \in W_{i,s},$$

and

(3.4) there is some $t < s$ such that:

    (a) $y > 2j, b_t, \quad u(\langle i,j \rangle, t),$

    (b) $y \in W_{j,t}$, and

    (c) $(A_s \cup B_s) \upharpoonright u(\langle i,j \rangle, t) = (A_t \cup B_t) \upharpoonright u(\langle i,j \rangle, t),$
       and $\{e\}_t^{A_t \cup B_t}(\langle i,j \rangle) = 0.$

(Note that condition (3.4)(c) implies that $u(\langle i,j \rangle, t) = u(\langle i,j \rangle, s)$.)

Choose the least $\langle i,j \rangle$ such that $P_{i,j}$ requires attention and then the least corresponding pair $\langle x,y \rangle$. Enumerate $\langle i,j \rangle$ into D. Enumerate A and B until the first number $z < u = u(\langle i,j \rangle, t) = u(\langle i,j \rangle, s)$ appears in $A \cup B$. (Namely, "speed up" the enumeration of A and B.) (If no such z appears, the construction never proceeds further. However, if e is the index obtained by the Recursion Theorem satisfying $D = \{e\}^{A \cup B}$, then z must appear.) If z appears in A, enumerate y in F_i. If z appears in B, enumerate x in E. We say that $P_{i,j}$ *receives attention*.

Lemma 1. *The sets E and F_i, $i \in \omega$, are coinfinite, $E \leq_T A$, $E \leq_T B$, and $F_i \leq_T A$.*

Proof. The sets are coinfinite by the conditions $x > 2i$, $y > 2j$ of (3.2) and (3.4)(a). The Turing reductions follow by the usual permitting method. For example, $E \leq_T A$ because

$$(\forall s)\,(\forall x)\, [A_s \upharpoonright x = A \upharpoonright x \implies E_{s+1} \upharpoonright x = E \upharpoonright x].$$

Exactly the same argument applies for $E \leq_T B$ and $F_i \leq_T A$.

Lemma 2. *If E is recursive, say $\overline{E} = W_i$, then*

    (i) F_i *is nonrecursive, and*

    (ii) $F_i \leq_T B$.

Proof of (i). If F_i is recursive, choose the least j such that $W_j = \overline{F}_i$. Choose s_0 such that no requirement $P_{m,n}$, with $\langle m,n \rangle < \langle i,j \rangle$, receives attention at any stage $s \geq s_0$. If $D(\langle i,j \rangle) = 1$, then $P_{i,j}$ received attention. Hence, we may assume $0 = D(\langle i,j \rangle) = \{e\}^{A \cup B}(\langle i,j \rangle)$. Choose $v \geq s_0$ such that for all $s \geq v$, $u(\langle i,j \rangle, s) = u(\langle i,j \rangle, v) > 0$, and $(A \cup B) \upharpoonright u = (A_v \cup B_v) \upharpoonright u$ where $u = u(\langle i,j \rangle, v)$. Now since B is nonrecursive, and, by Lemma 1, W_j is infinite, there exist $y > u$ and $t > v$ such that $y \in W_{j,t}$, and y is permitted by B at stage t, i.e., such that (3.4)(a), (b), and (c) hold for y and any $s > t$. But since A is nonrecursive, there exist infinitely many $x \in W_i$, satisfying (3.1), (3.2), and (3.3) for some $s > t$. Hence requirement $P_{i,j}$ receives attention, and either $E \cap W_i \neq \emptyset$ or $F_i \cap W_j \neq \emptyset$ contrary to hypothesis.

Proof of (ii). To determine whether $y \in F_i$, find the largest t such that $b_t \leq y$. Consider all j such that $y \in W_{j,t}$. For *each* such j find the least $s_j \geq t$ such that either:

$$(3.5) \qquad (A_{s_j} \cup B_{s_j}) \restriction u(\langle i,j \rangle, t) \neq (A_t \cup B_t) \restriction u(\langle i,j \rangle, t)$$

(in which case clause (3.4)(c) prevents any $P_{i,j}$ from later putting y into F_i), or else

$$(3.6) \qquad \text{requirement } P_{i,j} \text{ receives attention at stage } s_j.$$

In either case, $y \in F_i$ iff $y \in F_{i,s+1}$, where $s = \max\{s_j : y \in W_{j,t}\}$. Note that if (3.5) fails then (3.6) must hold, because as in part (i) there are infinitely many $x \in W_i$ satisfying (3.1), (3.2), and (3.3) and thus eligible to form a pair with y for $P_{i,j}$, so $P_{i,j}$ will receive attention. (Note that (3.4)(c) is used here because we need merely wait for the *first* stage when some $x \in W_{i,s}$ is ready to form a pair with y for an attack on $P_{i,j}$. It does so unless (3.4)(c) prevents this.) This argument to show $F_i \leq_T B$, say $F_i = \Psi^B$, is called *delayed permitting* since when $b_t \leq y$, we can discard the old computation $\Psi^B_{t-1}(y) \downarrow = 0$, but we *delay* redefining $\Psi^B_s(y)$ until the much later stage s.   ▯

3.2 Theorem (Lachlan [1966b]). *If* **a**, **b** *are r.e. degrees and* **d** *is a degree* $\leq$ **a** *and* $\leq$ **b**, *then there is an r.e. degree* **c** *such that* **d** $\leq$ **c**, **c** $\leq$ **a**, *and* **c** $\leq$ **b**.

Proof. Fix r.e. sets $A \in$ **a**, $B \in$ **b** and indices e and i, such that $\{e\}^A = \{i\}^B = D$ for some set $D \in$ **d**. Let $\{A_s\}_{s \in \omega}$ and $\{B_s\}_{s \in \omega}$ be recursive enumerations of A and B. For each x define a finite set C_x of "bad stages" as follows. If $\{e\}^{A_s}_s(x) = \{i\}^{B_s}_s(x) = y$ and at some later stage $t > s$, *neither* $\{e\}^{A_t}_t(x) = y$ *nor* $\{i\}^{B_t}_t(x) = y$, then enumerate all $v \leq s$ into C_x at stage t. Define $C = \bigoplus_x \{C_x : x \in \omega\}$, which is clearly r.e. because C_x is enumerated uniformly in x. Now $D \leq_T C$ because for each x we find some $s \in \overline{C_x}$ such that $\{e\}^{A_s}_s(x) \downarrow = \{i\}^{B_s}_s(x) = y$, and note that therefore $D(x) = y$. Note that $C \leq_T A$ because for each x we simply find t such that $\{e\}^{A_t}_t(x) \downarrow = y$ and $A_t \restriction u = A \restriction u$ where $u = u(A_t; e, x, t)$. Now $\{e\}^{A_v}_v(x) = y$ for all $v \geq t$ so no new s can enter C_x after stage t. The proof that $C \leq_T B$ is similar.   ▯

See Exercise 3.5 for results analogous to Theorem 3.2 about stronger reducibilities.

3.3 Corollary (Lachlan [1966b]—Yates [1966a]). *There are incomparable r.e. degrees* **a** *and* **b** *with no infimum. Hence, the r.e. degrees do not form a lattice. (See also Exercises 3.10 and 5.7.)*

Proof. Let **a** and **b** be any incomparable low r.e. degrees such that $\mathbf{a} \cup \mathbf{b} = \mathbf{0}'$. (These can be obtained by the Sacks Splitting Theorem or even the original

Friedberg-Muchnik Theorem as shown in Exercise VII.2.3.) Let $\mathbf{d}$ be any degree below both $\mathbf{a}$ and $\mathbf{b}$. Now $\mathbf{d}' = \mathbf{0}'$, since $\mathbf{a}' = \mathbf{0}'$, so the relativization to $\mathbf{d}$ of Theorem 3.1 produces $\mathbf{c} > \mathbf{d}$, such that $\mathbf{c} < \mathbf{a}$, $\mathbf{c} < \mathbf{b}$, and by Theorem 3.2 we may assume that $\mathbf{c}$ is r.e. Thus, $\mathbf{a}$ and $\mathbf{b}$ have no infimum in the upper semi-lattice of r.e. degrees or even in the upper semi-lattice of *all* degrees. ▯

On the basis of Theorem 3.1, Lachlan [1966b, p. 568] and Robinson [1971a, p. 313] suggested that perhaps r.e. degrees $\mathbf{a}$ and $\mathbf{b}$ such that $\mathbf{a} \cup \mathbf{b} = \mathbf{0}'$ can *never* have an infimum. This was refuted by Shoenfield and Soare [1978] and simultaneously by Lachlan [1980]. Indeed Lachlan obtained the following pleasing generalization of the Sacks Splitting Theorem VII.3.2.

3.4 Theorem (Lachlan Splitting Theorem) (Lachlan [1980]). *Let A be any nonrecursive r.e. set. There exist r.e. sets B_0, B_1, and C such that:*

(i) $C <_T A$;
(ii) $B_0 \cup B_1 = A$ *and* $B_0 \cap B_1 = \emptyset$;
(iii) $B_{i-1} \not\leq_T B_i \oplus C$, *for $i = 0, 1$; and*
(iv) $\deg(C) = \deg(B_0 \oplus C) \cap \deg(B_1 \oplus C)$.

Thus, for any r.e. degree $\mathbf{a} > \mathbf{0}$ there are incomparable r.e. degrees $\mathbf{b}_0 = \deg(B_0 \oplus C)$ and $\mathbf{b}_1 = \deg(B_1 \oplus C)$ which have supremum $\mathbf{a}$ and infimum $\mathbf{c} = \deg(C)$. The proof (which we omit) is interesting because it introduces a new method beyond those in §1 and §2 for preserving infimums. The idea is to guarantee that $\mathbf{b}_0 \cap \mathbf{b}_1 = \mathbf{c}$ by actively enumerating elements into C rather than by merely restraining elements from B_0 and B_1 as before.

In a different direction Ambos-Spies (Exercise XI.3.4) proves an interesting generalization of the Non-Diamond Theorem 3.1 using low degrees. As corollaries he proves that no low degree cups with half of a minimal pair to $\mathbf{0}'$, and that for every r.e. low degree $\mathbf{a}$ there exists $\mathbf{b} \in \mathbf{R}$ such that $\mathbf{a} \cap \mathbf{b}$ fails to exist.

3.5–3.12 Exercises

3.5. (a) Prove that the analogue of Theorem 3.2 holds with $\leq_T$ replaced by $\leq_{wtt}$ (as defined in Exercise V.2.16). Namely, if A and B are r.e., $D \leq_{wtt} A$, and $D \leq_{wtt} B$, then there exists an r.e. set $E \leq_{wtt} A, B$ such that $D \leq_{wtt} E$. *Hint.* To achieve $D \leq_{wtt} E$ replace C by the set $E = \mathrm{card}(\{ s : C_s \neq C_{s+1} \})$.

(b) Prove that the same result holds trivially with $\leq_T$ replaced by $\leq_m$ or $\leq_1$. (Fejer and Shore [ta] have shown it to be false for $\leq_{tt}$.)

3.6. Give an alternative proof of Theorem 3.1 using dominating functions instead of the Recursion Theorem. *Hint.* Fix e such that $\{ e \}^{A \oplus B}$ is total and $\{ e \}^{A \oplus B}(j) \geq \{ n \}(i, j)$ if $n, i \leq j$ and $\{ n \}(i, j)$ is defined. Define a partial recursive function $\{ n \}$ by setting $\{ n \}(i, j) = 1 + \{ e \}_s^{A_s \oplus B_s}(j)$ where

$s + 1$ is the first stage (if any) at which $P_{i,j}$ requires attention via some pair $\langle x, y \rangle$, in which case we *attach* the least such pair $\langle x, y \rangle$ to $P_{i,j}$. Now if $j \geq n$, then some $z < u(j, s)$ must later appear in either A or B, at which time we *act on* $\langle x, y \rangle$ just as before, enumerating either x in E or y in F_i. Ignoring those finitely many $P_{i,j}$ with $j < n$, each pair $\langle x, y \rangle$ attached to some $P_{i,j}$ is eventually acted upon, so the proof proceeds as before.

3.7 (Ambos-Spies).   Remove the Recursion Theorem from the proof of Theorem 3.1 as follows. Construct r.e. sets E^e and F_i^e for all $e, i \in \omega$ and meet the requirements:

$$P_{\langle e,i,j \rangle} : [D = \{ e \}^{A \oplus B} \text{ and } W_i, W_j \text{ are infinite}]$$

$$\implies [W_i \cap E^e \neq \emptyset \quad \text{or} \quad W_j \cap F_i^e \neq \emptyset].$$

Enumerate $\langle e, i, j \rangle$ into D as the attacker for $P_{\langle e,i,j \rangle}$ when (3.1)–(3.4) are satisfied, but now do not enumerate A and B until some $z \leq u$ is enumerated. Merely proceed with the construction and complete the attack when (and if) this later happens. (If $D = \{ e \}^{A \oplus B}$, this must eventually happen.) Since D is r.e., $D = \{ e \}^{A \oplus B}$ for some e. Prove that $P_{\langle e,i,j \rangle}$ is met for all i and j, and this e.

3.8.   Show that in the proof of Theorem 3.1 we can construct only *two* r.e. sets E and F, one of which has the desired degree **c**. *Hint.* Replace $P_{i,j}$ by the requirement $P'_{i,j} : E \cap W_i \neq \emptyset$ or $F \cap W_j \neq \emptyset$, which we attempt to meet for all i and j, $i < j$; replace the attacker $\langle i, j \rangle$ by j and $u(\langle i, j \rangle, s)$ by $u(j, s)$; allow D to be merely Δ_2^0 (rather than r.e.), since the attacker j may be inserted or extracted from D (according as $\{ e \}_s^{A_s \oplus B_s}(j) = 0$ or 1) whenever an attack is made on $P'_{i,j}$ for $i < j$. Since an attack is made on $P'_{i,j}$ at most once, each j is inserted in or extracted from D finitely often (once for each $i < j$), so D is Δ_2^0, and $D = \{ e \}^{A \oplus B}$ for some e. (This version is presented in detail in Soare [1980a, §6].)

3.9.   Show that the use of at least two sets E and F in Theorem 3.1 and Exercise 3.8 was necessary since there is no uniform procedure to obtain **c** from **a** and **b**. Namely, prove that there are no recursive functions f and g such that given nonrecursive r.e. sets $A = W_a$, $B = W_b$, and e such that $K = \{ e \}^{A \oplus B}$ then E is nonrecursive and $E = \{ f(a, b, e) \}^A = \{ g(a, b, e) \}^B$. *Hint.* Use the Recursion Theorem after first proving that for any *single* minimal pair requirement of the form

$$N_{f,g} : \{ f \}^A = \{ g \}^B = h \text{ total} \implies h \text{ is recursive},$$

we can construct nonrecursive r.e. sets A and B satisfying $N_{f,g}$ and such that $K \leq_T A \oplus B$. Whenever x is enumerated in K we enumerate some

$y \leq 3x$ in either A or B, thus ensuring $K \leq_T A \oplus B$ and allowing A and B to be coinfinite and simple. Define an (f, g)-expansionary stage analogously as we defined a 0-expansionary stage in §1. If $x \in K_{s+1} - K_s$ and s is not (f, g)-expansionary, we may assume one side (say A_s) still yields all computations achieved at the last (f, g)-expansionary stage $t < s$, and hence we enumerate x into the other set (say B). Simultaneously, we need to meet the positive requirements P_e of Theorem 1.2 to avoid the case $A \equiv_T K$, and $B \equiv_T \emptyset$ or vice versa.

3.10 (Jockusch). Let $\mathbf{M} = \{ \mathbf{a} : \mathbf{a} = \mathbf{0}$ or $\mathbf{a}$ is r.e. and is one half of a minimal pair $\}$. Let $\mathbf{M}^+$ denote the ideal of $\mathbf{R}$ generated by $\mathbf{M}$ (i.e., close $\mathbf{M}$ under finite sups and close downwards under $\leq$). Prove that $\mathbf{0}' \notin \mathbf{M}^+$. *Hint.* Consider r.e. sets A_1, A_2, A_3, B_1, B_2, B_3, such that $K \leq_T A_1 \oplus A_2 \oplus A_3$ and such that $\deg(A_i)$ and $\deg(B_i)$ form a minimal pair for $i = 1, 2, 3$. Build an r.e. set D as in Theorem 3.1 and assume we have an e such that $D = \{ e \}^{A_1 \oplus A_2 \oplus A_3}$. Construct r.e. sets E, F_i, $G_{i,j}$, and attempt to meet the requirements $P_{i,j,k}$: $W_i \cap E \neq \emptyset$ or $W_j \cap F_i \neq \emptyset$ or $W_k \cap G_{i,j} \neq \emptyset$. To attack $P_{i,j,k}$ find a stage t_0 at which

$$\{ e \}^{A_{t_0} \upharpoonright u}(\langle i, j, k \rangle) = 0 = D(\langle i, j, k \rangle),$$

where $A = A_1 \oplus A_2 \oplus A_3$. Now find $t_1 \geq t_0$ and $x_3 \in W_{k,t_1}$ such that $x_3 \geq u, 2k$, x_3 is permitted by B_3 at t_1, $W_{k,t_1} \cap G_{i,j,t_1} = \emptyset$, and $A_{t_1} \upharpoonright u = A_{t_0} \upharpoonright u$. Find $t_2 \geq t_1$ and $x_2 \in W_{j,t_2}$ such that $x_2 \geq u, 2j$, and x_2 is permitted by B_2 at t_2, $W_{k,t_2} \cap F_{i,t_2} = \emptyset$, and $A_{t_2} \upharpoonright u = A_{t_0} \upharpoonright u$. Find $t_3 \geq t_2$ and $x_1 \in W_{j,t_3}$, $x_1 \geq u, 2i$, and x_1 is permitted by B_1 at t_3, $W_{i,t_3} \cap E_{t_3} = \emptyset$, and $A_{t_3} \upharpoonright u = A_{t_0} \upharpoonright u$. Put $\langle i, j, k \rangle$ into D. Enumerate $A_1 \cup A_2 \cup A_3$ until a number $z < u$ appears. If z appears in A_1 (A_2, A_3) put x_1 (x_2, x_3) into E $(F_i, G_{i,j})$. Clearly, $E \leq_T A_1$, $E \leq_T B_1$, $F_i \leq_T A_2$, and $G_{i,j} \leq_T A_3$ by immediate permitting. If E is nonrecursive, we are done. If not, let $\overline{E} = W_i$. Now use the nonrecursiveness of B_1 to get $F_i \leq_T B_2$. If F_i is nonrecursive, we are done. Otherwise, let $\overline{F}_i = W_j$. Show that $G_{i,j} \leq_T B_3$ using the nonrecursiveness of B_1 and B_2. Thus, for each k if W_k is infinite then $P_{i,j,k}$ must be attacked and satisfied, so $W_k \cap G_{i,j} = \emptyset$. (See Exercise XI.3.4(b) for a related result and also see Theorem XIII.3.1 which asserts that surprisingly $\mathbf{M} = \mathbf{M}^+$.)

3.11 (Harrington). Prove that there is a low r.e. degree which is not in $\mathbf{M}^+$ as defined in Exercise 3.10. *Hint.* Apply the Sacks Splitting Theorem VII.3.2 to get K as the disjoint union of low r.e. sets A and B and apply Exercise 3.10. (See also results Exercise XIII.1.14, Theorem XIII.2.2, and Theorem XIII.3.1 which together imply this.)

3.12 (Jockusch [1981]).   Using the method of Theorem 3.1, give an easy direct proof of Corollary 3.3 that there are incomparable r.e. degrees **a** and **b** with no infimum in the r.e. degrees. *Hint.* Let $\{\Theta_e\}_{e\in\omega}$ and $\{(\Phi_e, \Psi_e)\}_{e\in\omega}$ enumerate all Turing reductions and all pairs of Turing reductions, respectively. Let $\{Z_e\}_{e\in\omega}$ be any acceptable numbering of the r.e. sets and $\{Z_{e,s}\}_{e,s\in\omega}$ a simultaneous recursive enumeration of $\{Z_e\}_{e\in\omega}$. Construct r.e. sets A, B, and $\{V_e\}_{e\in\omega}$ to meet the requirements

$$R_{\langle e,i\rangle} : \Phi_e(A) = Z_e = \Psi_e(B) \implies [V_e \leq_T A \ \& \ V_e \leq_T B \ \& \ V_e \neq \Theta_i(Z_e)].$$

Choose a witness $x = \langle e, i, k\rangle$ for requirement $R_{\langle e,i\rangle}$. Wait for a stage $s+1$ such that for u and v,

$$\Theta_i(Z_{e,s} \restriction u; x) = 0 \ \& \ (\forall y < u)\,[\Phi_{e,s}(A_s \restriction v; y)\downarrow = \Psi_{e,s}(B_s \restriction v; y) = Z_{e,s}(y)].$$

Put x into A_{s+1} and restrain $B_s \restriction v$ with priority $R_{\langle e,i\rangle}$. Wait for a stage $t > s$ such that either:

   (a) $(\exists y < u)\,[y \in Z_{e,t} - Z_{e,t-1}]$; or
   (b) some higher priority requirement receives attention; or
   (c) $(\forall y < u)\,[\Phi_{e,t}(A_t \restriction z; y)\downarrow]$.

If (a) holds then restrain $B_s \restriction v$ (thereby ensuring that $Z_e \neq \Psi_e(B)$); if (b) holds then start over with a new witness; if (c) holds and $\Phi_{e,t}(A_t) \restriction u \neq \Psi_{e,s}(B_s) \restriction u$, then restrain $A_t \restriction z$ and $B_s \restriction v$; if (c) holds and the equality holds, then put x into V_e and into B and restrain $A_t \restriction z$, thereby meeting $R_{\langle e,i\rangle}$. In the latter case, $V_e \leq_T B$ by simple permitting. Show by "delayed permitting" as in Lemma 2(ii) of Theorem 3.1 that $V_e \leq_T A$.

3.13 Notes. Related to Theorem 3.4, Ambos-Spies [1980, §6] considers which lattices can be embedded in the r.e. degrees by a map which takes the greatest element to $\mathbf{0}'$. He proves that any countable distributive lattice can be so embedded as well as certain nondistributive lattices such as N_5 of Diagram 2.1.

The Non-Diamond Theorem 3.1 first appeared in Lachlan [1966b] although the proof here is a simplification which arose from discussions between the author and Lachlan, Jockusch and Stob.

4.* Nonbranching Degrees

4.1 Definition.   Let **a** be an r.e. degree. Then **a** is called *branching* if there are r.e. degrees **b** and **c** different from **a** such that **a** is the infimum of **b** and **c**, and **a** is *nonbranching* otherwise.

For example, in §1 we used the minimal pair method to prove that $\mathbf{0}$ is branching, and it follows from Theorem 2.1, Exercise 2.3, or Exercise 2.4 that there are many other branching degrees. We now turn to nonbranching degrees. Such a degree was first constructed by Lachlan [1966b, p. 554]. Extensions of this nonbranching method played a key role in refuting the embedding conjecture [Lachlan-Soare, 1980] by proving that any r.e. degree which is the maximum element of a lattice of the form M_5 of Diagram 2.1 must be nonbranching. Fejer [1980] and [1983] combined the nonbranching degree construction with the Density Theorem to prove a density theorem for nonbranching degrees. Namely, for any r.e. degrees $\mathbf{c} < \mathbf{d}$ there is a nonbranching (r.e.) degree $\mathbf{a}$ such that $\mathbf{c} < \mathbf{a} < \mathbf{d}$. It follows that the nonbranching degrees generate (under $\cup$) all the nonzero r.e. degrees. This was the first nontrivial definable subset of the r.e. degrees shown to be dense and hence to generate the r.e. degrees. As suggested by Lerman, we give to the r.e. degrees the order topology where a typical subbasic open set has the form $\mathbf{R}(< \mathbf{a}) = \{\, \mathbf{b} : \mathbf{b} < \mathbf{a} \,\}$ or $\mathbf{R}(> \mathbf{a}) = \{\, \mathbf{b} : \mathbf{b} > \mathbf{a} \,\}$. The branching degrees together with $\mathbf{0}'$ are precisely the isolated points, and it follows from the Fejer Density Theorem above for nonbranching degrees that the Cantor-Bendixson rank of the r.e. degrees with this topology is 1 (see Exercise 4.5). See Notes 4.8 for Slaman's theorem that the branching degrees are also dense.

We present now the basic method for constructing nonbranching degrees. It is a finite injury argument and closely resembles the Friedberg-Muchnik method for satisfying a requirement of the form $B \neq \{e\}^A$, but it also requires the permitting method to ensure that B is recursive in certain sets $\widetilde{W}_i, \widetilde{W}_j$.

4.2 Theorem (Lachlan [1966b, p. 554]). *For any nonrecursive r.e. degree $\mathbf{c}$, there is a nonbranching (r.e.) degree $\mathbf{a} \not\geq \mathbf{c}$.*

Proof. We shall construct an r.e. set A and define $\mathbf{a} = \deg(A)$. To make $C \not\leq_T A$ we shall meet for each e the negative requirement

$$N_e : C \neq \{e\}^A,$$

which is done exactly as in Theorem VII.3.1, using the restraint function $r(e, s)$ defined there. For any r.e. set W_e, define $\widetilde{W}_e = W_e \oplus A$. To make A nonbranching we must ensure that if

(4.1) $$A <_T \widetilde{W}_i \quad \text{and} \quad A <_T \widetilde{W}_j,$$

then $\deg(A)$ is not the infimum of $\deg(\widetilde{W}_i)$ and $\deg(\widetilde{W}_j)$, namely, there is an r.e. set $B_{i,j}$ such that

(4.2) $$B_{i,j} \leq_T \widetilde{W}_i \quad \text{and} \quad B_{i,j} \leq_T \widetilde{W}_j;$$

and

(4.3) $B_{i,j} \not\leq_T A$.

For (4.3) we must meet for each e the requirement

(4.4) $R_{\langle e,i,j \rangle} : B_{i,j} \neq \{e\}^A$.

We sketch first the strategy for meeting a single such requirement R_n, $n = \langle e,i,j \rangle$, for fixed e, i, and j. For notational convenience we drop the subscripts i,j on $B_{i,j}$, and on the marker Γ_x described below.

We attempt to meet R_n just as in the usual Friedberg-Muchnik procedure. Namely, we:

(1) choose a fresh witness $x \in \omega^{[n]}$ (not restrained by any requirement of higher priority);

(2) wait for a stage s such that $\{e\}_s^{A_s}(x) \downarrow = 0$;

(3) define a *restraint function* $q(n, s+1) = u_x^s =_{\text{dfn}} u(A_s; e, x, s)$, and restrain with priority R_n any $y \leq q(n, s)$ from entering A;

(4) enumerate x in B_{s+1}, thereby guaranteeing that $B(x) \neq \{e\}^A(x)$.

The problem is that in order to ensure that $B \leq_T \widetilde{W_i}$, and $B \leq_T \widetilde{W_j}$, we must not put x into B unless we simultaneously put a certain *trace* y_x into A (and therefore into $\widetilde{W_i}$ and $\widetilde{W_j}$). However, doing so would be useless in preserving (2) unless $y_x > u_x^s$, else the computation $\{e\}_s^{A_s}(x) = 0$ would be destroyed. Thus we have, for each x, a "movable marker" Γ_x whose position at the end of stage s, Γ_x^s, denotes our current candidate for the trace y_x. The motion of Γ_x will satisfy:

(4.5) $\Gamma_x^{s+1} \neq \Gamma_x^s \implies (\exists z \leq x) [z \in W_{i,s+1} - W_{i,s}]$;

and

(4.6) $x \in B_{s+1} - B_s \implies \Gamma_x^{s+1} \in A_{s+1} - A_s$.

It follows from (4.5) that Γ_x moves only finitely often, so $\Gamma_x^\infty =_{\text{dfn}} \lim_s \Gamma_x^s < \infty$. Now by (4.5), $\lambda x[\Gamma_x^\infty]$ is a function recursive in W_i. Furthermore, $B \leq_T \widetilde{W_i}$ since $x \in B$ iff $x \in B_s$ where $s = \mu t[A_t \restriction (\Gamma_x^\infty + 1) = A \restriction (\Gamma_x^\infty + 1)]$ because of (4.6). We shall have $B \leq_T W_j$ by the usual simple permitting, namely

$$x \in B_{s+1} - B_s \implies W_{j,s+1} \restriction x \neq W_{j,s} \restriction x.$$

The hypothesis $A <_T \widetilde{W_i}$ is used to move Γ_x (according to (4.5)) to some element $y_x > u_x^s$ before performing the above Friedberg-Muchnik procedure, as we now explain.

At stage $s = 0$ we place (for each y) marker Γ_x on $\langle x, n \rangle$, the xth element of $\omega^{[n]}$, and we set $A_0 = B_0 = \emptyset$.

Stage $s + 1$.

   Step 1. For R_n find the least $x \in \omega^{[n]} - B_s$, $x \leq s$, (if such exists) such that:

$$(4.7) \qquad \{e\}_s^{A_s}(x) \downarrow = 0;$$

and

$$(4.8) \qquad W_{i,s+1} \restriction x \neq W_{i,s} \restriction x, \quad (\text{i.e., } W_i \text{ permits } x).$$

Now move Γ_x to the least $y \in \omega^{[n]} - A_s$ such that: $y > u_x^s = u(A_s; e, x, s)$; $y \geq \Gamma_x^s$; and y is not restrained with higher priority (i.e., $y > \max\{r(m, s), q(m, s)\}$ for all $m \leq n$). Also move markers Γ_z, $z > x$, in order to fresh elements of $\omega^{[n]} - A_s$.

   Step 2. We say that x is *eligible* if $x \leq s$; $x \in \omega^{[n]} - B_s$; (4.7) holds for x; and

$$(4.9) \qquad u(A_s; e, x, s) < \Gamma_x^{s+1}.$$

Choose the least eligible x (if such exists) such that for all $m \leq n$,

$$\Gamma_x^{s+1} > \max\{r(m, s), q(m, s)\},$$

and

$$(4.10) \qquad W_{j,s+1} \restriction x \neq W_{j,s} \restriction x.$$

If x exists we say that R_n *requires attention*. Enumerate Γ_x^{s+1} in A (and therefore in $\widetilde{W}_i$ and $\widetilde{W}_j$); define $q(n, s+1) = u(A_s; e, x, s)$; and enumerate x in B. In this case we say requirement R_n *receives attention* at stage $s + 1$.

   If R_n receives attention it remains satisfied and does not receive further attention unless it is later injured by a higher priority requirement. To see that R_n will eventually receive attention assume that $A <_{\mathrm{T}} \widetilde{W}_i$ and $A <_{\mathrm{T}} \widetilde{W}_j$ but $B = \{e\}^A$. For each x, let $u_x = \lim_s u(A_s; e, x, s)$, and $s_x = (\mu t)\,(\forall s \geq t)\,[u_x = u(A_s; e, s, x)]$. Since $\lambda x[s_x]$ is an A-recursive function the following set is A-r.e.:

$$U = \{x : x \in \omega^{[n]} \ \& \ (\exists s \geq s_x)\,[W_{i,s+1} \restriction x \neq W_{i,s} \restriction x]\,\}.$$

Now if U were finite then $W_i \leq_{\mathrm{T}} A$ since for almost every z, $z \in W_i$ iff $z \in W_{i,s_x}$ where $x = (\mu y > z)\,[y \in \omega^{[n]}]$. Hence, U is infinite; Step 1 is performed on each $x \in U$ at some stage $t_x \geq s_x$; and x is eligible at every stage $s \geq t_x$. Now since $\{t_x : x \in U\}$ is A-r.e., so is the set

$$V = \{x : x \in U \ \& \ (\exists s \geq t_x)\,[W_{j,s+1} \restriction x \neq W_{j,s} \restriction x]\,\}.$$

But V is infinite else $W_j \leq_T A$. Now for at most finitely many $x \in V$ is $\Gamma_x^{t_x}$ restrained from A with priority higher than R_n. Hence, R_n eventually receives attention via some $x \in V - B$, becomes satisfied, and remains satisfied thereafter (unless injured by a higher priority requirement R_m, $m < n$).

(Notice that we do not restrain $A \restriction u_x$ when Step 1 is performed and Γ_x is moved for the sake of R_n. Hence, lower priority requirements may enumerate elements $z < u_x$ into A. Thus x may become alternately eligible and ineligible many times before Step 2 is finally performed.)

This completes the description of the strategy for a single requirement. To handle all requirements simultaneously we restore the superscript e and subscripts i, j to $B_{i,j}^e$ and the superscripts e, i, j to $\Gamma_x^{e,i,j}$. At stage $s + 1$ perform Step 1 for each requirement R_n, $n \leq s$. At Step 2 choose the least n such that R_n requires attention and perform Step 2 on R_n.

Lemma. $(\forall n) \, [n = \langle e, i, j \rangle \ \& \ C \neq \{n\}^A$

$$\& \ [A <_T \widetilde{W}_i \ \& \ A <_T \widetilde{W}_j \implies B_{i,j}^e \neq \{e\}^A]$$
$$\& \ R_n \text{ receives attention at most finitely often}].$$

Proof. Fix n and assume for all $m < n$ that the above statements hold. Thus, the limits $r(m) = \lim_s r(m, s)$ and $q(m) = \lim_s q(m, s)$ exist for all $m < n$. Furthermore, R_n is injured at most finitely often and so is eventually satisfied forever. Thus, N_n is injured finitely often and is finally satisfied as in Chapter VII §3.   ▯

4.3–4.7 Exercises

4.3.   Show that the proof of Theorem 4.2 actually establishes that for every nonrecursive r.e. set C there exists an r.e. set A, $C \not\leq_T A$, such that for all i and $j \in \omega$,

$$[W_i \not\leq_T A \ \& \ W_j \not\leq_T A]$$
$$\implies (\exists \text{ r.e. } B_{i,j} \not\leq_T A) \, [B_{i,j} \leq_T A \oplus W_i \ \& \ B_{i,j} \leq_T W_j].$$

4.4 (Ambos-Spies [1980, §8]).   Construct a uniformly r.e. ascending sequence of r.e. degrees $\mathbf{a}_0 < \mathbf{a}_1 < \cdots$ which has no exact pair of r.e. degrees in the sense of Theorem VI.4.2.   *Hint.*   Construct r.e. sets A_n and $B_{i,j}$, for all $i, j, n \in \omega$, to meet the following requirements where $\hat{A}_n = \oplus \{A_m : m \leq n\}$:

$$S_{e,n} : A_{n+1} \neq \{e\}^{\hat{A}_n};$$
$$R_{i,j,n} : [W_i \not\leq_T \hat{A}_n \ \& \ W_j \not\leq_T \hat{A}_n] \implies$$
$$[B_{i,j} \leq_T A_0 \oplus W_i \ \& \ B_{i,j} \leq_T W_j \ \& \ B_{i,j} \not\leq_T \hat{A}_n].$$

The requirements $S_{e,n}$ are met with the usual Friedberg-Muchnik method by putting elements into A_{n+1} and restraining elements out of $\hat{A}_n$. Requirements $R_{i,j,n}$ are met as in the nonbranching construction of Theorem 4.2 by enumerating elements into A_0 to satisfy for every e the requirement $R_{i,j,n,e} : B_{i,j} \neq \{e\}^{\hat{A}_n}$.

4.5 (Fejer [1980]). Let $\mathbf{R}$ have the order topology given in the beginning of §4. For $\mathbf{a} \in \mathbf{R}$, let $\mathbf{R}(\leq \mathbf{a})$ denote $\{\mathbf{x} : \mathbf{x} \in \mathbf{R}$ and $\mathbf{x} \leq \mathbf{a}\}$, and similarly for $\mathbf{R}(\geq \mathbf{a})$.

(a) Prove by the Sacks Splitting Theorem VII.3.2 that for any r.e. degree $\mathbf{a} > \mathbf{0}$, $\mathbf{R}(\geq \mathbf{a})$ is an open set.

(b) Prove by induction on n that if $\mathbf{a} = \inf(\mathbf{b}_1, \mathbf{b}_2, \ldots, \mathbf{b}_n)$, where $n \geq 2$, $\mathbf{b}_i > \mathbf{a}$ and $\mathbf{b}_i$ is r.e. for every $i \leq n$, then $\mathbf{a}$ is a branching degree.

(c) Prove that $\mathbf{a}$ is branching iff $\mathbf{R}(\leq \mathbf{a})$ is an open set iff $\mathbf{a}$ is an isolated point.

(d) Prove that the branching degrees together with $\mathbf{0}'$ are precisely the isolated points.

(e) Let $\mathbf{R}^1$ be the first Cantor-Bendixson derivative of $\mathbf{R}$, namely $\mathbf{R}^1 = \mathbf{R} - \{$ the isolated points of $\mathbf{R}\}$. Assume the Fejer Density Theorem [Fejer, 1983] that given r.e. degrees $\mathbf{d} < \mathbf{c}$ there is a nonbranching degree $\mathbf{a}$, $\mathbf{d} < \mathbf{a} < \mathbf{c}$. Use this and (a) to show that $\mathbf{R}^1$ has no isolated points, so the Cantor-Bendixson process stops after one step.

4.6.° Use a permitting argument to show that for every nonzero r.e. degree $\mathbf{d}$, there is a nonbranching degree $\mathbf{a} \leq \mathbf{d}$.

4.7 (Ambos-Spies). Assuming the Fejer theorem [1983] that the nonbranching degrees are dense, prove that the lattice N_5 of Diagram 2.1 can be embedded in $\mathbf{R}$. *Hint.* Let $\mathbf{a}$ and $\mathbf{b}$ be a minimal pair. Choose $\mathbf{c}$ nonbranching with $\mathbf{0} < \mathbf{c} < \mathbf{a}$. Choose $\mathbf{d}$ such that $\mathbf{d} < \mathbf{a}$, $\mathbf{c} < \mathbf{d}$, and $\mathbf{d} < \mathbf{c} \cup \mathbf{b}$ since we know that $\mathbf{a} \cap (\mathbf{c} \cup \mathbf{b}) \neq \mathbf{c}$. Consider $\mathbf{0}$, $\mathbf{b}$, $\mathbf{c}$, $\mathbf{d}$, and $\mathbf{c} \cup \mathbf{b}$.

4.8 Notes. Fejer not only proved the density of nonbranching degrees [1983], but he also obtained a partial result on the density of the branching degrees by proving [1980] and [1982] that there is a branching degree above every low r.e. degree. Later Slaman [ta] proved the density of the branching degrees, using a $\emptyset''''$-priority construction like that described in Chapter XIV. The form of Theorem 4.2 first proved by Lachlan is the stronger form Exercise 4.6, which is also presented in Shoenfield [1971, p. 106].

5.* Noncappable Degrees

5.1 Definition. (i) An r.e. degree **a** is *cappable* (*caps*) if there exists an r.e. degree **b** > **0** such that **a** ∩ **b** = **0**, and **a** is *noncappable* (*n.c.*) otherwise.

(ii) Let **M** denote the class of cappable degrees (i.e., **0** together with those r.e. degrees which are halves of minimal pairs).

(iii) Let **NC** (= **R** − **M**) denote the class of noncappable r.e. degrees.

It is easy to see that **NC** forms a filter in **R** (see Exercise 5.4). In Theorem XIII.3.1 we shall see that **M** forms an ideal (namely **M** = **M**⁺ of Exercise 3.10) so that **M** and **NC** give the first algebraic decomposition of **R** into two disjoint classes definable in (**R**, ≤), one an ideal and the other a filter, indeed a strong filter. (See Definition II.1.14 for definitions of ideals, filters, and strong filters in an upper semi-lattice.)

To see that **NC** is nontrivial, we now give an easy and direct finite injury construction of an incomplete degree in **NC**. This construction is also useful because it can be combined with many others to investigate the degrees in **NC** and their relationship to the degrees in **R**. For example, it was suggested that because much pathology arises from the degrees in **M**, those in **NC** might be well-behaved and perhaps even satisfy Shoenfield's conjecture. This is disproved in Exercise 5.6 by constructing a branching degree in **NC**.

5.2 Theorem (Yates [1966a]).   *For every r.e. degree* **c** > **0** *there is a noncappable r.e. degree* **a** *such that* **c** ≰ **a**.

Proof. Let C be any nonrecursive r.e. set. We construct an r.e. set A to meet for all e and i the requirements:

$$N_e : \Phi_e(A) \neq C;$$

$$P_{\langle e,i \rangle} : W_e \text{ nonrecursive} \implies [A^{[e]} \leq_T W_e \ \& \ A^{[e]} \neq \overline{W}_i].$$

(Note that the $P_{\langle e,i \rangle}$ requirements guarantee that A is nonrecursive.) For N_e define the restraint function $r(e,s)$ as in Theorem VII.3.1. For $P_{\langle e,i \rangle}$ we enumerate some x of the form $x = \langle n,i,e \rangle$ into $A^{[e]}$ at stage $s+1$ if:

(5.1) $$A_s^{[e]} \cap W_{i,s} = \emptyset;$$

(5.2) $$x \in W_{i,s};$$

(5.3) $$W_{e,s+1} \restriction x \neq W_{e,s} \restriction x;$$

(5.4) $$(\forall j \leq \langle e,i \rangle) \, [r(j,s) < x];$$

and x is minimal with respect to these properties. Each $P_{\langle e,i \rangle}$ contributes at most one element to A, so N_e is injured at most finitely often and hence $r(e) = \lim_s r(e,s)$ exists as in Theorem VII.3.1. Condition (5.3) guarantees that $A^{[e]} \leq_T W_e$, and obviously $A^{[e]} \leq_T A$. Now if W_e is nonrecursive then the requirement $P_{\langle e,i \rangle}$ eventually receives attention (and remains satisfied) by the usual permitting proof as in Chapter V §3. (Namely, if $A^{[e]} = \overline{W}_i$, choose an increasing r.e. sequence of elements $x_1 < x_2 < \cdots$ in W_i of the form $\langle n,i,e \rangle$ such that $x_1 > r(j)$, for all $j \leq \langle e,i \rangle$. Choose t such that $r(j) = r(j,s)$ for all $s \geq t$ and all $j \leq \langle e,i \rangle$. Choose s_k minimal such that $s_k > t$ and $x_k \in W_{i,s_k}$. Now $W_{e,s_k} \upharpoonright x_k = W_e \upharpoonright x_k$ so W_e is recursive.)   ▯

5.3 Definition. (i) An r.e. degree $\mathbf{a}$ is *effectively noncappable* (*e.n.c.*) if there exist an r.e. set $A \in \mathbf{a}$ and recursive functions f, g and h such that $W_{f(e)} = \{ g(e) \}^A = \{ h(e) \}^{W_e}$, and if $\emptyset <_T W_e$ then $\emptyset <_T W_{f(e)}$.
    (ii) Let **ENC** denote the class of effectively noncappable (r.e.) degrees.

Notice that the construction in Theorem 5.2 produces an e.n.c. degree because we simply take $W_{f(e)} = A^{[e]}$. The e.n.c. degrees have apparently nicer properties than the n.c. degrees. For example, one can show that they form a strong filter (see Exercise 5.5) and that strictly below any e.n.c. degree there is another e.n.c. degree. In Theorem XIII.2.2 we shall show that every n.c. degree is e.n.c. This is particularly surprising because effective notions in recursion theory rarely coincide with their noneffective counterparts. For example, an r.e. set A is nonrecursive iff A is noncomplemented in $\mathcal{E}$, but A is creative iff A is *effectively* noncomplemented. Furthermore, there are four other apparently unrelated classes of r.e. degrees which coincide with **NC**. These results in Chapter XIII depend only upon the material through Chapter IX (except for the Low Non-Diamond Theorem XI.3.4), and the interested reader may proceed directly there, skipping Chapters X, XI and XII.

5.4–5.8 Exercises

5.4 (Ambos-Spies [1980, §1]).   Prove that the degrees **NC** form a filter in **R**, namely: (1) if $\mathbf{a} < \mathbf{b}$ and $\mathbf{a} \in$ **NC** then $\mathbf{b} \in$ **NC**; and (2) if $\mathbf{a}, \mathbf{b} \in$ **NC** and $\mathbf{c} = \mathbf{a} \cap \mathbf{b}$ exists then $\mathbf{c} \in$ **NC**.

5.5 (Ambos-Spies [1980]).   Prove that the e.n.c. degrees form a strong filter, namely, they are closed upwards, and if $\mathbf{a},\mathbf{b}$ are e.n.c then there exists $\mathbf{c} < \mathbf{a}, \mathbf{b}$ such that $\mathbf{c}$ is e.n.c. (Hence, using Theorem 5.2 for every e.n.c. $\mathbf{a}$ there exists an e.n.c. $\mathbf{b} < \mathbf{a}$.) *Hint.* Let f_1 and f_2 be the recursive functions as in Definition 5.3 for $A \in \mathbf{a}$ and $B \in \mathbf{b}$, respectively. Let $C = \oplus \{ W_{f_2(f_1(e))} : e \in \omega \}$ and prove that $f = f_2 \circ f_1$ satisfies Definition 5.3 for C.

5.6 (Ambos-Spies [1980, Theorem 1.1]).　Prove that there is an e.n.c. degree **a** which is branching. (Hence, the degrees **NC** do not satisfy Shoenfield's conjecture.) *Hint.* Combine the methods of Exercise 2.4 and Theorem 5.2 to construct r.e. sets A, B, and C such that A satisfies the requirements $P_{\langle e,i \rangle}$ as in Theorem 5.2 and also A, B, and C meet the requirements:

$$R_e : B \neq \{e\}^A \ \& \ C \neq \{e\}^A;$$

and

$$N_e : \{e\}^{B \oplus A} = \{e\}^{C \oplus A} = f \text{ total} \implies f \leq_T A.$$

Let the priority ranking of requirements be $N_0, P_0, R_0, N_1, \ldots$. For requirement R_e we have followers targeted for B and C as in the Friedberg-Muchnik method of VII §2. For N_e, define the restraint function $r(e, s)$ as in §1 with $B \oplus A$ and $C \oplus A$ in place of A and B, and allow a follower y of R_e to be enumerated at stage $s + 1$ only if $y > r(e, s)$. For $P_{\langle e,i \rangle}$ there are followers $x = \langle u, 0 \rangle$ targeted for A as in Theorem 5.2. The latter do not respect $r(e, s)$. However, whenever any follower x is appointed or enumerated, all followers of lower priority (which will be precisely those followers $y > x$) are cancelled. Furthermore, if x is appointed at stage s then $x > s$. These precautions guarantee that N_e is satisfied as in Theorem 2.1. The cancellation also preserves the A-computations necessary to meet requirement R_e. A follower x of $P_{\langle e,i \rangle}$ once appointed is later *realized* at stage $s + 1$ if there exists some $y > x$ of the form $\langle m, i, e + 1 \rangle$ such that y satisfies: (5.1) with $A^{[e+1]}$ in place of $A^{[e]}$; (5.2); and (5.3). In this case we enumerate both x and y in A_{s+1}. (The reason we need both x and y is that our branching technique for N_e requires x to be appointed early and realized much later. However, if W_e is recursive we may appoint infinitely many followers for $P_{\langle e,i \rangle}$, waiting in vain for one to satisfy (5.3). But then $P_{\langle e,i \rangle}$ would infinitely often cancel followers of lower priority.)

5.7° (Ambos-Spies [1984b, Corollary 1]).　Construct an r.e. degree **a**, $0 < \mathbf{a} < \mathbf{0}'$, which is *strongly noncappable*, namely, there are no r.e. degrees **b** and **d** such that $\mathbf{a} \mid \mathbf{b}$ and $\mathbf{a} \cap \mathbf{b} = \mathbf{d}$. *Hint.* Let C be any nonrecursive r.e. set. Construct an r.e. set A to meet the following requirements for all e, where $e = \langle e_0, e_1, e_2 \rangle$:

$$N_e : C \neq \{e\}^A;$$

$$P_e : A \neq \overline{W}_e;$$

$$R_e : [W_{e_0} = \{e_1\}^A \ \& \ A \not\leq_T W_{e_0} \ \& \ W_{e_2} \not\leq_T A]$$
$$\implies (\exists E_e) \, [E_e \leq_T W_{e_2} \ \& \ E_e \leq_T A \ \& \ E_e \not\leq_T W_{e_0}].$$

Meet N_e using the Sacks preservation strategy and restraint function $r(e, s)$ of Theorem VII.3.1. Meet R_e using the nonbranching method of Theorem 4.2. Namely, attempt to meet the requirement $R_{\langle e,i \rangle} : E_e \neq \{i\}^{W_{e_0}}$. When we enumerate x in E_e at stage $s+1$ for $R_{\langle e,i \rangle}$ we must enumerate a trace $f_e(x,s)$ into A to ensure that $E_e \leq_T A$. However, we do this only if there are computations: $\{i\}^{W_{e_0,s} \restriction u}(x) = 0$; and $\{e_1\}^{A_s \restriction v} \restriction u = W_{e_0,s} \restriction u$ so that by restraining $A_s \restriction v$ we can preserve the W_{e_0}-computation on x. This restraint $\tilde{r}(\langle e,i \rangle, s)$ imposed by $R_{\langle e,i \rangle}$ on $A_s \restriction u$ conflicts with putting the trace into A, so using $W_{e_0} <_T A$ we define $\gamma(x,0) = 0$,

$$\gamma(x, s+1) = \begin{cases} s+1 & \text{if } A_{s+1} \restriction x \neq A_s \restriction x, \\ \gamma(x,s) & \text{otherwise.} \end{cases}$$

Define

$$f_e(x,s) = \langle x, \max\{u(A_s; e, y, t) : y \leq \gamma(x,s) \ \& \ t \leq s\} \rangle.$$

Show that $\lim_s \gamma(x,s)$ exists and is recursive in A, and that if $\{e_1\}^A$ is total, then $f_e(x) = \lim_s f_e(x,s)$ exists and is recursive in A. We achieve $E_e \leq_T W_{e_2}$ by the usual permitting using that $W_{e_2} \not\leq_T A$. We say that $R_{\langle e,i \rangle}$ *requires attention* at stage $s+1$ iff there is an $x \in \omega^{[\langle e+1,i \rangle]}$ such that:

(5.5) $$\tilde{r}(\langle e,i \rangle, s) = 0;$$

(5.6) $$\{e\}_s^{W_{e_0,s}}(x) = 0 \ \& \ x \notin E_{e,s};$$

(5.7) $$\gamma(x,s) \geq u, \quad \text{where } u = u(W_{e_0,s}; i, x, s);$$

(5.8) $$W_{e_0,s} \restriction u = \{e_1\}_s^{A_s} \restriction u;$$

(5.9) $$f_e(x,s) \geq \max(\{r(k,s), \tilde{r}(j,s) : k \leq \langle e,i \rangle \text{ and } j < \langle e,i \rangle \});$$

and

(5.10) $$(\exists y < x)\,[y \in W_{e_2,s+1} - W_{e_2,s}].$$

Choose the highest priority requirement which requires attention at stage $s+1$ and let it act. Show that if the hypotheses of $R_{\langle e,i \rangle}$ hold then there is a stage at which $R_{\langle e,i \rangle}$ receives attention and is not later injured.

5.8° (Ambos-Spies [1984b, Theorem 2]). Given an r.e. degree $\mathbf{c} > \mathbf{0}$, prove that there are low strongly noncappable r.e. degrees $\mathbf{a}_0$ and $\mathbf{a}_1$ such that $\mathbf{c} \not\leq \mathbf{a}_0, \mathbf{a}_1$, and $\mathbf{a}_0 \cup \mathbf{a}_1 = \mathbf{0}'$.

Chapter X

The Lattice of R.E. Sets Under Inclusion

In Chapters VII–IX we have been studying the structure of the upper semi-lattice **R** of r.e. degrees, and the emphasis has been on Turing computations. In this chapter we study the lattice $\mathcal{E}$ of r.e. sets, and the emphasis will be on the algebraic structure of an r.e. set within the lattice of r.e. sets. In Chapter XI we combine these two approaches and consider the relationship between the algebraic structure of an r.e. set and its degree.

The two principal themes of this chapter are the splitting theorem methods of §2 and the *e*-state construction of a maximal set in §3. Variations and extensions of these methods give the remaining results of this chapter, including the classification in §7 of those Boolean algebras which are realized as the lattice of supersets of some r.e. set.

1. Ideals, Filters, and Quotient Lattices

Recall the Definitions II.1.14 of a lattice $\mathcal{L} = (L; \leq, \vee, \wedge, 0, 1)$, with greatest and least element 0 and 1 respectively, a distributive lattice, a Boolean algebra, and ideals and filters. Most of the lattices we consider will be sublattices or quotients of $\mathcal{N} = (2^\omega; \subseteq, \cup, \cap, \emptyset, \omega)$, the lattice of all subsets of natural numbers. Of special interest will be the lattice $\mathcal{E} \subseteq \mathcal{N}$ of r.e. sets and the Boolean algebra $\mathcal{R}$ of recursive sets, as defined in Definition II.1.15.

Notice that an ideal or filter itself forms a lattice under the induced operations. For example, the lattice $\mathcal{F}$ of finite sets forms an ideal in $\mathcal{R}$, $\mathcal{E}$, or $\mathcal{N}$, and the collection of cofinite sets $\mathcal{C}$ forms a filter in each. We have also seen (Exercises V.1.5 and V.2.8) that the simple sets together with $\mathcal{C}$ form a filter in $\mathcal{E}$, and likewise for the hypersimple sets. (The same is true for the hh-simple sets by Exercise 2.17.) If $\mathcal{L}$ is a lattice and $a \in \mathcal{L}$ then

$$I(a) = \{ b \in \mathcal{L} : b \leq a \},$$

and

$$D(a) = \{ b \in \mathcal{L} : b \geq a \}$$

are respectively the *principal ideal* and *principal filter* of $\mathcal{L}$ generated by a. An ideal (filter) is principal if and only if it is $I(a)$ ($D(a)$) for some $a \in \mathcal{L}$. Neither $\mathcal{F}$ nor $\mathcal{C}$ is principal.

Ideals and filters allow us to form quotient lattices, as the following theorem explains.

1.1 Theorem. *Let $\mathcal{L}$ be a distributive lattice, $I \subseteq \mathcal{L}$ ($D \subseteq \mathcal{L}$) an ideal (a filter). We say $a \equiv_I b$ ($a \equiv_D b$) if and only if for some $c, d \in I$ (in D) $a \vee c = b \vee d$ ($a \wedge c = b \wedge d$). Then $\equiv_I$ and $\equiv_D$ define congruence relations whose sets of equivalence classes form distributive lattices. These are denoted by $\mathcal{L}/I$ and $\mathcal{L}/^d D$ respectively, where $/^d$ indicates that the dual ideal (filter) is being used.*

Proof. Exercise 1.5.   ▯

It is easy to see that if $\mathcal{L}$ is a sublattice of $\mathcal{N}$, such that $\mathcal{L}$ is closed under symmetric difference, and I an ideal of $\mathcal{L}$ then $A \equiv_I B$ if and only if $A \triangle B \in I$, where $A \triangle B$ is the symmetric difference of A and B.

1.2 Definition. We let $\mathcal{L}^*$, $\mathcal{N}^*$, $\mathcal{E}^*$, and $\mathcal{R}^*$, denote the quotient lattices $\mathcal{L}/\mathcal{F}$, $\mathcal{N}/\mathcal{F}$, $\mathcal{E}/\mathcal{F}$, and $\mathcal{R}/\mathcal{F}$, where $\mathcal{F}$ is the ideal of finite sets. We also write $A =^* B$ in place of $A \equiv_{\mathcal{F}} B$, we let $A^* = \{ B : A =^* B \}$, and we write $A \subseteq^* B$ iff $A^* \leq B^*$, namely if $A - B$ is finite.

It is often more convenient to study the quotient lattice $\mathcal{E}^*$ rather than $\mathcal{E}$. An *automorphism* of a lattice $\mathcal{L}$ is a 1:1 map from $\mathcal{L}$ onto $\mathcal{L}$ which preserves the ordering of $\mathcal{L}$ (and hence preserves meets and joins).

1.3 Definition. A property P of r.e. sets (of r.e. sets modulo finite sets) is *invariant* or *lattice-theoretic* (*l.t.*) in $\mathcal{E}$ (in $\mathcal{E}^*$) if and only if it is invariant under all automorphisms of $\mathcal{E}$ (of $\mathcal{E}^*$). A property P is *elementary lattice theoretic* (*e.l.t.*) if there is a formula of one free variable (with no parameters) in the language $L(\leq, \vee, \wedge, 0, 1)$ which defines the class of those sets in $\mathcal{E}$ (in $\mathcal{E}^*$) having property P, where 0 and 1 are interpreted as the least and greatest elements respectively in the lattice.

Obviously, any e.l.t. property is also l.t. since any automorphism must preserve the inclusion relation. The converse is easily seen to be false since a cardinality argument proves that there are uncountably many l.t. properties of $\mathcal{E}^*$ (see Exercise 3.10(d)), while there are clearly only countably many e.l.t. properties. The properties of recursiveness and finiteness are e.l.t. in both $\mathcal{E}$ and $\mathcal{E}^*$ by the formulas:

(1.1) $$\mathrm{Rec}(x) \equiv (\exists y) \, [x \vee y = 1 \; \& \; x \wedge y = 0],$$

(1.2) $$\mathrm{Fin}(x) \equiv (\forall y) \, [y \leq x \implies \mathrm{Rec}(y)].$$

1.4 Definition. An r.e. set M is *maximal* if M^* is a coatom of $\mathcal{E}^*$, i.e., if $\overline{M}$ is infinite and there is no r.e. set W such that $|W \cap \overline{M}| = |\overline{W} \cap \overline{M}| = \infty$.

It is easy to show (see Exercise 1.6) that if P is a property of r.e. sets which is closed under finite differences then P is e.l.t. in $\mathcal{E}$ if and only if P is e.l.t. in $\mathcal{E}^*$. Since all the properties P we consider are closed under the finite differences, to prove P is e.l.t. in $\mathcal{E}$ it suffices to prove that P is e.l.t. in $\mathcal{E}^*$, which is usually easier and more elegant than proving it for $\mathcal{E}$. For example, simplicity is e.l.t. in $\mathcal{E}^*$ as is the property of A being maximal as demonstrated by the formulas:

$$(1.3) \qquad \mathrm{Sim}(x) \equiv x < 1 \ \& \ (\forall y) \, [y > 0 \implies x \wedge y > 0],$$

and

$$(1.4) \qquad \mathrm{Max}(x) \equiv x < 1 \ \& \ (\forall y) \, [x < y \implies y = 1].$$

Although hh-simplicity is not obviously e.l.t. from its original definition in Chapter V, the unexpected characterization here in Corollary 2.7 shows that hh-simplicity is indeed e.l.t. in $\mathcal{E}^*$ (and hence in $\mathcal{E}$).

The easiest way to prove that a property P is not e.l.t. is to prove that it is not l.t., which requires the method in Chapter XV for generating automorphisms of $\mathcal{E}$. For example, there we prove that hypersimplicity is not l.t. (Theorem XV.3.1), and that any two maximal sets lie in the same orbit (Theorem XV.4.6). Also we show that any automorphism of $\mathcal{E}^*$ is induced by an automorphism of $\mathcal{E}$ (Corollary XV.2.6). Hence, any property P of r.e. sets which is closed under finite differences is l.t. in $\mathcal{E}$ iff P is l.t. in $\mathcal{E}^*$, thereby further identifying $\mathcal{E}$ and $\mathcal{E}^*$ from an algebraic viewpoint.

1.5–1.9 Exercises

1.5. Prove Theorem 1.1. (This can also be found in Rogers [1967a, p. 225].)

1.6. Prove that if A is an infinite r.e. set and I is the principal ideal it generates in $\mathcal{E}$ then $I \cong \mathcal{E}$.

1.7. (a) Using (1.1) prove that the ideal $\mathcal{F}$ and the relation $A =^* B$ are each definable over $\mathcal{E}$.

(b) Prove that if a property P of r.e. sets is closed under finite differences (i.e., is well-defined on $\mathcal{E}^*$) then P is e.l.t. in $\mathcal{E}$ iff P is e.l.t. in $\mathcal{E}^*$. *Hint.* Given a formula Φ^* defining P over $\mathcal{E}^*$, obtain Φ defining P over $\mathcal{E}$ by replacing the atomic formulas $x = y$ using the definability in (a) of $A =^* B$.

1.8. Exhibit formulas which show that the properties of A being simple or maximal are e.l.t. in $\mathcal{E}$. (Use Exercise 1.7 and formulas (1.3) and (1.4).)

1.9.   Prove that $\mathcal{E}^*$ is not densely ordered iff $\mathcal{E}^*$ contains a maximal element. *Hint.* If $\mathcal{E}^*$ is not densely ordered choose $B, A \in \mathcal{E}$, $B^* < A^*$ and for no C, $B^* < C^* < A^*$, where $X^* < Y^*$ denotes that $X \subseteq^* Y$ but $Y \not\subseteq^* X$. Choose a 1:1 recursive function f such that $f(\omega) = A$ and prove that the *pullback* $D = f^{-1}(B)$ is maximal.

1.10 Notes.   Myhill [1956] was the first to suggest studying the lattices $\mathcal{E}$ and $\mathcal{E}^*$ and asked about the existence of maximal elements in $\mathcal{E}^*$. Friedberg [1958a] published three theorems on recursive enumeration: the Friedberg Splitting Theorem 2.1, the existence of a maximal set Theorem 3.3, and a recursive listing of the r.e. sets without repetition.

2. Splitting Theorems and Boolean Algebras

Although the Sacks Splitting Theorem VII.3.2 implies the weak version (asserting only (i) and (ii) below) of the following Friedberg Splitting Theorem, Friedberg's much simpler proof gives the stronger property (iii) which is needed in Lachlan's decision procedure [1968d] for the $\overrightarrow{\forall}\,\overrightarrow{\exists}$-sentences true in $\mathcal{E}^*$. (Of course, Friedberg's theorem does not imply the Sacks theorem because the former is concerned only with the algebraic properties of A_0 and A_1 and not the cone avoiding property of the latter, Theorem VII.3.2(ii).)

Given recursive enumerations $\{X_s\}_{s \in \omega}$ and $\{Y_s\}_{s \in \omega}$ of r.e. sets X and Y define the r.e. sets $X \setminus Y$ and $X \diagdown Y$ as in Definition II.2.9. Notice that $X \setminus Y = (X - Y) \cup (X \diagdown Y)$ as in Exercise II.2.10(b). Hence,

$$(2.1) \qquad\qquad X - Y \text{ non-r.e.} \implies X \diagdown Y \text{ is infinite},$$

because $X \setminus Y$ and $X \diagdown Y$ are both r.e. sets by Exercise II.2.10(a).

2.1 Friedberg Splitting Theorem.   *If B is any nonrecursive r.e. set there exist disjoint r.e. sets A_0 and A_1 such that*

  (i) $B = A_0 \cup A_1$, *and*
  (ii) A_0 *and* A_1 *are nonrecursive.*

Furthermore, if W is any r.e. set then

  (iii) $W - B$ *non-r.e.* $\implies W - A_i$ *non-r.e., for $i = 0, 1$.*

(Of course, (iii) implies (ii) by setting $W = \omega$.) Furthermore, r.e. indices for A_0 and A_1 can be found uniformly from an r.e. index for B.

Proof. Let f be a 1:1 recursive function with range B, and define $B_s = \{f(0), \ldots, f(s)\}$.   For (ii) we try to meet for all e, i the requirements

$R_{\langle e,i \rangle} : W_e \neq \overline{A}_i$. Using (2.1) our strategy automatically satisfies (iii), by satisfying:

$$W_e \setminus B \text{ infinite} \implies W_e \cap A_i \neq \emptyset, \quad \text{for } i = 0, 1,$$

because in (2.2), $f(s+1) \in W_e \setminus B$.

    Stage $s = 0$. Enumerate $f(0)$ in A_0.

    Stage $s + 1$. Choose the least $\langle e, i \rangle$ such that

$$(2.2) \qquad\qquad f(s+1) \in W_{e,s} \text{ and } W_{e,s} \cap A_{i,s} = \emptyset.$$

(Notice that $e < s$ by the definition of $W_{e,s}$.) Enumerate $f(s+1)$ in A_i (so $W_{e,s} \cap A_{i,s+1} \neq \emptyset$ and $R_{\langle e,i \rangle}$ is satisfied forever). Notice that $f(s+1) \in W_e \setminus B$. This is useful in some exercises like 2.19. If $\langle e, i \rangle$ fails to exist, then enumerate $f(s+1)$ in A_0.

    Let $A_i = \bigcup_s A_{i,s}$, $i = 0, 1$. Clearly A_0 and A_1 are disjoint r.e. sets which satisfy (i). To prove (iii), assume for a contradiction that W is a counterexample to (iii). Hence, $(W - A_i)$ is r.e. and

$$W - B = (W - A_i) - B \text{ non-r.e.} \implies (W - A_i) \setminus B \text{ infinite, by (2.1)},$$
$$\implies (W - A_i) \cap A_i \neq \emptyset, \text{ by construction,}$$

which is a contradiction. ☐

2.2 Definition. For a set $S \subseteq \omega$ define $\mathcal{E}(S)$, the lattice (under inclusion) of the r.e. sets restricted to S, to be $\{ W \cap S : W \text{ r.e.} \}$. For $W \in \mathcal{E}$, let W_S denote the set $W \cap S$ in $\mathcal{E}(S)$. A member A_S of $\mathcal{E}(S)$ is *complemented in* $\mathcal{E}(S)$ if there exists an r.e. set B such that $A_S \cup B_S = S$ and $A_S \cap B_S = \emptyset$, and A_S is *noncomplemented in* $\mathcal{E}(S)$ otherwise. (Note that $\mathcal{E}(S)$ is a Boolean algebra if and only if every member is complemented.)

Our goal is to give an algebraic characterization of the notions of hh-immune and hh-simple (see Definition V.2.1) by proving

2.3 Theorem (Morley-Soare). *An infinite Δ_2^0 set S is hh-immune if and only if $\mathcal{E}(S)$ is a Boolean algebra.*

Proof. By Lemma 2.4 and Exercise 2.10. ☐

2.4 Lemma (Lachlan). *If S is infinite (not necessarily in Δ_2^0) and $\mathcal{E}(S)$ is a Boolean algebra then S is hh-immune.*

Proof. If S is not hh-immune as witnessed by the array $\{ W_{f(n)} \}_{n \in \omega}$ of disjoint r.e. sets for some recursive function f, then A_S is noncomplemented in $\mathcal{E}(S)$ where we define the r.e. set $A = \bigcup \{ W_n \cap W_{f(n)} \}_{n \in \omega}$. If $B = W_n$ choose $x \in W_{f(n)} \cap S$. Now either $x \in A \cap B$ or $x \in \overline{A} \cap \overline{B}$, so B_S is not the complement of A_S in $\mathcal{E}(S)$. ☐

For the other direction of Theorem 2.3 we show in Exercise 2.9 that if S is Δ_2^0 then any noncomplemented element A_S of $\mathcal{E}(S)$ can be split uniformly into the disjoint union of noncomplemented elements B_S and C_S. (This uniformity which we stress in Theorem 2.5 and Exercise 2.9 is essential for doing the repeated splitting in Corollary 2.7 or Exercise 2.14.) Friedberg's theorem established this for $S = \omega$, and the next theorem establishes it for any co-r.e. set S. We give the original proof of the latter (which is similar to Friedberg's) because it gives the stronger property (iii) which is useful in decision procedures and which Exercise 2.9 does not give.

2.5 Owings Splitting Theorem. *Let $C \subseteq B$ be r.e. sets such that $B - C$ is not co-r.e. (i.e., B_S is noncomplemented in $\mathcal{E}(S)$ for $S = \overline{C}$). Then there exist disjoint r.e. sets A_0 and A_1 (whose indices may be obtained uniformly from those of B and C) such that*

(i) $B = A_0 \cup A_1$, *and*
(ii) $A_i - C$ *is not co-r.e., for $i = 0, 1$.*

(Namely $(A_i)_S$ is noncomplemented in $\mathcal{E}(S)$ for $S = \overline{C}$.) Furthermore, for any r.e. set W

(iii) $C \cup (W - B)$ *not r.e.* $\implies$ $C \cup (W - A_i)$ *not r.e., for $i = 0, 1$.*

(Clearly (iii) implies (ii) by setting $W = \omega$.)

Proof. Let f be a 1:1 recursive function with range B. Let $B_s = \{ f(0), f(1), \ldots, f(s) \}$, and $\{ C_s \}_{s \in \omega}$ any recursive enumeration of C. For (i) and (ii) we try to meet the requirements:

$$P_s : f(s) \in A_{0,s} \cup A_{1,s}, \quad s \in \omega;$$

$$R_{\langle e, i \rangle} : \overline{W}_e \neq A_i - C, \quad e \in \omega, \quad i = 0, 1,$$

and our strategy automatically guarantees (iii). For the sake of $R_{\langle e, i \rangle}$ we have a recursive function $g(e, i, s)$ which should be thought of as the position of a movable marker $\Gamma_{\langle e, i \rangle}$ at the end of stage s.

Stage $s = 0$. Enumerate $f(0)$ in A_0 and set $g(e, i, 0) = 0$ for all e and i.

Stage $s + 1$.

Step 1. If there is an $x \leq g(e, i, s)$ such that $x \in W_{e,s} \cap A_{i,s} \cap \overline{C}_s$, then set $g(e, i, s + 1) = g(e, i, s)$. Otherwise set $g(e, i, s + 1) = s + 1$.

Step 2. Let $y = f(s + 1)$. Choose the least $\langle e, i \rangle$ such that

$$y \in W_{e,s} \quad \text{and} \quad y \leq g(e, i, s),$$

and enumerate y in A_i. If no $\langle e, i \rangle$ exists, enumerate y in A_0.

Let $A_i = \bigcup_s A_{i,s}$, $i = 0, 1$. Clearly (i) holds. To prove (iii) suppose that for some e, i, and W, $C \cup (W - A_i) = W_e$. We must show that $C \cup (W - B)$ is r.e. Choose s' so large that for all $\langle e', i' \rangle$ less than $\langle e, i \rangle$ if $\lim_s g(e', i', s) < \infty$

then $g(e', i', s') = \lim_s g(e', i', s)$. Let z be the maximum of these values. (Note that $\lim_s g(e', i', s) < \infty$ if and only if $R_{\langle e', i' \rangle}$ is permanently satisfied.) Choose $s'' \geq s'$ such that $f(s) > z$ for all $s \geq s''$. Now $\lim_s g(e, i, s) = \infty$ because $W_e \cap (A_i - C) = \emptyset$. Define the r.e. set

$$V_e = \{ x : (\exists s \geq s'') [x \in W_{e,s} - B_s \ \& \ x \leq g(e, i, s)] \}.$$

Now $V_e \cap \overline{B} = W_e \cap \overline{B}$ since $\lim_s g(e, i, s) = \infty$. Hence, $V_e \cap \overline{B} = W \cap \overline{B}$ since $A_i, C \subseteq B$. Also $V_e \cap (B - C) = \emptyset$. If not suppose $x \in V_e \cap (B - C)$, say $x \in W_{e,s} - B_s$, $x \leq g(e, i, s)$, and $x = f(s+1)$ for some $s \geq s''$. Then by the construction at stage $s + 1$, x is chosen by $R_{\langle e, i \rangle}$ or by some $R_{\langle e', i' \rangle}$, for $\langle e', i' \rangle$ less than $\langle e, i \rangle$, such that $\lim_s g(e', i', s) = \infty$. But the latter implies that x cannot be a permanent witness for $R_{\langle e', i' \rangle}$ (else the limit is $< \infty$) so x must enter C at some stage after $s + 1$. Hence, $V_e \cap (B - C) = \emptyset$, and $V_e \cap \overline{B} = W \cap \overline{B}$ so $C \cup V_e = C \cup (W - B)$ is r.e. ∎

2.6 Definition. For any r.e. set A define the principal filter $\mathcal{L}(A) = \{ B : A \subseteq B \ \& \ B \in \mathcal{E} \}$.

Note that if A is r.e., then $\mathcal{L}(A) \cong \mathcal{E}(\overline{A})$, via the correspondence $W_e \cup A \leftrightarrow W_e \cap \overline{A}$. By Definition 1.2 we write $\mathcal{L}^*(A)$ and $\mathcal{E}^*(\overline{A})$ for the quotient lattices of $\mathcal{L}(A)$ and $\mathcal{E}(\overline{A})$ modulo the ideal $\mathcal{F}$ of finite sets.

2.7 Corollary (Lachlan). *Let C be a coinfinite r.e. set. Then C is hh-simple if and only if $\mathcal{E}(\overline{C})$ (or equivalently $\mathcal{L}(C)$) forms a Boolean algebra.*

Proof. ($\Longleftarrow$). By Lemma 2.4.

($\Longrightarrow$). Suppose $B \supseteq C$, B is r.e. and B is not complemented in $\mathcal{L}(C)$, i.e., $B - C$ is not co-r.e. Then repeatedly apply Theorem 2.5 to obtain a disjoint array $\{ W_{f(n)} \}_{n \in \omega}$ witnessing that $\overline{C}$ is not hh-immune. Namely apply 2.5 to B and C to obtain A_0 and A_1. Set $W_{f(0)} = A_0$, and apply 2.5 to A_1 and $C \cap A_1$ to obtain A_0^1 and A_1^1. Set $W_{f(1)} = A_0^1$ and apply 2.5 to A_1^1 and $C \cap A_1^1$, and so on. By the uniformity of 2.5, the sequence $\{ W_{f(n)} \}_{n \in \omega}$ is r.e., it is disjoint, and $W_{f(n)} \cap \overline{C} \neq \emptyset$ for all n. ∎

2.8 Corollary (Lachlan). *The property of being hh-simple is elementary lattice theoretic in $\mathcal{E}$.*

Proof. If $A \in \mathcal{E}$, then A is hh-simple iff

$$(\forall X \supseteq A) (\exists Y \supseteq A) [X \cap Y = A \ \& \ X \cup Y = \omega]. \quad ∎$$

In Exercise 2.9 we give a different way of proving Corollary 2.7 using the Sacks Splitting Theorem method in place of Theorem 2.5.

2.9–2.19 Exercises

2.9 (Morley-Soare). Prove the following generalization of the Sacks Splitting Theorem.

For any set S in Δ_2^0 and any r.e. set B, if B_S is a noncomplemented member of $\mathcal{E}(S)$ then there exist disjoint r.e. sets A^0 and A^1 whose indices may be found uniformly from that of B, such that:

  (i) $B = A^0 \cup A^1$;
  (ii) A_S^0 and A_S^1 are noncomplemented in $\mathcal{E}(S)$; and
  (iii) A^0 and A^1 are Turing incomparable.

Hint. Show that it suffices to meet the following requirements for $e \in \omega$, $i \in \{0,1\}$,

$$R_e^i : W_e^{A^i} \cap S \neq \overline{B} \cap S.$$

For example, if $A^0 \cap S = \overline{W}_j \cap S$ then $W_e^{A^1} = W_j \cap \overline{A}^1$ contradicts R_e^1. To meet these requirements let $\{\, S_s \,\}_{s \in \omega}$ be a recursive sequence of finite sets and that $S = \lim_s S_s$. Let f be a 1:1 recursive function with range B, and let $B_s = \{\, f(0), \ldots, f(s) \,\}$. Define the recursive functions for $i = 0, 1$

$$l(i, e, s) = \max\{\, x : x \le s \ \& \ (\forall y < x)\, [y \in S_s$$

$$\implies [y \in W_{e,s}^{A_s^i} \iff y \notin B_s]]\,\},$$

$$m(i, e, s) = \max\{\, l(i, e, t) : t \le s \,\},$$

$$r(i, e, s) = \max\{\, z : (\exists s)\, [x \le m(i, e, s) \ \& \ x \in W_{e,s}^{A_s^i} \ \& \ z \le u(A_s^i; e, x, s)\,]\,\}.$$

Enumerate $f(0)$ in A^0. At stage $s+1$, find the least $\langle e, i \rangle$ such that $f(s+1) \le r(i, e, s)$ and enumerate $f(s+1)$ in A^{1-i}. Show that each restraint function $r^i(e, s)$ is injured at most finitely often and each requirement R_e^i is met, else B_S is complemented.

2.10. Use Exercise 2.9 and the method of Corollary 2.7 to complete the proof of Theorem 2.3.

2.11. Prove that every maximal set is hh-simple.

2.12. Use Corollary 2.7 to show that a coinfinite r.e. set C is hh-simple if and only if for every r.e. set $B \supseteq C$ there is a recursive set R such that $B \cap \overline{C} = R \cap \overline{C}$.

2.13. If A and B are r.e., let $A < B$ denote that $A \subset B$ and $B - A$ is not co-r.e. Prove that if $A < B$ then $A < C < B$ for some r.e. set C.

2.14. Prove that if $C \subseteq B$ are r.e. sets and $B - C$ is hh-immune then $B - C$ is co-r.e. *Hint.* If not then B is noncomplemented in $\mathcal{E}(\overline{C})$ and $B - C$ can be continuously split using Theorem 2.5 as in the proof of Corollary 2.7.

2.15 (Robinson [1967a]).   Prove that a coinfinite r.e. set A is hh-simple if and only if for every disjoint weak array of finite sets $\{W_{f(n)}\}_{n\in\omega}$ there exists n, such that for all $m \geq n$, $|W_{f(m)} \cap \overline{A}| \leq m$. *Hint.* Assume that there exists a disjoint weak array $\{W_{f(n)}\}_{n\in\omega}$ violating this condition. Construct a disjoint weak array $\{W_{g(n)}\}_{n\in\omega}$ witnessing that A is not hh-simple according to Definition V.2.1. Attempt to put at least one element of $W_{f(n)} \cap \overline{A}$ (if such exists) into each set $W_{g(m)}$, $m \leq n$.

2.16 (Yates [1962]).   An infinite set B is *strongly hh-immune* (*shh-immune*) if there is no uniformly r.e. (u.r.e.) sequence $\{W_{f(n)}\}_{n\in\omega}$ of pairwise disjoint r.e. sets such that $W_{f(n)} \cap B \neq \emptyset$ for every n. (The difference between this and the Definition V.2.1 of hh-immune is that in the latter definition the r.e. sets $W_{f(n)}$ must all be finite.) Prove that if A is a coinfinite r.e. set then $\overline{A}$ is hh-immune iff $\overline{A}$ is shh-immune.

2.17.   Use Corollary 2.7 to prove that if A and B are hh-simple then $A \cap B$ is hh-simple. (Hence, the hh-simple sets and cofinite sets form a filter. By Exercises V.1.5 and V.2.8 the same is true for the simple and for the hypersimple sets.) *Hint.* Let A and B be hh-simple and C be r.e. Prove that $(A \cap B) \cup \overline{C}$ is r.e.

2.18.   Prove that if A is r.e., coinfinite, and not hh-simple, then $\mathcal{E}^*$ is embeddable in $\mathcal{L}^*(A)$. *Hint.* Let $\{S_n\}_{n\in\omega}$ witness that $\overline{A}$ is not hh-immune, and let

$$\widetilde{W}_n = \bigcup\{S_m : m \in W_n\}.$$

2.19 (Shore [1978a]).   An r.e. set B is *nowhere simple* if for every r.e. set C with $C - B$ infinite there is an infinite r.e. set $W \subseteq C - B$. Prove that every r.e. set B can be split into the disjoint union of nowhere simple r.e. sets A_0 and A_1. *Hint.* Modify the construction in Theorem 2.1 to meet the requirements,

$$R_{\langle e,i,n\rangle} : \text{ Put } n \text{ elements of } W_e \text{ into } A_i.$$

At stage s put $f(s)$ into A_i to meet the unsatisfied requirement of highest priority. To see that A_0 is nowhere simple fix e such that $W_e - A_0$ is infinite. Consider the cases where $W_e \cap A_1$ is infinite, and where $W_e \smallsetminus B$ is finite.

2.20 Notes.   Friedberg [1958a] proved Theorem 2.1. The next major advance was Lachlan's unexpected characterization Corollary 2.7 of the hh-simple sets [1968c] which he proved using a different method. After seeing a preprint of Lachlan's proof, Owings [1967, p. 174] proved Theorem 2.5 and showed how it could be used to derive Corollary 2.7. Later Morley and Soare [1975] proved Exercise 2.9 and Theorem 2.3.

3. Maximal Sets

Friedberg settled Myhill's question from Notes 1.10 by constructing a maximal set, i.e., a co-atom of $\mathcal{E}^*$. This was important for several reasons. Post had defined the concept of hh-simplicity but had been unable to construct an hh-simple set. Since maximal sets are clearly hh-simple, this gave the first (and still the easiest) proof of the existence of an hh-simple set. It is easy to modify the construction to make the maximal set Turing complete. Since maximal sets have the "thinnest" possible complements among r.e. sets, this dooms Post's program in Chapter V of finding a "thinness" property of the complement $\overline{A}$ of an r.e. set A which guarantees that A is incomplete. (Other properties in the style of Post which guarantee incompleteness have later been discovered; see Notes 3.13.) Finally, the existence of a maximal set proves by Exercise 1.9 that the lattice $\mathcal{E}^*$ is not densely ordered. (The closest we can come to a dense ordering is the property in Exercise 2.13.)

3.1 Definition. An infinite set C is *cohesive* if there is no r.e. set W such that $W \cap C$ and $\overline{W} \cap C$ are both infinite.

Thus, by Definition 1.4, an r.e. set A is maximal iff $\overline{A}$ is cohesive. Note that a cohesive set is hh-immune and that A is maximal if and only if A^* is a co-atom of $\mathcal{E}^*$. It is easy to give a noneffective construction of a cohesive set (see Exercise 3.4). To construct a maximal set we effectivize this strategy. To accomplish this we need the following technical device which will frequently be used.

3.2 Definition. The *e-state of x at stage s* is

$$\sigma(e, x, s) = \{\, i : i \leq e \ \& \ x \in W_{i,s} \,\}.$$

We identify each e-state with the string in $2^{<\omega}$ of length $e + 1$ which is the initial segment of its characteristic function. For fixed e, there are 2^{e+1} e-states which are ordered lexicographically by first differences. We read e-states from left to right so that, for example, the e-state 100 indicates that $x \in W_0$, $x \notin W_1$, and $x \notin W_2$. When σ is greater than τ in this ordering we say σ is a *higher* (also called *stronger*) e-state than τ and we write $\sigma > \tau$. For example $010 > 001$. For convenience we say $\sigma(-1, x, s) = \emptyset$ for all x and s. Define $\sigma(e, x) = \lim_s \sigma(e, x, s)$, which clearly exists.

3.3 Theorem (Friedberg [1958a]). *There exists a maximal set A.*

Proof. It suffices to construct an r.e. set A which meets for all e the requirements

$$N_e : |\overline{A}| \geq e,$$

and

$$P_e : \overline{A} \subseteq^* \overline{W}_e \text{ or } \overline{A} \subseteq^* W_e.$$

To construct A we have a sequence of markers $\{\Gamma_n\}_{n \in \omega}$, with a_n^s denoting the position of Γ_n at the end of stage s and such that

$$\overline{A}_s = \{ a_0^s < a_1^s < \cdots \}.$$

The strategy for meeting P_0 is to choose at stage $s+1$ the least e (if it exists) such that $a_e^s \notin W_{0,s}$ but $a_i^s \in W_{0,s}$ for some $i > e$, and to move Γ_e to a_i^s. (Namely, each marker moves to maximize (raise) its 0-state.) This strategy is clearly effective and satisfies P_0 because if $W_0 \cap \overline{A}$ is infinite then each marker eventually comes to rest on an element of W_0, and hence $\overline{A} \subseteq W_0$. The purpose of the e-state function is to combine the strategy for P_{i+1} with that for $P_{i'}$, $i' \leq i$. Namely, each marker Γ_e moves to maximize its e-state. Thus, each marker moves at most finitely often, but all markers Γ_i, $i \geq e$, respect requirement P_e. We construct A as follows.

Stage $s = 0$. Set $A_0 = \emptyset$ and $a_n^0 = n$ for all n.

Stage $s + 1$. Choose the least e such that for some $i, e < i \leq s$ and $\sigma(e, a_i^s, s) > \sigma(e, a_e^s, s)$. For this e choose the least such i. Move marker Γ_e to element a_i^s, and move markers Γ_j, $j > e$, preserving their order to elements a_k^s, $k > i$, thereby enumerating elements $a_e^s, a_{e+1}^s, \ldots, a_{i-1}^s$ into A. (If e fails to exist do nothing.)

Lemma 1. *For every e, marker Γ_e moves at most finitely often. (Hence, $\overline{A}$ is infinite.)*

Proof. Fix e and assume the lemma for all $i < e$. Choose s_0 such that for all $i < e$, $a_i^{s_0} = a_i =_{\text{dfn}} \lim_s a_i^s$. After stage s_0, Γ_e moves only to an element in a strictly higher e-state, but there are only 2^{e+1} different e-states. Hence Γ_e moves at most $2^{e+1} - 1$ times after stage s_0.

Lemma 2. $(\forall e) [\overline{A} \subseteq^* \overline{W}_e \text{ or } \overline{A} \subseteq^* W_e].$

Proof. Fix e and assume the lemma for all $i < e$. Choose $n \geq e$ such that for all $m \geq n$, $\sigma(e-1, a_m) = \sigma(e-1, a_n)$. Assume for a contradiction that $W_e \cap \overline{A}$ and $\overline{W}_e \cap \overline{A}$ are both infinite. Choose p, q such that $n < p < q$, $a_p \notin W_e$ and $a_q \in W_e$. Choose s so that for all $m \leq q$, $a_m = a_m^s$ and $\sigma(e, a_m) = \sigma(e, a_m, s)$. However, $\sigma(e, a_q) > \sigma(e, a_p)$ and hence $\sigma(p, a_q) > \sigma(p, a_p)$. Thus, some marker Γ_r, $r \leq p$, moves at stage $s+1$ contrary to the choice of s. □

3.4–3.12 Exercises

3.4. (a) Construct a cohesive set C as follows. Let $Z_{-1} = \omega$. Given Z_n, $n \geq -1$, let $Z_{n+1} = Z_n \cap W_n$ if the latter is infinite and $Z_n \cap \overline{W}_n$ otherwise.

Choose $c_0 \in Z_0$, $c_1 \in Z_1$ and $c_0 < c_1 \ldots$, and let $C = \{\, c_i \,\}_{i \in \omega}$. Show that C is cohesive.

(b) Show that any infinite set has a cohesive subset.

3.5. Show that ω is the disjoint union of a countable number of cohesive sets but not of a finite number of such sets.

3.6. (a) Combine the construction of Theorem 3.3 with that of Exercise V.1.6 to make A both maximal and effectively simple (and hence complete).

(b) Construct a complete maximal set A by enumerating a_n^s into A_{s+1} if $n \in K_{s+1} - K_s$, and also using the construction of Theorem 3.3.

3.7. We say that a partial function ψ *dominates* a partial function θ if for almost every n,

$$\theta(n) \downarrow \implies [\psi(n) \downarrow \ \& \ \theta(n) \leq \psi(n)].$$

Prove that if ψ dominates every p.r. function then $\emptyset' \leq_T \psi$. *Hint.* Consider the p.r. function $\theta(x) = \mu s[x \in K_s]$.

3.8 (Yates [1965]). Modify the proof of Theorem 3.3 to construct a maximal set A such that the principal function $p_{\overline{A}}$ dominates every p.r. function (and hence A is complete). *Hint.* Move marker Γ_n whenever necessary to arrange that $\varphi_e(n) \leq a_n$ for all $e \leq n$ such that $\varphi_e(n) \downarrow$, and simultaneously move Γ_n to maximize its n-state.

3.9. Show that for every infinite r.e. set A there is an r.e. set $B \subseteq A$ which is *maximal in* A in the sense that $A - B$ is cohesive. (Note that by Exercise 2.14, $A - B$ must be co-r.e.) *Hint.* Consider $f(M)$ for a 1:1 recursive function f with range A and a maximal set M.

3.10. A set A is *quasi-maximal* if A is the intersection of finitely many maximal sets.

(a) Prove that a coinfinite r.e. set A is quasi-maximal iff $\mathcal{L}^*(A)$ is finite.

(b) Prove that every quasi-maximal set is hh-simple.

(c) Prove that for each n if A_i, $1 \leq i \leq n$, are maximal sets such that $A_i^* \neq A_j^*$ for $i \neq j$, and $B_n = \bigcap\{\, A_i : i \leq n \,\}$ then $\mathcal{L}^*(B_n)$ is the Boolean algebra of cardinality 2^n.

(d) Use (c) to show that there are $2^{\aleph_0}$ properties P which are l.t. in $\mathcal{E}^*$. *Hint.* For every $f \in 2^\omega$ let property P_f include B_n iff $f(n) = 1$.

3.11 (Tennenbaum). Prove that there is a recursive linear ordering $<_L$ of the integers of order type $\omega + \omega^*$ (where ω^* is the order type of the negative integers) which has no infinite r.e. sequences which are either strictly increasing or strictly decreasing in $<_L$.

Hint. Construct the ordering by meeting for all $e \in \omega$ and $i < 2$, the requirements

$$R^i_{2e+1} : W_e \text{ infinite} \implies W_e \cap A_i \neq \emptyset$$

where A_0 is the initial segment of order type ω and $A_1 = \overline{A}_0$. Let L_s denote that portion of $<_L$ defined by the end of stage s, and let a^s_e, b^s_e be the positions of movable markers on L_s. Let $I^s_e = \{ x : a^s_e \leq_{L_s} x \leq_{L_s} b^s_e \}$. Suppose at the end of stage s we are given L-intervals I^s_e for $e \leq s$, such that $a^s_e <_L a^s_{e+1} <_L b^s_{e+1} <_L b^s_e$, for $e < s$, and such that at least all $x \in [0, 2s]$ have been assigned places in the L-ordering. However, there may be more elements in the L-ordering than $\{ a^s_e \}_{e \leq s}, \{ b^s_e \}_{e \leq s}$. If R_{2e} is unsatisfied at stage $s + 1$, and there exists $x \in I^s_{2e} \cap W_{e,s}$, meet R_{2e} by defining $a^{s+1}_{2e} = x$ and $b^{s+1}_{2e} = b^s_{2e}$, and by using fresh elements newly inserted in the L-ordering in the interval I^{s+1}_{2e} as end points for the intervals I^{s+1}_j, $e < j \leq s + 1$, making sure that $I^{s+1}_{j+1} \subseteq I^s_j$.

3.12. (a) Prove that there is no recursive ordering L of order type $\omega + \omega^*$ such that the initial segment A of order type ω is cohesive. (Hence, by symmetry the final segment of order type ω^* is not cohesive either.) *Hint.* As we enumerate the elements of L we color them red or blue so that the r.e. sets R and B of red (blue) elements satisfy the requirements $R_{2e} : |R \cap A| \geq e$ and $R_{2e+1} : |B \cap A| \geq e$.

(b) (D. A. Martin). Do part (a) with "hh-immune" in place of "cohesive." (See Jockusch [1968a, Corollary 4.5].)

3.13 Notes. Friedberg [1958a] first constructed a maximal set, and introduced the e-state function. The simplified construction in Theorem 3.3 is due to Yates [1965] who first constructed a complete maximal set as in Exercise 3.8. This shows that no property of "thinness" on the complement $\overline{A}$ of an r.e. set A can guarantee that A is incomplete. However, Marchenkov [1976a] suggested a different type of property in the style of Post which does guarantee incompleteness, namely being an η-maximal set which is also a lower cut in some recursive linear ordering of ω. An η-maximal set is one which is maximal when integers are replaced by equivalence classes under some r.e. equivalence relation. For a proof see Odifreddi [1981].

4. Major Subsets and r-Maximal Sets

If $A \subseteq B$ are r.e. sets and $B \neq^* A$ there are various ways A can be "close to" B. For example, A may be maximal in B in the sense of Exercise 3.9. Alternatively, A may be joined to ω by exactly the same r.e. sets joining B to ω modulo finite sets.

4.1 Definition. Given r.e. sets $A \subseteq B$, A is a *major subset* of B (written $A \subset_m B$) if $B - A$ is infinite and for every r.e. set W,

(4.1) $$\overline{B} \subseteq^* W \implies \overline{A} \subseteq^* W.$$

(Note that it is equivalent to replace (4.1) by the condition $\overline{B} \subseteq W \implies \overline{A} \subseteq^* W$.)

Note that if A and B are r.e. sets such that $A \subset_\infty B$ then $A \subset_m B$ iff there is no recursive set R such that $R \subseteq B$ and $R - A$ is infinite by Corollary II.1.10 and Exercise II.1.23. Major subsets are necessary for decision procedures for certain subclasses of the elementary theory of r.e. sets, and they give easy examples of nontrivial r-maximal sets, a property inspired by maximal sets.

4.2 Definition. (i) An infinite set C is *r-cohesive* if there is no recursive set R such that $R \cap C$ and $\overline{R} \cap C$ are both infinite.
   (ii) An r.e. set A is *r-maximal* if $\overline{A}$ is r-cohesive.

4.3 Proposition. *Assume $A \subset_\infty B$ and B is r-maximal. Then $A \subset_m B$ iff A is r-maximal.*

Proof. ($\implies$). Assume that $A \subset_m B$. To see that A is r-maximal, fix any recursive set R. Let $R = W_e$ and $\overline{R} = W_i$. Now $W_e \cup W_i = \omega \supseteq \overline{B}$, so either $W_e \cap \overline{B}$ or $W_i \cap \overline{B}$ is infinite, say the former. Then $\overline{B} \subseteq^* W_e$ by the r-maximality of B. Now $\overline{A} \subseteq^* W_e$ because $A \subset_m B$. Hence $W_i \cap \overline{A}$ is finite so R does not split $\overline{A}$.
   ($\impliedby$). Suppose A is r-maximal but $A \not\subset_m B$. Choose an r.e. set $W \supseteq \overline{B}$ such that $|\overline{W} \cap (B - A)| = \infty$. Choose any infinite recursive set $R \subseteq W$. Hence, R witnesses that A is not r-maximal. ▯

4.4 Corollary. *If B is r-maximal and $A \subset_m B$ then A is r-maximal but not maximal.*

Proof. Now A is r-maximal by Proposition 4.3 and not maximal since $A \subset_\infty B \subset_\infty \omega$. ▯

Clearly any maximal set B is r-maximal, so Corollary 4.4 gives an example of r-maximal sets which are not maximal, since by Theorem 4.6 any nonrecursive r.e. set B has a major subset.

4.5 Proposition. *If A is r-maximal and hh-simple then A is maximal.*

Proof. If A is r-maximal then $\mathcal{L}^*(A)$ contains no nontrivial complemented elements, but if A is hh-simple then every element of $\mathcal{L}^*(A)$ is complemented. Hence, $\mathcal{L}^*(A)$ is isomorphic to the two element Boolean algebra, and A is maximal. ▯

We cannot have r.e. sets $A \subset_m B$ such that A is maximal in B since the latter implies by Exercise 2.14 that $B-A$ is co-r.e. Thus, given a nonrecursive r.e. set B we cannot hope to construct $A \subset_m B$ by exactly the same proof as in §3, but surprisingly the same construction works if we merely modify the array with respect to which we measure the e-states.

4.6 Theorem (Lachlan). *For every nonrecursive r.e. set B there exists an r.e. set A such that $A \subset_m B$.*

Proof. We use a sequence of movable markers $\{\Gamma_n\}_{n\in\omega}$ as in Theorem 3.3, except that now $B - A = \{d_0 < d_1 < \cdots\}$, where d_n^s is the position of Γ_n at the end of stage s, and $d_n = \lim_s d_n^s$. We meet the requirements,

$$N_e : \text{ marker } \Gamma_e \text{ moves at most finitely often,}$$
$$P_e : \ \overline{B} \subseteq W_e \implies \overline{A} \subseteq^* W_e.$$

Define an r.e. array of finite sets $\{V_{e,s}\}_{s\in\omega}$ (which may be specified by the canonical indices in Definition II.2.4) as follows,

$$x \in V_{e,s} \equiv_{\mathrm{dfn}} x \in W_{e,s} \ \& \ (\forall y \leq x)\,[y \in W_{e,s} \cup B_s].$$

Let $V_e = \bigcup_s V_{e,s}$. The point of this definition is that

(4.2) $$\overline{B} \subseteq W_e \implies V_e = W_e,$$

but

(4.3) $$\overline{B} \not\subseteq W_e \implies V_e \text{ is finite.}$$

Now define the e-state function $\sigma(e, x, s)$ exactly as in Definition 3.2 but with $V_{e,s}$ in place of $W_{e,s}$. Choose a 1:1 recursive function f with range B and construct $A \subset_m B$ as follows.

Stage $s = 0$. Set $A_0 = \emptyset$ and $d_0^0 = f(0)$, and for any $e > 0$, $d_e^0 = -1$ (namely marker Γ_e is *unassigned*).

Stage $s + 1$. Let $y = f(s + 1)$. Choose the least e such that $d_e^s < y$; $-1 < d_{e-1}^s < y$; and either $\sigma(e, y, s) > \sigma(e, d_e^s, s)$ or $d_e^s = -1$. Set $d_e^{s+1} = y$, $d_i^{s+1} = -1$ for all $i > e$, and $d_i^{s+1} = d_i^s$ for all $i < e$, thereby enumerating into A_{s+1} all d_i^s, $i \geq e$, if $d_i^s > -1$. (If e fails to exist, enumerate y into A_{s+1}.)

Lemma 1. *Each marker Γ_e moves at most finitely often and is eventually permanently assigned to some $x > -1$. (Hence $B - A$ is infinite.)*

Proof. Fix e and assume that Γ_i has come to rest for each $i < e$. Now Γ_e will be assigned to some x because B is infinite, and thereafter will move at most $2^{e+1} - 1$ times, each time to an element newly enumerated in B.

Lemma 2. $(\forall e) [\overline{B} \subseteq W_e \implies \overline{A} \subseteq^* W_e].$

Proof. Fix e and assume the lemma by induction for all $i < e$. Let

$$\sigma = \{ i : i < e \ \& \ \overline{B} \subseteq W_i \}.$$

By (4.2) and (4.3) we have for $i < e$

$$i \in \sigma \implies V_i = W_i,$$

and

$$i \notin \sigma \implies V_i \text{ is finite.}$$

From these and the induction hypothesis we conclude that there is an m such that for all $n > m$, $\sigma(e-1, d_n) = \sigma$. Let $V_\sigma = \bigcap \{ V_i : i \in \sigma \}$. Now V_σ is r.e. and $\overline{B} \subseteq V_\sigma$ since $\overline{B} \subseteq V_i$ for every $i \in \sigma$. Assume $\overline{B} \subseteq W_e$. Then $V_e = W_e$ and the r.e. set $U =_{\mathrm{dfn}} V_\sigma \cap V_e$ contains $\overline{B}$. But B is not co-r.e., so $U \smallsetminus \overline{B}$ is infinite, namely

(4.4) $(\exists^\infty s) (\exists x) [x \in B_{s+1} - B_s \ \& \ \sigma(e-1, x, s) = \sigma \ \& \ x \in V_{e,s}].$

But if $\overline{V}_e \cap (B - A)$ is infinite we get a contradiction as in Lemma 2 of Theorem 3.3 by considering $d_p \in \overline{V}_e \cap (B - A)$, $m < p$, and noting that Γ_p will move to some element x which satisfies the matrix of (4.4). Hence, $(B - A) \subseteq^* V_e$.   ∎

4.7–4.12 Exercises

4.7.   Prove that if $A \subset_m B$ and $B \subset_m C$ then $A \subset_m C$.

4.8.   Let $A \subset_m B$, and let f be a 1:1 recursive function with range B. Define the *pullback* $\hat{A} = f^{-1}(A)$.
   (a) Prove that $\hat{A} \leq_T A$.
   (b) Show that $\hat{A}$ has no hh-simple (and therefore no maximal) superset.
Hint. Repeatedly use Theorem 2.5.

4.9 (Lerman).   For r-maximal sets A and B we write $A \sim B$ if $A \cup B \subset_\infty \omega$.
   (a) Prove that $\sim$ is an equivalence relation on r-maximal sets.
   (b) Prove that for A, B r-maximal, if $A \neq^* B$ then $A \sim B$ if and only if $A \cap B \subset_m A \cup B$.

4.10 Definition. (i) Given r.e. sets $A \subset_\infty B$, A is *small* in B (written $A \subset_s B$) if for every pair of r.e. sets X, Y, if $X \cap (B - A) \subseteq Y$, then $Y \cup (X - B)$ is r.e.
   (ii) A is *small* if $A \subset_s B \subset \omega$ for some r.e. set B. (The intuition is that $B - A$ is so large that if Y contains $X \cap (B - A)$, then Y contains enough of X so that $Y \cup (X - B)$ is r.e.)

4.11 (Stob [1979]).   Prove the following:

(a) If $A \subset B$ and A is recursive, then $A \subset_s B$.
(b) If $A \subset B$ and B is recursive then $A \subset_s B$.
(c) If $A \subset C \subset B$ and $A \subset_s C$, then $A \subset_s B$.
(d) If $A \subset C \subset B$ and $C \subset_s B$ then $A \subset_s B$.
(e) If $A \subset_s B$ and B is nonrecursive then $A \cup \overline{B}$ is not r.e.

4.12° (Lachlan [1968d]).   Show that if B is r.e. and nonrecursive then B has a small major subset A (written $A \subset_{sm} B$), namely $A \subset_s B$ and $A \subset_m B$. *Hint.* Construct $A \subset_\infty B$ as in Theorem 4.6 to meet requirements P_e there, and also requirements

$$N_i : Y_i \supseteq X_i \cap (B - A) \implies (X_i - B) \cup Y_i \text{ is r.e.}$$

where $\{ (X_i, Y_i) \}_{i \in \omega}$ is a listing of all pairs of r.e. sets. The priority ranking of requirements is $N_0, P_0, N_1, P_1, \dots$. Fix recursive enumerations $\{ X_{i,s} \}_{s \in \omega}$ and $\{ Y_{i,s} \}_{s \in \omega}$ of X_i and Y_i. To aid in meeting N_i we define by induction on s, $g(i, s)$ and $Z_{i,s}$ where g plays a role similar to the g in the Owings Splitting Theorem 2.5. For $s = 0$, set $g(i, 0) = 0$, $Z_{i,0} = \emptyset$. For $s + 1$ define

$$g(i, s + 1) = \begin{cases} (\mu x) \, [x \in (B_{s+1} - A_s) \cap (Z_{i,s} - Y_{i,s})] & \text{if such an } x \text{ exists,} \\ s + 1 & \text{otherwise.} \end{cases}$$

where $Z_{i,s+1} = \{ x : x \in X_{i,s+1} \ \& \ x \le g(i, s + 1) \} \cup Z_{i,s}$.

Elements in $(B_{s+1} - A_s) \cap (Z_{i,s+1} - Y_{i,s})$ are *restrained with priority* N_i from entering A_{s+1}. (Notice that only finitely many elements are permanently restrained by N_i because if Z_i is infinite then $\lim_s g(i, s) = \infty$.) Modify stage $s + 1$ of Theorem 4.6 by choosing the least e such that $d_e^s < y$ and for some $j \le e$:

$$\sigma(j - 1, y, s) = \sigma(j - 1, d_e^s, s);$$

$$\sigma(j, y, s) < \sigma(j, d_e^s, s);$$

and no d_k^s, $k \ge e$, is restrained by any N_i, $i \le j$.

Then set $d_e^{s+1} = y$ and enumerate into A_{s+1} (for the sake of P_j) all elements d_k^s, $e \le k$. (If there is no such e enumerate y into A_{s+1}.) Now each requirement P_j is met since each N_i permanently restrains at most finitely many elements. To verify N_i, fix i and set $\sigma = \{ e : e \le i \ \& \ \overline{B} \subseteq W_e \}$. Now $\overline{B} \subseteq V_\sigma$ so $\overline{A} \subseteq^* V_\sigma$. Choose x_0 such that $x \in A \cup V_\sigma$ for all $x \ge x_0$ and define

$$U_{i,s} = \{ x : x \in X_{i,s} - B_s \ \& \ x \le g(i, s)$$
$$\& \ (\forall y) \, [x_0 \le y \le x \implies y \in (A_s \cup V_{\sigma,s})] \}.$$

Now assume that $Y_i \supseteq X_i \cap (B - A)$. Then $\lim_s g(i, s) = \infty$, so $X_i - B \subseteq^* U_i$, and hence $X_i - B =^* U_i - B$. Now prove that $U_i \cup Y_i =^* (X_i - B) \cup Y_i$, so the latter is r.e. and N_i is satisfied. (To accomplish this note that we already have the inclusion $\,^* \supseteq$. To get the inclusion $\subseteq^*$ use the negative restraint imposed by N_i against elements $x \in U_{i,s} \cap (B_{s+1} - A_s)$ being enumerated in A before they appear in Y_i, noting that $U_{i,s} \subseteq Z_{i,s}$. This negative restraint will not be injured by any P_j, $j < i$, since every $x \in U_{i,s}$ is already in V_σ. Also $\lim_s g(i, s) = \infty$. Hence, $U_i \cap B \subseteq^* Y_i$, which establishes the inclusion $\subseteq^*$ desired above.)

4.13 Notes.   Lachlan proved Theorem 4.6 in [1968c] and then the stronger form Exercise 4.12 in [1968d] which he needed for his decision procedure for the $\overrightarrow{\forall}\,\overrightarrow{\exists}$-sentences true in $\mathcal{E}^*$. For a further discussion see Chapter XVI §2.

5. Atomless r-Maximal Sets

5.1 Definition.   A coinfinite r.e. set A is *atomless* if A has no maximal super-set.

We saw in Exercise 4.8 that atomless sets exist. An easy direct proof is given in Exercise 5.5. The r-maximal sets which we have encountered so far are not atomless since they are major subsets of maximal sets. We now construct atomless r-maximal sets and in the next section atomless hh-simple sets. The constructions are similar to that of the maximal set, but with markers arranged in certain nonlinear configurations and with the e-states measured with respect to certain new strong arrays in place of standard enumerations $\{W_{e,s}\}$.

For this theorem and later ones we wish to change our intuitive picture of movable markers. Instead of picturing the markers $\{\Gamma_n\}_{n\in\omega}$ as moving on elements, we think of the markers as stationary "windows" which have elements appearing in them, the elements changing (increasing in magnitude) from time to time. We still use d_n^s to denote the element associated with Γ_n at the end of stage s. The fundamental operation in which Γ_n was previously viewed as moving from d_n^s to d_{n+k}^s is now referred to as "Γ_n pulls d_{n+k}^s at stage $s+1$." The mathematical nature of the sequence $\{d_n^s\}_{n\in\omega}$ is unchanged but the picture is clearer in constructions such as this one where the markers are arranged in certain geometrical configurations (such as a square array or as the nodes on a tree) and the motion of elements is partially determined by their current position.

5.2 Definition (Robinson [1967a]). A sequence (not necessarily r.e.) of r.e. sets $\{H_n\}_{n\in\omega}$ forms a *tower* if $\bigcup_n H_n = \omega$ and for all n, $H_n \subset_\infty H_{n+1}$.

5.3 Lemma (Robinson [1967a]). *If A is an r.e. set and $\{H_n\}_{n\in\omega}$ is a tower such that $A \subset_\infty H_0$ and the requirement*

$$P_n : W_n \subseteq^* H_n \text{ or } \overline{A} \subseteq^* W_n$$

holds for every n, then A is atomless and r-maximal.

Proof. To see that A is atomless suppose that W_n is coinfinite and $W_n \supseteq A$. Then $W_n \subseteq^* H_n \subset_\infty H_{n+1} \subset_\infty \omega$, so H_{n+1} witnesses that W_n is not maximal. To see that A is r-maximal, suppose R is a recursive set, say $R = W_i$ and $\overline{R} = W_j$. If R splits $\overline{A}$, then the second clause of P_n fails for $n = i, j$, so $W_i \subseteq^* H_i$, $W_j \subseteq^* H_j$, and $\omega = W_i \cup W_j \subseteq^* H_k$, where $k = \max\{i, j\}$, contradicting $H_k \subset_\infty H_{k+1}$. $\blacksquare$

5.4 Theorem (Lachlan [1968c], Robinson [1967a]). *There exists an atomless r-maximal set A.*

Proof. We have markers $\{\Gamma_{\langle i,j\rangle}\}_{i,j\in\omega}$, arranged in a square array where d_n^s denotes the element associated with Γ_n at the end of stage s, $d_n =_{\mathrm{dfn}} \lim_s d_n^s$, and $\overline{A}_s = \{d_n^s\}_{n\in\omega}$, and $\overline{A} = \{d_n\}_{n\in\omega}$. We must meet for each n the requirements,

$$N_n : \quad d_n = \lim_s d_n^s < \infty,$$
$$P_n : \quad W_n \subseteq^* H_n \text{ or } \overline{A} \subseteq^* W_n,$$

and

$$R_n : \quad H_n \text{ is r.e.,}$$

where for each i we define the *columns*

$$C_i^s = \{d_{\langle i,j\rangle}^s\}_{j\in\omega},$$
$$C_i = \{d_{\langle i,j\rangle}\}_{j\in\omega},$$

and we define

$$H_n = A \cup \left(\bigcup\{C_i : i \le n\}\right).$$

To meet the requirements we define a strong array $\{U_{n,s}\}_{n,s\in\omega}$, and perform the Friedberg construction maximizing e-states measured with respect to $U_{n,s}$, but with a minor modification as explained below. We then define $U_n = \bigcup_s U_{n,s}$. Define

$$U_{n,s} = \{x : (\exists t \le s)\, [x = d_{\langle i,j\rangle}^t \ \& \ x \in W_{n,t} \ \& \ n < i]\},$$

and

$$\sigma(e, x, s) = \{\, i : i \leq e \ \& \ x \in U_{i,s} \,\}, \text{ the } e\text{-state of } x \text{ at stage } s.$$

Stage $s = 0$. Define $d_n^0 = n$.

Stage $s + 1$. Find the least e such that for some i, $e < i \leq s$, $\sigma(e, d_i^s, s) > \sigma(e, d_e^s, s)$. Choose $i > e$, $i \leq s$ and with $\sigma(e, d_i^s, s)$ as large as possible and define $d_e^{s+1} = d_i^s$. We say Γ_e *pulls* d_i^s. (If i fails to exist, do nothing.) Enumerate d_k^s, $e \leq k \leq s$, $k \neq i$, into A. (The enumeration into A of all these elements, rather than just d_k^s, $e \leq k < i$, is necessary for technical reasons and is referred to as a *dump*.) Set $d_j^{s+1} = d_j^s$, for all $j < e$, and $d_{e+k}^{s+1} = d_{s+k}^s$, for all $k > 0$.

The purpose of the dump is the following. If $z \in W_{k,s}$, then $z < s \leq d_s^s$, so either $z < d_e^s$, or $z = d_e^{s+1}$, or else $z \in A_{s+1}$. Thus, it follows as we see in Lemma 3 that if Γ_e pulls an element $y = d_i^s$ at stage $s + 1$, then y must have improved its e-state at stage s. (See also the explanation for (6.5).)

Lemma 1. *For every* n, $d_n = \lim_s d_n^s$ *exists and* $\overline{A} = \{\, d_n \,\}_{n \in \omega}$.

Proof. The proof that $\lim_s d_n^s$ exists is by induction on n exactly as in Lemma 1 of Theorems 3.3 or 4.6.

Lemma 2. $(\forall n) [W_n \subseteq^* H_n \text{ or } \overline{A} \subseteq^* W_n]$.

Proof. The proof is by induction on n. Note that U_0 is infinite if and only if $\overline{A} \subseteq U_0$. (If U_0 is infinite then $\overline{A} \subseteq U_0$ by the definition of U_0 and the construction of A. The converse is immediate since $\overline{A}$ is infinite by Lemma 1.) Now fix $n > 0$ and suppose for all $m < n$ that

(5.1) $U_m \cap \overline{A}$ is finite or $\overline{A} \subseteq^* U_m$.

Let $\sigma = \{\, m : m < n \ \& \ \overline{A} \subseteq^* U_m \,\}$, the true $(n-1)$-state of (almost all elements of) $\overline{A}$, and let $U_\sigma = \bigcap \{\, U_m : m \in \sigma \,\}$. If $U_n \cap U_\sigma =^* \emptyset$, then $U_n \cap \overline{A} =^* \emptyset$ and $W_n \subseteq^* H_n$, by the definition of U_n. But if $U_n \cap U_\sigma$ is infinite, then by construction $\overline{A} \subseteq^* U_n$. Thus, (5.1) holds for $m = n$.

Lemma 3. $(\forall e) [H_e \text{ is r.e.}]$.

Proof. Fix e. Let $\sigma = \{\, i : i \leq e \ \& \ \overline{A} \subseteq^* U_i \,\}$. Choose n such that $\sigma(e, d_m) = \sigma$ for all $m \geq n$. Choose s such that $d_m = d_m^s$ for all $m \leq n$. Define an r.e. set $\hat{H}_e$ by enumerating $y = d_k^t$ into $\hat{H}_e$ at stage $t + 1 > s$ just if $n \leq k \leq t$;

(5.2) $(\exists i)\, (\exists j)\, [k = \langle i, j \rangle \ \& \ i \leq e]$;

(5.3) $(\forall m) [n \leq m \leq k \implies \sigma(e, d_m^t, t) = \sigma]$,

and

(5.4) $(\forall m \leq k) \, [d_m^t = d_m^{t+1}].$

We claim that if $y = d_k^t$ is enumerated in $\hat{H}_e$ at stage $t + 1$ then either
$y = d_k$ or $y \in A$. If $y \neq d_k$, choose the first stage $v > t + 1$ such that some
marker Γ_m, $m < k$, pulls some element z. Now for all $p \leq e$, $y \in U_{p,t}$ if and
only if $y \in U_{p,v}$ by (5.3), (5.2), the choice of s, and the definition of U_p. For
all $p > e$, $y \in U_{p,t}$ iff $y \in U_{p,v}$ by (5.2), the choice of s, and the definition
of U_p. But by (5.4), Γ_m does not pull y at stage $t + 1$ so it does not pull y at
stage v. Hence, $y \neq z$ so $y \in A_v$ by the dump property of the construction,
since $y = d_t^t \leq d_{v-1}^{v-1}$. Hence, by the claim, $\hat{H}_e \cup A =^* H_e$ so H_e is r.e.   ▯

5.5–5.10 Exercises

5.5 (Martin [1963]).   Give an easy direct proof that there exists an atomless
r.e. set A by meeting for each e the requirements

$$P_e : [A \subseteq W_e \ \& \ \overline{A} \not\subseteq^* W_e] \implies (\exists^\infty n) \, [|F_n \cap \overline{W}_e| \geq 2].$$

where $\{F_n\}_{n \in \omega}$ is a fixed disjoint strong array of finite sets with union ω
such that for each n, $|F_n| = n + 1$. *Hint.* At stage $s + 1$, for each e and
n, $e < n \leq s$, such that $|F_n - (W_{e,s} \cup A_s)| = 1$, let $x(n, e)$ be the unique
$y \in F_n - (W_{e,s} \cup A_s)$, and enumerate $x(n, e)$ into A. (See also Exercise XI.2.8
where it is proved that in addition A can have an arbitrary high r.e. degree.)

5.6.   Show that the set A of Theorem 5.4 has no hh-simple superset.

5.7.   (a) Using the proof of Lemma 3 show that in Theorem 5.4 we have
automatically ensured that $A \cup C_i$ is r.e. for any column $C_i = \{d_{\langle i,j \rangle}\}_{j \in \omega}$.
(Indeed let R be any recursive set for which there exists an n such that if
$\langle i, j \rangle \in R$ then $i \leq n$. Prove that $A \cup \{d_k : k \in R\}$ is r.e.)
   (b) Conclude that $\mathcal{L}^*(A)$ does not satisfy the formula $(\exists x) \, \text{Sim}(x)$, defined
in (1.3) of §1.

5.8.   Modify the construction in Theorem 5.4 to produce an atomless r-max-
imal set A such that $\mathcal{L}^*(A)$ satisfies $(\exists x) \, \text{Sim}(x)$. (Conclude from Exercises
5.7 and 5.8 that there exist atomless r-maximal sets A_1 and A_2 such that
$\mathcal{L}^*(A_1)$ and $\mathcal{L}^*(A_2)$ do not satisfy the same first order sentences.)   *Hint.*
Ensure that H_0 is simple in $\mathcal{L}^*(A)$ by meeting for each n the requirement

$$Q_n : |W_n \cap \overline{A}| = \infty \implies |W_n \cap (H_0 - A)| = \infty.$$

Conflict with the P_n requirements of Theorem 5.4 is resolved by the priority
ranking $P_0, Q_0, P_1, Q_1, \ldots$.

5.9.   Prove that if B is r-maximal and $A \subset_{sm} B$ (as defined in Exercise 4.12) then $\mathcal{L}^*(A)$ satisfies $(\exists x)\, \mathrm{Sim}(x)$. (Note that A is r-maximal by Proposition 4.3.)

5.10 (Lachlan [1968c, Theorem 8]).   Give an alternative construction of an atomless r-maximal set A in the style of Exercise 5.5. Let $\{\,F_j\,\}_{j \in \omega}$ be a disjoint array of finite sets such that $|F_j| = 2^j \cdot j$. Let $\hat{F}_{j,s} = F_j - A_s$. For each $i < j$, at the first stage when $W_{i,s}$ contains at least half of the members of $\hat{F}_{j,s}$, enumerate the remaining members of $\hat{F}_{j,s}$ into A_{s+1}. Finally, let the markers of Theorem 3.3 move on the sets $\hat{F}_{j,s}$ in place of integers to maximize e-states where the e-state of $\hat{F}$ at stage s is

$$\sigma(e, \hat{F}, s) = \{\, i : i \le e\ \&\ \hat{F} \subseteq W_{i,s}\,\}.$$

Prove that A is r-maximal and has no hh-simple superset. *Hint.* To see that A is r-maximal assume that a recursive set R splits $\overline{A}$. Let $W_e = R$ and $W_i = \overline{R}$, where $e < i$. Show that if $e \in \sigma(e, \hat{F}_j)$ for infinitely many j, then this is true for almost all j, and similarly for i.

5.11 Notes.   Atomless r-maximal sets were constructed by Robinson [1967a] and independently by Lachlan [1968c], although the present construction is not the same as either. It was invented by John Norstad and appears in Lerman and Soare [1980b, Theorem 2.15] in connection with an extension of Lachlan's decision procedure for a fragment of the elementary theory of $\mathcal{E}^*$.

6. Atomless hh-simple Sets

In Corollary 2.7 we saw that a coinfinite r.e. set A is hh-simple if and only if $\mathcal{L}(A)$ forms a Boolean algebra. It is natural to ask exactly which Boolean algebras are so realized. First it is easy to show (see Exercise XV.2.12) that the isomorphism type of $\mathcal{L}^*(A)$ determines that of $\mathcal{L}(A)$ so it suffices to classify the former. We now construct an r.e. set A such that $\mathcal{L}^*(A)$ is a countable atomless Boolean algebra. In §7 we modify the method to show that any Boolean algebra with sufficiently effective presentation is isomorphic to $\mathcal{L}^*(B)$ for some r.e. set B.

6.1 Theorem (Lachlan [1968c]).   *There exists an atomless hh-simple set A.*

6.2 Corollary.   *There exists an r.e. set A such that $\mathcal{L}^*(A)$ is a countable atomless Boolean algebra.*

Proof. For A as in Theorem 6.1, if the Boolean algebra $\mathcal{L}^*(A)$ contains an atom C^* then its complement in $\mathcal{L}^*(A)$ is a coatom D^* so D is a maximal superset of A, contrary to A being atomless.   ▯

Proof of Theorem 6.1. Fix an effective 1:1 coding f from $2^{<\omega}$ onto ω such that if $lh(\alpha) < lh(\beta)$ then $f(\alpha) < f(\beta)$. Let $\alpha_i =_{\mathrm{dfn}} f^{-1}(i)$. For each $i \in \omega$ we have a marker Γ_i placed at node α_i in the binary tree $2^{<\omega}$. For each $\alpha \in 2^{<\omega}$ of length e define the *e-region* of nodes $R_\alpha = \{\beta : \alpha \subseteq \beta\}$, and the corresponding *cones* of elements,

$$C_\alpha^s = \{d_i^s : \alpha_i \in R_\alpha\},$$

and

$$C_\alpha = \{d_i : \alpha_i \in R_\alpha\},$$

where, as usual, d_i^s is the integer x associated with marker Γ_i at the end of stage s, $d_i^s = -1$ if no such x exists, $d_i = \lim_s d_i^s$, and $\overline{A} = \{d_i\}_{i \in \omega}$. We shall arrange that $d_i^s > -1$ if and only if $i \leq s$.

Lemma 1. *It suffices to construct A to meet the following requirements,*

$$P_e : \ lh(\alpha) = e \implies [C_\alpha \cap W_e =^* \emptyset \ or \ C_\alpha \subseteq^* W_e],$$

and

$$N_\alpha : \ A \cup C_\alpha \ is \ r.e.$$

Proof. Suppose that $A \subseteq W_e$ and W_e is maximal. Then $\overline{W}_e$ is infinite, so for some α with $lh(\alpha) = e$, $C_\alpha \not\subseteq^* W_e$ so $C_\alpha \cap W_e =^* \emptyset$ by P_e. Choose $\beta \supset \alpha$. Then $B = A \cup C_\beta \cup W_e$ is r.e. by N_β and $W_e \subset_\infty B \subset_\infty \omega$, contradicting the maximality of W_e. Thus, A is atomless.

Now suppose $A \subseteq W_e$. To conclude that A is hh-simple it suffices by Corollary 2.7 to show that $A \cup \overline{W}_e$ is r.e. Define

$$V_e = A \cup \left(\bigcup \{C_\alpha : lh(\alpha) = e \ \& \ W_e \cap C_\alpha =^* \emptyset\}\right).$$

Now V_e is r.e. by the requirements N_α for $lh(\alpha) = e$, and $V_e =^* A \cup \overline{W}_e$ by P_e. This concludes the proof of Lemma 1.

In the construction of A the condition allowing us to take positive action is referred to as: "Γ_i *wants to pull* d_j^s *for (the sake of)* P_e *at stage* $s + 1$," and is defined as the conjunction of the following conditions:

(6.1) $e \leq i;$

(6.2) $i < j;$

(6.3) $\sigma(e - 1, d_i^s, s) = \sigma(e - 1, d_j^s, s)$, but $d_i^s \notin W_{e,s}$ and $d_j^s \in W_{e,s};$

and

(6.4) α_i and α_j lie in the same e-region.

(The e-states $\sigma(e, x, s)$ are defined as usual as in Definition 3.2.)

An integer which has not yet been assigned to any marker or enumerated in A is said to be *fresh*.

Stage $s = 0$. Set $d_0^0 = 0$.

Stage $s + 1$.

Step 1 (a) (Pulling). Choose the least i, then the least e, and then the least j such that $e, i, j \leq s$, and such that Γ_i wants to pull d_j^s for P_e. Define $d_i^{s+1} = d_j^s$. We say Γ_i *pulls* d_j^s.

(b) (Dumping). For each k, $i \leq k \leq s$ and $k \neq j$, enumerate (dump) d_k^s into A. (If no such e, i, j exist then set $d_k^{s+1} = d_k^s$ for all $k \leq s$.)

Step 2. For each marker Γ_k, $k \leq s + 1$, not now assigned an element of $\overline{A}$, assign in increasing order of k the first fresh element $> s+1$. Enumerate into A_{s+1} all fresh elements $\leq s + 1$.

This ends the construction.

The purpose of the extra dump in step 1(b) (which enumerates into A the elements d_k^s, $j < k \leq s$, as well as those d_k^s, $i \leq k < j$, which would be normally enumerated) is to ensure the following property needed to prove Lemma 3.

(6.5)   If $d_i^{s+1} \neq d_i^s$ and $i \leq k \leq s$, then $y = d_k^s$ is enumerated in A at stage $s + 1$ unless y is the element pulled.

It follows that an element y can be pulled for the sake of P_e only when it has just changed e-state since if some Γ_i wants to pull it at that time, then y is either immediately pulled by $\Gamma_{i'}$, some $i' \leq i$, or y is dumped into A because $\Gamma_{i'}$ pulls some $z \neq y$. An appropriate modification of this fact will play a key role in §7, particularly in Lemma 2. (Note that when a fresh element y is assigned to a marker at some stage $s + 1$, the e-state of y is $\emptyset$ because $y > s+1$. Hence y does not want to be pulled until after it has been assigned to a marker and has then changed its e-state.)

Lemma 2.   *For every n, $d_n = \lim_s d_n^s$ exists, and $\overline{A} = \{ d_n \}_{n \in \omega}$.*

Proof. The first part follows by induction on n as in Lemma 1 of Theorem 3.3 or 4.6. The second part follows because every number is eventually either assigned to a marker or is enumerated in A.

Lemma 3.   *For every e, requirement P_e is met.*

Proof. Fix e and assume P_i holds for all $i < e$. Fix α of length e. Since almost all elements $x \in C_\alpha$ are in the same $(e - 1)$-state σ, we can choose $n > e$ such that for all $m \geq n$, $\sigma(e - 1, d_m) = \sigma$ if $d_m \in C_\alpha$. Suppose that W_e splits C_α. Choose $j > i > n$ such that $\sigma(e, d_i) = \sigma$, $\sigma(e, d_j) = \sigma \cup \{ e \}$,

$\alpha \subset \alpha_i$ and $\alpha \subset \alpha_j$. Choose s such that $d_k^s = d_k$ and $\sigma(e, d_k, s) = \sigma(e, d_k)$ for all $k \leq j$. Then at stage $s+1$, Γ_i wants to pull d_j^s for P_e, so $d_k^{s+1} \neq d_k^s$ for some $k \leq j$ contrary to the choice of s.

Lemma 4. *For every* α, $A \cup C_\alpha$ *is r.e.*

Proof. Fix α and let $e = lh(\alpha)$. By Lemma 3 for each β of length e there is an e-state σ_β such that $\sigma(e, x) = \sigma_\beta$ for almost all $x \in C_\beta$. Choose n, $e \leq n$, $\alpha_n \in R_\alpha$, such that for all $m \geq n$,

(6.6) $$(\exists \beta)\,[lh(\beta) = e \ \& \ d_m \in C_\beta],$$

and

(6.7) $$(\forall \beta)\,[[lh(\beta) = e \ \& \ d_m \in C_\beta] \implies \sigma(e, d_m) = \sigma_\beta].$$

Choose v such that for all $m \leq n$,

(6.8) $$d_m = d_m^v \text{ and } \sigma(e, d_m) = \sigma(e, d_m^v, v).$$

Define the r.e. set $\hat{C}_\alpha$ by enumerating $y = d_k^t$ in $\hat{C}_\alpha$ at stage $t + 1 > v$ just if $k \geq n$, $\alpha \subseteq \alpha_k$, $\sigma(e, y, t) = \sigma_\alpha$, and no Γ_i wants to pull y at stage $t + 1$. We claim that $A \cup \hat{C}_\alpha =^* A \cup C_\alpha$ so the latter is r.e. Clearly, $C_\alpha \subseteq^* \hat{C}_\alpha$ since every $d_m \in C_\alpha$, $m \geq n$, is eventually enumerated in $\hat{C}_\alpha$.

To see that $\hat{C}_\alpha \subseteq C_\alpha \cup A$, we show that no $y = d_k^t$ enumerated in $\hat{C}_\alpha$ at stage $t + 1$ can later leave the e-region R_α without being enumerated in A. Suppose to the contrary that $s + 1 > t + 1$ is the least stage $> t + 1$ where y enters a different e-region. Then Γ_i pulls $y = d_j^s$ for the sake of $P_{e'}$ at $s + 1$ for some i, j, e' such that $i < j \leq k$, and $e' < e$ by (6.4). Also $d_m^s = d_m^t$ for all $m < j$ by (6.5) else $y \in A_s$. In particular, $d_i^s = d_i^t = x$ say.

Now by the definition of $\hat{C}_\alpha$, Γ_i did not want to pull y for any $P_{e''}$, $e'' \leq e'$ at $t + 1$, so $\sigma(e', x, t) \geq \sigma(e', y, t)$, but it did at $s + 1$, so $\sigma(e', x, s) < \sigma(e', y, s)$ by (6.3). Hence $\sigma(e, y, s) > \sigma(e, y, t) = \sigma_\alpha$ so by (6.7) Γ_n wants to pull y at stage $s + 1$. Thus, for some $m \leq n$, Γ_m pulls some element at $s + 1$, contrary to (6.8). ∎

6.3 Notes. Lachlan's original construction [1968c] for proving Theorem 6.1 measured e-states with respect to a more complicated array $\{V_{e,s}\}$ in place of $\{W_{e,s}\}$ and allowed an element y to be pulled for P_e only at a stage when y had just increased its e-state. The present variation as suggested by M. Stob accomplishes the same thing using the extra dumping to achieve (6.5). This considerably simplifies the proof in §7.

7*. Σ_3 Boolean Algebras Represented as Lattices of Supersets

7.1 Definition. A countable Boolean algebra $\mathcal{B} = (\{\, b_i \,\}_{i \in \omega}, \leq, \cup, \cap, ')$ is a Σ_3 *Boolean algebra* if there are recursive functions f and g and a Σ_3 relation R such that $b_i \cup b_j = b_{f(i,j)}$, $b_i \cap b_j = b_{g(i,j)}$ and $b_i \leq b_j$ if and only if $R(i,j)$. (A given element of $\mathcal{B}$ may appear more than once in the list $\{\, b_i \,\}_{i \in \omega}$.)

If A is hh-simple then $\mathcal{L}^*(A)$ is a Σ_3 Boolean algebra by Theorem II.1.9 and Exercise IV.1.12. We now prove that as A ranges through all hh-simple sets, $\mathcal{L}^*(A)$ ranges through all Σ_3 Boolean algebras.

7.2 Theorem (Lachlan [1968c]). *Let $\mathcal{B}$ be any Σ_3 Boolean algebra. Then there is an hh-simple set A such that $\mathcal{L}^*(A)$ is isomorphic to $\mathcal{B}$.*

Proof. Let $\mathcal{B} = (B, \subseteq, \cup, \cap, ')$ be a Σ_3 Boolean algebra with zero element $\emptyset$. A sequence $\{\, b_0, b_1, \dots \,\}$ of members of $\mathcal{B}$ is said to *generate* $\mathcal{B}$ if its closure under the operations of union, intersection and complementation contains every member of $\mathcal{B}$. A mapping F from $2^{<\omega}$ to $\{\, 0, 1 \,\}$ is an *associate* of $\mathcal{B}$ if there is a generating sequence $\{\, b_0, b_1, \dots \,\}$ such that $F(\alpha) = 0$ if and only if $b_\alpha = \emptyset$ where

$$(7.1) \qquad b_\alpha =_{\text{dfn}} \left(\bigcap \{\, b_i : \alpha(i) = 1 \,\} \right) \cap \left(\bigcap \{\, b_i' : \alpha(i) = 0 \,\} \right).$$

Now each member of $\mathcal{B}$ is of the form $\bigcup \{\, b_\alpha : \alpha \in S \,\}$ for some finite set $S \subseteq 2^{<\omega}$. In particular,

$$b_i = \bigcup \{\, b_\alpha : lh(\alpha) = i + 1 \text{ and } \alpha(i) = 1 \,\}.$$

It is easy to see that if $\mathcal{B}^0$ and $\mathcal{B}^1$ are two countable Boolean algebras with a common associate F then $\mathcal{B}^0$ and $\mathcal{B}^1$ are isomorphic. (Let $\{\, b_i^0 \,\}, \{\, b_i^1 \,\}$ be the corresponding generating sequences and simply map b_i^0 to b_i^1.)

The sequence $B = \{\, b_0, b_1, \dots \,\}$ clearly generates $\mathcal{B}$. From the recursive functions for union and intersection we can find recursive functions $\hat{f}$ and $\hat{g}$ such that

$$b_{\hat{f}(\alpha)} = \bigcap \{\, b_i : \alpha(i) = 1 \,\}, \text{ and } b_{\hat{g}(\alpha)} = \bigcup \{\, b_i : \alpha(i) = 0 \,\}.$$

Hence, we have

$$F(\alpha) = 0 \iff b_{\hat{f}(\alpha)} \cap b_{\hat{g}(\alpha)}' = 0 \iff b_{\hat{f}(\alpha)} \leq b_{\hat{g}(\alpha)}.$$

Since the third relation is Σ_3, so is the first, so by Corollary IV.3.7 for approximating Σ_3 relations there is a recursive sequence of r.e. sets $\{\, X_{\alpha,y} : \alpha \in 2^{<\omega} \ \& \ y \in \omega \,\}$ such that

$$(7.2) \qquad \begin{aligned} F(\alpha) = 0 &\implies (\text{a.e. } y) \, [X_{\alpha,y} = \omega] \\ \& \ F(\alpha) = 1 &\implies (\forall y) \, [X_{\alpha,y} =^* \emptyset]. \end{aligned}$$

Now define the recursive sequence of r.e. sets $\{\, Z_\alpha : \alpha \in 2^{<\omega} \,\}$ by

$$Z_\alpha = \bigcup \{\, X_{\beta,y} : \beta \subseteq \alpha \;\&\; y \le lh(\alpha) \,\}.$$

Note that $Z_\beta \subseteq Z_\alpha$ if $\beta \subseteq \alpha$ and that if $F(\beta) = 0$ then there exists y such that $|Z_\alpha| = \infty$ for all $\alpha \supseteq \beta$, $lh(\alpha) \ge y$. Thus the sets Z_α have the following properties which we shall need:

(7.3) $$F(\alpha) = 0 \iff (\text{a.e. } \gamma \supseteq \alpha)\, [Z_\gamma = \omega]$$

and

(7.4) $$Z_\alpha =^* \emptyset \implies (\forall \beta \subseteq \alpha)\, [Z_\beta =^* \emptyset].$$

(The proof of (7.3) ($\Longleftarrow$) uses the fact that if $F(\alpha) = 1$ and $e > lh(\alpha)$ then there is some $\beta \supseteq \alpha$, $lh(\beta) = e$, such that $F(\beta) = 1$ also.) Identify each node α_i with its code number i as fixed in Theorem 6.1 and write Z_i in place of Z_{α_i}. Let $\{\, Z_{i,s} \,\}_{i,s\in\omega}$ be a simultaneous recursive enumeration of the sets $\{\, Z_i \,\}$.

We adopt all the notation, conventions and the requirements P_e, N_α of Theorem 6.1. In addition we assume $W_{e,s}$ has been defined as in Exercise I.3.11 so that as in I (3.3) we have

(7.5) $$(\forall s)\, (\exists \text{ at most one } \langle e, x \rangle)\, [x \in W_{e,s+1} - W_{e,s}].$$

The construction here of A is very similar to that in Theorem 6.1 except that the above "kicking set" Z_i is associated with Γ_i, and whenever a new element appears in Z_i at stage s, we enumerate (kick) d_i^s into A. This is to ensure that $d_i = \lim_s d_i^s$ exists if and only if Z_i is finite so that F is an associate of $\mathcal{L}^*(A)$.

The fact that $d_i = \lim_s d_i^s$ does not always exist necessitates a few modifications in the original construction. First we shall allow Γ_i to pull d_j^s only if $|Z_{i,s}| < j$. This ensures that if d_i is undefined (i.e., Z_i is infinite) then there is no $j > i$ such that Γ_i pulls d_j^s infinitely often. (Otherwise, the infinite pulling by Γ_i might cause d_j to be undefined even though Z_j may be finite). Secondly, whenever $d_i^{s+1} \neq d_i^s$ for any reason we enumerate (dump) into A all elements $d_j^s \le d_s^s$ (except possibly d_i^{s+1}) such that $|Z_{i,s}| < j$, i.e., all elements $d_j^s \le d_s^s$ which Γ_i might later want to pull. This guarantees the appropriate analogue of (6.5) which we need in Lemma 3.

We say that Γ_i *wants to pull* d_j^s *for (the sake of)* P_e at stage $s+1$ if (6.1) through (6.4) hold and also

(7.6) $$|Z_{i,s}| < j.$$

  Stage $s = 0$. Set $d_0^0 = 0$.
  Stage $s + 1$.

Step 1. (Pulling) (Same as Theorem 6.1 Step 1(a).) Choose the least i, then the least e, and then the least j such that $e, i, j \leq s$, and such that Γ_i wants to pull d_j^s for P_e. Define $d_i^{s+1} = d_j^s$, and enumerate d_i^s into A. We say Γ_i pulls d_j^s.

Step 2. (Kicking) For each k, $k \leq s$ and $k \neq j$ such that $Z_{k,s+1} \neq Z_{k,s}$, enumerate (kick) d_k^s into A.

Step 3. (Dumping) For all $p \leq s$, do the following in increasing order of p. If d_p^s has been enumerated in A at stage $s + 1$ and $|Z_{p,s}| < q \leq s$ enumerate d_q^s into A unless $q = j$ of Step 1.

Step 4. Same as Step 2 of Theorem 6.1.

This ends the construction.

Lemma 1. *For all j, $d_j = \lim_s d_j^s$ exists if and only if Z_j is finite, and $\overline{A} = \{ d_j \}_{j \in \omega}$.*

Proof. Fix j, and assume the lemma by induction for all $i < j$. If Z_j is infinite then Step 2 applies to d_j^s for infinitely many s, so $\lim_s d_j^s$ does not exist. If Z_j is finite then Step 2 applies to d_j^s at most finitely often, say never after stage u. By the inductive hypothesis choose $v > u$ such that for all $i < j$, either $d_i^v = d_i$, or $|Z_{i,v}| \geq j$. Now by (7.6) no Γ_i can pull d_j^s at any stage $s > v$, so Γ_j changes its assigned element at most 2^{j+1} times after stage v and hence $\lim_s d_j^s$ exists.

Lemma 2. *If any Γ_i wants to pull $y = d_j^s$ for P_e at stage $s+1$ then $\sigma(e, y, s) \neq \sigma(e, y, s - 1)$ and hence y is pulled by $\Gamma_{i'}$ for some $i' \leq i$.*

Proof. The proof is by induction on s. Assume the lemma is true for all stages $t \leq s$. Now consider stage $s + 1$ and suppose that Γ_i wants to pull $y = d_j^s$ for P_e at stage $s + 1$. Hence, $|Z_{i,s-1}| \leq |Z_{i,s}| < j$. Assume for a contradiction that $\sigma(e, y, s) = \sigma(e, y, s - 1)$. Also $\sigma(e, y, s) \neq \emptyset$ so y was not just placed as a fresh element on Γ_j at the end of stage s. Hence, $y = d_k^{s-1}$ for some $k \geq j$. Furthermore, $d_i^s = d_i^{s-1}$ else y would have been enumerated in A at stage s during Step 3. Hence, Γ_i wanted to pull y at stage s. By the induction hypothesis y is pulled at stage s by $\Gamma_{i'}$ for some $i' \leq i$, contradicting $y = d_j^s$. Hence, $\sigma(e, y, s) \neq \sigma(e, y, s - 1)$. But at most one y changes its e-state at each stage, so j of Step 1 is unique. Thus, y is pulled at stage $s + 1$.

Lemma 3. $lh(\alpha) = e \implies [C_\alpha \cap W_e =^* \emptyset \text{ or } C_\alpha \subseteq^* W_e]$, *where*

$$C_\alpha = \{ d_i : \alpha \subseteq \alpha_i \text{ and } d_i \text{ exists} \}.$$

Proof. The proof is like that of Theorem 6.1 Lemma 3 except that we consider only those m such that d_m exists. Choose e, i, and j as in Theorem 6.1

Lemma 3 except that we now require in addition that d_i and d_j exist and that $j > |Z_i|$. (Since d_i exists, Z_i is finite by Lemma 1 so such a j exists for each appropriate i.) Now since $y = d_j$ has a higher e-state than d_i, Γ_i wants to pull y at all sufficiently large stages. Hence, by Lemma 2, y is eventually pulled away from Γ_j, contradicting $y = d_j$.

Lemma 4. *For every α, $A \cup C_\alpha$ is r.e.*

Proof. Fix α and let $e = lh(\alpha)$. Choose σ_β, and n as in Theorem 6.1 Lemma 4 except that now d_n must exist, and (6.6) and (6.7) only apply to those d_m which exist. (Now n exists because if d_m exists for only finitely many $\alpha_m \in R_\alpha$ then C_α is finite and clearly $A \cup C_\alpha$ is r.e.) Choose v satisfying (6.8) for all $m \leq n$ such that d_m exists, and such that $|Z_{n,v}| = |Z_n| = p$ say and

$$(7.7) \qquad (\forall m \leq p)\,[d_m \text{ exists} \implies d_m^v = d_m].$$

(Since d_n exists, Z_n is finite for Lemma 1, so v exists.) Using v define $\hat{C}_\alpha$ as before. The proof is now the same as in Theorem 6.1 Lemma 4 because although we no longer have (6.5) we do know that if Γ_i pulls $y = d_j^s$ at stage $s+1 > t+1$ then $|Z_{i,s}| < j$ and therefore $d_i^s = d_i^t$ because otherwise y would have been dumped because of Step 3.

We claim as before that no $y = d_k^t$ enumerated in $\hat{C}_\alpha$ at some stage $t+1$ can later leave the e-region R_α without being eventually enumerated in A. Suppose to the contrary that $s+1 > t+1$ is the least stage $> t+1$ where y enters a different e-region. Then Γ_i pulls $y = d_j^s$ for the sake of $P_{e'}$ at $s+1$ for some i, j, e' such that $i < j \leq k$ and $e' < e$. But since $d_i^s = d_i^t$, and Γ_i did not want to pull y at stage t, y must have increased its e-state between stages $s+1$ and $t+1$. Also $|Z_n| < j$ since otherwise y is eventually enumerated in A by (7.7). Hence, Γ_n wants to pull y at stage $s+1$, so by Lemma 2 some Γ_m, $m \leq n$, does pull y at stage $s+1$. But if $\lim_s d_m^s$ does not exist then y later enters A, and if $\lim_s d_m^s$ exists this contradicts the assumption that $d_m^v = d_m$ for $m \leq n$ such that d_m exists.

Using Lemma 4 it is easy to construct a sequence $\{B_i^*\}_{i \in \omega}$ of members of $\mathcal{E}^*$ which generates $\mathcal{L}^*(A)$ and which has F as associate. Namely, for each $\alpha \in 2^{<\omega}$ the set $B_\alpha = A \cup C_\alpha$ is r.e. by Lemma 4. Now B_α plays the role of b_α in (7.1), so we can let

$$B_i = \bigcup \{ B_\alpha : lh(\alpha) = i+1 \ \& \ \alpha(i) = 1 \}.$$

7.3 Notes. No algebraic characterization analogous to Theorem 7.2 is known for r.e. sets such that $\mathcal{L}^*(A)$ is not a Boolean algebra, but since $\mathcal{E}^*$ is embeddable in any such infinite $\mathcal{L}^*(A)$ by Exercise 2.18, any characterization is likely to be difficult.

Chapter XI

The Relationship Between the Structure and the Degree of an R.E. Set

In Chapters I–IX we have studied the properties of the r.e. degrees **R**, particularly the classes $\mathbf{H}_n$ and $\mathbf{L}_n$ of high$_n$ and low$_n$ degrees studied in Exercise IV.4.5, Exercise VII.1.6, Corollary VIII.3.4, and Corollary VIII.3.5. In Chapter X we studied the algebraic properties of the r.e. sets as a distributive lattice $\mathcal{E}$ under inclusion. In this chapter we bring together these two branches by studying the relationship between the algebraic structure of an r.e. set and its Turing degree (i.e., its "information content"). We think of the high (low) degrees as having high (low) information content.

The maximal set produced in Theorem X.3.3 required a complicated construction and may be viewed as being complex from an algebraic and computational complexity viewpoint. In §1 we show that maximal sets must have high degree (and therefore high information content), and in §2 that every high r.e. degree contains a maximal set. Hence, deg$\{\, M : M$ maximal $\} = \mathbf{H}_1$. Furthermore, these results also hold with maximality replaced by other properties such as hh-simplicity, being a major subset and many others.

In contrast we show in §3 that low r.e. sets and degrees behave much like recursive sets both in $\mathcal{E}$ and in **R**. For example, $\mathcal{L}(A) \cong \mathcal{E}$ for every coinfinite low r.e. set (not merely for every recursive set), and the Sacks Splitting Theorem holds over any low r.e. degree (and not merely over **0**). Let **A** denote the class of degrees of coinfinite r.e. sets which are atomless (have no maximal superset). In §4 we continue the progression begun in §1 and §2 by proving Shoenfield's theorem that $\mathbf{A} \supset \overline{\mathbf{L}}_2$. In §5 we strengthen Robinson's result of §3 that $\mathbf{A} \cap \mathbf{L}_1 = \emptyset$ by proving Lachlan's theorem that $\mathbf{A} \cap \mathbf{L}_2 = \emptyset$, and hence that $\mathbf{A} = \overline{\mathbf{L}}_2$.

1. Martin's Characterization of High Degrees in Terms of Dominating Functions

Post's program described in Chapter V was to produce coinfinite r.e. sets A with ever "sparser" complements $\overline{A}$ in an effort to find a sparseness condition on $\overline{A}$ that would guarantee incompleteness of A. Yates [1965] proved the impossibility of this program by constructing a complete maximal set, since the complement of a maximal set is the sparsest possible co-r.e. set. Recall that Yates achieved this (Exercise X.3.8) by constructing a maximal set M such that the principal function $p_{\overline{M}}$ of $\overline{M}$ dominates every p.r. function, since by Exercise X.3.7 any function that dominates every *partial* recursive function must be complete.

Knowing the latter fact Tennenbaum [1961] suggested just the opposite of Post's program, namely that if M is maximal then $\overline{M}$ must be so sparse that $p_{\overline{M}}$ dominates every p.r. function and thus M is complete. Sacks [1964b], however, refuted Tennenbaum's program by constructing an incomplete maximal set using the infinite injury method of Chapter VIII. Martin [1966b] and independently Tennenbaum [1962] salvaged the true kernel of Tennenbaum's program by showing that if M is maximal then $p_{\overline{M}}$ dominates all *total* recursive functions (see Proposition 1.2). By analogy to Exercise X.3.7 Martin [1966b] then connected this domination property to the high degrees, and used it to show that the high r.e. degrees are precisely the degrees of maximal sets (see Theorems 1.3 and 2.3).

1.1 Definition.   A function f is *dominant* if f dominates every (total) recursive function.

1.2 Proposition (Martin [1966b]—Tennenbaum [1962]).   *If M is a maximal set then the principal function $p_{\overline{M}}$ of $\overline{M}$ is dominant. (Namely, every maximal set is dense simple as defined in Exercise 1.11.)*

Proof.   Suppose to the contrary that $p_{\overline{M}}$ fails to dominate some recursive function f. By Corollary V.2.4 f does not dominate $p_{\overline{M}}$. Hence, there are infinitely many "crossover points" x such that $f(x) < p_{\overline{M}}(x)$ but $p_{\overline{M}}(x+1) \leq f(x+1)$, and hence card $(\overline{M} \cap I_x) \geq 2$ where $I_x = [f(x), f(x+1)]$. Let $\{M_s\}_{s \in \omega}$ be a recursive enumeration of M. Let $\psi(x, s) = \mu y[y \in I_x - M_s]$ if y exists, and let $\psi(x, s)$ be undefined otherwise. Now give a recursive enumeration $\{A_s\}_{s \in \omega}$ of an r.e. set A such that $M \subset_\infty A \subset_\infty \omega$ by enumerating z in A_{s+1} if $z \in M_s$ or $z \in \{\psi(x, s) : x \leq s\}$.   ☐

1.3 Theorem (Martin [1966b]).   *A set A satisfies $\emptyset'' \leq_T A'$ iff there is a dominant function $f \leq_T A$. (Note that A need not be of r.e. degree or even recursive in $\emptyset'$.)*

Proof.   By Theorem IV.3.2 we know that Tot $\equiv_T \emptyset''$. Hence, by the Limit Lemma III.3.3 relativized to A, we have $\emptyset'' \leq_T A'$ iff there is an A-recursive $\{0, 1\}$-valued function $g(x, s)$ such that $\lim_s g(x, s) = \chi_{\text{Tot}}$, the characteristic function of Tot.

($\Longrightarrow$). Assume $\emptyset'' \leq_T A'$. Given $g(x, s)$ as above we define a dominant function $f \leq_T A$ as follows.

Stage s. (To define $f(s)$). For all $e \leq s$ define

$$t(e) = (\mu t \geq s) \, [(\forall x \leq s) \, [\varphi_{e,t}(x) \downarrow] \quad \text{or} \quad g(e, t) = 0],$$

and define

$$f(s) = \max\{\, t(e) : e \leq s \,\}.$$

Note that $t(e)$ exists because if φ_e is not total then $g(e, t) = 0$ for almost every (a.e.) t. If φ_e is total then $g(e, t) = 1$ for a.e. t and hence $f(s) > \varphi_e(s)$ for a.e. s.

($\Longleftarrow$). Assume $f \leq_T A$ is dominant. Define an A-recursive function $g(x, s)$ such that $\lim_s g(x, s) = \chi_{\text{Tot}}$ as follows:

$$g(x, s) = \begin{cases} 1 & \text{if } (\forall z \leq s) \, [\varphi_{x, f(s)}(z) \downarrow]; \\ 0 & \text{otherwise.} \end{cases}$$

Note that if φ_x is total then so is

$$\psi_x(y) = (\mu s) \, (\forall z \leq y) \, [\varphi_{x,s}(z) \downarrow].$$

Thus, f dominates ψ_x and $g(x, s) = 1$ for a.e. s. If φ_x is not total then $\varphi_x(y)$ and hence $\psi_x(y)$ diverge for some y, and therefore $g(x, s) = 0$ for all $s \geq y$. ▯

Recall the high and low sets and degrees $\mathbf{H}_n$ and $\mathbf{L}_n$, defined in Definition IV.4.2 and Definition VII.1.3.

1.4 Corollary. *An r.e. degree* $\mathbf{a}$ *is high* $(\mathbf{a} \in \mathbf{H}_1)$ *iff there is a dominant function* f *such that* $\deg(f) \leq \mathbf{a}$. ▯

1.5 Corollary. *Every maximal set has high degree.*

Proof. Apply Proposition 1.2 and Corollary 1.4. ▯

It is interesting to see that not only maximal sets must be high but many other types of sets considered in Chapter X, such as hh-simple sets, r-maximal sets, and major subsets must also be high. However, unlike maximal sets, not all of these need have a dominant complement so we must first replace Proposition 1.2 by a more flexible fact relating structural properties to high sets.

1.6 Lemma. *Let A be a coinfinite r.e. set that is not high and $h(x)$ any strictly increasing recursive function. Then A has a recursive enumeration $\{A_s\}_{s \in \omega}$ that is h-diagonally correct in the sense that*

(1.1) $$(\exists^\infty s) \, [a^s_{h(s)} = a_{h(s)}],$$

where $\overline{A}_s = \{\, a_0^s < a_1^s < \cdots \}$ *and* $\overline{A} = \{\, a_0 < a_1 < \cdots \}$.

Proof. Choose any index i such that $W_i = A$. Let $B_s = W_{i,s}$, $\overline{B}_s = \{\, b_0^s < b_1^s < \cdots \}$, and $g(x) = \mu s[b_{h(x)}^s = a_{h(x)}]$. Now $g \leq_T A$ and hence by Corollary 1.4, g fails to dominate some recursive function f. Define $A_s = B_{f(s)}$. Hence, $a_x^s = b_x^{f(s)}$ for all x. Now if $f(s) \geq g(s)$ then $a_{h(s)}^s = b_{h(s)}^{f(s)} = b_{h(s)}^{g(s)} = a_{h(s)}$. Thus (1.1) holds.   ▯

(See the proof of Theorem 5.1 for another application of Lemma 1.6.)

1.7 Theorem (Martin [1966b]).   *If A is hh-simple or r-maximal then A is high. (Indeed if A is finitely strongly hypersimple, as defined in Exercise 1.8, then A is high.)*

Proof. Suppose A is not high. Apply Lemma 1.6 with $h(s) = s$ to get a recursive enumeration $\{A_s\}_{s\in\omega}$ of A satisfying (1.1). We shall construct a uniformly recursive (see Definition II.2.5(ii)) sequence $\{V_n\}_{n\in\omega}$ of mutually disjoint finite sets such that $V_n \cap \overline{A} \neq \emptyset$ for all n. The sequence $\{V_n\}_{n\in\omega}$ itself witnesses that A is not hh-simple. Since the sequence is uniformly recursive we can define the recursive set $R = \bigcup_n V_{2n}$ which witnesses that A is not r-maximal.

We construct the sequence $\{V_n\}_{n\in\omega}$ by stages letting V_n^s denote the elements enumerated by the end of stage s, and $V^s = \bigcup\{\, V_n^s : n \leq s\}$. At each stage $s + 1$ the element a_s^s will be enumerated in exactly one set V_n. Hence, $|V^s| = s$ and the sequence $\{V_n\}_{n\in\omega}$ is uniformly recursive because the recursive function $f(s) = a_s^s$ is strictly increasing. Let $V_n^0 = \emptyset$ for all n.

Stage $s + 1$. Define

$$n = (\mu m \leq s)\,[V_m^s \cap \{\, a_i^s : i < s\} = \emptyset],$$

and enumerate a_s^s in V_n^{s+1}. (Note that n exists because $V_s^s = \emptyset$.)

Lemma. $(\forall n)\,[V_n \cap \overline{A} \neq \emptyset$ and V_n is finite].

Proof. Fix n and assume the lemma for all $m < n$. Choose stage r such that $V_m^r = V_m$ for all $m < n$. Let $t = (\mu s > r)\,[a_s^s = a_s]$. Now either there is some $x < t$ such that $a_x^t = a_x$ is already in V_n^t or else $a_t^t = a_t$ is enumerated in V_n^{t+1}. In either case $V_n^{t+1} \cap \overline{A} \neq \emptyset$ and $V_n^s = V_n^{t+1}$ for all $s \geq t + 1$.   ▯

The proof of Theorem 1.7 actually establishes more than the statement of the theorem. Young [1966] and D. A. Martin [1966b] introduced a new class of r.e. sets called *finitely strongly hypersimple (fsh-simple)* which includes both hh-simple and r-maximal sets and to which the proof of 1.7 exactly applies. (See Exercise 1.8.) Martin also introduced the notion of *dense simple sets* (see Exercise 1.11) since Proposition 1.2 actually shows that maximal sets are dense simple. Young [1966] also introduced the strongly hypersimple sets. The following diagram summarizes the relationships between these classes of simple sets, as well as those studied in Chapter X. Note that "$P \to Q$" indicates that property P implies property Q, and the lack of an arrow indicates that the implication fails. Definitions, inclusions, and the lack of reverse inclusions are given in the exercises below.

$$\text{maximal} \overset{(1)}{\nearrow} \quad r\text{-maximal} \xrightarrow{(2)} \text{sh-simple} \xrightarrow{(3)} \text{finitely strongly h-simple}$$

$$\underset{(4)}{\searrow} \quad \text{quasi-maximal} \xrightarrow{(5)} \text{hh-simple} \xrightarrow{(7)} \text{dense simple} \xrightarrow{(8)} \text{h-simple}$$

with vertical arrows $(6) \uparrow$, $(9) \downarrow$, and $(10) \downarrow$ to simple

Diagram 1.1

1.8–1.24 Exercises

1.8. An infinite set B is *strongly hyperimmune* (*sh-immune*) if there is no uniformly recursive (see Definition II.2.5) sequence $\{V_n\}_{n \in \omega}$ of mutually disjoint recursive sets such that $V_n \cap B \neq \emptyset$, for all n, and $V = \bigcup_n V_n$ is recursive. (This differs from hh-immune and h-immune of Definition V.2.1 primarily because in the former we give a Σ_1^0 index for V_n, in the latter a canonical index for V_n, and here a characteristic index for V_n as defined in Definitions II.2.1 and II.2.4.) An r.e. set A is *strongly hypersimple* (*sh-simple*) if $\overline{A}$ is sh-immune. Furthermore, B is *finitely strongly hyperimmune* (*fsh-immune*) and A is *finitely strongly hypersimple* (*fsh-simple*) if we add to the above conditions that each V_n be finite.

(a) Prove that an infinite set B is sh-immune (fsh-immune) iff there is no uniformly r.e. (u.r.e.) sequence $\{W_{f(n)}\}_{n \in \omega}$ of mutually disjoint r.e. finite sets such that: $W_{f(n)} \cap B \neq \emptyset$ for all n; and $\bigcup_n W_{f(n)} = \omega$.

(b) Prove that a coinfinite r.e. set A is sh-simple (fsh-simple) iff there is no u.r.e. sequence $\{W_{f(n)}\}_{n \in \omega}$ of mutually disjoint r.e. (finite) sets such that: $W_{f(n)} \cap \overline{A} \neq \emptyset$ for all n; and $\overline{A} \subseteq \bigcup_n W_{f(n)}$.

(c) D. A. Martin [1966b, p. 306] called a coinfinite r.e. set A *supersimple* if there is no uniformly recursive sequence $\{V_n\}_{n \in \omega}$ of mutually disjoint finite sets such that $V_n \cap \overline{A} \neq \emptyset$, for all n. (Namely, the condition above that $V = \bigcup_n V_n$ be recursive is removed.) Prove that A is supersimple iff A is fsh-simple. *Hint* (Lerman-Shore). Let $\{V_n\}_{n \in \omega}$ witness that A is not supersimple. Let $g(n) = \sum_{i=1}^{n} i = \frac{n(n+1)}{2}$. Define

$$U_n = \bigcup \{V_i : g(n) \leq i < g(n+1)\} - \{0, 1, \ldots, n-1\}.$$

(Hence, $U_0 = V_0$, $U_1 = V_1 \cup V_2 - \{0\}$, etc.) Prove that $\{U_n\}_{n \in \omega}$ witnesses that A is not fsh-simple.

1.9. In Diagram 1.1 prove the implications (2), (3), (6) and (9). Note that implications (1), (4), (5) and (10) follow from Exercise X.3.10 and the definitions in Chapters V and X. (See Exercise 1.11 for implications (7) and (8).)

1.10 (Robinson [1967a]). Let B be an infinite set. Prove that the principal function p_B is dominant iff for every disjoint strong array $\{F_n\}_{n \in \omega}$ (of finite sets),

$$(1.2) \qquad\qquad (\exists m)\,(\forall n \geq m)\,[|F_n \cap B| \leq n].$$

Hint. If there is such an array not satisfying (1.2) it is easy to show that p_B is not dominant. Conversely, assume that p_B fails to dominate some strictly increasing recursive function f. Define a strictly increasing recursive function g by $g(0) = f(0)$, and $g(n+1) = f(g(n) + n + 1)$. Let $F_n = (g(2n), g(2n+2)]$, and show that $\{F_n\}_{n \in \omega}$ violates (1.2). If $p_B(x) \leq f(x) > g(1)$ find y such that $g(y) < f(x) \leq g(y+1)$ and prove that $g(y-1) < x$. Prove that

$$\text{card } (B \cap (g(y-1), g(y+1)]) \geq y.$$

Next consider separately the cases $y = 2n$ and $y = 2n + 1$, to see that there are at least n elements of B in either F_{n-1} or F_n.

1.11 (Martin). A coinfinite r.e. set A is *dense simple* (Martin [1966b]) if the principal function $p_{\overline{A}}$ is dominant. (For example, Proposition 1.2 shows that maximal sets are dense simple.) We now prove implications (7) and (8) of Diagram 1.1 so that by Exercise 1.9 all the implications are established.

(a) Prove that dense simple sets are h-simple. (See Theorem V.2.3.)

(b) (Martin [1963, p. 273]). Prove that hh-simple sets are dense simple. *Hint.* If A is not dense simple let $\{F_n\}_{n \in \omega}$ be as in Exercise 1.10 violating (1.2) for $B = \overline{A}$. Construct a disjoint weak array $\{V_n\}_{n \in \omega}$ that witnesses that $\overline{A}$ is not hh-immune by attempting whenever possible to see that for each n, every V_m, $m < n$, contains an element of $F_n - A_s$. (Note that by Exercise X.2.16 it does not matter whether V_n is finite or not.)

One can also give a direct proof of Exercise 1.11 (without using Exercise 1.10) by choosing a strictly increasing function f such that $p_{\overline{A}}(n) < f(n)$ for infinitely many n. Construct the array $\{V_i\}_{i \in \omega}$ in stages and let

$$V^s = \bigcup \{V_i^s : i \leq s\},$$

the elements enumerated in the array by the end of state s. At stage $s + 1$ choose the least $i \leq s$ such that

$$(\exists n) \, [2i < n \leq s \;\&\; V_i^s \cap \overline{A}_s \upharpoonright f(n) = \emptyset$$
$$\&\; (\exists x) \, [f(2i) \leq x < f(n) \;\&\; x \in \overline{A}_s \cap \overline{V}^s]].$$

Enumerate the least such x into V_i^{s+1}. Prove that for every i, V_i is finite and $V_i \cap \overline{A} \neq \emptyset$.

1.12 (a) (Martin [1966b, Theorem 3]).   Show that the proof of Theorem 1.7 establishes that every fsh-simple set is high.

(b) Show that each class of r.e. sets in Diagram 1.1, except for the simple and h-simple sets, is contained in the high sets.

(c) Prove that the implications (8) and (9) of Diagram 1.1 cannot be reversed. *Hint.* Use Theorem V.2.5.

1.13 (Lachlan and Robinson).   Prove that there is an r-maximal set without any dense simple superset. Conclude that there should be no other arrows in Diagram 1.1 pointing to "dense simple" except those shown.

Hint. Show that in the construction in Exercise X.5.10 of an atomless r-maximal set if $F_j \cap \overline{A} \neq \emptyset$ then $|F_j \cap \overline{A}| \geq j$ and use Exercise 1.10.

1.14 (Robinson [1967a]).   Prove that we can modify the proof of Theorem X.5.4 to make A dense simple. *Hint.* See Exercise X.3.8. This gives an example of an r-maximal (and therefore sh-simple) set that is dense simple but has no hh-simple superset. Hence, the implications (6) and (7) of Diagram 1.1 cannot be reversed, and there is no arrow from r-maximal to hh-simple. (The latter also follows from Corollary X.4.4 and Proposition X.4.5.)

1.15 (Martin [1966b, p. 306]).   Prove that there is a dense simple set A that is not fsh-simple. *Hint.* Use movable markers $\{\Gamma_n\}_{n \in \omega}$ so that Γ_n comes to rest on a_n, where $\overline{A} = \{a_0 < a_1 < \cdots\}$. Move Γ_n beyond $\varphi_i(n)$ if $i \leq n$ and $\varphi_i(n) \downarrow$. Let V_n be the set of positions of Γ_n. To make $\{V_n\}_{n \in \omega}$ uniformly recursive move Γ_n at stage $s + 1$ only to an element $> \max\{V^s\}$.

1.16.   Young [1966, p. 80] has shown that there is a set which is fsh-simple but not sh-simple, so that implication (3) of Diagram 1.1 cannot be reversed. Prove that there are no further arrows that can be added to Diagram 1.1. *Hint.* Use Exercises 1.12, 1.13, 1.14, 1.15, 1.16 and results from Chapters V and X.

1.17 (Robinson [1967a]).   Prove that a coinfinite r.e. set A is sh-simple iff for every disjoint weak array $\{W_{f(n)}\}_{n \in \omega}$ whose union contains $\overline{A}$ there exists m such that for all $n \geq m$, $W_{f(n)} \cap \overline{A}$ is finite. (Compare this with similar characterizations Exercise 1.10 and Exercise X.2.15 of dense simple and hh-simple sets.) *Hint.* Use Exercise 1.8(b).

1.18.° Let $A \subset_m B$ and let $\hat{A}$ be the pullback of A as defined in Exercise X.4.8.

(a) (Robinson [1968, p. 344]). Prove that $\hat{A}$ is dense simple (and therefore high), and is not contained in any sh-simple set. (Indeed Stob and Herrmann have observed that $\hat{A}$ is not contained in any fsh-simple set.)

(b) (Stob [1979, Theorem I.2.5] and Herrmann [1978, p. 191]). Prove that $\hat{A}$ is a major subset of some r.e. set.

1.19 (Jockusch [1973a]). Prove that if $A \subset_m B$ then A is high. *Hint.* Let $c(n) = (\mu s) \, (\exists x > n) \, [x \in B_s - A]$. Now $c \leq_T A$ so if A is not high then c fails to dominate some recursive function f. Let $R = \bigcup_{n=0}^{\infty}(B_{f(n)} - [0, n])$. Prove that $\overline{R}$ violates $A \subset_m B$.

1.20 (Robinson [1968]). Let $\{A_s\}_{s \in \omega}$ be a (recursive) enumeration of an r.e. set A. The *computation function* $c_A(x)$ of this enumeration is defined by

$$c_A(x) = (\mu s) \, [A_s \restriction x = A \restriction x].$$

We say that A is *enumeration dominant* (*e-dominant*) if it has an enumeration whose computation function is dominant.

(a) Let $c_A(x)$, $\{A_s\}_{s \in \omega}$, and A be as above, and $g(x) \leq_{wtt} c_A(x)$. (See the definition in Exercise V.2.16 of $\leq_{wtt}$.) Prove that if $c_A(x)$ is not dominant then neither is $g(x)$.

(b) Prove that if an r.e. set is e-dominant via the computation function for *one* enumeration then it is e-dominant via the computation function for *every* enumeration.

(c) Prove that if A is e-dominant, B is r.e., and $A \leq_{wtt} B$, then B is e-dominant.

(d) Prove that every *wtt*-complete set is e-dominant.

1.21 (Robinson [1968]). Prove that every r.e. degree contains an r.e. set that is not e-dominant. (Also see Exercise 2.7 for e-dominant sets.)

1.22 (Robinson [1968]). If $\{A_s\}_{s \in \omega}$ is an enumeration of A, define the *deficiency subset* (of A) for this enumeration to be

$$B = \{\, x : (\exists s) \, (\exists y < x) \, [x \in A_s \ \& \ y \in A - A_s] \,\},$$

and the *deficiency set*

$$D = \{\, s : A_s \subseteq A_{s-1} \cup B \,\}.$$

(Note that if $|A_{s+1} - A_s| = 1$, for all s, then D is the same as the Dekker deficiency set of Theorem V.2.5.)

(a) Prove that an infinite r.e. set A is e-dominant iff every deficiency subset B of A is a major subset of A.

(b) Prove that if $\{A_s\}_{s\in\omega}$ enumerates an e-dominant set A then the corresponding deficiency set D is dense simple and is not contained in any sh-simple set.

(c) Prove that if $\{A_s\}_{s\in\omega}$ enumerates a creative set then D is an atomless coinfinite r.e. set.

1.23° (Robinson [1968]).   Prove that there is a maximal set that is e-dominant and one that is not e-dominant.

1.24 (Cooper [1972c]).   (a) If $B \leq_T \emptyset'$ prove that B is hh-immune iff B is shh-immune. (Adapt the proof of Exercise X.2.16.)

(b) Show that the hypothesis in Martin 1.7 that A is r.e. is unnecessary. Namely, prove that if B is hh-immune and $B \leq_T \emptyset'$ then B is high. *Hint* (Jockusch).   Assume B is hh-immune.   Relativize the Arslanov criterion Theorem V.5.1 and Exercise V.5.8(c) to $\emptyset'$. In order to show that $\emptyset'' \leq_T B'$ it suffices to find a function $h \leq_T B'$ such that $h(x) \neq \varphi_x^{\emptyset'}(x)$ for all x. Define the recursive function

$$g(x, s) = \begin{cases} \varphi_{x,s}^{K_s}(x) & \text{if defined;} \\ 0 & \text{otherwise.} \end{cases}$$

For each x define the disjoint weak array $\{V_n\}_{n\in\omega}$ by $V_n = \{s : g(x, s) = n\}$. Since B is shh-immune, $V_n \subseteq \overline{B}$ for some n. Use B' to compute such an n, and then note that $n \neq \varphi_x^K(x)$.

1.25 Notes.   Exercise 1.14 is an improvement suggested by Robinson [1967a] of a theorem of Lachlan and Robinson, while Exercise 1.13 uses the construction of Lachlan [1968c] and the characterization of Robinson [1967a]. Robinson [1968] called the e-dominant sets of Exercise 1.20 *high* and the non-e-dominant sets *low*, but it has since become standard to use the high and low terminology with the meaning defined here in Definition IV.4.2.

1.26 Open Questions (Jockusch).   Does every cohesive set C (or indeed every shh-immune set) satisfy $\emptyset'' \leq_T C'$? This is known to be false for hh-immune sets B unless we add the hypotheses $B \leq_T \emptyset'$ as in Exercise 1.24.

2. Maximal Sets and High R.E. Degrees

By Corollary 1.5 every maximal set has high degree. In this section we wish to prove that conversely every high r.e. degree **d** contains a set that is maximal (and therefore lies in every class shown in Diagram 1.1). Let $D \in \mathbf{d}$ be an r.e. set. In Exercise V.3.6 we saw how to use the *Yates permitting method* to construct an r.e. set $A \leq_T D$ such that A satisfies certain *finitary* positive requirements that suffice to guarantee that A is simple, or hypersimple. To guarantee that A is maximal we must meet certain *infinitary* positive requirements P_e as in Theorem X.3.3 where each P_e may cause an *infinite* set S_e of elements to be enumerated in A, for example if P_e is the requirement $S_e \subseteq^* A$. We can satisfy this new requirement P_e providing that: (1) S_e is recursive; and (2) D is high. The method of achieving this is called the *Martin permitting method* and will be illustrated in the next two theorems, which are preparation for the main result, Theorem 2.3.

2.1 Theorem.   *If A is a high r.e. set then there is an r.e. set $D \equiv_T A$ and a recursive enumeration $\{ D_s \}_{s \in \omega}$ of D whose computation function $c_{\overline{D}}(x)$ is dominant where $\overline{D} = \{ d_0 < d_1 < \cdots \}$, $\overline{D}_s = \{ d_0^s < d_1^s < \cdots \}$, and $c_{\overline{D}}(x) = (\mu s) \, [d_x^s = d_x]$.*

Proof. Since A is high there exists a dominant function $f \leq_T A$ by Theorem 1.3. Let $f = \{ e \}^A$, and $\{ A_s \}_{s \in \omega}$ be a recursive enumeration of A. Define $g(x, s) = \{ e \}_s^{A_s}(x)$ if the latter is defined and $g(x, s) = 0$ otherwise. Let $c_g(x)$ be the least modulus function for g, namely

$$c_g(x) = (\mu s) \, (\forall t \geq s) \, [g(x, t) = g(x, s)].$$

Now $c_g \leq_T A$ by the Modulus Lemma III.3.2, and c_g is dominant because if c_g fails to dominate some recursive function h then f fails to dominate the recursive function $g(x, h(x)) + 1$. We define $\{ D_s \}_{s \in \omega}$ by stages. Let $D_0 = \emptyset$.

    Stage $s + 1$.

        Step 1. (To ensure $c_g(x) \leq c_{\overline{D}}(x)$). If $g(x, s+1) \neq g(x, s)$ and $x \leq s$ enumerate d_x^s in D_{s+1}. (Note that D_{s+1} is finite.)

        Step 2. (To ensure $A \leq_T D$). If $x \in A_{s+1} - A_s$, enumerate d_x^s in D_{s+1}.

Since f is total $\lim_s g(x, s)$ exists; hence $\lim_s d_x^s$ exists, and $c_{\overline{D}}(x)$ is defined for all x. Now $c_{\overline{D}}$ is dominant by Step 1. Secondly, $D \leq_T A$ because for each x if we choose $s > c_g(x)$ such that $A_{s+1} \restriction x+1 = A \restriction x+1$ then $d_x^s = d_x$. Finally, $A \leq_T D$ because if $d_x^s = d_x$ then $x \in A$ iff $x \in A_s$.   ▯

(Note that in Theorem 2.1 it was essential that f be total so that $\lim_s g(x, s)$ exists for all x, else $|\overline{D}| < \infty$. This is used in the proof of Theorem XII.4.13.) The easiest application of the Martin permitting method is to prove the following variation of the Thickness Lemma VIII.1.1.

2.2 Theorem. *If* **d** *is a high r.e. degree and* B *is a piecewise recursive r.e. set then there is a thick r.e. subset* A *of* B *such that* $\deg(A) \leq$ **d**.

Proof. Choose an r.e. set $D \in$ **d** and an enumeration $\{D_s\}_{s \in \omega}$ of D with $c_{\overline{D}}$ dominant as in Theorem 2.1. Let $\{B_s\}_{s \in \omega}$ be a recursive enumeration of B. If $x \in B_{s+1} - A_s$ enumerate x in A_{s+1} iff $d_x^{s+1} \neq d_x^s$. Clearly $A \subseteq B$ and $A \leq_T D$. We claim that every positive requirement $P_e : A^{[e]} =^* B^{[e]}$ is satisfied. If not choose e such that $B^{[e]} - A^{[e]}$ is infinite. Since $B^{[e]}$ is recursive we can define the recursive function

$$h(x) = \begin{cases} (\mu s)\,[x \in B_s] & \text{if } x \in B^{[e]}, \\ 0 & \text{otherwise.} \end{cases}$$

Now for all $x \in B^{[e]} - A^{[e]}$, $d_x^{h(x)} = d_x$ and hence $c_{\overline{D}}(x) \leq h(x)$. Thus if $B^{[e]} - A^{[e]}$ is infinite then $c_{\overline{D}}(x)$ is not dominant. ▯

Given D with $c_{\overline{D}}$ dominant we now wish to construct a maximal set $A \leq_T D$ by using the maximal set construction and notation of Theorem X.3.3 to meet for each e the positive requirement

$$P_e : \overline{A} \subseteq^* \overline{W}_e \quad \text{or} \quad \overline{A} \subseteq^* W_e.$$

We have a movable marker Γ_e that rests on a_e^s at the end of stage s; Γ_e may move at stage $s+1$ to some a_i^s, $e < i$, in order to maximize its e-state; when Γ_e moves we enumerate a_e^s into A_{s+1}. Now to ensure that $A \leq_T D$ we allow marker Γ_e to move and hence allow $a_e^s \in A_{s+1} - A_s$ only if $d_e^{s+1} \neq d_e^s$ as in Theorem 2.2. To see that this permitting still allows requirement P_e to be satisfied, consider $e = 0$ and define $h(x)$ to be the least s such that either: $x \in A_s \cup W_{0,s}$; or $x = a_e^s$ and Γ_e wants to move to some $a_i^s \in W_{0,s}$, $e < i$, in order to maximize its 0-state. If $W_0 \cap \overline{A}$ is infinite then h is total and recursive. But $c_{\overline{D}}(x)$ dominates $h(x)$ so almost all markers are permitted by D to move to elements in W_0 and hence $\overline{A} \subseteq^* W_0$. Note that once a marker wants to move to maximize its 0-state it continues to have this desire until it is fulfilled.

2.3 Theorem (Martin [1966b]). *If* **d** *is a high r.e. degree then there is a maximal set* A *of degree* **d**.

2.4 Definition. If $\mathcal{C} \subseteq \mathcal{E}$ let $\deg(\mathcal{C}) = \{\deg(W) : W \in \mathcal{C}\}$.

2.5 Corollary. *Let* $\mathcal{M}$ *be the class of maximal sets. Then* $\deg(\mathcal{M}) = \mathbf{H}_1$. *Furthermore, for each class* $\mathcal{C}$ *of r.e. sets shown in Diagram 1.1, except the classes of simple and hypersimple sets,* $\deg(\mathcal{C}) = \mathbf{H}_1$.

Proof. Use Exercise 1.12, Theorem 2.3 and the implications in Diagram 1.1. ▯

Proof of Theorem 2.3. Choose an r.e. set $D \in \mathbf{d}$ and $\{D_s\}_{s \in \omega}$ and $c_{\overline{D}}(x)$ as in Theorem 2.1. Construct A by stages using the method and notation of Theorem X.3.3. Let $\overline{A}_s = \{a_0^s < a_1^s < \cdots\}$, and $\overline{A} = \{a_0 < a_1 < \cdots\}$. For $s = 0$, set $A_0 = \emptyset$ so $a_n^0 = n$ for all n. If $x, y \in \overline{A}_s$, $x < y$, $x = a_i^s$, and $\sigma(i, x, s) < \sigma(i, y, s)$, we say x is *attracted to* y at stage $s + 1$ for the sake of P_i.

  Stage $s + 1$.
    Step 1 (to satisfy requirement P_i). Choose the least i such that

(2.1) $$(\exists j > i)\, [\sigma(i, a_i^s, s) < \sigma(i, a_j^s, s)],$$

and

(2.2) $$(\exists y \le a_i^s)\, [d_y^{s+1} \ne d_y^s].$$

Let σ_1 be the highest i-state such that $\sigma(i, a_j^s, s) = \sigma_1$ for some j, $i < j \le s$. Choose the minimum j, $i < j \le s$, such that $\sigma(i, a_j^s, s) = \sigma_1$. Enumerate a_k^s in A_{s+1} for all $k \ne j$, $i \le k \le s$. Move marker Γ_i to a_j^s, and marker Γ_{i+k} to a_{s+k}^s for all $k > 0$. (Notice that at the end of this enumeration no $x \ge a_i^{s+1}$ is attracted to any $y > x$.) If there is no such i do nothing.
    Step 2 (to ensure $D \le_T A$). Choose $z = (\mu y)\, [d_y^{s+1} \ne d_y^s]$. If some $x \le a_{z+1}^s$ was enumerated in A_{s+1} under Step 1 do nothing. Otherwise, enumerate a_z^s in A_{s+1} unless $\sigma(z, a_{z+1}^s, s) < \sigma(z, a_z^s, s)$, in which case enumerate a_{z+1}^s into A_{s+1}. If z fails to exist do nothing.

Remark. Under Step 2 we enumerate whichever of a_z^s and a_{z+1}^s has the lower z-state, and hence we guarantee that if any $x \le a_z^s$ is attracted to some $y \in \{a_z^s, a_{z+1}^s\}$ *before* performing Step 2 then x is still attracted to some remaining element $y' \in \{a_z^s, a_{z+1}^s\} - A_{s+1}$ *after* performing Step 2.

Lemma 1. *For every e, marker Γ_e moves at most finitely often. (Hence, $\overline{A}$ is infinite.)*

Proof. Choose t such that $d_e^t = d_e$. Thus, no marker Γ_i, $i \le e$, moves at any stage $s \ge t$ because of Step 2, but only because of Step 1. Now the proof of Theorem X.3.3 Lemma 1 applies to show that Γ_e moves at most finitely often after stage t.

Lemma 2. $A \equiv_T D$.

Proof. First $D \le_T A$ since by Step 2 if $a_{x+1}^s = a_{x+1}$, then $d_x^s = d_x$. Second, $A \le_T D$ since if $d_x^s = d_x$ then $x \in A$ iff $x \in A_s$.

Lemma 3. $(\forall e) [\overline{A} \subseteq^* \overline{W}_e \ or \ \overline{A} \subseteq^* W_e]$.

Proof. Fix e and assume the lemma for all $i < e$. Choose $n \geq e$ such that for all $m \geq n$, $\sigma(e-1, a_m) = \sigma(e-1, a_n) = \sigma_0$. Let $\sigma_1 = \sigma_0 \cup \{e\}$. Choose s_0 such that for all $m \leq n$, $a_m^{s_0} = a_m$ and $\sigma(e, a_m, s_0) = \sigma(e, a_m)$.

Assume for a contradiction that $W_e \cap \overline{A}$ and $\overline{W}_e \cap \overline{A}$ are both infinite. Define a recursive function $h(x)$ not dominated by $c_{\overline{D}}(x)$ as follows. For $x \leq a_n$ let $h(x) = 0$. For $x > a_n$ define

$$h(x) = (\mu s > s_0) \, [x \in A_s \vee \sigma(e, x, s) > \sigma_0 \vee$$
(2.3)
$$(\exists i) \, [x = a_i^s \ \& \ \sigma(e, a_i^s, s) = \sigma_0 \ \& \ (\exists j > i) \, [\sigma(e, a_j^s, s) \geq \sigma_1]]].$$

(If the third clause of (2.3) holds then $x = a_i^s$ is attracted to $y = a_j^s$ for the sake of P_e.) We wish to show that if

(2.4)
$$h(x) \text{ is defined by the third clause of (2.3)}$$
$$\& \ x \in \overline{A} \ \& \ \lim_s \sigma(e, x, s) = \sigma_0,$$

then

(2.5)
$$d_x^{h(x)} = d_x,$$

and $c_{\overline{D}}(x)$ fails to dominate $h(x) + 1$. But if $x \in \overline{W}_e \cap \overline{A}$ then (2.4) must hold for x since $W_e \cap \overline{A}$ is infinite. Hence, since $\overline{W}_e \cap \overline{A}$ is infinite $c_{\overline{D}}(x)$ is not dominant contrary to hypothesis.

Fix x and assume that (2.4) holds for x. To prove (2.5) it suffices to prove

(2.6)
$$(\forall s \geq h(x)) \, (\exists k) \, [x < a_k^s \ \& \ \sigma_1 \leq \sigma(e, a_k^s, s)],$$

namely for all $s \geq h(x)$, x is attracted to some $y > x$ for the sake of requirement P_e. Now (2.6) implies (2.5) because otherwise $d_x^{s+1} \neq d_x^s$ for some $s \geq h(x)$, so for some $z \leq x$, $z \in A_{s+1} - A_s$ by (2.1) and (2.2), and thus either x enters A_{s+1} or x is the new position of the least marker Γ_p which moves at stage $s + 1$. The former cannot occur by (2.4). The latter cannot occur because suppose x is the new position of the least marker Γ_p which moves at $s + 1$. Since $s + 1 > h(x) \geq s_0$, we must have $p > n \geq e$. By (2.6) there exists k such that $x < a_k^s$ and $\sigma_1 \leq \sigma(e, a_k^s, s)$. Since $\sigma(e, x) = \sigma_0$, we have $\sigma(p, x, s) < \sigma(p, a_k^s, s)$, and therefore x cannot be the new position of Γ_p which would have preferred to move to a_k^s instead of to x.

It remains only to prove (2.6) by induction on $s \geq h(x)$. Now (2.6) clearly holds for $s = h(x)$ via $k = j$ in (2.3) because we are assuming (2.4) for x. Fix $s \geq h(x)$, and assume a_k^s satisfies (2.6) for s but not for $s + 1$.

Case 1. $a_k^s \in A_{s+1} - A_s$ by Step 1. Then some least marker Γ_p moves at stage $s+1$ to an element a_q^s for some p and q, $n < p < q \leq s$. Now $p \leq k$ and $k \neq q$ since $a_k^s \in A_{s+1}$. But $x < a_p^s$ because $x \notin A_{s+1}$. Thus, $x < a_p^s \leq a_k^s$, and $a_p^s < a_q^s$. Now the p-state (and hence the e-state) at s of a_q^s must have been at least as high as that of a_k^s since Γ_p moved to a_q^s rather than to a_k^s. Thus, (2.6) holds at $s + 1$ via element $y = a_q^s = a_p^{s+1}$.

   Case 2. $a_k^s \in A_{s+1} - A_s$ by Step 2. The remark at the end of Step 2 shows that (2.6) still holds for x at $s + 1$ via either $y = a_{k+1}^s$ or $y = a_{k-1}^s$. (Namely, let $z = (\mu y)\ [d_y^{s+1} \neq d_y^s]$. Suppose $z = k$. Then the z-state, and hence the e-state, of a_{k+1}^s is at least as high as that of a_k^s so (2.6) holds at $s + 1$ via $y = a_{k+1}^s$. Suppose $z = k - 1$. Then $a_k^s \in A_{s+1}$ implies $\sigma(k - 1, a_k^s, s) < \sigma(k - 1, a_{k-1}^s, s)$. Hence, $x < a_{k-1}^s$ and $\sigma(e, a_{k-1}^s, s) \geq \sigma_1$, so (2.6) holds via $y = a_{k-1}^s$.)

   This completes the proof of (2.6), the proof of Lemma 3, and the proof of the theorem.   ▯

2.6 Remark.   We visualize this construction as follows. If $h(x) = s$ via the third clause of (2.3) then we assign an *e-symbol* $*_e$ to x indicating that x *wants* to enter A for the sake of P_e as soon as D permits x. By (2.6) x continues to want to enter at all stages $s \geq h(x)$. To satisfy a finitary positive requirement P_e using Yates permitting (for example $P_e : W_e$ infinite $\implies W_e \cap A \neq \emptyset$, so A is simple) it suffices to show that at least *one* element x with an e-symbol $*_e$ is eventually enumerated in A. To satisfy an infinitary positive requirement with Martin permitting we must show that almost every x that receives $*_e$ is eventually permitted (or loses $*_e$ because $\sigma(e, x, s) = \sigma_1$). The crucial point is that the positive requirements P_e must be such that we can define a recursive function $h(x)$ as in (2.3) in order to force D to permit almost every element retaining the symbol $*_e$. Namely, P_e must be ready to act with the starred element at all stages $s \geq h(x)$.

2.7–2.18 Exercises

2.7° (Robinson [1968]).   Prove that every high r.e. degree **d** contains an e-dominant r.e. set D as defined in Exercise 1.20. (For degrees of non-e-dominant sets see Exercise 1.21.) (Thus, if we had proved Exercise 2.7 first in place of Theorem 2.1 we could have chosen D and $\{D_s\}_{s\in\omega}$ in Theorems 2.2 and 2.3 so that merely $c(x)$ (rather than $c_{\overline{D}}(x)$) is dominant where $c(x) = (\mu s)\ [D_s \restriction x = D \restriction x]$. Then we could have used the permitting condition $D_{s+1} \restriction x \neq D_s \restriction x$ as in Chapter V instead of $d_x^{s+1} \neq d_x^s$ as used here.) *Hint.* Choose an r.e. set $A \in \mathbf{d}$. It suffices to construct $D \leq_T A$, D e-dominant, since then $A \oplus D$ is e-dominant by Exercise 1.20(c). Let $\{A_s\}_{s\in\omega}$, f, and $g(x, s)$ be as in Theorem 2.1. Let $D = \bigcup_s D_s$ where

$$D_s = D_{s-1} \cup \{x : g(x, s) \neq g(x, s + 1) \ \& \ (\forall y < 3x)\ [\{(x)_0\}_s(y)\downarrow]\}.$$

Clearly $D \leq_T A$. Use c_g dominant to prove that c_D is dominant.

2.8 (a) (Martin [1967b]). Prove that for any high r.e. degree **d** there is an atomless r.e. set A of degree $\leq$ **d**. *Hint.* Combine Exercise X.5.5 with the method of Theorem 2.3 to obtain $A \leq_T D$. When the element $x = x(n, e)$ wants to enter A for P_e assign the e-symbol $*_e$ to x and enumerate x later in A_{s+1} if $d_x^{s+1} \neq d_x^s$. Show that if W_e is a maximal superset of A then almost every element x with $*_e$ enters A. Analogously as in Lemma 3 of Theorem 2.2 define the recursive function,

$$h(x) = (\mu s)\,[x \in A_s \cup W_{e,s} \vee [x \in F_n \ \& \ |F_n - (W_{e,s} \cup A_s)| = 1]].$$

(b) Modify the proof of (a) to show that $A \equiv_T D$. *Hint.* Choose F_n so that $|F_n| = n+2$ and enumerate some element of F_n into A if $n \in D_{s+1} - D_s$.

(c) Construct $A \in$ **d** so that A has no dense simple superset. (See Exercises 1.13 and X.5.10.)

2.9 (Jockusch-Soare [1972a]). (This is another example of the "maximum degree principle" of Exercise V.5.6. See also Exercises 2.10, 2.11 and Notes 2.19.) Prove that Exercise 2.8(b) was unnecessary because the set $A \leq_T D$ in 2.8(a) automatically satisfies $A \equiv_T D$.

Hint. Let $\{A_s\}_{s \in \omega}$ and $\{D_s\}_{s \in \omega}$ be the enumerations of A and D. Let F_n be as in Exercise X.5.5. Let $m(x) = (\mu s)\,[s \in D_s]$ if $x \in D$, and $m(x)$ diverge otherwise. As in Theorem V.5.1 use the Recursion Theorem to define a recursive function $h(x)$ by

$$W_{h(x)} = \begin{cases} F_{h(x)} - \{y\}, \text{ where } y = (\mu z)\,[z \in F_{h(x)} - A_{m(x)}] & \text{if } x \in D, \\ \emptyset & \text{otherwise.} \end{cases}$$

Let $g(x) = \max(F_{h(x)})$, and

$$r(x) = (\mu s)\,[A_s \restriction g(x) = A \restriction g(x)].$$

As usual $r \leq_T A$ because g is recursive. Let $\hat{D} = \{x : x \in D - D_{r(x)}\}$. If $\hat{D}$ is finite then $D \leq_T A$. If $\hat{D}$ is infinite prove that $D \leq_T A$ as follows. Clearly, $\hat{D}$ is r.e. in A. Fix any number y. To determine whether $y \in D$, find $x \in \hat{D}$ such that $y < h(x)$. Let s be the least stage such that $|W_{h(x),s}| = |F_{h(x)}| - 1$. Since $x \in \hat{D}$, $m(x) > r(x)$ so $W_{h(x)} = F_{h(x)} - \{z\}$, for some $z \in \overline{A}$. Show that $y \in D$ iff $y \in D_s$.

2.10.° Let $A \leq_T B$ be the semicreative set constructed recursive in some nonrecursive r.e. set B according to the method in the hint of Exercise V.3.5. Prove that $B \leq_T A$ automatically as in Exercise 2.9.

2.11° (Jockusch-Soare) [1972a]. (a) Prove that Step 2 in the proof of Theorem 2.3 is unnecessary, namely that if $A \leq_T D$ is constructed using Step 1 alone then $D \leq_T A$ automatically as in Exercises 2.9 and V.5.6. *Hint.* Let

$m(x) = (\mu s)\,[x \in D_s]$ if $x \in D$ and $m(x)$ is undefined otherwise. Define a recursive function h by:

$$W_{h(x)} = \begin{cases} \{\, a^{m(x)}_{h(x)+2i} : 0 < i \leq h(x) + 1 \,\} & \text{if } x \in D, \\ \emptyset & \text{otherwise.} \end{cases}$$

(b) Prove that the maximal set A of Theorem X.3.3 is automatically complete (without the Yates modification in Exercise X.3.8).

2.12° (Lachlan [1968c, p. 27]).   Prove that the construction of the r.e. set A in Theorem X.7.2 (or more simply Theorem X.6.1) can be carried out in any high r.e. degree.

2.13° (Lerman [1971a]).   Show that for any high r.e. degree **d** there is an atomless r-maximal set $A \in \mathbf{d}$.

2.14° (Lerman [1971a]).   Prove that if **d** is a high r.e. degree and B is a nonrecursive r.e. set then there is a major subset A of B such that $A \in \mathbf{d}$. Conclude by Exercise 1.19 that $\deg(\mathcal{C}) = \mathbf{H}_1$ where $\mathcal{C} = \{\, A : (\exists X)\,[A \subset_m X]\,\}$. *Hint.* Combine the construction of Theorem X.4.6 with the permitting method of Theorem 2.3. Let V_e be as in Theorem X.4.6 and $U_e = V_e \backslash B$. The markers $\{\Gamma_e\}_{e \in \omega}$ resting on elements of $B_s - A_s$ maximize e-states measured with respect to $\{U_n\}_{n \in \omega}$. Make sure that $\Gamma_e^s \leq \Gamma_e^{s+1}$ for all $e \in \omega$.

2.15° (Cooper [1974b]).   Prove that if **d** is a high r.e. degree then there is a minimal pair **a**, **b** of r.e. degrees each $\leq \mathbf{d}$. *Hint.* Let $D \in \mathbf{d}$ be as in Theorem 2.1. Construct A and B using the pinball machine method sketched in Exercise IX.1.7. In addition divide each gate G_e into an upper level G_e^+ and a lower level G_e^-. Let $x_1, x_2, \ldots$ be the elements arriving at G_e. When x_n arrives at G_e, say at stage s_n, x_n enters G_e^+ and waits for a stage $t_n \geq s_n$ such that $r(e, t_n) < x$, at which time x_n drops to G_e^-, and we define $f_e(s_n) = t_n$. Now x_n waits for $v \geq t_n$ such that $d_{s_n}^{v+1} \neq d_{s_n}^v$, at which time x_n drops to G_{e-1}. Define $f_e(z) = 0$ if $z \notin \{s_n\}_{n \in \omega}$, and note that f_e is recursive if $\{x_n\}_{n \in \omega}$ is infinite. If x_n remains permanently at G_e^- then $f(s_n) > c_{\overline{D}}(s_n)$. Hence G_e^- cannot have infinitely many permanent residents. Also if follower x lies above hole H_e and then becomes realized at stage s, x must wait for $t \geq s$, $d_x^{t+1} \neq d_x^t$, before x may leave hole H_e and drop to gate G_e. Since D permitting of x is required before x can leave any hole or gate, D can compute the final position within the pinball machine of every element x. Hence $A, B \leq_{\mathrm{T}} D$.

2.16 (Martin [1967b]).   Let A be a coinfinite r.e. set which is not high. Prove that A has a coinfinite r.e. superset B such that the principal function $p_{\overline{B}}$ dominates every partial recursive function, and so in particular B is dense simple. *Hint.* Use Lemma 1.6.

2.17. (a) (Martin and Miller [1968]). Let $A \leq_T \emptyset'$, and $\{A_s\}_{s \in \omega}$ be a uniformly recursive array of recursive sets such that $A(x) = \lim_s A_s(x)$. Prove that if A is nonrecursive then there is no recursive function f that dominates c_A where $c_A(x) = (\mu s \geq x) [A_s \upharpoonright x = A \upharpoonright x]$. *Hint.* To compute $A(x)$ look for the least n such that $A_s(x) = A_{f(n)}(x)$ for all $s \in [n, f(n)]$.)

   (b) Use (a) to show that every nonzero degree below $\mathbf{0}'$ is hyperimmune.

   (c)$^\circ$ (Soare). Show that this is false if we incorrectly define

$$c_A(x) = (\mu s) [A_s \upharpoonright x = A \upharpoonright x].$$

(Note that A is merely Δ_2 not Σ_1.)

2.18° (Posner and Robinson [1981]). If $\mathbf{0} < \mathbf{c} \leq \mathbf{0}'$ and $\mathbf{0} < \mathbf{b} \leq \mathbf{0}'$ then there is a degree $\mathbf{a}$ with $\mathbf{a} \cup \mathbf{b} = \mathbf{0}'$ and $\mathbf{c} \not\leq \mathbf{a}$. *Hint.* Use Exercise 2.17. Note that these degrees are *not* necessarily r.e.

2.19 Notes. The device in Step 2 of Theorem 2.3 for ensuring $D \leq_T A$ is due to Lachlan and makes the proof considerably simpler than Martin's original proof [1966b], which required substages at each stage.

   The *maximum degree principle* asserts very roughly that any r.e. set A being constructed to have a certain property P automatically has the highest degree not explicitly violating P. For example, sets constructed with weak negative requirements tend to be complete, such as K, Post's simple set S (Theorems V.1.3 and V.4.1), the join $A \oplus B$ of the Friedberg-Muchnik sets A and B (Exercise VII.2.3), and the maximal set (Exercise 2.11(b)). See also Arslanov's Completeness Criterion, Theorem V.5.1. The observation that permitting constructions of an r.e. set $A \leq_T D$ often lead automatically to $D \leq_T A$ (such as Exercises 2.9, 2.10, 2.11, and V.5.6) was made in Jockusch and Soare [1972a, p. 615]. It would be interesting to formulate and prove a precise version of the principle.

   A great deal of information on non-r.e. degrees below $\mathbf{0}'$, including Exercises 2.17 and 2.18 and Martin permitting below high (non-r.e.) degrees, may be found in Posner [1980].

2.20 Open Questions. If $\mathbf{d}$ is any high r.e. degree and B any nonrecursive r.e. set, is there an r.e. set $A \in \mathbf{d}$ such that $\mathcal{L}^*(A) \cong \mathcal{L}^*(B)$. The results of this section, especially Theorem 2.3 and Exercises 2.12, 2.13, and 2.14 suggest a positive answer. Maass, Shore and Stob [1981] have shown that the statement is false with the stronger conclusion "A is automorphic to B." Soare [1982a] showed that $\mathcal{L}^*(A) \cong \mathcal{E}^*$ if A is low.

3. Low R.E. Sets Resemble Recursive Sets

In §1 we saw that sets with certain complicated or unusual structural properties (such as being maximal, hh-simple or a major subset) *had* to be high,
and in §2 that *any* high r.e. degree contained an r.e. set with such a property.
The point of this section is that *low* r.e. sets and degrees tend to have very
uniform structure that resembles that of recursive sets. (We have already
seen some partial evidence of this in Theorem 1.7 and Exercise 2.16 for r.e.
sets that are *not* high.)

The Sacks Splitting Theorem (VII.3.2) implied that any r.e. degree $\mathbf{b} > \mathbf{0}$
splits over $\mathbf{0}$. In Theorem 3.2 we prove the Robinson Splitting Theorem,
which asserts that moreover $\mathbf{b}$ splits over any *low* r.e. degree $\mathbf{c} < \mathbf{b}$. (Lachlan's
Nonsplitting Theorem [1975a] (known as the "monster" theorem) asserts that
this is not true in general when $\mathbf{c}$ is not low.) In Exercise 3.4 we establish a
generalization above low r.e. degrees for Lachlan's Non-Diamond Theorem.
In Exercise 3.5 we show that every coinfinite low r.e. set C has a maximal
superset. Indeed it is possible to get a much stronger result by showing that
$\mathcal{L}(C) \cong \mathcal{E}$ for every such C (see Notes 3.7). This gives a very strong structural
resemblance between low r.e. sets and recursive sets.

3.1 Lemma. *If C is a low set (not necessarily r.e.) then*

$$(3.1) \qquad X =_{\mathrm{dfn}} \{ j : (\exists n \in W_j)\, [D_n \subseteq \overline{C}] \} \leq_{\mathrm{T}} \emptyset'.$$

Proof. Clearly X is Σ_1^C, so $X \leq_{\mathrm{T}} C'$. If C is low then $C' \leq_{\mathrm{T}} \emptyset'$, so
$X \leq_{\mathrm{T}} \emptyset'$. ▯

(Note that an index for the Turing reduction in (3.1) is not uniform in
an r.e. index for C. We also need a *lowness index* for C, i.e., an index e such
that $C' = \{ e \}^{\emptyset'}$. See Exercise XII.4.24.)

3.2 Theorem (Robinson Splitting Theorem) (Robinson [1971a, Corollary 9]).
*Let $\mathbf{b}$ and $\mathbf{c}$ be r.e. degrees such that $\mathbf{c} < \mathbf{b}$, and $\mathbf{c}$ is low. Then there exist
incomparable low r.e. degrees $\mathbf{a}_0$ and $\mathbf{a}_1$ such that $\mathbf{b} = \mathbf{a}_0 \cup \mathbf{a}_1$, and $\mathbf{a}_i > \mathbf{c}$,
for $i < 2$.*

Proof. Let $B \in \mathbf{b}$ and $C \in \mathbf{c}$ be r.e. sets, such that $B \subseteq 2\omega$, the even numbers,
and $C \subseteq 2\omega + 1$, the odd numbers. We shall recursively enumerate r.e. sets
A_0 and A_1 such that

$$(3.2) \qquad A_0 \cup A_1 = B \text{ and } A_0 \cap A_1 = \emptyset,$$

and

$$(3.3) \qquad B \not\leq_{\mathrm{T}} A_i \oplus C, \qquad \text{for } i < 2.$$

Let $\mathbf{a}_i = \deg(A_i \oplus C)$. Now (3.2) guarantees that $\mathbf{b} = \mathbf{a}_0 \cup \mathbf{a}_1$ by Proposition VII.3.3, and (3.3) guarantees $\mathbf{a}_i < \mathbf{b}$ and hence $\mathbf{c} < \mathbf{a}_i < \mathbf{b}$, for $i < 2$. That $\mathbf{a}_i$ is low can be ensured by modifying the following construction or by modifying the remark at the end of the proof of Theorem VII.3.2. Note that $A_i \cup C \equiv_T A_i \oplus C$, so we use the former from now on.

Let $\{ B_s \}_{s \in \omega}$ and $\{ C_s \}_{s \in \omega}$ be recursive enumerations of B and C such that $B_0 = \emptyset$ and $|B_{s+1} - B_s| = 1$ for all s. Let $g(e, s)$ be a recursive function such that $\lim_s g(e, s)$ is the characteristic function of the set in (3.1). As in the Sacks Splitting Theorem VII.3.2 we shall give enumerations $\{ A_{i,s} \}_{s \in \omega}$, $i \in \{ 0, 1 \}$, satisfying the single positive requirement

$$(3.4) \qquad P : x \in B_{s+1} - B_s \implies [x \in A_{0,s+1} \text{ or } x \in A_{1,s+1}],$$

and the negative requirements for $i < 2$ and all e,

$$N_{\langle e, i \rangle} : B \neq \{ e \}^{A_i \cup C}.$$

Fix e, i, and x. Intuitively, we use the lowness of C to help to "C-certify" a computation $\{ e \}^{(A_{i,s} \cup C_s) \upharpoonright u}(x)$ as follows. Let $D_n = \overline{C}_s \upharpoonright u$. Enumerate n into an r.e. set V that we shall build during the construction. By the Recursion Theorem we may assume that we have in advance an index j such that $V = W_j$. Find the least $t \geq s$ such that either $D_n \cap C_t \neq \emptyset$ (in which case the computation is obviously incorrect), or $g(j, t) = 1$, in which case we C-certify the computation (and guess that it is C-correct). It may later happen that we were wrong and $C \upharpoonright u \neq C_s \upharpoonright u$, but this happens at most finitely often, by (3.1) and the definition of g. Since we are really using g as an "oracle" to inquire whether $D_n \subseteq \overline{C}$ for the *current* $D_n = \overline{C}_s \upharpoonright u$, it is very important that there are no previous elements $m \in V_s$ unless $D_m \cap C_s \neq \emptyset$. Thus, whenever a C-certified computation first becomes A-invalid (by $A_t \upharpoonright u \neq A_s \upharpoonright u$), we abandon the old r.e. set V and start with a new version of V and hence a new index j, such that $W_j = V$.

This C-certification process is best formalized by transforming the functional $\{ e \}_s^X(y)$ to a recursive functional $\widetilde{\Phi}(X; e, y, s)$ that resembles $\widehat{\Phi}$ of §1 of Chapter VIII. When we have fixed e and i, for notational convenience we let

$$(3.5) \qquad \widetilde{\Phi}_s(x) \qquad \text{denote} \qquad \widetilde{\Phi}(A_{i,s} \cup C_s; e, x, s),$$

and

$$(3.6) \qquad u_x^s \qquad \text{denote} \qquad u(A_{i,s} \cup C_s; e, x, s).$$

If $\widetilde{\Phi}_{s-1}(x) \downarrow$ but $\widetilde{\Phi}_s(x) \uparrow$ we say that the $\langle e, i, x \rangle$-computation $\widetilde{\Phi}_{s-1}(x)$ becomes *A-invalid* if

$$(3.7) \qquad (\exists z < u_x^{s-1}) \, [z \in A_s - A_{s-1}]$$

and otherwise becomes C-*invalid*.

Fix e, i, x, and s, and define $\widetilde{\Phi}_s(x)$ as follows. Assume we are given $A_{i,t}$ and C_t, $t \leq s$. Assume that

$$(3.8) \qquad \{e\}_s^{A_{i,s} \cup C_s}(x) \downarrow = y,$$

and

$$(3.9) \qquad \neg(\exists z < u_x^{s-1})\, [z \in (A_{i,s} \cup C_s) - (A_{i,s-1} \cup C_{s-1})],$$

(namely that $\widehat{\Phi}_e(A_{i,s} \cup C_s; e, x, s)$ is defined where $\widehat{\Phi}$ is as in §VIII.1).

Let $D_n = \overline{C}_s \upharpoonright u_x^s$. Enumerate n into $V_s^{e,i,x}$. Let v be the greatest stage $< s$ at which an $\langle e, i, x\rangle$-computation became A-invalid, and $v = 0$ if no such stage exists. By the Recursion Theorem choose j such that $W_j = \bigcup\{V_t^{e,i,x} : v < t\}$. Find the least $t \geq s$ such that either

$$(3.10) \qquad D_n \cap C_t \neq \emptyset,$$

or

$$(3.11) \qquad g(j,t) = 1.$$

If (3.11), define $\widetilde{\Phi}_s(x) = y$. Otherwise $\widetilde{\Phi}_s(x)\uparrow$. Define as usual the recursive functions

(length fn.) $\quad \widetilde{\ell}(e,i,s) = \max\{\, x : (\forall y < x)[B_s(y) = \widetilde{\Phi}(A_{i,s} \cup C_s;\ e, y, s)]\,\}$,

(restraint function) $\quad \widetilde{r}(e,i,s) = \max\{\, u(A_{i,s} \cup C_s;\ e, x, s) : x \leq \widetilde{\ell}(e,i,s)$
$$\& \ \ \widetilde{\Phi}_s(A_{i,s} \cup C_s;\ e, x, s)\downarrow \},$$

and

(injury set) $\quad \widetilde{I}(e,i) = \{\, x : (\exists s)\, [x \in A_{i,s+1} - A_{i,s}\ \&\ x \leq \widetilde{r}(e,i,s)]\,\}$.

Stage $s = 0$. Define $A_{i,0} = \emptyset$, for $i < 2$.

Stage $s + 1$. Given $A_{i,t}$, $t \leq s$, define $\widetilde{\ell}(e,i,s)$ and $\widetilde{r}(e,i,s)$ as above. Let $x \in B_{s+1} - B_s$. Choose $\langle e', i'\rangle$ to be the least $\langle e, i\rangle$ such that $x \leq \widetilde{r}(e,i,s)$. Enumerate x in $A_{1-i'}$.

To prove that the construction succeeds we must prove by induction on $\langle e, i\rangle$ that

$$(3.12) \qquad \widetilde{I}(e,i) \text{ is finite},$$

$$(3.13) \qquad B \neq \{e\}^{A_i \cup C},$$

and

(3.14) $\tilde{r}(e, i) =_{\mathrm{dfn}} \lim_s \tilde{r}(e, i, s)$ exists and is finite.

Fix $\langle e, i \rangle$ and assume (3.12), (3.13), and (3.14) for all $\langle j, k \rangle < \langle e, i \rangle$. By (3.14) choose s_0 such that $\tilde{r}(j, k) = \tilde{r}(j, k, s)$ for all $s \geq s_0$, and $\langle j, k \rangle < \langle e, i \rangle$. Define $r = 1 + \max\{ \tilde{r}(j, k) : \langle j, k \rangle < \langle e, i \rangle \}$.

Lemma 1. $\tilde{I}_{e,i}$ is finite.

Proof. Choose $s_1 > s_0$ such that $B \upharpoonright r = B_{s_1} \upharpoonright r$. Now $N_{\langle e, i \rangle}$ is never injured after stage s_1.

Lemma 2. $B \neq \{ e \}^{A_i \cup C}$.

Proof. Assume to the contrary that $B = \{ e \}^{A_i \cup C}$. Then $\lim_s \tilde{\ell}(e, i, s) = \infty$. We shall show that $B \leq_T C$ contrary to hypothesis. To C-recursively compute $B(p)$ for $p \in \omega$, find some $s > s_1$ such that $\tilde{\ell}(e, i, s) > p$ and each computation $\tilde{\Phi}_s(x)$, $x \leq p$, is C-correct, namely, $C_s \upharpoonright u_x^s = C \upharpoonright u_x^s$. It follows by induction on $t \geq s$ as in Lemma 1 of Theorem VII.3.1 that

$$(\forall t \geq s) \, [\tilde{\ell}(e, i, t) > p \; \& \; \tilde{r}(e, i, t) \geq \max\{ u_x^t : x \leq p \}],$$

and hence that for all $t \geq s$, $\tilde{\Phi}_t(p) = \{ e \}^{A_i \cup C}(p) = B(p)$.

Lemma 3. $\tilde{r}(e, i) =_{\mathrm{dfn}} \lim_s \tilde{r}(e, i, s)$ exists and is finite.

Proof. By Lemma 2, choose $p = (\mu x) \, [B(x) \neq \{ e \}^{A_i \cup C}(x)]$. Choose $s_2 \geq s_1$ sufficiently large so that for all $s \geq s_2$,

(3.15) $(\forall x < p) \, [\tilde{\Phi}_s(x) \downarrow = \{ e \}^{A_i \cup C}(x)],$

and

(3.16) $(\forall x \leq p) \, [B_s(x) = B(x)].$

  Case 1. $\{ e \}^{A_i \cup C}(p) \downarrow \neq B(p)$. Choose $s_3 \geq s_2$ such that for all $s \geq s_3$, $\tilde{\Phi}_s(p) \downarrow = q \neq B(p)$. Hence, for all $s \geq s_3$, $\tilde{\ell}(e, i, s) = \tilde{\ell}(e, i, s_3)$ and $\tilde{r}(e, i, s) = \tilde{r}(e, i, s_3)$.
  Case 2. $\{ e \}^{A_i \cup C}(p) \uparrow$. We shall find a stage v such that for all $s \geq v$, $\tilde{\Phi}_s(p) \uparrow$. Hence, for all $s \geq v$, $\tilde{r}(e, i, s) = \tilde{r}(e, i, v)$.
  Note that if $\tilde{\Phi}_s(p) \downarrow$ for any $s \geq s_2$ then $\tilde{r}(e, i, s) \geq u_p^s$ so by induction on $t \geq s$, the computation $\tilde{\Phi}_t(p) = \tilde{\Phi}_s(p)$ holds as long as it remains C-valid. Let s' be the least t such that no $\langle e, i, p \rangle$-computation becomes A-invalid at any stage $\geq t$. By the Recursion Theorem choose j such that $W_j = \bigcup \{ V_s^{e,i,p} : s \geq s' \}$. Since $\{ e \}^{A_i \cup C}(p) \uparrow$, any computation $\tilde{\Phi}_s(p)$, $s > s'$,

becomes C-invalid at some stage $t > s$, at which time $D_m \cap C_t \neq \emptyset$ for every $m \in V_t^{e,i,p}$. Hence, $\lim_s g(j,s) = 0$ by (3.1). Choose $v > s_2$ such that $\widetilde{\Phi}_v(p)\uparrow$, and $g(j,s) = 0$ for all $s \geq v$. We claim that $\widetilde{\Phi}_s(p)\uparrow$ for all $s \geq v$. Suppose $s > v$, $\widetilde{\Phi}_{s-1}(p)\uparrow$ and $\widetilde{\Phi}_s(p)\downarrow$. Then we enumerate $n \in V_s^{e,i,p}$, where $D_n = \overline{C}_s \upharpoonright u_p^s$, and we choose the least $t \geq s$ satisfying (3.10) or (3.11). But (3.11) cannot occur by the choice of v, so (3.10) occurs and $\widetilde{\Phi}_s(p)\uparrow$.   ☐

3.3–3.6 Exercises

3.3.   Use the method at the end of the proof of Theorem VII.3.2 to show that the degrees $\mathbf{a}_i$, $i < 2$, of Theorem 3.2 are automatically low.

3.4 (Low Non-Diamond Theorem, Ambos-Spies [1984a]).   (This generalizes the Lachlan Non-Diamond Theorem IX.3.1.)

   (a) Prove that for r.e. degrees $\mathbf{a}_0, \mathbf{a}_1, \mathbf{b}_0, \mathbf{b}_1$ if $\mathbf{0}' = \mathbf{a}_0 \cup \mathbf{a}_1$ and $\mathbf{b} = \mathbf{b}_0 \cup \mathbf{b}_1$ is low then there are no r.e. degrees $\mathbf{c}_i \nleq \mathbf{b}_i$ such that $\mathbf{a}_i \cap \mathbf{c}_i \leq \mathbf{b}_i$, $i = 0, 1$. *Hint.* Choose disjoint r.e. sets $A_i \in \mathbf{a}_i$, $B_i \in \mathbf{b}_i$, $C_i \in \mathbf{c}_i$, $i = 0, 1$. Assume $B = B_0 \cup B_1$ is low and $K \leq_T A = A_0 \cup A_1$. As in Theorem IX.3.1 enumerate r.e. sets $E, F_0, F_1, F_2, \ldots$ attempting to make $E \leq_T C_0, A_0$, $F_n \leq_T C_1, A_1$, and to meet the following requirements for all i and j,

$$P_{i,j} : E \neq \{i\}^{B_0} \quad \text{or} \quad F_i \neq \{j\}^{B_1}.$$

Choose a witness pair (x, y) for $P_{i,j}$ as before such that $\{j\}^{B_{1,s} \upharpoonright u}(y)\downarrow = 0$ and $\{i\}_s^{B_{0,s} \upharpoonright u}(x)\downarrow = 0$, but before attacking $P_{i,j}$ by inserting $\langle i, j\rangle$ into D use the lowness of B and the $\emptyset'$-oracle to test whether $B_s \upharpoonright u = B \upharpoonright u$ as in the proof of Theorem 3.2.

   (b) Use the method of Exercise IX.3.10 to replace $i \leq 1$ in (a) by $i \leq n$ for any $n \geq 1$.

   (c) In (b) by setting $\mathbf{a}_1 = \mathbf{b}_1$ low, and $\mathbf{b}_0 = \mathbf{0}$, prove that if $\mathbf{a}_1 \in \mathbf{L}_1$ and $\mathbf{d} = \mathbf{a}_0 \in \mathbf{M}^+$ then $\mathbf{a}_1 \cup \mathbf{d} < \mathbf{0}'$. (In particular, if an r.e. degree $\mathbf{d}$ nontrivially *caps* to $\mathbf{0}$ then $\mathbf{d}$ cannot be *cupped* to $\mathbf{0}'$ by any low r.e. degree. The converse will be proved in Theorem XIII.4.2.)

   (d) Prove that if $\mathbf{a} \in \mathbf{L}_1$ and $\mathbf{a} \neq \mathbf{0}$ then there exists an r.e. degree $\mathbf{b}$ such that $\mathbf{a} \cap \mathbf{b}$ does not exist.

3.5 (Robinson [1966a, Theorem 5.3]).   Prove that if C is a coinfinite low r.e. set then C has a maximal superset A. *Hint.* We only use the weaker hypothesis that $\overline{C}$ is semi-low (see Exercise IV.4.6), namely that

$$(3.17) \qquad\qquad \{j : W_j \cap \overline{C} \neq \emptyset\} \leq_T \emptyset'.$$

Let $g(j,s)$ be a recursive function such that $\lim_s g(j,s)$ is the characteristic function of the set in (3.17). Construct A as in Theorem X.3.3 with markers

$\{\Gamma_e\}_{e\in\omega}$ such that Γ_e rests on a_e^s at the end of stage s and Γ_e comes to a limit on a_e where $\overline{A} = \{a_0 < a_1 < \cdots\}$. However, now if Γ_0 wants to move to an element $x > a_e^s$ at stage $s+1$ we first enumerate x in $V_{0,s+1}$ (as in Theorem 3.2) and use the Recursion Theorem to choose an index j such that $W_j = \bigcup_t V_{0,t}$. Now choose the least $t \geq s$ such that either: (1) $x \in C_t$; or (2) $g(j,t) = 1$. If (2) then move Γ_0 to the now C-certified element x.

Note that

(3.18) $(\exists x)\, [x \in V_{0,s+1} - V_{0,s}] \implies V_{0,s} \subseteq C_s,$

so that in effect g is answering (perhaps finitely often incorrectly) the question "Is $x \in \overline{C}$?" for the current candidate x and not for some previous element $y \in V_{0,s}$. In the full construction we have for each marker Γ_e and e-state σ a set V_e^σ. When Γ_e wants to move at stage $s+1$ to an element $x > a_e^s$, where $\sigma(e,x,s) = \sigma$, enumerate x in $V_{e,s+1}^\sigma$. At each stage $s+1$ perform the above procedure for each marker Γ_e and each e-state σ in increasing order of e and decreasing order of σ until some Γ_e is C-certified to move. To guarantee the analogue of (3.18) for V_e^σ choose $j(e,\sigma)$ such that

$$W_{j(e,\sigma)} = \bigcup\{V_{e,t}^\sigma : t > t_e\}$$

where t_e is the greatest stage $\leq s$ at which some marker Γ_i, $i < e$, moved.

3.6.   Let C be an r.e. set. (See also Exercise IV.4.7.)
   (a) Prove that $\overline{C}$ is semi-low iff there is a recursive function f such that for all j:

$$W_j \cap \overline{C} = W_{f(j)} \cap \overline{C};$$

and

$$W_j \cap \overline{C} = \emptyset \implies W_{f(j)} \text{ is finite.}$$

   (b) Prove that C is low iff there is a recursive function f such that for all j:

$$W_j \cap \{n : D_n \subseteq \overline{C}\} = W_{f(j)} \cap \{n : D_n \subseteq \overline{C}\};$$

and

$$W_j \cap \{n : D_n \subseteq \overline{C}\} = \emptyset \implies W_{f(j)} \text{ is finite.}$$

3.7 Notes. Theorem 3.2 appears in Robinson [1971a, Corollary 9], and Exercise 3.5 in Robinson [1966a, Theorem 5.3] and [1966b], although the proofs here are simpler. The method used here was worked out in Soare [1982a, §2], where a complete proof of Exercise 3.5 is given. This method also was used in Soare and Stob [1982], and in Ambos-Spies [1984a].

   Exercise 3.5 was considerably generalized by Soare [1982a], as we discuss in Theorem XVI.1.1. The characterization in Exercise 3.6 appears in Soare

[1977] where it is shown that an r.e. set C is *nonspeedable* (a certain computational complexity property in the style of M. Blum) iff $\overline{C}$ is semi-low. A survey of some of these computational complexity properties of r.e. sets and their relation to the promptly simple sets of Chapter XIII may be found in Soare [1982b]. Roughly, an r.e. set C is *nonspeedable* iff C has a fastest program almost everywhere (modulo some recursive function). If C fails to have a *single* fastest program then M. Blum suggested asking whether C has a computable sequence of programs cofinal in the running times of all programs for C. Bennison and Soare [1978] showed that an r.e. set C has this property iff $\overline{C}$ is *semi-low$_{1.5}$*, namely $\{\, e : W_e \cap \overline{A} \text{ infinite}\,\} \leq_m \text{Inf}$, a strictly weaker property than $\overline{C}$ semi-low. They then extended Exercise 3.5 by showing that such a set C has a maximal superset. Maass [1983] combined this method with an extension of the automorphism machinery in [Soare 1982a] to prove that a coinfinite r.e. set C satisfies $\mathcal{L}^*(C) \cong^{\mathrm{eff}} \mathcal{E}\,{}^*$ iff $\overline{C}$ is semi-low$_{1.5}$, a result that generalizes Exercise 3.5 and the two results above (see Theorem XVI.1.2). The general intent of these results is roughly that an r.e. set C has low information content (e.g., $\overline{C}$ is semi-low or semi-low$_{1.5}$) iff C is easy to compute (from a computational complexity viewpoint) iff C has uniform algebraic structure in $\mathcal{E}\,{}^*$ (and sometimes in $\mathbf{R}$) resembling that of the recursive sets.

Robinson [1971a] proved that if $\mathbf{a}$ and $\mathbf{b}$ are r.e. degrees and $\mathbf{a} < \mathbf{b}$ then any countable partially ordered set is embeddable in the r.e. degrees between $\mathbf{a}$ and $\mathbf{b}$ with joins preserved whenever they exist (see Exercise VIII.4.10). He also proved that if $\mathbf{a}$ is low then there is such an embedding that also preserves greatest and least elements whenever they exist. By Theorem XVI.4.6 it is necessary to require the hypothesis that $\mathbf{a}$ is low.

4. Non-Low$_2$ R.E. Degrees Contain Atomless R.E. Sets

Let $\mathcal{M}$, $\mathcal{H}$, and $\mathcal{D}$ denote the classes of maximal sets, hh-simple, and dense simple sets, respectively. By the results in §1 and §2, $\deg(\mathcal{M}) = \deg(\mathcal{H}) = \deg(\mathcal{D}) = \mathbf{H}_1$. For any class $\mathcal{C}$ of r.e. sets let $\mathcal{C}^{\#}$ denote the class of coinfinite r.e. sets that have no superset in $\mathcal{C}$. Martin showed (see Exercises 2.8(c) and 2.16) that $\deg(\mathcal{D}^{\#}) = \mathbf{H}_1$ also. Let $\mathbf{A} = \deg(\mathcal{M}^{\#})$, the degrees of atomless sets. We have seen in Exercise 2.8(a) that $\mathbf{H}_1 \subseteq \mathbf{A}$, and in Exercise 3.5 that $\mathbf{A} \cap \mathbf{L}_1 = \emptyset$. Lachlan [1968a] improved the latter by showing that $\mathbf{A} \cap \mathbf{L}_2 = \emptyset$, namely that any coinfinite low$_2$ r.e. set has a maximal superset (see Theorem 5.1). Shoenfield [1976] completed the classification of $\mathbf{A}$ by showing that $\overline{\mathbf{L}}_2 \subseteq \mathbf{A}$, and hence that $\mathbf{A} = \overline{\mathbf{L}}_2$. Since his proof actually shows that every non-low$_2$ r.e. degree $\mathbf{d}$ contains a coinfinite r.e. set A with no hh-simple superset, this establishes that $\deg(\mathcal{H}^{\#}) = \overline{\mathbf{L}}_2$ as well.

4.1 Theorem (Shoenfield [1976]). *For every non-low$_2$ r.e. degree **d** there is a coinfinite r.e. set $A \in$ **d** such that A has no hh-simple superset. Hence, $\overline{\mathbf{L}}_2 \subseteq$ **A**. Furthermore, for any r.e. set $B \in$ **d** we may take A to be the Dekker deficiency set of B.*

Proof. Choose any r.e. set $B \in$ **d** and any 1:1 recursive function f with range B. Let A be the Dekker deficiency set of B with respect to f, namely

$$A = \{\, s : (\exists t > s)\, [f(t) < f(s)]\,\}.$$

Recall from Theorem V.2.5 that A is r.e., h-simple, and $A \equiv_T B$. We shall show that if A has a hh-simple superset H then $B'' \leq_T \emptyset''$, so B is not low$_2$.

Let P denote $\mathrm{Tot}^B = \{\, e : W_e^B = \omega\,\}$. By Theorem IV.3.1 relativized to B, P is Π_2^B-complete, and there is a recursive function g such that

(4.1) $e \in P \implies W_{g(e)}^B = \omega \implies (\forall x)\,(\exists u)\,[x \in W_{g(e)}^{B \restriction u}],$

and

(4.2) $e \notin P \implies W_{g(e)}^B =^* \emptyset \implies (\text{a.e. } x)\neg(\exists u)\,[x \in W_{g(e)}^{B \restriction u}].$

Clearly $P \in \Pi_3$ because $P \in \Pi_2^B$ and B is r.e. If A has a hh-simple superset H then we shall build an r.e. class of r.e. sets $\{\, V_{e,k}\,\}_{e,k \in \omega}$ such that for all e,

(4.3) $e \in P \iff V_e$ is cofinite,

where $V_e = \bigcup_k V_{e,k}$. Hence, $P \in \Sigma_3$ because Cof is Σ_3-complete by Theorem IV.3.5. Therefore, $P \in \Delta_3$, so $P \leq_T \emptyset''$ by Post's Theorem IV.2.2. Hence, $B'' \leq_T \emptyset''$ since P is Π_2^B-complete.

Let H be a hh-simple superset of B. Let H^s, B^s, V_e^s, and $V_{e,k}^s$ denote the set of elements enumerated in H, B, V_e, and $V_{e,k}$, respectively, by the end of stage s. Let $H^0 = V_e^0 = V_{e,k}^0 = \emptyset$, for all e and k.

Stage $s + 1$. Find the least $k \leq s$ such that there exists $t \leq s$ such that:

(4.4) $t \notin V_e^s;$

(4.5) $(\exists u < f(t))\,[k \in W_{g(e),s}^{B^s \restriction u}];$

and

(4.6) $V_{e,k}^s \subseteq H^s.$

Enumerate the least such t in $V_{e,k}$. (If k and t fail to exist do nothing.) This completes the construction of $\{\, V_{e,k}\,\}_{e,k \in \omega}$.

Note that $|V_{e,k} - H| \leq 1$ for each e and k by (4.6).

Lemma. $(\forall e) [e \in P \Longleftrightarrow V_e \text{ is cofinite}]$.

Proof. ($\Longrightarrow$). Fix e, and assume $e \in P$. Since the sets $\{V_{e,k}\}_{k \in \omega}$ are disjoint and H is hh-simple, there is a k such that $V_{e,k} \subseteq H$. (By Exercise X.2.16 it is unnecessary that every $V_{e,k}$ be finite.) Since $e \in P$, $(\exists u) [k \in W_{g(e)}^{B \restriction u}]$ by (4.1). Choose t_0 such that $f(t) > u$ for every $t \geq t_0$. Now every $t \geq t_0$ is put into V_e because choose $s \geq t$ such that: $k \in W_{g(e),s}^{B_s \restriction u}$; (4.6) holds; and

$$(\forall z < t) [u < f(z) \implies z \in V_e^s].$$

Now at stage $s + 1$, either t is put into $V_{e,k}$, into $V_{e,k'}$ for some $k' < k$, or else $t \in V_e^s$ already.

($\Longleftarrow$). Assume $e \notin P$. Then by (4.2) $W_{g(e)}^B$ is finite, and

(4.7) $(\exists k_0) (\forall k \geq k_0) \neg (\exists u) [k \in W_{g(e)}^{B \restriction u}]$.

Now for all k, $|V_{e,k} - H| \leq 1$, by (4.6). We claim that $V_{e,k} - H = \emptyset$ for all $k \geq k_0$, and hence $V_e \subseteq^* H$ (indeed $V_e \subseteq^* A$), so V_e is coinfinite. Suppose t is enumerated in $V_{e,k}$, $k \geq k_0$, at some stage $s + 1 > t$. Then (4.5) holds for t, s, k, and u. However, by (4.7), $B \restriction u \neq B_s \restriction u$ so $f(z) \leq u < f(t)$ for some $z > t$. Hence t later enters A, but $A \subseteq H$ so $t \in H$.   ⬚

4.2–4.3 Exercises

4.2 (Lachlan).   Let A be hh-simple, $A \subseteq B$ an r.e. set, $\mathbf{a} = \deg(A)$ and $\mathbf{b} = \deg(B)$. Prove that there is an r.e. degree $\mathbf{c}$ such that $\mathbf{a} = \mathbf{b} \cup \mathbf{c}$. (See also Exercise 4.3 and Corollary XII.3.8.) *Hint.* Use Corollary X.2.7.

4.3 (Lachlan [1968a]).   Let A be a coinfinite r.e. set that is not hh-simple, and let $\mathbf{a} = \deg(A)$. Let $\mathbf{b}$ be an r.e. degree $\geq \mathbf{a}$. Prove that there exists an r.e. superset $B \supseteq A$ of degree $\mathbf{b}$. (See also Exercise 4.2 and Corollary XII.3.8.) *Hint.* Choose a u.r.e. array $\{V_e\}_{e \in \omega}$ of r.e. sets such that $e < \min(V_e)$, for all e and witnessing A not hh-simple. Choose any r.e. set $C \in \mathbf{b}$. Let $\psi(e, s) = (\mu x) [x \in V_{e,s} - A_s]$. Let $B = A \cup \{\psi(e, s) : e \in C \ \& \ s \in \omega\}$.

4.4 Notes.   After seeing Exercises 3.5 and 2.8 Lachlan [1968a] improved both by a factor of a jump by showing that $\mathbf{A} \cap \mathbf{L_2} = \emptyset$ and that $\mathbf{H_2} \subseteq \mathbf{A}$. Shoenfield's proof of Theorem 4.1, the stronger fact that $\overline{\mathbf{L_2}} \subseteq \mathbf{A}$, is much simpler. There are no further known connections between $\mathbf{H_2}$ and lattice definable classes of r.e. sets but it would be very interesting to find some. Cohen and Jockusch [1975] showed that Post's simple set S in Theorem V.1.3 is atomless. D. Miller [1981b] used the method of Theorem 4.1 to prove that every η-maximal semirecursive set is low$_2$. These sets were introduced by Marchenkov [1976] in an attempt to give a solution to Post's problem in the style of Post by finding a structural property on $\overline{A}$ that guarantees that an r.e. set A is incomplete. (See Odifreddi [1981, Theorem 2.11].) Robinson [1968] used deficiency sets in a way similar to that in Theorem 4.1. (See Exercise 1.22.)

4.5 Open Questions. A class $\mathbf{C}$ of r.e. degrees is *invariant* if $\mathbf{C} = \deg(\mathcal{C})$ for some class $\mathcal{C} \subseteq \mathcal{E}$ invariant under Aut $\mathcal{E}$, the automorphisms of $\mathcal{E}$. By Corollaries 2.5 and 5.2, $\mathbf{H}_1$ and $\overline{\mathbf{L}}_2$ are invariant. Of course, $\overline{\mathbf{L}}_0 = \mathbf{R}^+ = \mathbf{R} - \{\mathbf{0}\}$ is invariant since $\mathbf{R}^+ = \deg(\mathcal{C})$ for $\mathcal{C}$ the class of nonrecursive sets (or also the class of simple sets). Trivially, $\mathbf{L}_0 = \{\mathbf{0}\}$ is invariant since it contains exactly the recursive sets. Harrington has shown (see Chapter XV) that the creative sets form an orbit and therefore $\mathbf{H}_0 = \{\mathbf{0}'\}$ is invariant. Are there any other invariant classes of the form $\mathbf{H}_n$, $\mathbf{L}_n$, $\overline{\mathbf{H}}_n$ or $\overline{\mathbf{L}}_n$? In particular, the sequence $\overline{\mathbf{L}}_0, \mathbf{H}_1, \overline{\mathbf{L}}_2$ suggests that perhaps $\mathbf{H}_{2n}$ and $\overline{\mathbf{L}}_{2n+1}$ are invariant for every n, but there is no further evidence for this. Lerman and Soare [1980a] refuted a conjecture that every invariant class is of one of the above forms. They introduced the class of d-simple sets (a property arising from studying automorphisms of $\mathcal{E}$) and showed that the class $\mathbf{C}$ of their degrees splits $\mathbf{L}_1$. Maass, Shore, and Stob [1981] improved this by introducing an invariant class $\mathbf{SP\overline{H}}$ (consisting of non-hh-simple sets with a certain splitting property) and then showed that $\mathbf{SP\overline{H}}$ splits $\mathbf{H}_n$ and $\mathbf{L}_n$ for every n. Ambos-Spies, Jockusch, Shore and Soare [1984] showed that $\mathbf{SP\overline{H}} = \mathbf{PS}$ the degrees of promptly simple sets studied in Chapter XIII, so that this class is definable both in $\mathcal{E}$ and in $\mathbf{R}$; forms a filter in $\mathbf{R}$; and its complement is an ideal. Which other classes of r.e. degrees are invariant? In particular, is $\overline{\mathbf{L}}_1$ invariant? If so perhaps one can find a property in the style of Theorem 4.1 to show it. If not, one might try to use automorphisms of $\mathcal{E}$ (as in Chapter XV) to show

$$(\forall A)\,[\deg(A) \in \mathbf{L}_2 - \mathbf{L}_1 \implies (\exists \Phi \in \text{Aut } \mathcal{E})\,[\deg(\Phi(A)) \in \mathbf{L}_1]].$$

Let $\mathcal{C}$ be the class of coinfinite r.e. sets with no r-maximal superset. Classify $\deg(\mathcal{C})$ in the $\mathbf{H}_n$, $\mathbf{L}_n$ hierarchy.

5.* *Low$_2$ R.E. Degrees Do Not Contain Atomless R.E. Sets*

R. W. Robinson showed (Exercise 3.5) that every coinfinite low$_1$ r.e. set has a maximal superset and hence that $\mathbf{A} \cap \mathbf{L}_1 = \emptyset$. Lachlan improved this one jump level by replacing $\mathbf{L}_1$ by $\mathbf{L}_2$. This allows us to complete the classification of $\mathbf{A}$ as exactly $\overline{\mathbf{L}}_2$. Recall from §4 that for any class $\mathcal{C}$ of r.e. sets we let $\mathcal{C}^\#$ denote the class of coinfinite r.e. sets that have no superset in $\mathcal{C}$.

Block B_0		Block B_1			Block B_j	
current residents of B_0	0-states	current residents of B_1	1-states		current residents of B_j	j-states

The motion of a given element x — — — — — — — — — — — — — — — →
is upward and to the left

Diagram 5.1. Arrangement of Blocks $\{\,B_e\,\}_{e\in\omega}$

Block B_0:

current residents of B_0	0-states
x_1, x_2	$\sigma = 1$
x_3	$\sigma = 0$

Block B_1:

current residents of B_1	1-states
	$\sigma = 11$
x_4, x_5, x_6	$\sigma = 10$
x_7	$\sigma = 01$
x_8	$\sigma = 00$

Block B_j:

current residents of B_j	j-states
x_{11}, x_{12}	$11\cdots 11$
x_{13}	$11\cdots 10$
	$\vdots$
	$\vdots$
	$\vdots$
	$\vdots$
x_{14}, x_{15}	$00\cdots 01$
x_{16}	$00\cdots 00$

5.1 Theorem (Lachlan [1968a]).   *If A is a coinfinite r.e. set which is low$_2$ $(A'' \equiv_T \emptyset'')$ then A has a maximal superset M. Hence, $\mathbf{A} \cap \mathbf{L}_2 = \emptyset$.*

5.2 Corollary. $\mathbf{A} =_{\mathrm{dfn}} \deg(\mathcal{M}^{\#}) = \deg(\mathcal{H}^{\#}) = \overline{\mathbf{L}}_2$.

Proof. Apply Theorems 4.1 and 5.1.   ▯

Proof of Theorem 5.1. We shall give a recursive construction where at stage s we define a finite set M_s, the elements enumerated in M by the end of stage s. By the Recursion Theorem we may assume that we have an index for the recursive sequence $\{ M_s \}_{s \in \omega}$ given before the construction.

For each e we define the e-states σ in Definition X.3.2 to be strings $\sigma \in 2^{<\omega}$, $lh(\sigma) = e + 1$, with the ordering $<$ defined there. If $\tau < \sigma$ we say σ is *greater* (*higher*) *than* τ.

In Theorem X.3.3 we approximated the complement of the maximal set using movable markers. In § X.5, § X.6, and § X.7 we viewed the markers as fixed "windows" through which the elements passed. In this proof we expand the windows to *blocks* $\{ B_j \}_{j \in \omega}$ where each block is divided into 2^{j+1} rows, one corresponding to each j-state σ, and the rows are arranged in decreasing order according to the ordering of the j-states. (See Diagram 5.1.) The blocks $B_0, B_1, B_2, \ldots$ are arranged in order from left to right. At the end of stage $s + 1$ each row σ of Block B_j may contain at most finitely many elements x of $\overline{M}_s$ satisfying $\sigma(j, x, s) = \sigma$. The motion of the elements within the blocks is up and to the left. Namely, after x enters Block B_j, x may move up among the rows of B_j as x increases in j-state; x may move left to some Block B_k, $k < j$; or x may be enumerated in A or M in which case x is removed (forever) from the blocks. The permanent residents of the blocks will constitute exactly $\overline{M}$.

Let $B_{j,s}$ represent the residents of block B_j at the end of stage s, and $B_{j,\infty}$ the permanent residents of B_j. We shall show that $B_{j,\infty}$ is nonempty and finite for all j, and that $\overline{M} = \bigcup_{j \in \omega} B_{j,\infty}$. Let $\sigma(e, x) = \lim_s \sigma(e, x, s)$ which always exists.

We use the hypothesis that A is low$_2$ as follows. First, since A is not high$_1$, A has a recursive enumeration $\{ A_s \}_{s \in \omega}$ such that

(5.1) $(\exists^{\infty} s) \, [a_s^s = a_s],$

where $\overline{A}_s = \{ a_0^s < a_1^s < \cdots \}$, and $\overline{A} = \{ a_0 < a_1 < \cdots \}$, by Lemma 1.6 with $h(s) = s$. Every stage s such that $a_s^s = a_s$ is called a *true stage*.

For each $e \in \omega$ and e-state σ we define an A-r.e. set U_σ^A as follows. If s is a true stage and

(5.2) $k \leq \operatorname{card} \{ x : x \in \overline{M}_{s+1} \ \& \ x \leq a_s^s \ \& \ \sigma(e, x, s) = \sigma \},$

enumerate k into U_σ^A. Since A is low$_2$, $\Pi_2^A \subseteq \Sigma_3$ (by Exercise IV.4.5(a)). Hence by Corollary IV.3.7 there is a u.r.e. array of sets $\{ Z^{y,\sigma} \}_{y\in\omega,\, \sigma\in 2^{<\omega}}$ such that for all σ and y:

(5.3) $$|U_\sigma^A| = \infty \implies \text{(a.e. } y) \, [Z^{y,\sigma} = \omega];$$

and

(5.4) $$|U_\sigma^A| < \infty \implies (\forall y) \, [Z^{y,\sigma} \text{ is finite}].$$

For each y and σ define the r.e. set

$$V^{y,\sigma} = \bigcup \{ Z^{y,\tau} : \tau \subseteq \sigma \ \& \ (\exists k \geq 0) \, [\sigma = \tau^\frown 0^k] \}.$$

Technically, the A-r.e. sets U_σ^A are defined *during* the construction since they depend upon the sequence $\{ M_s \}_{s\in\omega}$ being defined. However, by the Recursion Theorem, we may assume that we have an index for the sequence $\{ M_s \}_{s\in\omega}$ and hence for the sequence $\{ V^{y,\sigma} : y \in \omega, \ \sigma \in 2^{<\omega} \}$ given *before* the construction. Fix a simultaneous recursive enumeration $\{ V_s^{y,\sigma} \}_{s\in\omega}$ of the sequence $\{ V^{y,\sigma} \}$ such that at most one set $V^{y,\sigma}$ receives a new element at each stage.

(The intuition is roughly the following. An element x is *true* if $x \in \overline{A}$. The goal of block B_j is to put a *true* element x of j-state σ on row σ for the highest possible row σ. If U_σ^A is infinite then there are infinitely many true elements in j-state σ available to B_j. Whenever a new element appears at stage $s + 1$ in $Z^{j,\sigma}$, and hence in $V^{j,\sigma}$, this "oracle" is telling B_j that U_σ^A is infinite and so it is safe to "dump" into M_{s+1} all elements x of B_j less desirable than elements in e-state σ, namely all elements in B_j of j-state $\nu < \sigma$. We perform this dump only if B_j contains an element on some row $\tau \geq \sigma$ so that B_j will not become empty. Further intuitive remarks appear after the proof.)

In the following construction of M, an element x is *unused* at a certain time during stage $s + 1$ if $x \leq a_s^s$, and x has not yet been enumerated into M or into $B_{j,s+1}$ for any $j \leq s$.

Construction of M.

Stage $s = 0$. Define $M_0 = \emptyset = B_{j,0}$ for all j.

Stage $s + 1$. Enumerate into M all elements $x \in A_s$ such that $x \leq a_s^s$. For each j, $0 \leq j \leq s$, perform substage j and then pass to the terminating substage $j = s + 1$.

Substage j.

Step 1. (Dump rows $\nu < \rho$). If there is a j-state ρ such that $V_{s+1}^{j,\rho} \neq V_s^{j,\rho}$ then enumerate into M all unused x such that $x \in B_{j,s}$ and $\sigma(j,x,s) < \rho$.

Step 2. (Update). Enumerate into $B_{j,s+1}$ all unused $x \in B_{j,s}$.

Step 3. (Add-One). If there exists an unused x such that

$$\sigma(j, x, s) > \sigma(j, y, s)$$

for all y so far enumerated in $B_{j,s+1}$ then first choose the greatest j-state for such an x and then the least x in this j-state and enumerate x into $B_{j,s+1}$.

Substage $s + 1$. (Terminating substage). Enumerate into M all unused x. Define $B_{e,s+1} = \emptyset$ for all $e > s$.

This ends the construction.

The construction immediately yields the following three facts for all s and x:

(5.5) $$x \in B_{j,s+1} \implies x \in \{ a_0^s < a_1^s < \cdots < a_s^s \};$$

(5.6) $$x \in M_{s+1} \implies x \leq a_s^s \leq a_s;$$

(5.7) $$[s \text{ a true stage and } x \in B_{j,s+1}] \implies x \in \overline{A};$$

(5.8) $$x \in B_{j,s} \implies [x \in M_{s+1} \lor (\exists k \leq j)\,[x \in B_{k,s+1}].$$

For each j define

$$\sigma_j = \begin{cases} \text{greatest } j\text{-state } \sigma \text{ such that } |V^{j,\sigma}| = \infty & \text{if such } \sigma \text{ exists,} \\ \text{least } j\text{-state} & \text{otherwise.} \end{cases}$$

Lemma 1. (Stable Block Lemma). *For every j, there is a stage s_j such that:*

(i) $(\forall t \geq s_j)\,(\forall x)\,[[x \in B_{j,t} \ \& \ \sigma_j \leq \sigma(j, x, t) \ \& \ x \in \overline{A}]$
$\implies (\forall v \geq t)\,[x \in B_{j,v}]],$
(ii) $(\exists x)\,[x \in \overline{A} \ \& \ \sigma_j \leq \sigma(j, x, s_j) \ \& \ (\forall t \geq s_j)\,[x \in B_{j,t}]],$
(iii) $(\forall t \geq s_j)\,(\forall x)\,[x \in B_{j,t} \implies \sigma_j \leq \sigma(j, x, t)],$
(iv) $(\forall t \geq s_j)\,(\forall x)\,[x \in B_{j,t+1} - B_{j,t} \implies y \in A].$

Proof. Assume Lemma 1 for all $e < j$. We shall define r_1 (respectively, r_2, r_3, r_4) such that part (i) (part (ii), (iii), (iv)) holds for all $t \geq r_1$ (respectively, $t \geq r_2$, $t \geq r_3$, $t \geq r_4$), and then we shall choose $s_j > \max\{ r_1, r_2, r_3, r_4 \}$. Choose $r_1 \geq \max\{ s_0, \ldots, s_{j-1} \}$ such that $V^{j,\sigma} = V_{r_1}^{j,\sigma}$ for all $\sigma > \sigma_j$. Now

(5.9) $$(\forall t \geq r_1)\,[(\bigcup_{e<j} B_{e,t}) \cap \overline{A} \subseteq \{ a_0, \ldots, a_{r_1-1} \}],$$

by (iv) and (5.5).

First we claim that (i) holds for all $t \geq r_1$. Fix $t \geq r_1$ and x satisfying the hypothesis of (i). Now x is not enumerated into M at the beginning of stage

$t + 1$ because $x \in \overline{A}$. Next x is not enumerated into $B_{e,t+1}$ at any substage $e < j$ because (iv) holds for all $e < j$. Furthermore,

$$V_{t+1}^{j,\rho} \neq V_t^{j,\rho} \implies \rho \leq \sigma_j \leq \sigma(j, x, t)$$

because $r_1 \leq t$. Hence, x is not dumped into M under Step 1 at substage j. Thus, $x \in B_{j,t+1}$.

Secondly we choose $x \in \overline{A}$ and r_2 such that (ii) holds for all $t \geq r_2$.

Case 1. V^{j,σ_j} is finite. Thus, σ_j is the lowest j-state. Let r_2 be a true stage $\geq r_1$. Then $a_{r_2}^{r_2} = a_{r_2} \notin M_{r_2}$ by (5.6); a_{r_2} is not enumerated into M at the beginning of stage $r_2 + 1$ because $a_{r_2} \in \overline{A}$; and for all $e < j$, $a_{r_2} \notin B_{e,r_2+1}$ by (5.9) because $a_{r_2} \geq a_{r_1}$. Hence, at substage j there is an unused x. Thus, there exists $x \in B_{j,r_2+1}$, and $x \in \overline{A}$ since r_2 is a true stage. Finally, $\sigma_j \leq \sigma(j, x, r_2)$ because σ_j is the lowest j-state.

Case 2. V^{j,σ_j} is infinite. Then $Z^{j,\sigma}$ is infinite for some e-state σ, where $e \leq j$ and $\sigma_j = \sigma^\frown 0^{j-e}$. Hence, U_σ^A is infinite and

$$\limsup_{s \text{ a true stage}} \text{card}\{x : x \in \overline{M}_{s+1} \ \& \ x \leq a_s^s \ \& \ \sigma(e, x, s) = \sigma\} = \infty.$$

Note that if $\sigma(e, x, s) = \sigma$ then $\sigma_j \leq \sigma(j, x, s)$. Choose a true stage $r_2 \geq r_1$ such that

$$\text{card}\{x : x \in \overline{M}_{r_2+1} \ \& \ x \leq a_{r_2}^{r_2} \ \& \ \sigma_j \leq \sigma(j, x, r_2)\} > r_1.$$

Since $|(\bigcup_{e<j} B_{e,r_2+1}) \cap \overline{A}| \leq r_1$ at stage $r_2 + 1$, substage j, there exists an unused x such that $\sigma_j \leq \sigma(j, x, r_2)$. Hence, there exists $x \in B_{j,r_2+1}$ such that $\sigma_j \leq \sigma(j, x, r_2)$, and since r_2 is a true stage, $x \in \overline{A}$ by (5.7).

Regardless of which case holds, we have by (i) that $x \in B_{j,t}$ for all $t > r_2 \geq r_1$.

Thirdly, we choose $r_3 > r_2$ such that (iii) holds for all $t \geq r_3$. We can assume that $\sigma_j >$ the lowest j-state (since otherwise (iii) is trivially true). Therefore, V^{j,σ_j} is infinite. Choose $r_3 > r_2$ such that $V_{r_3+1}^{j,\sigma_j} \neq V_{r_3}^{j,\sigma_j}$. Then

$$(\forall x \in B_{j,r_3+1}) \ [\sigma_j \leq \sigma(j, x, r_3)].$$

Since $r_3 > r_2$,

$$(\exists x) \ [\sigma_j \leq \sigma(j, x, r_3) \ \& \ (\forall t \geq r_3) \ [x \in B_{j,t}]]$$

by the proof above for r_2. Hence, the only new elements which can enter B_j at stages $t > r_3$ must have j-state $> \sigma_j$.

Fourth, we define r_4 such that (iv) holds for all $t \geq r_4$. Define

$$\tau = \max\{\sigma(j, y, s) : (\exists s \geq r_3) \ [y \in B_{j,s+1} \cap \overline{A}]\}.$$

By (ii), $\tau \geq \sigma_j$. Choose x and a stage $r_4 \geq r_3$ such that $\sigma(j, x, r_4) = \tau$ and $x \in B_{j,r_4+1} \cap \overline{A}$. Then by (i), $x \in B_{j,t}$ for all $t > r_4$. Now suppose some $y \in B_{j,t+1} - B_{j,t}$ where $t > r_4$ and $\sigma(j, y, t) > \tau$. Then $y \in A$ by our choice of τ. Thus Lemma 1 is proved by setting $s_j = r_4$.

Lemma 2. *For all j, $B_{j,\infty}$ is finite and nonempty.*

Proof. By Lemma 1 parts (ii) and (iv).

Lemma 3. $\overline{M} = \bigcup_{j<\infty} B_{j,\infty}$.

Proof. ($\subseteq$). Suppose $x \in \overline{M}$. Choose a true stage s such that $x \leq a_s^s = a_s$. Then $x \in B_{j,s+1}$ for some j (else $x \in M_{s+1}$). Choose the least k such that $x \in B_{k,t}$ for some stage $t > s$. Then either x remains in block B_k forever or x is enumerated in M by (5.8). Hence, $x \in B_{k,\infty}$.

($\supseteq$). If $x \in B_{j,\infty}$ for some j, then there exists a stage s such that $x \in B_{j,t}$ for all $t \geq s$. Hence, $x \notin M_t$ for all $t \geq s$. Therefore, $x \notin M$.

Lemma 4. *M is maximal.*

Proof. By Lemmas 2 and 3, $\overline{M}$ is infinite. Suppose M is not maximal. Choose the least e such that $\overline{M} \cap W_e$ and $\overline{M} \cap \overline{W}_e$ are both infinite. Let σ be the $(e-1)$-state such that $\sigma(e-1, x) = \sigma$ for a.e. $x \in \overline{M}$. Let $\sigma_1 = \sigma \,\hat{}\, 1$ and $\sigma_0 = \sigma \,\hat{}\, 0$. Clearly, $U_{\sigma_1}^A$ is infinite. Hence,

$$(5.10) \qquad\qquad (\exists k)\,(\forall j \geq k)\,[Z^{j,\sigma_1} \text{ is infinite}].$$

For each $j \geq k$ of (5.10) consider $V^{j,\sigma_1\hat{}0^{j-e}}$. Since $Z^{j,\sigma_1} \subseteq V^{j,\sigma_1\hat{}0^{j-e}}$, these sets are all infinite. Hence, $\sigma_j \geq \sigma_1\hat{}0^{j-e}$. Therefore, no permanent residents of B_j have e-state σ_0. By Lemmas 2 and 3 $|\overline{M} \cap \overline{W}_e| < \infty$.

This completes the proof of Theorem 5.1.   ∎

To better understand the intuition of this proof consider only a single r.e. set W_0. Let σ_1 be the high 0-state, $\sigma_1 = 1$, and σ_0 be the low 0-state $\sigma_0 = 0$. Imagine that each block B_j has only two rows, one for each of σ_0 and σ_1. An element x is *true* if $x \in \overline{A}$. The most conservative strategy for B_j is to fill both of his rows σ_i, $i = 0, 1$, with an element x_i in state σ_i, but then W_0 may split $\overline{M}$. The most aggressive strategy is to discard (dump) $x = b_s(j, \sigma_0)$ whenever $y = b_s(j, \sigma_1)$ is defined, but then each such y may later enter A so B_j will have no permanent resident. Hence, a more discriminating strategy for filling rows and discarding present residents is needed.

Each B_j would really like to know whether $|U_{\sigma_1}^A| = \infty$, since if so then there will be infinitely many *true* elements in state σ_1 available to each B_j. This is a Π_2^A question and is therefore Σ_3 since A is low₂. Let $\{Z^{j,\sigma}\}$ satisfy (5.3) and (5.4) for $\sigma = \sigma_0, \sigma_1$. Think of Z^{j,σ_1} as an "oracle" for B_j which "tells" B_j that $U_{\sigma_1}^A$ is infinite whenever Z^{j,σ_1} receives a new element, and which "lies" at most finitely often by (5.3) and (5.4).

As an aid to the imagination, we picture each B_j as a person (of either sex) and the integers $x \in \omega$ as members of the opposite sex (m.o.s.). The latter are divided into blondes (if $x \in \sigma_1$) and brunettes (if $x \in \sigma_0$). Brunettes may change (once) to blondes but never conversely. Each B_j prefers blondes over brunettes, but the primary goal is to choose as a permanent companion at least one m.o.s. x who is true (i.e., $x \in \overline{A}$). At substage j, B_j will discard a brunette resident if: (1) B_j now has a blonde resident; and (2) the oracle of B_j for blondes, Z^{j,σ_1}, receives a new element, thereby telling B_j that there will be infinitely many true blondes available to B_j. B_j will always attempt to fill the blonde row σ_1 whether or not the brunette row σ_0 contains a resident, but B_j will never add a *new* brunette if B_j currently has a blonde. In filling an empty blonde row, however, B_j must be very cautious, since the set of true blondes, although infinite, may be very elusive. B_j first looks to see whether B_j's own brunette row σ_0 contains some blonde x (since a brunette may change to a blonde even while in B_j). If so B_j moves x to the blonde row σ_1. If not B_j searches for the next available (unused) blonde.

Of course, if $|U_{\sigma_1}^A| < \infty$ then B_j may only succeed in finding a permanent brunette resident y but no permanent blonde resident x. This, however, does not deter B_j from diligently adding an infinite sequence $\{x_n\}_{n \in \omega}$ of temporary blonde residents (in preference to y) each of whom later proves to be untrue. (Oh well, some of the best things in life are ephemeral.) Meanwhile, since B_j's blonde oracle set Z^{j,σ_1} is finite, B_j is careful never to discard the true brunette y during a visit of one of the transient blondes x_n.

5.3 Notes. The original proof of Theorem 5.1 appeared in Lachlan [1968a]. The present style and format using the blocks and the sets $V^{y,\sigma}$ were introduced by the author in a first draft of this section. After reading an earlier draft of this chapter S. Ahmad suggested further simplifications and corrections which we have incorporated here.

Chapter XII

Classifying Index Sets of R.E. Sets

In Chapter IV we saw how the arithmetical hierarchy gave us a precise measure for classifying index sets and yielded considerably more information than the primitive methods of Chapters I and II, which used only the s-m-n Theorem. In this chapter we carry the classification of index sets much further by combining the arithmetical hierarchy methods of Chapter IV with the priority methods developed in Chapters VII, VIII, X, and XI. Interesting in their own right, these index set results in addition often have unexpected and important applications. For example, the Yates Index Set Theorem 1.5 yields a very short and elegant proof of the Sacks Density Theorem VIII.4.1. More recently Herrmann [ta] has used index sets in his proof of the undecidability of the elementary theory of $\mathcal{E}$.

We begin in §1 by proving the Yates Index Set Theorem. This enables us to show that if A is r.e. then $G(A) =_{\mathrm{dfn}} \{\, x : W_x \equiv_{\mathrm{T}} A \,\}$ is Σ_3^A-complete. We continue this topic in §2 by obtaining the Yates classification of $G(\leq A)$, $G(\geq A)$ and $G(|\ A)$ but replacing Yates's original proof by newer and more powerful results which yield more information.

In §3 we relate uniformly r.e. (u.r.e.) sequences of r.e. sets to Σ_3 index sets and to low$_2$ r.e. degrees. Closely related results by Jockusch are presented on degrees in which the recursive functions (sets) are uniform (respectively subuniform). The latter concepts of uniform and subuniform have recently proved useful in studying degrees of models of arithmetic.

In §4 we present some elegant recent results with unexpectedly short proofs by Schwarz, Solovay, and Jockusch for classifying index sets of high$_n$ and low$_n$ sets, as well as the intermediate sets of Corollary VIII.3.5 introduced by Lachlan, Martin, and Sacks. Surprisingly, one easily obtains from these general theorems that Max $=_{\mathrm{dfn}} \{\, x : W_x$ is maximal $\}$ is Π_4-complete, and $\{\, x : W_x$ atomless $\}$ is Π_5-complete, the latter having been a long open question until this method appeared.

In §5 we continue the study of fixed point theorems which began with Kleene's Recursion Theorem II.3.1 and continued with Arslanov's Fixed Point Theorems V.5.1 and V.5.5. We show that for any function $f \leq_{\mathrm{T}} \emptyset^{(2)}$ there is *Turing fixed point* n such that $W_n \equiv_{\mathrm{T}} W_{f(n)}$.

In §6 we generalize both the Recursion Theorem and the Arslanov Completeness Criterion to all levels in the arithmetical hierarchy and even $\emptyset^{(\omega)}$.

The material in sections 4, 5, and 6 is relatively recent, having been proved in 1982–1985, and much of it has not previously appeared in print.

1. Classifying the Index Set $G(A) = \{\, x : W_x \equiv_T A \,\}$

1.1 Definition.   If A is an r.e. set let $G(A) = \{\, x : W_x \equiv_T A \,\}$.

In this section we prove that $G(A)$ is Σ_3^A-complete, and we prove an important index set theorem by Yates which gives a short and elegant proof of the Sacks Density Theorem VIII.4.1.

1.2 Lemma (Yates [1966b]).   *If A is r.e. then $G(A) \in \Sigma_3^A$.*

Proof. We use the methods of Chapter IV. Note that

$$(1.1) \quad x \in G(A) \iff (\exists e)\,(\exists i)\,[W_x = \{e\}^A \ \& \ A = \{i\}^{W_x}],$$

$$(1.2) \qquad\qquad \iff (\exists e)\,(\exists i)\,[\{e\}^A \text{ is total } \ \& \ \{i\}^{\{e\}^A} \text{ is total}$$
$$\& \ (\forall y)\,[[y \in W_x \iff \{e\}^A(y) = 1]$$
$$\& \ [y \in A \iff \{i\}^{W_x}(y) = 1]]].$$

The trick is now to replace the last line by the following:

$$(1.3) \qquad\quad y \in A \iff (\exists s)\,(\exists \sigma)\,[\{i\}_s^\sigma(y) = 1$$
$$\& \ (\forall z < lh(\sigma))\,[\{e\}_s^A(z)\!\downarrow = \sigma(x)]].$$

Now the right hand side of line (1.3) is in Σ_1^A form so (1.2) can easily be put into Σ_3^A form.   ▯

The following representation lemma for a Σ_3^A set S is the main tool of this section.

1.3 Representation Theorem (Yates [1966b]).   *Let A be an r.e. set. For any set $S \in \Sigma_3^A$ there is a u.r.e. sequence of r.e. sets $\{V_k\}_{k\in\omega}$ which is uniformly recursive in A, and such that for all k,*

$$(1.4)$$
$$k \in S \implies (\exists e_0)\,[(\forall e \geq e_0)\,[V_k^{[e]} \equiv_T A] \ \& \ (\forall e < e_0)\,[V_k^{[e]} \text{ is recursive}]],$$

and

$$(1.5) \qquad\qquad k \notin S \implies (\forall e)\,[V_k^{[e]} \text{ is recursive}].$$

1.4 Lemma. *If A is r.e. and $R \in \Pi_2^A$ then there is a u.r.e. sequence of r.e. sets $\{\, U_i \,\}_{i \in \omega}$ which is uniformly recursive in A, and such that for all i,*

$$(1.6) \qquad\qquad i \in R \implies U_i \equiv_T A,$$

and

$$(1.7) \qquad\qquad i \notin R \implies U_i \text{ is recursive.}$$

Furthermore, an index for the sequence $\{\, U_i \,\}_{i \in \omega}$ may be found uniformly from a Π_2^A-index for R.

Proof of Theorem 1.3 (using Lemma 1.4). Fix $S \in \Sigma_3^A$. By Theorem IV.3.2 relativized to A choose $R \in \Pi_2^A$ such that for all k,

$$(1.8) \quad k \in S \implies (\exists e_0)\, [(\forall e \geq e_0)\, [\langle k, e \rangle \in R] \ \& \ (\forall e < e_0)\, [\langle k, e \rangle \notin R]],$$

and

$$(1.9) \qquad\qquad k \notin S \implies (\forall e)\, [\langle k, e \rangle \notin R].$$

By Lemma 1.4 choose a u.r.e. sequence $\{\, U_{\langle k,e \rangle} \,\}_{k,e \in \omega}$ uniformly recursive in A and satisfying (1.6) and (1.7) for R. Define $V_k = \bigoplus_{e \in \omega} U_{\langle k,e \rangle}$. $\quad\blacksquare$

Proof of Lemma 1.4. Fix $R \in \Pi_2^A$. By Theorems IV.3.2 and III.1.5 there is a recursive function f such that for all i,

$$(1.10) \qquad i \in R \implies W_{f(i)}^A = \omega \implies (\forall y)\, [y \in W_{f(i)}^A],$$

and

$$(1.11) \qquad i \notin R \implies W_{f(i)}^A \text{ is a finite initial segment of } \omega,$$
$$\implies (\exists y_0)\, [(\forall y \geq y_0)\, [y \notin W_{f(i)}^A]$$
$$\& \ (\forall y < y_0)\, [y \in W_{f(i)}^A]].$$

Define for each i and y the set

$$L_{i,y} = \{\, u : (\exists s \geq u)\, [y \notin W_{f(i),s}^{A_s \restriction u}] \,\}.$$

Clearly, $\{\, L_{i,y} \,\}_{i,y \in \omega}$ is uniformly r.e. and uniformly recursive in A. Note that

$$(1.12) \qquad\qquad y \in W_{f(i)}^A \implies L_{i,y} \text{ is finite,}$$

and

$$(1.13) \qquad\qquad y \notin W_{f(i)}^A \implies L_{i,y} = \omega.$$

Define

$$U_i = \{ \langle x, y \rangle : y \in A \vee x \in \bigcup_{z \leq y} L_{i,z} \}.$$

Note that $U_i \leq_T A$ uniformly in i because $\{ L_{i,y} \}$ is uniformly recursive in A. Now by (1.10) and (1.12),

$$i \in R \implies (\forall y) \, [y \in W^A_{f(i)}],$$
$$\implies (\forall y) \, [L_{i,y} \text{ is finite}],$$
$$\implies U_i \equiv_T A,$$

because $y \in \overline{A}$ iff $(\exists x) \, [\langle x, y \rangle \in \overline{U}_i]$. Next by (1.11), (1.12), and (1.13),

$$i \notin R \implies (\exists y_0) \, [(\forall y \geq y_0) \, [y \notin W^A_{f(i)}] \ \& \ (\forall y < y_0) \, [y \in W^A_{f(i)}]],$$
$$\implies (\exists y_0) \, [(\forall y \geq y_0) \, [L_{i,y} = \omega] \ \& \ (\forall y < y_0) \, [L_{i,y} \text{ is finite}]],$$
$$\implies U_i \text{ is recursive;}$$

because for $y \geq y_0$, $U_i^{[y]} = \omega^{[y]}$, and for $y < y_0$ if $y \in A$ then $U_i^{[y]} = \omega^{[y]}$ and if $y \notin A$ then $U_i^{[y]}$ is finite.   ▯

The following important results are now easy consequences of the Representation Theorem 1.3.

1.5 Index Set Theorem (Yates [1966b]). *Given r.e. sets C and D such that $D <_T C$ and $S \in \Sigma_3^C$ there is a recursive function $g(k)$ such that for all k,*

(i) $D \leq_T W_{g(k)} \leq_T C$, *and*
(ii) $k \in S \iff W_{g(k)} \equiv_T C$.

Proof. Fix $S \in \Sigma_3^C$ and $\{V_k\}_{k \in \omega}$ the sequence for S given by Theorem 1.3. For each k define the r.e. set B_k by $B_k^{[0]} = \{ \langle x, 0 \rangle : x \in D \}$, and $B_k^{[e+1]} = \{ \langle x, e+1 \rangle : x \in V_k^{[e]} \}$. (Note that $B_k \leq_T C$, for all k, because $V_k \leq_T C$ and $D \leq_T C$.) For each k, apply the Strong Thickness Lemma VIII.2.3(ii) to B_k and to C to find the r.e. subset $A_k \subseteq B_k$, $A_k \leq_T B_k \leq_T C$, such that A_k satisfies the conclusion of VIII.2.3(ii). Now for each k, $D \leq_T A_k$ because $A_k^{[0]} =^* B_k^{[0]} \equiv_T D$ by Lemma VIII.2.3(ii) with $e = 0$. By the uniformity of VIII.2.3 there is a recursive function $g(k)$ such that $W_{g(k)} = A_k$. Now if $k \notin S$ then $B_k^{[e]}$ is recursive for all $e > 0$ by (1.5) whence $C \not\leq_T A_k$ by VIII.2.3(i). If $k \in S$ then by (1.4) and the definition of B_k choose e such that

$$B_k^{[e]} \equiv_T C \ \& \ (\forall j) \, [0 < j < e \implies B_k^{[j]} \text{ is recursive}].$$

Hence, $B_k^{[<e]} \equiv_T D <_T C$. Therefore, by VIII.2.3(ii), $A_k^{[e]} =^* B_k^{[e]} \equiv_T C$, so that $C \leq_T A_k$. Thus, $A_k \equiv_T C$ because $A_k \leq_T B_k \leq_T C$.   ▯

1.6 Corollary (Yates). *If C is r.e. then $G(C)$ is Σ_3^C-complete.*

Proof. First $G(C) \in \Sigma_3^C$ by Lemma 1.2. Now if C is recursive apply Corollary IV.3.6. Otherwise, apply Theorem 1.5 with $D = \emptyset$. ▯

Recall that in Definition IV.3.3 we defined Comp $= \{\, x : W_x \equiv_T \emptyset' \,\} = G(\emptyset')$. In Theorem IV.3.4 we showed that $(\Sigma_3, \Pi_3) \leq_1 (\text{Cof}, \text{Comp})$ and hence that $A \leq_1$ Comp for any Π_3 set A. We now improve this by showing that Comp is Σ_4-complete.

1.7 Corollary. Comp *is Σ_4-complete.*

Proof. Apply Corollary 1.6 with $C = \emptyset'$ and note that $\Sigma_3^{\emptyset'} = \Sigma_4$ by Post's Theorem IV.2.2. ▯

We have as a corollary of Theorem 1.5 an elegant proof of the Sacks Density Theorem VIII.4.1.

1.8 Density Theorem (Sacks). *If D and C are r.e. and $D <_T C$ then there exists an r.e. set A such that $D <_T A <_T C$.*

Proof. Let $S = \{\, x : W_x \equiv_T D \,\}$. Then by Corollary 1.6, $S \in \Sigma_3^D$ and a fortiori $S \in \Sigma_3^C$. Apply Theorem 1.5 to find a recursive function $g(x)$ such that $D \leq_T W_{g(x)} \leq_T C$ and $W_x \equiv_T D$ if and only if $W_{g(x)} \equiv_T C$. By the Recursion Theorem II.3.1 choose n such that $W_n = W_{g(n)}$. Hence, $D <_T W_{g(n)} <_T C$. ▯

1.9–1.14 Exercises

1.9 (Yates [1969]). There is no u.r.e. sequence of r.e. sets $\{\, A_n \,\}_{n \in \omega}$ which is uniformly of degree $\leq \mathbf{0}'$ and such that for every r.e. degree $\mathbf{d}$ there is exactly one n such that $\mathbf{d} = \deg(A_n)$. (In particular the r.e. *degrees cannot* be recursively enumerated without repetition, although Friedberg [1958] showed that this *can* be done for the r.e. *sets*.) *Hint.* Assume to the contrary that $\{\, A_n \,\}_{n \in \omega}$ is such a u.r.e. sequence. Let $\{\, B_n \,\}$ be the r.e. sequence obtained from $\{\, A_n \,\}$ by omitting the unique set A_j of degree $\mathbf{0}'$. Thus,

$$G(\emptyset') = \{\, x : (\forall n)\, [W_x \not\leq_T B_n] \,\}.$$

Show that then $G(\emptyset') \in \Pi_4$ thereby violating Corollary 1.7 that $G(\emptyset')$ is Σ_4-complete.

1.10 (Yates [1966b]). (a) Prove that if C is r.e., $C \neq \emptyset$, $C \neq \omega$, then the index set $G_m(C) =_{\text{dfn}} \{\, x : W_x \equiv_m C \,\}$ is Σ_3-complete. *Hint.* $G_m(C) \in \Sigma_3$ by Exercise IV.3.13. Now prove Σ_3-completeness. The recursive case follows from Exercise I.4.25 and Corollary IV.3.6. For C nonrecursive choose $S \in \Sigma_3$ and let g satisfy Corollary IV.3.7 for $A = S$. For each x define the r.e. set

$$B_x = \{\, \langle z, e \rangle : z \in C \cap W_{g(x,e)} \,\}.$$

Apply the proof of the Strong Thickness Lemma VIII.2.3(ii) to C and B_x to obtain $W_{f(x)} = A_x \subseteq B_x$, and show $x \in S$ iff $f(x) \in G_m(C)$. Abbreviating A_x by A, we must slightly modify that earlier proof so that if $\langle z, e \rangle \in A_{s+1} - A_s$, then for all $e' > e$ and $z' \le z$, $\langle z', e' \rangle \notin A$ unless $\langle z', e' \rangle \in A_s$ already. This ensures that if $x \in S$, and say e is minimal such that $W_{g(e,x)}$ is infinite, then not only is $A^{[e]} \equiv_m C$ but also $A^{[>e]}$ is recursive so that $A \le_m C$.

   (b) Prove that $\{\, x : W_x \text{ is creative} \,\}$ is Σ_3-complete. (See Exercise 1.12 for a generalization.)

1.11.   Prove that if A is r.e. then $G_{tt}(A) =_{\text{dfn}} \{\, x : W_x \equiv_{tt} A \,\}$ is Σ_3-complete.

1.12.$^{\diamond\diamond}$ (Herrmann [ta]).   If A is r.e., infinite, and coinfinite then

$$G_1(A) =_{\text{dfn}} \{\, x : W_x \equiv_1 A \,\}$$

is Σ_3-complete. (This generalizes Exercise 1.10(b).)

1.13 (Young).   Show that any two maximal sets are either m-incomparable or have the same m-degree.

1.14 (D. A. Martin and C. E. M. Yates).   Prove that the maximal sets of degree $\mathbf{0}'$ lie in infinitely many m-degrees. *Hint.* Apply Exercise 1.13. Suppose for a contradiction that they lie in finitely many m-degrees, whose classes of r.e. sets we write as: $C_1, C_2, \ldots, C_k$. Define a recursive function f such that for all x,

$$W_x \in \bigcup_{i \le k} C_i \implies W_{f(x)} \text{ is recursive,}$$

and

$$W_x \notin \bigcup_{i \le k} C_i \implies W_{f(x)} \text{ is maximal and of degree } \mathbf{0}'.$$

Apply the Recursion Theorem to f to obtain a contradiction.

2. Classifying the Index Sets $G(\le A)$, $G(\ge A)$ and $G(|A)$

By analogy with Definition 1.1 we make the following definitions.

2.1 Definition. Let A be an r.e. set, and $\mathbf{a} = \deg(A)$. Define

$$G(\leq \mathbf{a}) = G(\leq A) = \{\, x : W_x \leq_T A \,\},$$

$$G(\geq \mathbf{a}) = G(\geq A) = \{\, x : A \leq_T W_x \,\},$$

and

$$G(|\mathbf{a}) = G(|A) = \{\, x : W_x \mid_T A \,\}.$$

Upper bounds for these three index sets as Σ_3^A, Σ_4, and Π_4, respectively, are easily established in Corollary 2.3, and in Exercises 2.7 and 2.6. We now prove that (except in the trivial cases of $A \equiv_T \emptyset$ or $A \equiv_T \emptyset'$ where $G(\geq \emptyset) = \omega = G(\leq \emptyset')$ and $G(|\emptyset) = \emptyset = G(|\emptyset')$ these are the best possible bounds.

We could quickly classify the latter two index sets by first proving Corollary 2.4, which is an easy corollary of Theorem 1.3 and the Thickness Lemma VIII.2.3, as we sketch in Exercise 2.8. However, all three index sets can be classified using the following powerful theorem whose proof uses the standard infinite injury methods of Chapter VIII. The statement is due to Kallibekov and the proof to Stob as explained in Notes 2.17. The interesting new idea in this argument lies in the proof of line (2.6) in Lemma 3.

2.2 Theorem (Kallibekov-Stob). *Let C and D be r.e. sets such that*

$$\emptyset <_T C <_T \emptyset'$$

and $C \leq_T D$. Let $S \in \Sigma_3^D$. Then there is a u.r.e. sequence of r.e. sets $\{A_k\}_{k\in\omega}$ such that for all k,

$$k \in S \implies A_k \equiv_T D,$$

$$k \notin S \implies A_k \mid_T C.$$

Proof. Let $\{V_k\}_{k\in\omega}$ be as in the Representation Theorem 1.3 satisfying (1.4) and (1.5) with D in place of A. Let $\{V_{k,s}\}_{k,s\in\omega}$, $\{C_s\}_{s\in\omega}$, $\{D_s\}_{s\in\omega}$, and $\{K_s\}_{s\in\omega}$ be recursive enumerations of $\{V_k\}_{k\in\omega}$, C, D, and K, respectively. We shall construct A_k. It will be clear that the construction is uniform in k so that $\{A_k\}_{k\in\omega}$ is u.r.e.

If $k \notin S$, to guarantee $A_k \mid_T C$ we meet for all e the requirements

$$N_e : C \neq \Phi_e(A_k),$$

and

$$P_e : A_k \neq \Phi_e(C).$$

These requirements and the strategies for meeting them are exactly as in the Density Theorem VIII.4.1. Namely, to meet P_e, if $\Phi_{e,s}(C_s)$ agrees with $A_{k,s}$ on long initial segments we attempt to code K into $A_k^{[2e+1]}$. Thus, if

$\Phi_e(C) = A_k$, then $A_k \equiv_T K$ so $K \leq_T C$ contrary to the hypothesis on C. This coding strategy is arranged so that $A_k^{[2e+1]}$ is recursive for all e so as not to interfere with N_e, which is handled as in Chapter VIII.

We also attempt to arrange that $A_k^{[2e]} \equiv_T V_k^{[e]}$. Hence, if $k \in S$ then $D \leq_T A_k$ by (1.4). If $k \notin S$, this additional action does not conflict with N_e since then $V_k^{[e]}$ is recursive for every e by (1.5). If $k \in S$, however, we do not succeed in making $A_k^{[2e]} \equiv_T V_k^{[e]}$ for every e, but it suffices to succeed on one of the cofinitely many e such that $V_k^{[e]} \equiv_T D$. The trick in this case will be to show by (2.6) that $A_k \leq_T D$.

Let $\hat{l}(e, s)$, $\hat{r}(e, s)$, and $\hat{l}_e$ be as in Lemma VIII.1.1 but with A_k in place of A. Define $\hat{R}(e, s) = \max\{\hat{r}(i, s) : i \leq e\}$, and $\hat{R}(-1, s) = 0$ for all s. Analogously to Theorem VIII.4.1 define the recursive function

$$\hat{l}^C(e, s) = \max\{x : (\forall y < x)\, [A_{k,s}(y) = \hat{\Phi}_{e,s}(C_s; y)]\}.$$

Construction of A_k.

   Stage $s + 1$.

      Step 1. Enumerate $\langle x, 2e \rangle$ in $A_{k,s+1}$ iff $\langle x, e \rangle \in V_{k,s}$ and

$$\hat{R}(e - 1, s) < \langle x, 2e \rangle.$$

   Step 2. Enumerate $\langle x, t, 2e + 1 \rangle$ in $A_{k,s+1}$ iff $x \in K_s$, $t \leq s$, $\hat{R}(e - 1, s) < \langle x, t, 2e + 1 \rangle$, and $x < \hat{l}^C(e, v)$ for all v, $t \leq v \leq s$. (This is the same as the Sacks coding strategy of Theorem VIII.4.1.)

Lemma 1. *If $V_k^{[e]}$ is recursive (and hence $V_k^{[j]}$ is recursive for all $j \leq e$) then for all $j \leq e$:*

(2.1) $A_k^{[2j]}$ *is recursive,*

(2.2) $\Phi_j(C) \neq A_k$,

(2.3) $A_k^{[2j+1]}$ *is recursive,*

(2.4) $\Phi_j(A_k) \neq C$.

Proof. Suppose that (2.1)–(2.4) hold for all $i < j$. We show that they hold for j. Now by (2.4) and the Window Lemma VIII.2.2,

$$r =_{\text{dfn}} \liminf_s \hat{R}(j - 1, s) < \infty.$$

Thus, if $\langle x, 2j \rangle > r$ then $\langle x, 2j \rangle \in A_k$ iff $\langle x, j \rangle \in V_k$ and hence (2.1) is proved. For (2.2) assume to the contrary that $\Phi_j(C) = A_k$. Then we contradict the hypothesis that $C <_T K$ by C-recursively computing K as in Theorem VIII.4.1 Lemma 2. (Namely, C-recursively compute the modulus $M(x) = (\mu s) (\forall t \geq s) [l^C(j,t) > x]$. Now by Step 2 of the construction, if $x > r$, then $x \in K$ iff $\langle x, M(x), 2j+1 \rangle \in A_k$. Hence, $K \leq_T A_k \leq_T C$.) Next (2.3) follows from (2.2) as in Theorem VIII.4.1 Lemma 3, but with A, C, D, and e replaced by A_k, K, C, and j, respectively. Finally, note that the injury set $\hat{I}_j \subseteq A_k^{[<2j+2]}$ and hence $\hat{I}_j \leq_T A_k^{[<2j+2]} \equiv_T \emptyset$, as in Lemma VIII.1.1. Thus, $C \neq \Phi_j(A_k)$ by the Injury Lemma VIII.2.1, thereby proving (2.4).

Lemma 2. *If $k \notin S$ then $A_k \mid_T C$.*

Proof. By (1.5), and Lemma 1 (2.2) and (2.4), $\Phi_j(C) \neq A_k$ and $\Phi_j(A_k) \neq C$ for all j.

Lemma 3. *If $k \in S$ then $A_k \equiv_T D$.*

Proof. If $k \in S$ then by (1.4) there is a least number e_0 so that $V_k^{[e_0]} \equiv_T D$. For $i < e_0$, $V_k^{[i]}$ is recursive so (2.1)–(2.4) hold for all $i < e_0$ by Lemma 1. By (2.4) and the Window Lemma VIII.2.2, $\liminf_s \hat{R}(e_0 - 1, s) < \infty$ so that $A_k^{[2e_0]} \equiv_T V_k^{[e_0]} \equiv_T D$.

It remains only to show that $A_k \leq_T D$. Since $C \leq_T A_k$ there is a least integer e_1, necessarily $e_1 \geq e_0$, such that $\Phi_{e_1}(A_k) = C$. We shall show that $A_k \leq_T D$ by showing that

$$(2.5) \qquad\qquad A_k^{[\leq 2e_1+1]} \leq_T D,$$

and

$$(2.6) \qquad\qquad A_k^{[>2e_1+1]} \leq_T D.$$

To prove (2.5), note that if $e < e_1$ then $\Phi_e(A_k) \neq C$ so that $\liminf_s \hat{R}(e, s) < \infty$. Thus for $e \leq e_1$, $A_k^{[2e]} \equiv_T V_k^{[e]} \leq_T D$, $\Phi_e(C) \neq A_k$, and $A_k^{[2e+1]}$ is recursive by the proofs of (2.1)–(2.3) in Lemma 1. (Note that the proof of (2.3) requires only (2.2), not (2.1).) This proves (2.5).

To prove (2.6), notice that $\hat{I}_{e_1} \leq_T A_k^{[\leq 2e_1+1]} \equiv_T D$ so that we can D-recursively decide when a computation of the form $\Phi_{e_1,s}(A_{k,s}; x)$ will never again be injured. Thus, since $\Phi_{e_1}(A_k) = C$, given y we can D-recursively find a stage s so that $y < \hat{r}(e_1, s)$ where $\hat{r}(e_1, s)$ is protecting a length of agreement $\hat{l}(e_1, s)$ which will never decrease after stage s. Thus, if $y = \langle x, t, e \rangle$, $e > 2e_1 + 1$, then $y \in A_k$ iff $y \in A_{k,s}$ because $y < \hat{r}(e_1, v)$ for all $v \geq s$. This proves (2.6). ∎

2.3 Corollary (Yates).   *If C is r.e. and $C <_T \emptyset'$ then $G(\leq C)$ is Σ_3^C-complete.*

Proof.   First $G(\leq C) \in \Sigma_3^C$ by the same method as in Lemma 1.2. Next $G(\leq C)$ is Σ_3^C-complete by Theorem 2.2 with $D = C$. (The case $C \equiv_T \emptyset'$ is uninteresting because then $G(\leq C) = \omega$.)   ▯

2.4 Corollary (Yates).   *If C is r.e. and $\emptyset <_T C <_T \emptyset'$ then $(\Sigma_4, \Pi_4) \leq_1$ $(G(\emptyset'), G(|C))$, and hence $G(|C)$ is Π_4-complete.*

Proof.   Apply Theorem 2.2 with $D = \emptyset'$. (Note that $\Sigma_3^{\emptyset'} = \Sigma_4$.)   Also $G(|C) \in \Pi_4$ by Exercise 2.6.   ▯

2.5 Corollary (Yates).   *If C is r.e. and $\emptyset <_T C$ then $G(\geq C)$ is Σ_4-complete.*

Proof.   First $G(\geq C) \in \Sigma_4$ by Exercise 2.7. If $C <_T \emptyset'$, let $D = \emptyset'$ and apply Theorem 2.2. If $C \equiv_T \emptyset'$ then $G(\geq C) = G(\emptyset')$ is Σ_4-complete by Corollary 1.7.   ▯

2.6–2.16 Exercises

2.6 (Yates).   Prove that if A is r.e. then $G(|A) \in \Pi_4$. *Hint.* Use the same method as in the reduction of the matrix in Lemma 1.2.

2.7 (Yates).   Prove that if A is r.e. then $G(\geq A) \in \Sigma_4$. *Hint.* See the method in Lemma 1.2.

2.8.   Give an easy direct proof of Corollary 2.4 using the Strong Thickness Lemma VIII.2.3 instead of Theorem 2.2. *Hint.* Fix $S \in \Sigma_4 = \Sigma_3^K$ and apply Theorem 1.3 to K to produce a u.r.e. sequence $\{V_k\}_{k\in\omega}$ satisfying (1.4) and (1.5) (with K in place of A). Choose an r.e. set $E \mid_T C$. Choose a u.r.e. sequence $\{B_k\}_{k\in\omega}$ such that for all k, $B_k^{[0]} \equiv_T E$ and $B_k^{[e+1]} \equiv_T V_k^{[e]}$. Apply Lemma VIII.2.3 to C and $\{B_k\}_{k\in\omega}$ to obtain a u.r.e. sequence $\{A_k\}_{k\in\omega}$ such that if $k \in S$ then $A_k \equiv_T K$ but if $k \notin S$ then $A_k \mid_T C$.

2.9 (Yates).   Let $\emptyset <_T C \leq_T \emptyset'$ and let $D_1 \leq_T D_2 \leq_T D_3 \leq_T \cdots$ be a u.r.e. sequence (of r.e. sets) such that $C \not\leq_T D_i$ for all i. Prove that if $S \in \Sigma_4$ then there is a u.r.e. sequence $\{A_k\}_{k\in\omega}$ such that for all k and i, $D_i \leq_T A_k$ and

$$k \in S \implies A_k \equiv_T \emptyset',$$

and

$$k \notin S \implies C \not\leq_T A_k.$$

Hint. The proof is the same as in Exercise 2.8 except that for each k we define an r.e. set B_k such that $B_k^{[2e]} \equiv_T V_k^{[e]}$ and $B_k^{[2e+1]} \equiv_T D_e$.

2.10 (Yates). Let $\emptyset <_T C <_T \emptyset'$ and let $D_1 \leq_T D_2 \leq_T D_3 \leq_T \cdots$ be a u.r.e. sequence such that $C \not\leq_T D_i$ for all i. Prove that there exists an r.e. set $A \mid_T C$ such that $D_i \leq_T A$ for all i. *Hint.* Set $S = G(\leq C) \in \Sigma_4$. Apply Exercise 2.9 to S to obtain $\{A_k\}_{k \in \omega}$. Use the Recursion Theorem to find n such that $W_n = A_n$, and set $A = A_n$.

2.11. Prove a different version of Theorem 2.2 with the additional hypothesis "$C <_T D$", and the additional conclusion "$\{A_k\}_{k \in \omega}$ is uniformly recursive in D." *Hint.* In the proof of 2.2 replace all instances of "K" by "D". The point is that D can now compute which elements want to enter $A_k^{[2e+1]}$ as well as which elements want to enter $A_k^{[2e]}$ (subject as usual to the restraint function). The proof that D can now compute $A_k^{[2e]}$ uniformly in e is similar to the proof of Theorem VIII.4.1 Lemma 4.

2.12. Prove that for any nonrecursive r.e. set D,

$$(\Sigma_3^D, \Pi_3^D) \leq_1 (G(D), G(< D)).$$

Hint. Choose an r.e. set C, $\emptyset <_T C <_T D$ and apply Exercise 2.11.

2.13 (Slaman). Show that in Theorem 2.2 we may remove the hypothesis "$C \leq_T D$", and still obtain the same conclusion as in 2.2. *Hint.* Let $\{V_k\}_{k \in \omega}$ be as in the proof of 2.2. By the proofs of Theorem 1.3 and Lemma 1.4 there is a recursive function f such that for all k and e

$$(2.7) \qquad\qquad V_k^{[e]} \equiv_T D \implies W_{f(e,k)}^D = \omega;$$

and

$$(2.8) \qquad\qquad V_k^{[e]} \equiv_T \emptyset \implies W_{f(e,k)}^D =^* \emptyset.$$

Define the recursive function

$$q(e, s) = \max\{x : (\forall y \leq x) \, [y \in W_{f(e,k),s}^{D_s} \And D_s \restriction u_y = D_{s-1} \restriction u_y]\},$$

for $u_y = u(D_{s-1}; f(e, k), y, s - 1)$. Let $Q(e, s) = \max\{q(i, s) : i \leq e\}$. Note that

$$(2.9) \quad W_{f(e,k)}^D =^* \emptyset \implies [Q(e) =_{\text{dfn}} \liminf_s Q(e, s) < \infty \And Q \leq_T D],$$

while

$$(2.10) \qquad\qquad W_{f(e,k)}^D = \omega \implies \liminf_s Q(e, s) = \infty,$$

and $m \leq_T D$ where $m(x) = (\mu s) \, (\forall t \geq s) \, [Q(e, t) \geq x]$. The point is that the hypothesis "$C \leq_T D$" was used only in Lemma 3 to obtain e_1 such that $\liminf_s \hat{R}(e_1, s) = \infty$. Now we simply choose e_0 as in Lemma 3 and note that $\liminf_s Q(e, s) = \infty$ by (2.2) and (2.10), and $m \leq_T D$ which suffices to prove Lemma 3.

Therefore in the construction we would like to require that if $y = \langle x, 2e \rangle$ in Step 1 (or $y = \langle x, t, 2e + 1 \rangle$ in Step 2) then $y > Q(e - 1, s)$ in addition to $y > \hat{R}(e - 1, s)$. The problem is that $\liminf_s Q(e, s)$ is achieved on the true stages of the enumeration $\{ D_s \}_{s \in \omega}$ and $\liminf_s \hat{R}(e, s)$ on the true stages of $\{ A_{k,s} \}_{s \in \omega}$ so that $\liminf_s P(e, s)$ need not exist where $P(e, s) = \max\{ \hat{R}(e, s), Q(e, s) \}$.

The solution is to use the pinball method of Chapter VIII §5. Let an element y which wants to enter A_k for a positive requirement of Theorem 2.2 drop from Hole H_e if $y = \langle x, e \rangle$, for some x. (See Chapter VIII Diagram 5.1.) Divide the gate G_e, $e \geq 0$, into two subgates G_e^1 and G_e^2. Allow y to pass through G_e^1 at stage $s + 1$ if $y > Q(e, s)$ and through G_e^2 at stage $s + 1$ if $y > \hat{R}(e, s)$. Enumerate y in A_k if y reaches gate G_{-1}. Use the method of proof in Chapter VIII §5 to show that this construction succeeds in enabling us to prove Lemmas 1 and 2 of Theorem 2.2. Prove Lemma 3 using $Q(e_0, s)$ in place of $\hat{R}(e_1, s)$.

2.14. Suppose D and C are r.e. sets such that $\emptyset <_T D <_T C <_T \emptyset'$ and $D' \equiv_T C'$. Prove that

$$(\Sigma_3^C, \Pi_3^C) = (\Sigma_3^D, \Pi_3^D) \leq_1 (G(D), G(|_T C) \cap G(> D)).$$

Hint. Note that $D' \equiv_T C'$ ensures that $\Sigma_3^C = \Sigma_3^D$. Apply the proof of Exercise 2.13 but code D on $A_k^{[0]}$ for each k so that $D \leq_T A_k$ for all k. (Note that without the hypothesis that $D' \equiv_T C'$ we can still get the same conclusion but without the first equality.)

2.15. Fix any r.e. set A such that $\emptyset <_T A <_T \emptyset'$.

(a) Prove that $G(< A) \in [\Sigma_3^A \ \& \ \Pi_3^A]$. *Hint.* Use Corollary 2.3 and Lemma 1.2.

(b) (Slaman). Prove that: (i) $\Sigma_3^A \leq_1 G(< A)$; and $\Pi_3^A \leq_1 G(< A)$. *Hint.* For (i) use Exercise 2.14 with $A = C$, noting that $G(D) \subseteq G(< C)$. Note that given C, with $\emptyset <_T C <_T \emptyset'$, we can choose D such that $\emptyset <_T D <_T C$ and $D' \equiv_T C'$ by Theorem VIII.4.3. For (ii) use Exercise 2.12 with $A = D$.

(c) Prove that: (i) $\Sigma_4 \leq_1 G(> A)$; and (ii) $\Pi_3^A \leq_1 G(> A)$. *Hint.* For (i) use Corollary 2.4, and for (ii) use Exercise 2.14 with $A = D$.

2.16. Call an r.e. degree **a** *m-topped* if there is a maximum among the r.e. m-degrees in **a**.

(a) (Jockusch [1972a]). Prove that any incomplete m-topped r.e. degree **a** is low$_2$. *Hint.* Use the index set classifications of $\{ e : W_e \leq_m A \}$ and $\{ e : W_e \leq_T A \}$.

(b)° (Downey and Jockusch).   No m-topped (or even tt-topped) r.e. degree is low.

(c)°° (Downey and Jockusch).   There exists a nonzero m-topped r.e. degree **a** with $\mathbf{0} < \mathbf{a} < \mathbf{0}'$. (In fact for every low r.e. degree **b** there exists an m-topped r.e. degree **a** with $\mathbf{b} \leq \mathbf{a} < \mathbf{0}'$.)

2.17 Notes. Kallibekov [1971, Theorem 1] announced Theorem 2.2 and proposed a new and ingenious method for doing priority arguments which has also been used by Kinber [1977]. Unfortunately, Kallibekov's proof contained an error which does not appear to be reparable. Stob then first tried to give a standard infinite injury proof of Theorem 2.2 in the style of Chapter VIII but encountered an obstacle to showing $A_k \leq_T D$ which he overcame [1982b] by the unusual device to prove (2.6) in Lemma 3. Slaman in Exercise 2.13 then showed how the hypothesis "$C \leq_T D$" could be removed. The results Corollaries 2.3, 2.4, and 2.5 and Exercises 2.6, 2.7, 2.9, and 2.10 appeared in Yates [1969], but with different proofs. The proofs sketched in Exercises 2.8, 2.9, and 2.10 appeared in Soare [1976, §6].

Kallibekov [1971] used Theorem 2.2 to prove that if **d** is an r.e. degree such that $\mathbf{d} < \mathbf{0}'$ and $\mathbf{d}'' = \mathbf{0}'''$ (i.e., $\mathbf{d} \in \mathbf{H_2}$) then **d** contains an infinite antichain of r.e. m-degrees and **d** does not contain an r.e. m-degree maximal among its r.e. m-degrees. (Kallibckov also proved this for tt-degrees in place of m-degrees.)   Moreover, Degtev [1972b] showed that *every* nonzero r.e. degree contains an infinite antichain of r.e. tt-*degrees*. Kobzev [1976] showed that every nonzero r.e. degree contains an infinite antichain of r.e. tt-degrees. Also Kobzev [1973b] showed that every nonzero r.e. degree contains infinitely many minimal r.e. m-degrees.

3. Uniform Enumeration of R.E. Sets and Σ_3 Index Sets

Let C be a class of r.e. sets and $G(C) = \{\, x : W_x \in C \,\}$. In this section we study some results about those classes C such that $G(C) \in \Sigma_3$. This is a particularly interesting and useful property because it is equivalent to C being u.r.e. (providing $\mathcal{F} \subseteq C$).

3.1 Theorem (Yates [1969]).   *Let C be a class of r.e. sets containing $\mathcal{F}$, the class of all finite sets. Then the following are equivalent:*

   (i) C *is uniformly r.e.,*
   (ii) C *is uniformly of degree $\leq \mathbf{0}'$,*
   (iii) $G(C) \in \Sigma_3$.

Proof. (i) $\implies$ (ii). This is immediate since $K_0 \equiv \emptyset'$ and K_0 is 1-complete.

(ii) $\implies$ (iii). Choose a function $f \leq_T \emptyset'$ such that if A_n is the set with characteristic function $\lambda x[f(n, x)]$ then $C = \{ A_n \}_{n \in \omega}$. Hence

$$k \in G(C) \iff (\exists n)\, (\forall x)\, [x \in W_k \iff f(n, x) = 1].$$

The matrix is recursive in $\emptyset'$ and hence Δ_2 by Post's Theorem IV.2.2 so $G(C) \in \Sigma_3$.

(iii) $\implies$ (i). If $G(C) \in \Sigma_3$, choose g as in Corollary IV.3.7 and define $B_{\langle x,y \rangle} = W_x \cap W_{g(x,y)}$. If $x \in G(C)$ then $B_{\langle x,y \rangle} = W_x$ for a.e. y, and is finite for the remaining y. If $x \notin G(C)$ then $B_{\langle x,y \rangle}$ is finite for all y. Thus, $\{ B_n \}_{n \in \omega}$ is u.r.e. and equals C, since $C \supseteq \mathcal{F}$. $\quad \square$

It follows from Theorem 3.1 that the class of recursive sets is u.r.e., although we have already seen this in Exercise II.2.11. Not so obvious is that this is also true for the class of r.e. sets of degree $\leq \mathbf{a}$ if $\mathbf{a}$ is low$_2$. Indeed the following result establishes that $\mathbf{a}$ being low$_2$ is also a necessary condition.

3.2 Corollary (Yates). *If $\mathbf{a} < \mathbf{0}'$ is an r.e. degree then the following are equivalent:*

(i) the class of r.e. sets of degree $\leq \mathbf{a}$ is u.r.e.,
(ii) the class of r.e. sets of degree $\leq \mathbf{a}$ is uniformly of degree $\leq \mathbf{0}'$,
(iii) $\mathbf{a}^{(2)} = \mathbf{0}^{(2)}$ (i.e., $\mathbf{a}$ is low$_2$).

Proof. (i) $\iff$ (ii) by Theorem 3.1.

(ii) $\iff$ (iii). Note that (ii) iff $G(\leq \mathbf{a}) \in \Sigma_3$ by Theorem 3.1, iff $\Sigma_3 = \Sigma_3^{\mathbf{a}}$ by Corollary 2.3. But $\Sigma_3 = \Sigma_1^{\mathbf{0}^{(2)}}$ and $\Sigma_3^{\mathbf{a}} = \Sigma_1^{\mathbf{a}^{(2)}}$ by Theorems IV.2.2 and IV.4.1. Hence $\Sigma_3 = \Sigma_3^{\mathbf{a}}$ iff $\mathbf{a}^{(2)} = \mathbf{0}^{(2)}$. $\quad \square$

In contrast to Corollary 3.2 we see in Corollary 3.5 below that there is no r.e. degree $\mathbf{a} < \mathbf{0}'$ such that the r.e. sets of degree $\leq \mathbf{a}$ are uniformly of degree $\leq \mathbf{a}$. Theorem 3.4 relates nicely to Martin's Theorem XI.1.3 on dominant functions and high degrees. Also the notions introduced in Definition 3.3 have proved useful in other areas of recursion theory and models of arithmetic, for example Knight, Lachlan and Soare [1984].

3.3 Definition (Jockusch). If f is a binary function then f_e denotes $\lambda n[f(e, n)]$. If C is a class of (unary) functions and $\mathbf{a}$ is a degree, C is called $\mathbf{a}$-*uniform* ($\mathbf{a}$-*subuniform*) if there is a binary function f of degree $\leq \mathbf{a}$ such that

$$C = \{ f_e : e \in \omega \} \quad \text{(respectively, } C \subseteq \{ f_e : e \in \omega \}\text{)}.$$

3.4 Theorem (Jockusch [1972a]).   *If* **a** *is any degree, then statements* (i)–(iv) *are equivalent:*

(i) $\mathbf{a}' \geq \mathbf{0}''$;

(ii) the recursive functions are **a**-uniform;

(iii) the recursive functions are **a**-subuniform;

(iv) the recursive sets are **a**-uniform.

If **a** *is r.e., then* (i)–(iv) *are each equivalent to* (v):

(v) the recursive sets are **a**-subuniform.

Proof. The implications (ii) $\implies$ (iii), (ii) $\implies$ (iv), and (iv) $\implies$ (v) are immediate.

(i) $\implies$ (ii). By Theorem XI.1.3 choose a dominant function g of degree $\leq$ **a**. Define $f(\langle e,i \rangle, x) = \varphi_{e,i+g(x)}(x)$ if $\varphi_{e,i+g(y)}(y) \downarrow$ for all $y \leq x$ and $f(\langle e,i \rangle, x) = 0$ otherwise. Now either $f_{\langle e,i \rangle} = \varphi_e$ a total function, or $f_{\langle e,i \rangle}$ is finitely nonzero. In either case $f_{\langle e,i \rangle}$ is recursive. If φ_e is total then $g(x)$ dominates $c(x) = (\mu s) [\varphi_{e,s}(x) \downarrow]$, so $\varphi_e = f_{\langle e,i \rangle}$ for some i.

(iii) $\implies$ (i). Let $f(e, x)$ be a function of degree $\leq$ **a** such that every recursive function is an f_e. Define $g(x) = \max\{ f_e(x) : e \leq x \}$. Then g is dominant so $\mathbf{a}' \geq \mathbf{0}''$ by Theorem XI.1.3.

(iv) $\implies$ (i). By Theorem IV.3.2 and Exercise IV.3.8 we have

$$(\mathrm{Tot}, \overline{\mathrm{Tot}}) \leq_m (\mathrm{Tot}, \overline{\mathrm{Ext}})$$

via some recursive function g. Assume f has degree $\leq$ **a** and that the f_e's are exactly the recursive characteristic functions. Then for all e,

$$e \in \mathrm{Tot} \iff (\exists i) [f_i \text{ extends } \varphi_{g(e)}],$$
$$\iff (\exists i) (\forall x) (\forall y) (\forall s) [\varphi_{g(e),s}(x) = y \implies f_i(x) = y].$$

Thus $\mathrm{Tot} \in \Sigma_2^A$. Hence, $\mathrm{Tot} \in \Delta_2^A$ so $\mathbf{0}'' \leq \mathbf{a}'$ by Theorem IV.4.1.

(v) $\implies$ (i). (The following resembles the proof that the recursive functions are not uniformly recursive.) Assume that **a** is r.e. but (i) is false and $f(e, x)$ is any function of degree $\leq$ **a**. We must construct a $\{0, 1\}$-valued recursive function $h \neq f_e$ for all e. Since $\deg(f) \leq \mathbf{0}'$ there is a recursive function $\hat{f}(e, x, s)$ such that $f(e, x) = \lim_s \hat{f}(e, x, s)$ and a modulus function $m(e, x)$ for $\hat{f}$ which has degree $\leq$ **a** by the Modulus Lemma III.3.2. Let $p(x) = \max\{ m(e, \langle e, x \rangle) : e \leq x \}$. Since $\deg(p) \leq$ **a** and (i) fails there is a recursive function $q(x)$ which $p(x)$ fails to dominate. Define $h(\langle e, x \rangle) = 1 \dot{-} \hat{f}(e, \langle e, x \rangle, q(x))$. Then h is a recursive function and $h(\langle e, x \rangle) \neq f_e(\langle e, x \rangle)$ whenever $x \geq e$ and $q(x) \geq p(x)$. (By Exercise 3.9 the hypothesis **a** r.e. is necessary in the proof of (v) $\implies$ (i).)   ▯

3.5 Corollary (Jockusch).   *If* $\mathbf{a} < \mathbf{0}'$ *is r.e. then the class of r.e. sets of degree* $\leq \mathbf{a}$ *is not* $\mathbf{a}$-*uniform.*

Proof. Assume $\mathbf{a}$ is a counterexample. Then the recursive sets are $\mathbf{a}$-subuniform so $\mathbf{a}' = \mathbf{0}''$ by (v) $\Longrightarrow$ (i) of Theorem 3.4. However, since the r.e. sets of degree $\leq \mathbf{a}$ are $\mathbf{a}$-uniform, they are $\mathbf{0}'$-uniform and so $\mathbf{a}'' = \mathbf{0}''$ by Corollary 3.2.   ∎

The next theorem continues the topic of Σ_3 index sets begun in Theorem 3.1, and yields the interesting Corollary 3.8 which relates to Exercises XI.4.2 and XI.4.3 about degrees of r.e. supersets of an r.e. set.

3.6 Theorem (Stob [1982b]).   *If* S *is a* Σ_3 *set such that*

$$(3.1) \qquad k \in S \implies W_k \text{ is not an infinite, coinfinite recursive set,}$$

then there is a simple set A *such that*

$$(3.2) \qquad [k \in S \ \& \ W_k \text{ nonrecursive}] \implies W_k \not\leq_T A.$$

3.7 Corollary (Kinber [1977, Theorem 1′]).   *If* T *is a* Π_2 *set and* ψ *a p.r. function such that if* $k \in T$ *then* $\psi(k)\!\downarrow$ *and* $W_{\psi(k)}$ *is nonrecursive, then there is a nonrecursive r.e. set* A *such that if* $k \in T$ *then* $W_{\psi(k)} \not\leq_T A$.

Proof. Apply Theorem 3.6 to $S = \{\,\psi(k) : k \in T\,\}$ which is Σ_3.   ∎

(By Theorem V.4.1, Post's simple set is complete regardless of which recursive enumeration $\{W_{e,s}\}_{e,s\in\omega}$ of the r.e. sets is used in the construction. In contrast Jockusch and Soare [1973] showed that the *hypersimple* set H constructed by Post [1944] may be complete or incomplete depending upon *which* enumeration $\{W_{e,s}\}_{e,s\in\omega}$ is used, and they asked whether H could have *any* nonzero r.e. degree for some enumeration. Kinber [1977] used Corollary 3.7 to negatively answer this question.)

3.8 Corollary (Stob [1982b]).   *If* B *is simple, then there is a nonrecursive r.e. set* A *such that if* $B \subseteq C \neq^* \omega$ *then* $C \not\leq_T A$. *Thus, a coinfinite r.e. set* B *is simple iff* B *does not have a superset of every r.e. degree.*

Proof. Let $S = \{\,e : B \subseteq W_e\,\}$. Then S is a Π_2 set satisfying the hypothesis of Corollary 3.7 since B is simple iff no infinite, coinfinite superset is recursive. Hence, 3.7 gives an r.e. set A such that if $B \subseteq W_e$ and W_e is nonrecursive then $W_e \not\leq_T A$.   ∎

Proof of Theorem 3.6. Fix $S \in \Sigma_3$ satisfying (3.1). Since Inf $\in \Pi_2$ we may assume $S \subseteq$ Inf by replacing if necessary S by $S \cap$ Inf. By Corollary IV.3.7 we fix a u.r.e. sequence $\{V_k\}_{k\in\omega}$ such that for all k,

$$(3.3) \qquad\qquad k \in S \implies (\exists e)\,(\forall i \geq e)\,[V_k^{[i]} = \omega],$$

and

$$(3.4) \qquad\qquad k \notin S \implies (\forall e)\,[V_k^{[e]} \text{ is finite}].$$

It suffices to construct A coinfinite and meeting for all e and k the requirements:

$$P_e : W_e \text{ infinite} \implies A \cap W_e \neq \emptyset;$$

$$N_{\langle e,k \rangle} : [V_k^{[e]} \text{ infinite and } W_k \text{ nonrecursive}] \implies \Phi_e(A) \neq W_k.$$

To see that the requirements $N_{\langle e,k\rangle}$ guarantee (3.2) note that if $k \in S$ and W_k is nonrecursive then for almost every e, $V_k^{[e]} = \omega$ by (3.3), and hence $\Phi_e(A) \neq W_k$ by $N_{\langle e,k\rangle}$. Hence, $W_k \not\leq_T A$ since there are infinitely many indices for every Turing reduction.

We meet $N_{\langle e,k\rangle}$ using the preservation strategy of Theorem VII.3.1 but we must take care to avoid too much restraint if W_k is cofinite. Define the recursive functions:

$$l(\langle e, k \rangle, s) = \max\{\, x : (\forall y < x)\,[\Phi_{e,s}(A_s; x) = W_{k,s}(x)]\,\};$$

$$m(\langle e, k \rangle, s) = \begin{cases} l(\langle e, k \rangle, s) & \text{if } m(\langle e, k \rangle, s - 1) < \\ & \qquad \min\{\, |V_{k,s}^{[e]}|, |\overline{W}_{k,s} \upharpoonright l(\langle e, k \rangle, s)|\,\} \\ m(\langle e, k \rangle, s - 1) & \text{otherwise.} \end{cases}$$

$$r(\langle e, k \rangle, s) = \max\{\, u(A_s; e, x, s) : x \leq m(\langle e, k \rangle, s)\,\}$$

Note that l is the usual "length of agreement" function, but that m is modified to increase only as $|V_k^{[e]}|$ and $|\overline{W}_k|$ increase. Note also that if the first clause in the definition of m holds then $m(\langle e, k \rangle, s - 1) < l(\langle e, k \rangle, s)$.

Stage $s + 1$. Choose the least i such that $W_{i,s} \cap A_s = \emptyset$, and

$$(\exists x)\,[x \in W_{i,s} \;\&\; 2i < x \;\&\; (\forall j < i)\,[r(j,s) < x]].$$

Enumerate the least such x into A, and say that P_i *receives attention* at stage $s + 1$.

Lemma. *For each pair $\langle e, k \rangle$,*

   (i) $N_{\langle e,k\rangle}$ *is satisfied;*
   (ii) $\lim_s m(\langle e, k \rangle, s) < \infty$;
   (iii) $\lim_s r(\langle e, k \rangle, s) < \infty$.

Proof. Fix $\langle e, k \rangle$. Choose s_0 so that no P_i, $i \le \langle e, k \rangle$ receives attention at any stage $s \ge s_0$. Note that for $s \ge s_0$, no $x \le r(\langle e, k \rangle, s)$ enters A at stage $s + 1$, so if $\Phi_{e,s}(A_s; x) = y$ and $x \le m(\langle e, k \rangle, s)$ then for all $t \ge s$, $\Phi_{e,t}(A_t; x) = y$ and $u(A_t; e, x, t) = u(A_s; e, x, s)$. Thus, (ii) implies (iii) because if $m = \lim_s m(\langle e, k \rangle, s)$ then

$$\lim_s r(\langle e, k \rangle, s) = \max\{ u(A; e, x) : x \le m \}.$$

Hence, it suffices to prove (i) and (ii). If $V_k^{[e]}$ is finite then (i) and (ii) are obvious by the definitions. Thus, we may assume the $V_k^{[e]}$ is infinite so $k \in S$ by (3.3) and (3.4).

Case 1. W_k is nonrecursive.

Then $|\overline{W}_k| = \infty$ so if $N_{\langle e, k \rangle}$ is not met and $\Phi_e(A) = W_k$ then $\lim_s l(\langle e, k \rangle, s) = \infty$, $\lim_s m(\langle e, k \rangle, s) = \infty$, and we can show that W_k is recursive contrary to hypothesis. (To recursively compute whether $x \in W_k$, find $s \ge s_0$ such that $x < m(\langle e, k \rangle, s)$, and $\Phi_{e,s}(A_s; e, x, s) \downarrow = y$. Now $x \in W_k$ iff $y = 1$.) Hence, (i) and therefore (ii) both hold.

Case 2. W_k is cofinite.

Note that (i) holds immediately. Assume (ii) fails. Then

$$\lim_s m(\langle e, k \rangle, s) = \infty.$$

Choose $t > s_0$ such that $m(\langle e, k \rangle, t) > m(\langle e, k \rangle, t - 1) > |\overline{W}_k|$. Let $m = m(\langle e, k \rangle, t)$. Then $m = l(\langle e, k \rangle, t)$, and hence

$$(\forall x < m)\, [\Phi_{e,t}(A_t; x) = W_{k,t}(x)].$$

Since $t > s_0$, each of the computations $\Phi_{e,t}(A_t; x)$ is permanent. But $m > |\overline{W}_k|$ so there must exist $x < m$, $x \notin W_{k,t}$ but $x \in W_{k,v}$ for some $v > t$. Hence,

$$0 = \Phi_{e,v}(A_v; x) \ne W_{k,v}(x) = 1,$$

so this disagreement will prevent either $l(\langle e, k \rangle, s)$ or $m(\langle e, k \rangle, s)$ from increasing at any stage $s > v$.

This completes the proof of the lemma. Each positive requirement P_i is satisfied by the construction and part (iii) of the lemma. Hence the theorem is proved. ∎

3.9–3.10 Exercises

3.9 (Jockusch). Show that the hypothesis **a** r.e. in the proof of (v) $\Longrightarrow$ (i) of Theorem 3.4 was necessary by proving that there is a degree **a** such that

$\mathbf{a}' = \mathbf{0}'$ and the recursive sets are $\mathbf{a}$-subuniform. *Hint.* Apply the Low Basis Theorem VI.5.12 to the Π_1^0 class $\mathcal{C} \subseteq 2^{<\omega}$ defined by

$$f \in \mathcal{C} \iff \text{rng } f \subseteq \{0,1\}$$
$$\&\ (\forall e)\,(\forall x)\,[\varphi_e(x)\!\downarrow\ \implies\ f(\langle e, x\rangle) = \min\{\,1, \varphi_e(x)\,\}],$$

to obtain some $f \in \mathcal{C}$ of low degree.

3.10 (Jockusch [1972a]).   Show that if $\mathbf{a}$ and $\mathbf{b}$ are r.e. degrees, $\mathbf{b} \leq \mathbf{a}$, and $\mathbf{b} < \mathbf{0}'$ then the following three statements are equivalent:

   (i) the r.e. sets of degree $\leq \mathbf{b}$ are $\mathbf{a}$-subuniform;
   (ii) $\mathbf{b}'' = \mathbf{a}' = \mathbf{0}''$;
   (iii) there is an r.e. sequence of r.e. sets which is uniformly of degree $\leq \mathbf{a}$ and consists exactly of the r.e. sets of degree $\leq \mathbf{b}$.

4. Classifying the Index Sets of the $High_n$, Low_n, and Intermediate R.E. Sets

4.1 Definition. (i) Let $\emptyset^{(\omega+1)} = (\emptyset^{(\omega)})'$, the jump of $\emptyset^{(\omega)}$.

   (ii) A set A is $\Sigma_{\omega+1}$ $(\Pi_{\omega+1})$ if $A \in \Sigma_1^{\emptyset^{(\omega)}}$ $(\Pi_1^{\emptyset^{(\omega)}})$.

   (iii) A $\Sigma_{\omega+1}$ $(\Pi_{\omega+1})$ set A is $\Sigma_{\omega+1}$-*complete* $(\Pi_{\omega+1}$-*complete*) if $\emptyset^{(\omega+1)} \leq_1 A$ (respectively $\overline{\emptyset^{(\omega+1)}} \leq_1 A$). (By the same remarks as in Definition IV.2.1 and Exercises IV.2.6 and IV.2.7, it makes no difference whether we use "$\leq_m$" or "$\leq_1$" in the definition of $\Sigma_{\omega+1}$-complete and $\Pi_{\omega+1}$-complete.)

4.2 Definition. (i) $High_n = \{\,x : W_x \text{ is } high_n\,\}$,
   (ii) $Low_n = \{\,x : W_x \text{ is } low_n\,\}$,
   (iii) $Low_{<\omega} = L_{<\omega} = \bigcup_n Low_n$ and $High_{<\omega} = H_{<\omega} = \bigcup_n High_n$,
   (iv) $Int = \{\,x : W_x \text{ is intermediate}\,\} = \omega - (L_{<\omega} \cup H_{<\omega})$.

Since $A^{(0)} = A$, it is clear that $Low_0 = Rec$, and $High_0 = Comp = \{\,x : W_x \equiv_T \emptyset'\,\}$, as also defined in Definition IV.3.3. We have seen in Theorem IV.3.4 that $(\Sigma_3, \Pi_3) \leq_1 (Rec, Comp) = (Low_0, High_0)$ and in Corollary 1.7 that $Comp = High_0$ is Σ_4-complete. We now generalize these results by proving that for all $n \geq 0$, $High_n$ is Σ_{n+4}-complete, and

$$(\Sigma_{n+3}, \Pi_{n+3}) \leq_1 (Low_n, High_n).$$

We also prove that $L_{<\omega}$ and $H_{<\omega}$ are each $\Sigma_{\omega+1}$-complete and that Int is $\Pi_{\omega+1}$-complete. We begin by establishing some easy upper bounds.

   In particular, it follows that Low_1 is Σ_4-complete and $High_1$ is Σ_5-complete. These results are important because many times we are given an r.e. set

W_x and we would like to know whether W_x is low$_1$ since many constructions are easier if we are working over a low$_1$ r.e. set or degree. (See Theorem XI.3.2, Exercises XI.3.4 and XI.3.5, and Notes XI.3.7 for low$_1$, and Theorem XI.5.1 for low$_2$.) However, we usually need not only know that W_x is low$_1$ but also its *lowness index* (see Exercise 4.24).

4.3 Theorem. (i) Low$_n \in \Sigma_{n+3}$, *uniformly in n;*
　(ii) High$_n \in \Sigma_{n+4}$, *uniformly in n;*
　(iii) L$_{<\omega} \in \Sigma_{\omega+1}$ *and* H$_{<\omega} \in \Sigma_{\omega+1}$; *and*
　(iv) Int $\in \Pi_{\omega+1}$.

Proof. Note that (iii) follows from (i) and (ii), and (iv) follows from (iii), so it suffices to prove (i) and (ii).
　(i)

$$x \in \text{Low}_n \iff W_x^{(n)} \leq_T \emptyset^{(n)}$$
$$\iff (\exists e)\,(\forall y)\,[[y \in W_x^{(n)} \ \& \ \{e\}^{\emptyset^{(n)}}(y) = 1]$$
$$\vee \ [y \notin W_x^{(n)} \ \& \ \{e\}^{\emptyset^{(n)}}(y) = 0]],$$
$$\iff \exists\forall\,[[\Sigma_{n+1} \ \& \ \Sigma_{n+1}] \vee [\Pi_{n+1} \ \& \ \Sigma_{n+1}]],$$
$$\iff \exists[\Pi_{n+2} \ \& \ \Pi_{n+2}],$$
$$\iff \Sigma_{n+3}.$$

　(ii)

$$x \in \text{High}_n \iff \emptyset^{(n+1)} \leq_T W_x^{(n)},$$
$$\iff (\exists e)\,[\{e\}^{W_x^{(n)}} \text{ is total} \ \& \ \cdots],$$
$$\iff (\exists e)\,[\Pi_2^{W_x^{(n)}} \ \& \ \cdots],$$
$$\iff \exists[\Pi_{n+3} \ \& \ \cdots], \text{ because } W_x^{(n)} \leq_T \emptyset^{(n+1)},$$
$$\iff \Sigma_{n+4}. \ \square$$

4.4 Theorem (Schwarz [1982]). *For all $n \geq 0$, High$_n$ is Σ_{n+4}-complete, uniformly in n.*

Proof. We wish to define a recursive function $f(n,k)$ such that

$$\emptyset^{(n+4)} \leq_m \text{High}_n$$

via $\lambda k[f(n,k)]$. In Corollary 1.7 we showed that $H_0 = \text{Comp}$ is Σ_4-complete. By relativizing this proof to $\emptyset^{(n)}$ and using the relativized *s-m-n* Theorem we obtain a *recursive* function $g(n,k)$ such that $\text{Comp}^{\emptyset^{(n)}}$ is Σ_{n+4}-complete via $\lambda k[g(n,k)]$, namely for all n and k,

$$(4.1) \qquad k \in \emptyset^{(n+4)} \iff W_{g(n,k)}^{\emptyset^{(n)}} \equiv_T \emptyset^{(n+1)}.$$

Let $q(x)$ be the recursive function from the Sacks Jump Theorem satisfying line VIII (3.6). Hence, by Corollary VIII.3.6, with $Y = \emptyset$ we have for all x and n,

$$(4.2) \qquad \emptyset <_T W_{q^n(x)} <_T \emptyset' \ \& \ W^{(n)}_{q^n(x)} \equiv_T W^{\emptyset^{(n)}}_x \oplus \emptyset^{(n)}.$$

Define $f(n,x) = q^n(g(n,x))$. Hence, by (4.1) and (4.2) for all n and k,

$$k \in \emptyset^{(n+4)} \iff W^{(n)}_{f(n,k)} \equiv_T \emptyset^{(n+1)}.$$

Note that the proof establishes that for all $Y \subseteq \omega$ and $n, k \in \omega$,

$$k \in Y^{(n+4)} \iff \left(W^Y_{f(n,k)}\right)^{(n)} \equiv_T Y^{(n+1)}. \quad \Box$$

Schwarz [1982] then proved Corollary 4.7 by virtually the same proof as in Theorem 4.4 but with $(\Sigma_3, \Pi_3) \leq_1$ (Rec, Comp) used in the base case. Jockusch later used Schwarz's method to get a more general result, Corollary 4.10, on n-jump closed sets. The following theorem is slightly more general than Schwarz's Corollary 4.7 because in the case $k \in Y^{(n+3)}$ we replace the lower bound $Y^{(n)}$ of (4.6) by the lower bound $(W^Y_j)^{(n)}$ of (4.4).

4.5 Theorem (Jockusch).   *There is a 1:1 recursive function $f(n, j, k)$ such that for all sets $Y \subseteq \omega$ and integers n, j, k,*

$$(4.3) \qquad\qquad\qquad Y \leq_T W^Y_{f(n,j,k)},$$

$$(4.4) \qquad\qquad k \in Y^{(n+3)} \implies (W^Y_{f(n,j,k)})^{(n)} \equiv_T (W^Y_j)^{(n)},$$

and

$$(4.5) \qquad\qquad k \notin Y^{(n+3)} \implies (W^Y_{f(n,j,k)})^{(n)} \equiv_T Y^{(n+1)}.$$

Proof. By relativizing Theorem IV.3.4 to $Y^{(n)}$ we get a recursive function $g(n, k)$ such that for all n,

$$\left(\Sigma_3^{Y^{(n)}}, \Pi_3^{Y^{(n)}}\right) \leq_1 \left(\mathrm{Rec}^{Y^{(n)}}, \mathrm{Comp}^{Y^{(n)}}\right) = \left(\mathrm{Low}_0^{Y^{(n)}}, \mathrm{High}_0^{Y^{(n)}}\right)$$

via $\lambda k[g(n,k)]$. Hence, for all n and k,

$$(4.6) \qquad\qquad k \in Y^{(n+3)} \implies W^{Y^{(n)}}_{g(n,k)} \equiv_T Y^{(n)},$$

and

$$(4.7) \qquad\qquad k \notin Y^{(n+3)} \implies W^{Y^{(n)}}_{g(n,k)} \equiv_T Y^{(n+1)}.$$

Define the recursive function h by

$$W_{h(n,k,j)}^{Y^{(n)}} = (W_j^Y \oplus Y)^{(n)} \oplus W_{g(n,k)}^{Y^{(n)}} \geq_T Y^{(n)}.$$

Let $f(n,k,j) = q^n(h(n,k,j))$. Now (4.3), (4.4), and (4.5) follow from (4.6), (4.7), and Corollary VIII.3.6.   ▯

4.6 Corollary (Jockusch).   *For any r.e. set A there is a recursive function $g(n,k)$ such that for all n and k,*

$$k \in \emptyset^{(n+3)} \implies W_{g(n,k)}^{(n)} \equiv_T A^{(n)},$$

$$k \notin \emptyset^{(n+3)} \implies W_{g(n,k)}^{(n)} \equiv_T \emptyset^{(n+1)}.$$

Namely, $(\Sigma_{n+3}, \Pi_{n+3}) \leq_m (G^n(A), G^n(\emptyset'))$ *where we define*

$$G^n(X) = \{\, e : W_e^{(n)} \equiv_T X^{(n)} \,\}.$$

Proof. In Theorem 4.5 set $Y = \emptyset$, and $g(n,k) = f(n,j,k)$ for some j such that $W_j = A$.   ▯

4.7 Corollary (Schwarz [1982]).   *For all $n \geq 0$,*

$$(\Sigma_{n+3}, \Pi_{n+3}) \leq_1 (\text{Low}_n, \text{High}_n),$$

uniformly in n and hence Low_n is Σ_{n+3}-complete.

Proof. In Theorem 4.5 set $Y = \emptyset$ and $W_j^\emptyset = \emptyset$.   ▯

4.8 Corollary. (i) (Jockusch). $A =_{\text{dfn}} \{\, x : W_x \text{ is atomless} \,\}$ *is Π_5-complete.*
   (ii) Sim $=_{\text{dfn}} \{\, x : W_x \text{ is simple} \,\}$ *is Π_3-complete. (See Exercise V.1.8.)*
   (iii) HSim $=_{\text{dfn}} \{\, x : W_x \text{ is hypersimple} \,\}$ *is Π_3-complete. (See Exercise V.2.10.)*

Proof. (i) For each y let $W_{g(y)}$ be the Dekker deficiency set of W_y (as defined in Theorem V.2.5 but using the standard enumeration $\{\, W_{y,s} \,\}_{s \in \omega}$ from Exercise I.3.11 and omitting repetitions). By Corollary 4.7 with $n = 2$,

$$(\Sigma_5, \Pi_5) \leq_1 (\text{Low}_2, \text{High}_2)$$

via some recursive function h. Thus, $(\Sigma_5, \Pi_5) \leq_1 (\overline{A}, A)$ via $g \circ h$ by Theorem XI.4.1 and Corollary XI.5.2. Thus, $\Pi_5 \leq_1 A$. Furthermore, $A \in \Pi_5$, and hence A is Π_5-complete, because

$$x \in A \iff x \notin \text{Cof} \ \& \ \neg(\exists y) [y \in \text{Max} \ \& \ W_x \subseteq W_y],$$

but $\text{Max} \in \Pi_4$ by Exercise 4.21.
   (ii) and (iii). By Corollary 4.7 with $n = 0$, $(\Sigma_3, \Pi_3) \leq_1 (\text{Rec}, \text{Comp})$, via some h. Hence $(\Sigma_3, \Pi_3) \leq_1 (\text{Rec}, \text{HSim})$ via $g \circ h$. By the methods of Chapters IV and V it is easy to show that Sim, HSim $\in \Pi_3$.   ▯

4.9 Definition. (i) A set C is *n-jump closed* if whenever $x \in C$ and $W_x^{(n)} \equiv_T W_y^{(n)}$ then $y \in C$.

(ii) C is *ω-closed* if C is n-jump closed for every n.

4.10 Corollary (Jockusch). *Suppose $C \neq \emptyset$ and $C \neq \omega$.*

(i) *If C is n-jump closed, then either $\emptyset^{(n+3)} \leq_1 C$ or $\emptyset^{(n+3)} \leq_1 \overline{C}$.*
(ii) *If C is ω-closed then $\emptyset^{(\omega)} \leq_1 C$.*

Proof. To prove (i) assume $C \cap \mathrm{Comp} = \emptyset$, and fix $j \in C$. Then

$$(\Sigma_{n+3}, \Pi_{n+3}) \leq_1 (C, \overline{C})$$

by Corollary 4.6. Note that (ii) follows from the uniformity in (i). ▯

4.11 Corollary (Solovay). $\emptyset^{(\omega)} \leq_1 \mathrm{L}_{<\omega}$, $\emptyset^{(\omega)} \leq_1 \mathrm{H}_{<\omega}$, *and* $\emptyset^{(\omega)} \leq_1 \mathrm{Int}$. ▯

4.12 Theorem (Jockusch). *Theorem IV.3.4 is uniform in the sense that there exist recursive functions f and g such that $(\emptyset^{(3)}, \overline{\emptyset^{(3)}}) \leq_1 (\mathrm{Cof}, \mathrm{Comp})$ via f and for all x,*

$$x \notin \emptyset^{(3)} \implies \{g(x)\}^{W_{f(x)}} = \chi_K,$$

$$x \in \emptyset^{(3)} \implies W_{f(x)} \equiv_T \emptyset,$$

and

$$\mathrm{compat}(\chi_K, \{g(x)\}^{W_{f(x)}}),$$

where χ_K is the characteristic function of K, even though if $x \in \emptyset^{(3)}$ the function $\{g(x)\}^{W_{f(x)}}$ will only be partial, indeed with finite domain.

Proof. In the proof of Theorem IV.3.4, the procedure $\psi(y) = \{g(x)\}^{W_{f(x)}}(y)$ first finds the least stage $s(y)$ such that $b_{x,y}^{s(y)} = b_{x,y}$ and then gives output 1 or 0 according as $y \in K_{s(y)}$ or not. If $x \notin \emptyset^{(3)}$ then $s(y)$ exists for all y so ψ is total. If $x \in \emptyset^{(3)}$ then $\psi(y)$ is defined for only a finite initial segment of ω, but if $\psi(y)$ is defined then $\psi(y) = K(y)$ by the construction of $W_{f(x)}$. ▯

4.13 Theorem (Schwarz). (i) $(\Sigma_4, \Pi_4) \leq_1 (\mathrm{Low}_1, \mathrm{Max})$ *where*

$$\mathrm{Max} =_{\mathrm{dfn}} \{x : W_x \text{ is maximal}\}.$$

(ii) *Max is Π_4-complete.*
(iii) *For each of the other five classes C of types of simple sets in Chapter XI Diagram 1.1 (excluding the classes simple, hypersimple, and quasimaximal) $\{x : W_x \in C\}$ is Π_4-complete. This includes the classes: r-maximal, strongly hypersimple, finitely strongly hypersimple, hyperhypersimple, and dense simple.*

Proof. Parts (ii) and (iii) follow immediately from (i) and from Exercise 4.21 because for each such class C, $\deg(C) \subseteq \mathbf{H}_1$ by Exercise XI.1.12(b), so $\deg(C) \cap \mathbf{L}_1 = \emptyset$, and $\mathcal{M} \subseteq C$ where $\mathcal{M}$ is the class of maximal sets. Thus, for C the class of quasi-maximal sets we also have $\Pi_4 \leq_1 \{ x : W_x \in C \}$, but this class is not Π_4 but rather Σ_5-complete (see Lempp [1986]). The other classes mentioned are all in Π_4 form (see Exercise 4.21) except for simple and hypersimple which are Π_3-complete (see Exercises V.1.8 and V.2.10).

(i) (Jockusch). By Corollary 4.7 with $n = 1$, let

$$(\Sigma_4, \Pi_4) \leq_1 (\text{Low}_1, \text{High}_1)$$

via some recursive function $f(k)$. (The idea is now to apply Martin's reduction Theorem XI.2.3 from a high$_1$ r.e. set D to a maximal set $M \equiv_T D$ but first some uniformities are needed.) Relativize Theorem 4.12 to $\emptyset'$ and invoke Theorem III.1.5 to obtain recursive functions g and F such that for all k,

$$(4.8) \qquad k \notin \emptyset^{(4)} \implies \emptyset^{(2)} = \{ g(k) \}^{W_{F(k)}^{\emptyset'}},$$

$$(4.9) \qquad k \in \emptyset^{(4)} \implies W_{F(k)}^{\emptyset'} \equiv_T \emptyset',$$

and

$$(4.10) \qquad \text{compat}(\emptyset^{(2)}, \{ g(k) \}^{W_{F(k)}^{\emptyset'}}).$$

Define the recursive function $f(k) = q(F(k))$ where q satisfies Corollary VIII.3.6. Hence, $W_{F(k)}^{\emptyset'} \equiv_T W_{f(k)}'$, and the Turing reductions are uniform in k, so we may replace $W_{F(k)}^{\emptyset'}$ by $W_{f(k)}'$ in (4.8)–(4.10). It suffices to obtain a function h_k such that for all k,

$$(4.11) \qquad h_k \text{ is total,}$$

$$(4.12) \qquad h_k \leq_T W_{f(k)}, \text{ uniformly in } k,$$

and

$$(4.13) \qquad k \notin \emptyset^{(4)} \implies h_k \text{ is dominant,}$$

because then we simply apply Martin's construction in Theorems XI.2.1 and XI.2.3 with inputs $W_{f(k)}$ and h_k to obtain an r.e. set $A_k \equiv_T W_{f(k)}$ which is maximal if h_k is dominant. Fix k and define $h_k(x)$ as follows. From (4.8)–(4.10), the Limit Lemma III.3.3 applied to $\{ g(k) \}^{W_{f(k)}'(x)}$, and the fact that $\emptyset^{(2)} \equiv_T \text{Tot}$, there is a total function $m(k, x, s)$ recursive in $W_{f(k)}$ such that

$$(4.14) \qquad k \notin \emptyset^{(4)} \implies (\forall x) [\text{Tot}(x) = \lim_s m(k, x, s)],$$

(4.15) $[k \in \emptyset^{(4)}$ & $\psi_k(x) = \lim_s m(k, x, s)] \implies \mathrm{compat}(\psi_k(x), \mathrm{Tot}(x))$,

even though $\psi_k(x)$ may not be total if $k \in \emptyset^{(4)}$. Now obtain h_k as in the proof of Theorem XI.1.3 ($\implies$) with A, $g(x, s)$ and f replaced by $W_{f(k)}$, $m(k, x, s)$ and h_k. Note that (4.15) ensures that in XI.1.3 the value $t(e)$ exists even if $k \in \emptyset^{(4)}$. Hence, h_k is total for all k, whether $k \in \emptyset^{(4)}$ or not.   ▯

(Note that in the proof of Theorem 4.13 we need h_k total in order to apply the construction of Theorem XI.2.3. If not suppose that $h_k(x_0)\uparrow$. Then for each $x \geq x_0$ we may have $m(k, x, s) \neq m(k, x, s + 1)$ for infinitely many s. But at each such stage $s + 1$, x is permitted to enter $W_{f(k)}$ according to the construction in Theorem XI.2.3 so we might get $W_{f(k)} \equiv_T \emptyset'$ even if $k \in \emptyset^{(4)}$. This is why we are so careful to use (4.10) and Theorem 4.12.)

Solovay answered Schwarz's question about the classification of Int with the following theorem which improves Corollary 4.11. The proof resembles Sacks's proof of Corollary VIII.3.5 that Int $\neq \emptyset$.

4.14 Theorem (Solovay).   Int *is* $\Pi_{\omega+1}$*-complete, and* $\mathrm{H}_{<\omega}$ *and* $\mathrm{L}_{<\omega}$ *are each* $\Sigma_{\omega+1}$*-complete.*

Proof. By Theorem 4.3, Int $\in \Pi_{\omega+1}$ and $\mathrm{L}_{<\omega}, \mathrm{H}_{<\omega} \in \Sigma_{\omega+1}$. Let S be any $\Sigma_{\omega+1}$-complete set such as $\emptyset^{(\omega+1)}$. In Exercises 4.15 and 4.16 we establish the following two facts which we need,

(4.16)   ($\exists$ recursive function h) ($\forall y$) $[y \in S \iff (\exists n) [h(y, n) \in \emptyset^{(n)}]]$,

(4.17) $(\exists i_0) (\forall n) [\{i_0\}^{\emptyset^{(n)}}(0) = n]$.

Given x, y, and Y define a recursive function $g(x, y)$ by enumerating $W^Y_{g(x,y)}$ as follows.

Step 1. Attempt to compute $\{i_0\}^Y(0)$. Let n be the output (if any). (Thus, if $Y = \emptyset^{(j)}$ then $n = j$.) If there is no output let $W^Y_{g(x,y)} = \emptyset$. Otherwise, do the following.

Step 2. See whether $h(y, n) \in Y$.

   Case 1. If so, define $W^Y_{g(x,y)} = Y'$.

   Case 2. If not, define $W^Y_{g(x,y)} = W^Y_{q(x)}$, for $q(x)$ as in Corollary VIII.3.6.

Now by the Relativized Recursion Theorem III.1.6 with parameters there is a recursive function $p(y)$ such that

(4.18) $W^Y_{p(y)} = W^Y_{g(p(y),y)}$.

We claim that for all y,

(4.19) $y \in S \implies (\exists n) [(W^\emptyset_{p(y)})^{(n)} \equiv_T \emptyset^{(n+1)}]$,

and

(4.20) $\qquad\qquad y \notin S \implies W_{p(y)}^{\emptyset}$ is intermediate.

For every j, define $\mathbf{d}_j = \deg(W_{p(y)}^{\emptyset^{(j)}})$. If $y \in S$ choose the least n such that $h(y, n) \in \emptyset^{(n)}$. Now for all $j \leq n$, $(W_{p(y)}^{\emptyset})^{(j)} \equiv_T W_{p(y)}^{\emptyset^{(j)}}$. Hence,

(4.21) $\qquad\qquad (\forall j < n) \, [\mathbf{0}^{(j)} < \mathbf{d}_j < \mathbf{0}^{(j+1)} \ \& \ \mathbf{d}_j' = \mathbf{d}_{j+1}]$

by Step 2 Case 2 and the properties of q in Corollary VIII.3.6. But for $j = n$, $\mathbf{d}_n = \mathbf{0}^{(n+1)}$, by Step 2 Case 1, and thus $\mathbf{d}_0^{(n)} = \mathbf{0}^{(n+1)}$ by (4.21). Hence $\mathbf{d}_0 \in \mathbf{H}_n$. If $y \notin S$, then $(\forall n) \, [h(y, n) \notin \emptyset^{(n)}]$, so (4.21) holds for all n, and thus $W_{p(y)}^{\emptyset}$ is intermediate. Note that (4.19) and (4.20) establish that $(\Sigma_{\omega+1}, \Pi_{\omega+1}) \leq_1 (\mathrm{H}_{<\omega}, \mathrm{Int})$ and hence that $\mathrm{H}_{<\omega}$ is $\Sigma_{\omega+1}$-complete and Int is $\Pi_{\omega+1}$-complete.

To prove that $\mathrm{L}_{<\omega}$ is $\Sigma_{\omega+1}$-complete we set $W_{g(x,y)} = Y$ in Step 2 Case 1. The rest of the proof is the same and establishes that

$$(\Sigma_{\omega+1}, \Pi_{\omega+1}) \leq_1 (\mathrm{L}_{<\omega}, \mathrm{Int}). \qquad \square$$

4.15–4.26 Exercises

4.15. Prove line (4.16), by first proving the following facts.
(i) There is a recursive function $f(u, v)$ such that for all u and v,

$$[D_u \subseteq \emptyset^{(\omega)} \ \& \ D_v \cap \emptyset^{(\omega)} = \emptyset] \iff f(u, v) \in \emptyset^{(\omega)}.$$

(ii) There is a recursive function $r(i, j, x)$ such that

$$(\forall x) \, (\forall i, j)_{i<j} [x \in \emptyset^{(i)} \iff r(i, j, x) \in \emptyset^{(j)}].$$

Hint. By the definition of $\emptyset^{(\omega+1)}$ we have $y \in \emptyset^{(\omega+1)} \iff \{y\}^{\emptyset^{(\omega)}}(y) \downarrow$. Hence, by (i) there is a recursive function $g(y)$ such that

(4.22) $\qquad y \in \emptyset^{(\omega+1)} \iff (\exists \langle a, b \rangle) \, [\langle a, b \rangle \in W_{g(y)} \ \& \ a \in \emptyset^{(b)}].$

Use (ii) and $W_{g(y)}$ to define a recursive function $k(y)$ such that:

$$W_{k(y)} = \{ \langle u_1^y, v_1^y \rangle, \langle u_2^y, v_2^y \rangle, \dots \} \text{ is infinite;}$$

$v_j^y < v_{j+1}^y$ for all j; and (4.22) holds with $k(y)$ in place of $g(y)$. Let $\{e_0\}^Y(z)$ be undefined for all Y and z, so that $e_0 \notin \emptyset^{(n)}$ for any n. Define $h(y, n) = u_j^y$ if $n = v_j^y$ for some j and $= e_0$ otherwise. (See also the similar proof of Lemma 1 in Theorem 6.2.)

4.16. (a) Prove line (4.17). *Hint.* Define $\{e_1\}^Y = \lambda z[0]$. For $n \geq 1$ define $\{e_{n+1}\}^Y = \lambda z[0]$ if $e_n \in Y$, and $\lambda z[\uparrow]$ otherwise. Note that $e_n \in \emptyset^{(m)}$ exactly for $1 \leq n \leq m$. Note that the function $g(n) = e_n$ is recursive by the *s-m-n* Theorem. To compute $\{i_0\}^Y(0)$ search for the least $n > 0$ such that $e_n \notin Y$ and give output $n - 1$.

(b) Prove

$$(4.23) \qquad (\exists i_0)\,(\exists e_1)\,(\forall n)\,(\forall Y)\,[\{i_0\}^{(Y-\{e_1\})^{(n)}}(0) = n].$$

Hint. Define e_1 as in part (a). The proof for $Z = Y - \{e_1\}$ is the same as for $Z = \emptyset$ in (a).

4.17 (Jockusch). (a) Prove that the class $\mathcal{C}$ of ω-closed sets C is a $\Pi_1^{0,\emptyset^{(\omega)}}$ class (as defined in Definition VI.5.11 and relativized to $\emptyset^{(\omega)}$).

(b) Apply the Low Basis Theorem VI.5.13 and Corollary 4.10(ii) to conclude that $\mathcal{C}$ has a nontrivial member A low over $\emptyset^{(\omega)}$ (i.e., $A' \equiv_T \emptyset^{(\omega+1)}$).

(c) Show that unlike Theorem 4.14 the conclusion of Corollary 4.10(ii) cannot be strengthened to read "$\emptyset^{(\omega+1)} \leq_T C$."

4.18 (Jockusch). A disjoint pair (A_0, A_1) of $\emptyset^{(\omega)}$-r.e. sets is $\emptyset^{(\omega)}$-*effectively inseparable* if the definition in Exercise II.4.13 holds with $\emptyset^{(\omega)}$-r.e. and $\emptyset^{(\omega)}$-recursive in place of r.e. and recursive, respectively. Let $\mathcal{D}$ be the class of sets D which separate some such fixed disjoint pair. Prove that if C is any nontrivial ω-closed set then $(\exists D \in \mathcal{D})\,[D \leq_m C]$. *Hint.* Let A_0 and A_1 be disjoint, r.e. in $\emptyset^{(\omega)}$, and $\emptyset^{(\omega)}$-effectively inseparable. For $i = 0, 1$ choose a recursive function h_i which satisfies (4.16) with A_i in place of S. Let C be ω-closed and nontrivial. Choose $e_0 \in C$ and $e_1 \notin C$. Modify the definition in Step 2 Case 1 of Theorem 4.14 as follows. Define

$$W_{g(x,y)}^Y = \begin{cases} W_{z(i,n)}^Y & \text{if } [h_i(y,n) \in Y \ \& \ h_{1-i}(y,n) \notin Y], \\ W_{q(x)}^Y & \text{otherwise,} \end{cases}$$

where $z(i, n)$ is chosen so that $W_{z(i,n)}^{\emptyset^{(n)}} = W_{e_i}^{(n)}$. Proceed as in the proof of Theorem 4.14.

4.19 (Jockusch). For $A \subseteq \omega$ define the ω-*closure* of A,

$$\mathrm{Cl}_\omega(A) = \{e : (\exists i \in A)\,(\exists n)\,[W_i^{(n)} \equiv_T W_e^{(n)}]\}.$$

Note that $\mathrm{Cl}_\omega(A)$ is the smallest ω-closed set containing A.

(a) Prove that if A is r.e. in $\emptyset^{(\omega)}$, then $\mathrm{Cl}_\omega(A)$ is also r.e. in $\emptyset^{(\omega)}$.

(b) Prove that if A is r.e. in $\emptyset^{(\omega)}$ and $\mathrm{Cl}_\omega(A)$ is neither $\emptyset$ nor ω then $\mathrm{Cl}_\omega(A) \equiv_T \emptyset^{(\omega+1)}$. *Hint.* Use (a) and Exercise 4.16 to show that $\mathrm{Cl}_\omega(A)$ is $\emptyset^{(\omega+1)}$-complete.

4.20 (Jockusch).   (a) Prove that if $C \subseteq 2^\omega$ is any nonempty Π_1^0 class and D separates a pair of effectively inseparable sets then $(\exists C \in \mathcal{C})\, [C \leq_T D]$. *Hint.* Fix a recursive tree T as in Definition VI.5.11 such that $[T] = C$. Let $T_\sigma = \{ \tau \in T : \sigma \subseteq \tau \}$. Show that one can D-recursively compute a function $f(\sigma)$ such that if $[T_\sigma] \neq \emptyset$ then $[T_{\sigma^\frown \langle f(\sigma) \rangle}] \neq \emptyset$ also.

   (b) Relativize (a) to $\emptyset^{(\omega)}$ and combine with Exercise 4.18 to prove that for $\mathcal{C}$ in Exercise 4.17 and $\mathcal{D}$ in Exercise 4.16 the upward closure of $\deg(\mathcal{C})$ equals the upward closure of $\deg(\mathcal{D})$.

4.21.   (a) Prove that Max $\in \Pi_4$.

   (b) For each of the remaining five classes $\mathcal{C}$ of simple sets shown in Diagram 1.1 of Chapter XI, excluding the simple, hypersimple, and quasi-maximal sets, prove that $\{ x : W_x \in \mathcal{C} \} \in \Pi_4$. This includes the classes: r-maximal, strongly hypersimple, finitely strongly hypersimple, hyperhyper-simple, and dense simple.

4.22 (Schwarz).   Prove that $(\Sigma_4, \Pi_4) \leq_1 (\text{Low}_1, \text{NSemilow})$ where

$$\text{NSemilow} = \{ x : \overline{W}_x \text{ is not semi-low} \}.$$

Hint. Use Corollary 4.7 with $n = 1$, and Exercise IV.4.10. (This result also follows from Theorem 4.13 because if A is maximal then $\overline{A}$ is not semi-low.)

4.23.   Give a direct proof that Max is Π_4-complete without using Theorem 4.13. *Hint.* (D. A. Martin). Let S be a Π_4-complete set. Then there is a recursive function f such that $k \in S$ iff $(\forall i)\, (\exists j)\, W_{f(k,i,j)}$ is infinite. We may assume that $W_{f(k,i,0)}$ is finite and nonempty for all k and i, and that $W_{f(k,i,j)} \subseteq W_{f(k,i,j')}$ for $j' \geq j$. Construct an r.e. set A_k such that if $k \in S$ then A_k is maximal and otherwise A_k is not finitely strongly hypersimple. Fix k from now on, and drop the subscript k from A_k.

   Use the maximal set construction of Theorem X.3.3, where $\overline{A}_s = \{ a_0^s < a_1^s < \cdots \}$ except that as in Theorem X.7.2 $\lim_s a_n^s$ may not exist for every n. At stage $s = 0$, set $a_n^0 = 2n$. An element x is *fresh* at stage $s+1$ if $x \notin A_s \cup \{ a_n^s : n \in \omega \}$.

   Stage $s + 1 = 2t + 1$. If $W_{f(k,i,j),s+1} \neq W_{f(k,i,j),s}$, enumerate $a_{\langle i,j \rangle}^s$ into A_{s+1}. This is like the "kicking" of Theorem X.7.2 Step 2. (By Convention II.2.6 there is at most one such $\langle i, j \rangle$.) Let $a_{\langle i,j \rangle}^{s+1}$ be the least fresh integer and $a_p^{s+1} = a_p^s$ if $p \neq \langle i, j \rangle$.

   Stage $s + 1 = 2t + 2$. Find the least $n = \langle i, j \rangle$ such that some $m > n$, $m = \langle i', j' \rangle$ satisfies:

   (a) $|W_{f(k,i,j),s}| \leq m$,
   (b) $\sigma(i, a_n^s, s) < \sigma(i, a_m^s, s)$,
   (c) $\sigma(i, a_m^s, s - 1) \neq \sigma(i, a_m^s, s)$ or a_m^s was fresh at stage s.

If n exists enumerate a_n^s into A_{s+1}, choose the least m corresponding to n, let $a_n^{s+1} = a_m^s$, and let a_m^{s+1} be the least fresh integer. Let $a_p^{s+1} = a_p^s$ for all $p \neq n$, and for all p if n fails to exist. (Condition (c) is like condition X (7.6) of Theorem X.7.2 to ensure that if $\lim_s a_n^s$ fails to exist then there is no $m > n$ such that infinitely often the element a_m^s under marker m is pulled to the position of marker n.)

Prove in order the following facts:

(1) $a_{\langle i,j \rangle} = \lim_s a_{\langle i,j \rangle}^s$ exists iff $W_{f(k,i,j)}$ is finite;

(2) $(\forall i)\, [a_{\langle i,0 \rangle}$ exists$]$, and hence $|\overline{A}_k| = \infty$ whether $k \in S$ or not;

(3) $(\forall i)\,(\exists \sigma_0)$ (a.e. n) $[[n = \langle i', j' \rangle$ & $i \leq i'$
$$\implies \sigma(i, a_n) =_{\mathrm{dfn}} \lim_s \sigma(i, a_n, s) = \sigma_0]];$$

(4) $k \in S \implies A_k$ is maximal;

(5) $k \notin S \implies A_k$ is not finitely strongly hypersimple.

For (5) choose the least i such that $(\forall j)\, W_{f(k,i,j)}$ is finite. Let σ_0 satisfy (3). For every j define

$$V_j = \{\, x : (\exists s)\, [x = a_{\langle i,j \rangle}^s = a_{\langle i,j \rangle}^{s+1} \ \& \ \sigma(i, x, s) = \sigma_0]\,\}.$$

Show that the sets V_j are finite, disjoint, meet $\overline{A}_k$ and are uniformly recursive.

4.24. If an r.e. set A is low_1, e is a *lowness index* for A if $A' = \{e\}^{\emptyset'}$. For each $n \geq 1$ define
$$P_n = \{\, \langle x, e \rangle : W_x^{(n)} = \{e\}^{\emptyset^{(n)}} \,\}.$$
Show that P_n is Π_{n+2}-complete. Hence, in particular P_1 is Π_3-complete, thereby saving one quantifier over the Σ_4 definition of Low_1. *Hint.* To prove that $P_n \in \Pi_{n+2}$ note that

$$\langle x, e \rangle \in P_n \iff \{e\}^{\emptyset^{(n)}} \text{ is total}$$
$$\& \ (\forall y)\, [[y \in W_x^{(n)} \ \& \ \{e\}^{\emptyset^{(n)}}(y) = 1]$$
$$\vee\ [y \notin W_x^{(n)} \ \& \ \{e\}^{\emptyset^{(n)}}(y) = 0]],$$
$$\iff [\Pi_2^{\emptyset^{(n)}} \ \& \ (\forall y)\, [[\Sigma_{n+1} \ \& \ \Sigma_{n+1}] \vee [\Pi_{n+1} \ \& \ \Sigma_{n+1}]]].$$

4.25. If an r.e. set A is high_1, e is a *highness index* for A if $\emptyset'' = \{e\}^{A'}$. Let

$$Q_n = \{\, \langle x, e \rangle : \emptyset^{(n+1)} = \{e\}^{W_x^{(n)}} \,\}.$$

Show that Q_n is Π_{n+3}-complete. *Hint.* To see that $Q_n \in \Pi_{n+3}$ note that

$$\langle x, e \rangle \in Q_n \iff \{e\}^{W_x^{(n)}} \text{ is total } \& \ (\forall y)\, [\cdots];$$
$$\iff \Pi_2^{W_x^{(n)}} \ \& \ \forall y [\cdots];$$
$$\iff [\Pi_{n+3} \ \& \ \Pi_{n+3}], \text{ because } W_x^{(n)} \leq_T \emptyset^{(n+1)}.$$

4.26. Prove that $(\Sigma_4, \Pi_4) \leq_1 (\text{Low}_1, Q_1) \leq_1 (\text{Low}_1, \text{Max})$, for Q_1 defined in Exercise 4.25.

4.27 Notes. Schwarz [1982] and [ta] proved Theorem 4.4 and then Corollary 4.7 with essentially the same proof, but with $(\Sigma_3, \Pi_3) \leq_1 (\text{Rec}, \text{Comp})$ as the base step. Schwarz raised the question of the classification of Int. After seeing Schwarz's results Solovay (private communication) proved Theorem 4.14. Independently, Jockusch (private communication) expanded Schwarz's method to obtain results 4.5, 4.6, 4.8, 4.10, 4.17, 4.18, 4.19 and 4.20. In the late 1960's Lachlan, Martin, R. W. Robinson and Yates each proved Corollary 4.13(ii) that Max is Π_4-complete (probably by the method of Exercise 4.23) although no proof was ever published. Tulloss [1971] wrote out a more complicated proof that Max is Π_4-complete, and he also obtained the upper bounds in Exercise 4.21. Tulloss's proof was simplified by D. A. Martin (private communication) to the version presented in Exercise 4.23. Schwarz [1982] used this method of 4.23 in a more complicated form to obtain Theorem 4.13 and Exercise 4.22. Jockusch then found the elegant proof here of Theorem 4.13(i) using Corollary 4.7 and Theorem XI.2.3. Let $\text{QMax} = \{ x : W_x \text{ is quasi-maximal} \}$. It is easy to show that $\text{QMax} \in \Sigma_5$ using Exercise 4.21(a). The question of whether QMax is Σ_5-complete had been open since the thesis of Tulloss [1971]. This has been proved by Lempp [1986]. Furthermore, Lempp extended the notions of maximal and quasi-maximal to α-atomic and α-quasi-atomic, and he proved the corresponding index sets to be $\Pi_{2\alpha+2}$-complete and $\Sigma_{2\alpha+3}$-complete for α any recursive ordinal (see Lempp [1986]).

5. Fixed Points up to Turing Equivalence

The Recursion Theorem II.3.2 asserted that every recursive function f has a fixed point n satisfying $W_n = W_{f(n)}$. In Corollary V.5.2 this was extended to every function f of r.e. degree $< \mathbf{0}'$. Moving up one level we saw in Exercise V.5.5 that every function $f \leq_T \emptyset'$ has a *-fixed point n satisfying $W_n =^* W_{f(n)}$. We now show that every function $f \leq_T \emptyset^{(2)}$ has a *Turing fixed point*, namely some n satisfying $W_n \equiv_T W_{f(n)}$. This is an immediate corollary of Theorem 5.1. Both Theorem 5.1 and Corollary 5.2 are generalized in Theorems 6.2 and 6.3, respectively.

5.1 Theorem (Arslanov).   *If ψ is a function partial recursive in $\emptyset^{(2)}$ then there is a recursive function g such that for all $e \in \text{dom } \psi$, $W_{g(e)} \equiv_T W_{\psi(e)}$.*

5.2 Corollary (Arslanov).   *If f is a total function recursive in $\emptyset^{(2)}$ then there is an integer n such that $W_n \equiv_T W_{f(n)}$.*

Proof of Corollary 5.2. By Theorem 5.1 and the Recursion Theorem II.3.2. ▯

Proof of Theorem 5.1. Since ψ is p.r. in $\emptyset^{(2)}$ we can apply the Limit Lemma III.3.3 twice to obtain a total recursive function $\hat{f}$ such that

$$\psi(e) = \lim_y \lim_s \hat{f}(e, y, s)$$

for all $e \in \mathrm{dom}\ \psi$, and such that $\lim_s \hat{f}(e, s, y)$ exists for all e and y. For each e, we shall construct uniformly in e an r.e. set A_e, and hence there is a recursive function g such that $W_{g(e)} = A_e$. Furthermore, we shall arrange that $A_e \equiv_T W_{\psi(e)}$ for all $e \in \mathrm{dom}\ \psi$. The reductions between A_e and $W_{\psi(e)}$, however, will not be uniform in e.

We define $A_e = \oplus_y A_e^y$ where the yth "row" $A_e^y = \bigcup_s B_{e,y,s}$, for $B_{e,y,s}$ constructed as follows.

Stage $s = 0$. For each e and y, set $B_{e,y,0} = \emptyset$.

Stage $s > 0$. Let $B_{e,y,s}$ contain all elements x satisfying:

(i) $x \in W_{\hat{f}(e,y,s),s}$;
(ii) (vertical condition) $(\forall j \leq x)\ [\hat{f}(e, y, s) = \hat{f}(e, y + j, s)]$; and
(iii) (horizontal condition) $(\forall t)_{x \leq t \leq s}[\hat{f}(e, y, t) = \hat{f}(e, y, t + 1)]$.

This ends the construction. (Note that $B_{e,y,s}$ is finite because

$$B_{e,y,s} \subseteq W_{\hat{f}(e,y,s),s}$$

by (i).)

Fix $e \in \mathrm{dom}\ \psi$. Define y_0 and s_0 by

$$y_0 = (\mu y)\ (\forall z \geq y)\ [\lim_s \hat{f}(e, z, s) = \psi(e)],$$

and

$$s_0 = (\mu s)\ (\forall t \geq s)\ [\hat{f}(e, y_0, t) = \psi(e)].$$

(We think of all rows A_e^y, $y \geq y_0$, as *good* and rows A_e^y, $y < y_0$, as *bad*. We shall see that $A_e^y =^* W_{\psi(e)}$ for $y = y_0$ (and indeed for every good row A^y) while condition (ii) ensures that $A_e^y =^* \emptyset$ for each bad row. Condition (iii) ensures that $A_e \leq_T W_{\psi(e)}$.)

Lemma 1. $(\forall y < y_0)\ [A_e^y$ is finite$]$.

Proof. Fix $y < y_0$. By definition of y_0, we have some y_1, $y < y_1 \leq y_0$, such that $\lim_s \hat{f}(e, y_1, s) \neq \lim_s \hat{f}(e, y, s)$. Choose s such that $\hat{f}(e, z, t) = \lim_s \hat{f}(e, z, s)$ for all $t \geq s$ and $z \leq y_1$. Let $k = y_1 - y$. Choose any $t \geq s$. Now any $x \geq k$ fails by (ii) to enter $B_{e,y,t}$ so $B_{e,y,t} \subseteq [0, k)$, and hence A_e^y is finite since $A_e^y \subseteq \bigcup_{t < s} B_{e,y,t} \cup [0, k)$.

Lemma 2. $A_e^{y_0} =^* W_{\psi(e)}$ (and hence $W_{\psi(e)} \leq_T A_e$).

Proof. First $A_e^{y_0} \subseteq^* W_{\psi(e)}$ since for every $s \geq s_0$, $\hat{f}(e, y_0, s) = \psi(e)$ so $B_{e,y_0,s} \subseteq W_{\psi(e)}$ by (i). To see that $W_{\psi(e)} \subseteq^* A_e^{y_0}$ note that no $x \geq s_0$ is restrained from any $B_{e,y_0,t}$ by (iii). Thus, for any $x \geq s_0$, $x \in W_{\psi(e)}$, choose any t such that $x \in W_{\psi(e),t}$ and $\hat{f}(e, y_0, t) = \hat{f}(e, y_0 + j, t)$ for all $j \leq x$. Then $x \in B_{e,y_0,t}$ so $x \in A_e^{y_0}$. (Such t exists since all good rows A_e^y have $\lim_t \hat{f}(e, y, t) = \psi(e)$.)

Lemma 3. $A_e \leq_T W_{\psi(e)}$.

Proof. By Lemma 1, $\bigcup_{y < y_0} A^y =^* \emptyset$ so it suffices to prove that for every $y \geq y_0$, $A_e^y \leq_T W_{\psi(e)}$ uniformly in y. Fix $y \geq y_0$. To test whether $x \in A_e^y$ see whether $x \in W_{\psi(e)}$.

Case 1. $x \in W_{\psi(e)}$. Choose the least $s > s_0$ such that $x \in W_{\psi(e),s}$ and (ii) is satisfied for x (i.e., $\hat{f}(e, y, s) = \hat{f}(e, y + j, s)$ for all $j \leq x$). If $x \notin B_{e,y,s}$ it must be that (iii) restrains x from entering $B_{e,y,s}$, but then (iii) ensures $x \notin B_{e,y,t}$ for any $t > s$. Hence, $x \in A_e^y$ iff $x \in \bigcup_{t \leq s} B_{e,y,t}$.

Case 2. $x \notin W_{\psi(e)}$. Then $x \in A_e^y$ only if $x \in B_{e,y,s}$ for some s such that $\hat{f}(e, y, s) \neq \psi(e)$. Choose any $s > x$ such that $\hat{f}(e, y, s) = \psi(e)$. (Such s exists because $y \geq y_0$.) If $x \in B_{e,y,t}$ for any $t > s$ then $\hat{f}(e, y, v) \neq \psi(e)$ for some minimal v, $s < v \leq t$. But then (iii) prevents $x \in B_{e,y,p}$ for any $p \geq v$.

5.3 Notes. Theorem 5.1 appeared in Arslanov [1981]. The proof here is a modification of a version by L. Welch [1981]. Theorem 5.1 can also be proved using the tree method of Chapter XIV. This involves more machinery but perhaps gives a clearer insight into the proof.

6. *A Generalization of the Recursion Theorem and the Completeness Criterion to All Levels of the Arithmetical Hierarchy*

The Kleene Recursion Theorem II.3.2 asserted that every recursive function has a *fixed point* e such that $W_e = W_{f(e)}$. In Exercise V.5.5 and Corollary 5.2, respectively, we have seen that at the higher levels $n = 1, 2$, every function $f \leq_T \emptyset^{(1)}$ ($f \leq_T \emptyset^{(2)}$) also has a "fixed point" e such that $W_e =^* W_{f(e)}$ (respectively, $W_e \equiv_T W_{f(e)}$). In this section we prove that for any $n \in \omega$, every function $f \leq_T \emptyset^{(n)}$ has a fixed point e such that $W_e \sim_n W_{f(e)}$ for an appropriate definition of certain equivalence relations $\sim_n$, $n \in \omega$, on r.e. sets. This is accomplished as in Theorem 5.1 for the case $n = 2$ by proving that for every $f \leq_T \emptyset^{(n)}$ there is a *recursive* function g such that for all x,

$W_{f(x)} \sim_n W_{g(x)}$. Indeed we state and prove a version of the latter for *partial* functions $\psi \leq_T \emptyset^{(n)}$ (in place of a *total* $f \leq_T \emptyset^{(n)}$) in order to prove Theorem 6.4 and Corollary 6.5.

Similarly, the Arslanov Completeness Criterion V.5.1 asserted that an r.e. set A is complete iff there is a function $f \leq_T A$ such that $W_{f(x)} \neq W_x$. This was generalized in Exercise V.5.10 for the level $n = 1$. Here, we extend this completeness criterion to all levels $\emptyset^{(n)}$ of the arithmetical hierarchy using the equivalence relations $\sim_n$, for all $n \in \omega$. All the results here can be proved not only for $\alpha = n \in \omega$ but also for $\alpha = \omega$ where we view ω as an ordinal (as well as the set of integers), and we let "$\alpha \leq \omega$" denote "$\alpha \in \omega$ or $\alpha = \omega$".

6.1 Definition. We define for each $\alpha \leq \omega$ the following equivalence relations on subsets of ω.

(i) $A \sim_0 B$ if $A = B$,
(ii) $A \sim_1 B$ if $A =^* B$,
(iii) $A \sim_2 B$ if $A \equiv_T B$,
(iv) $A \sim_{n+2} B$ if $A^{(n)} \equiv_T B^{(n)}$, for $n \in \omega$,
(v) $A \sim_\omega B$ if $(\exists n)\, [A \sim_n B]$.

6.2 Recursive Index Simulation Theorem.   *For every $\alpha \leq \omega$ if ψ is a function partial recursive in $\emptyset^{(\alpha)}$ then there is a (total) recursive function g such that*

$$(\forall x \in \text{dom } \psi)\, [W_{g(x)} \sim_\alpha W_{\psi(x)}].$$

6.3 Generalized Fixed Point Theorem.   *For every $\alpha \leq \omega$,*

$$f \leq_T \emptyset^{(\alpha)} \implies (\exists e)\, [W_e \sim_\alpha W_{f(e)}].$$

Proof of Theorem 6.3. For $\alpha = 0$ this is the Kleene Recursion Theorem II.3.2. The case of $0 < \alpha \leq \omega$ follows immediately from Theorem 6.2 and the case $\alpha = 0$.   ∎

(Cases $\alpha = 1$ and 2 are Exercise V.5.5(b) and Corollary 5.2, respectively, and were proved by Arslanov, Nadirov, and Solov'ev [1977] and Arslanov [1981], and Arslanov [1981]. The cases $3 \leq \alpha < \omega$ were proved by Jockusch, Lerman, Soare, and Solovay [ta].)

Proof of Theorem 6.2. We consider the case of α, $0 \leq \alpha \leq \omega$.
    Case $\alpha = 0$. Define

$$W_{g(x)} = \begin{cases} W_{\psi(x)} & \text{if } \psi(x)\downarrow, \\ \emptyset & \text{otherwise.} \end{cases}$$

    Case $\alpha = 1$. This is Exercise V.5.5(a), using the function g from the case $\alpha = 0$ in case ψ is partial.

Case $\alpha = 2$. This is Theorem 5.1, again using the function g from the case $\alpha = 0$ in case ψ is partial.

Case $3 \leq \alpha < \omega$. (Jockusch). Fix $n \geq 1$ and let $\psi \leq_T \emptyset^{(n+2)}$ be given. We want to find a recursive function g such that

$$(6.1) \qquad (\forall x \in \text{ dom } \psi) \, [W_{g(x)}^{(n)} \equiv_T W_{\psi(x)}^{(n)}].$$

First from ψ we fix a partial function $\theta \leq_T \emptyset^{(n+2)}$ such that $\text{dom } \theta = \text{dom } \psi$, and

$$(6.2) \qquad (\forall x \in \text{ dom } \psi) \, [W_{\theta(x)}^{\emptyset^{(n)}} = W_{\psi(x)}^{(n)}].$$

By Theorem 5.1, relativized to $\emptyset^{(n)}$ and applied to θ, we obtain a total function $g_1 \leq_T \emptyset^{(n)}$ such that for all x in dom θ,

$$(6.3) \qquad W_{g_1(x)}^{\emptyset^{(n)}} \equiv_T W_{\theta(x)}^{\emptyset^{(n)}}.$$

By Exercise III.1.20, we may replace $g_1(x)$ in (6.3) by a (total) *recursive* function $g_2(x)$. Now define the recursive function $g(x) = q^n(g_2(x))$, where $q(x)$ is the recursive function produced by the Sacks Jump Theorem satisfying Corollary VIII.3.6. Now for all $x \in$ dom ψ,

$$(6.4) \qquad W_{g(x)}^{(n)} \equiv_T W_{g_2(x)}^{\emptyset^{(n)}} = W_{g_1(x)}^{\emptyset^{(n)}} \equiv_T W_{\theta(x)}^{\emptyset^{(n)}} = W_{\psi(x)}^{(n)},$$

by Corollary VIII.3.6, Exercise III.1.20, (6.3), and (6.2), respectively. This establishes (6.1).

Case $\alpha = \omega$. Let ψ be partial recursive in $\emptyset^{(\omega)}$. The following lemma gives a convenient representation of ψ and the proof is similar to the proof in Exercise 4.15 of line (4.16).

Lemma 1.   *There is a recursive function $h(x, y, n)$ such that for all x and y,*

$$\psi(x) = y \iff (\exists n \geq y) \, [h(x, y, n) \in \emptyset^{(n)}].$$

Proof. Let g_1 be a recursive function such that for all x and y,

$$\psi(x) = y \iff (\exists \sigma \in W_{g_1(x,y)}) \, [\sigma \subset \emptyset^{(\omega)}].$$

Let h_1 and h_2 be recursive functions such that for all σ,

$$\sigma \subseteq \emptyset^{(\omega)} \iff h_1(\sigma) \in \emptyset^{(h_2(\sigma))}.$$

Let h_3 be a recursive function such that for all i, j, k with $k \geq j$,

$$i \in \emptyset^{(j)} \iff h_3(i, j, k) \in \emptyset^{(k)}.$$

Fix a_0 such that $a_0 \notin \emptyset^{(n)}$ for all n. Define $h(x, y, n)$ as follows. Let $n = \langle m, s \rangle$, and let σ be the string in $2^{<\omega}$ with code number m. If $h_2(\sigma) < n$

and $\sigma \in W_{g_1(x,y),s}$ define $h(x,y,n) = h_3(h_1(\sigma),h_2(\sigma),n)$. Otherwise, let $h(x,y,n) = a_0$.   ▯

Lemma 2. $(\exists i_0)\,(\forall n)\,[\{\,i_0\,\}^{\emptyset^{(n)}}(0) = n]$.

Proof. By Exercise 4.16.   ▯

It is obvious from Post's Theorem IV.2.2 that

$$(6.5) \qquad (\exists \text{ recursive function } r)\,(\forall i)\,(\forall n)\,[W^{\emptyset^{(n)}}_{r(i,n)} = W^{(n)}_i].$$

Let h, i_0 and r be as above. Let $q(x)$ be the recursive function from the Sacks Jump Theorem satisfying Corollary VIII.3.6.

Define the recursive function f as follows. Given x and z, let $f(x,z)$ be an index such that for all Y,

$$(6.6) \qquad W^Y_{f(x,z)} = \begin{cases} \emptyset & \text{if } \{\,i_0\,\}^Y(0)\uparrow, \\ W^Y_{r(y_0,n)} & \text{if } \{\,i_0\,\}^Y(0)\downarrow = n \ \& \\ & \quad y_0 = (\mu y \le n)\,[h(x,y,n) \in Y], \\ W^Y_{q(z)} & \text{if } \{\,i_0\,\}^Y(0) = n \\ & \quad \& \ \neg(\exists y \le n)\,[h(x,y,n) \in Y]. \end{cases}$$

By the Relativized Recursion Theorem III.1.6 with parameters there is a recursive function g such that

$$(6.7) \qquad (\forall x)\,(\forall Y)\,[W^Y_{f(x,g(x))} = W^Y_{g(x)}].$$

Claim. $(\forall x \in \text{dom } \psi)\,[W_{\psi(x)} \sim_\omega W^\emptyset_{g(x)}]$.

Proof. Fix $x \in \text{dom } \psi$, and say $\psi(x) = y$. Choose

$$n = (\mu i \ge y)\,[h(x,y,i) \in \emptyset^{(i)}].$$

It suffices to show that

$$(6.8) \qquad \left(W^\emptyset_{g(x)}\right)^{(n)} \equiv_T W^{(n)}_y.$$

For $i < n$,

$$W^{\emptyset^{(i)}}_{g(x)} = W^{\emptyset^{(i)}}_{f(x,g(x))} = W^{\emptyset^{(i)}}_{q(g(x))},$$

by (6.7) and (6.6). Hence, for $i < n$,

$$\left(W^{\emptyset^{(i)}}_{g(x)}\right)' = \left(W^{\emptyset^{(i)}}_{q(g(x))}\right)' \equiv_T W^{\emptyset^{(i+1)}}_{g(x)} \oplus \emptyset^{(i+1)},$$

and

$$\emptyset^{(i)} \le_T W^{\emptyset^{(i)}}_{g(x)}.$$

Thus, by induction on $i \leq n$ we have for all $i \leq n$,

$$(6.9) \qquad\qquad W_{g(x)}^{\emptyset^{(i)}} \equiv_T \left(W_{g(x)}^{\emptyset} \right)^{(i)}$$

For $i = n$, we have

$$W_{g(x)}^{(n)} \equiv_T W_{g(x)}^{\emptyset^{(n)}} = W_{f(x,g(x))}^{\emptyset^{(n)}} = W_{r(y,n)}^{\emptyset^{(n)}} = W_y^{(n)},$$

by (6.9), (6.7), (6.6) and (6.5), respectively. This proves (6.8) and hence the case $\alpha = \omega$. ▯

Similarly, we now wish to obtain a generalization of the Arslanov Completeness Criterion Theorem V.5.1 in terms of $\sim_n$ using Theorem 6.2. However, it will be convenient to prove first the following general theorem. The following theorem is similar to a result by Arslanov [1985a, Theorem 2].

6.4 Theorem.   *Let $\equiv_E$ be an equivalence relation on r.e. sets. Let C and A be any sets such that $C \leq_T A <_T C'$ and A is r.e. in C. Assume that for any ψ partial recursive in C there is a (total) function $g \leq_T A$ such that*

$$(6.10) \qquad\qquad (\forall z \in \mathrm{dom}\ \psi)\ [W_{g(z)} \equiv_E W_{\psi(z)}].$$

Then for any function $f \leq_T A$ there exists e such that $W_e \equiv_E W_{f(e)}$.

6.5 Corollary (Generalized Completeness Criterion).   *Fix $\alpha \leq \omega$. Suppose $\emptyset^{(\alpha)} \leq_T A$ and A is r.e. in $\emptyset^{(\alpha)}$. Then*

$$A \equiv_T \emptyset^{(\alpha+1)} \iff (\exists f \leq_T A)\ (\forall x) \neg [W_{f(x)} \sim_\alpha W_x].$$

Proof of Corollary 6.5. ($\implies$). Trivial since $\{ x : W_x \sim_\alpha \emptyset \} \leq_T \emptyset^{(\alpha+1)}$.

($\impliedby$). Assume $A <_T \emptyset^{(\alpha+1)}$. In Theorem 6.4 let $C = \emptyset^{(\alpha)}$ and let $\equiv_E$ be $\sim_\alpha$. Theorem 6.2 guarantees the hypothesis of Theorem 6.4, and hence 6.4 produces an e such that $W_{f(e)} \sim_\alpha W_e$. ▯

(Note that in Corollary 6.5 the case $\alpha = 0$ is Theorem V.5.1 (Arslanov [1981]), and the case $\alpha = 1$ is Exercise V.5.10 (Arslanov [1981]). The case $\alpha = 2$ is due to Arslanov [1981, Theorem 10]. The case $3 \leq \alpha \leq \omega$ is due to Jockusch.)

Proof of Theorem 6.4. Suppose for a contradiction that $f \leq_T A$ and

$$(\forall e)\ [W_e \not\equiv_E W_{f(e)}].$$

It suffices to define $h \leq_T A$ such that

$$(\forall e)\ [h(e) \neq \{ e \}^C(e)],$$

because then by Theorem V.5.1 and Exercise V.5.8, each relativized to C, we can conclude that $C' \equiv_T A$ contrary to hypothesis.

Since $\lambda e[\{e\}^C(e)]$ is partial recursive in C, there exists by hypothesis an A-recursive function g with $W_{g(e)} \equiv_E W_{\{e\}^C(e)}$ for all e such that $\{e\}^C(e)$ converges. Since f has no $\equiv_E$ fixed points, $W_{f(g(e))} \not\equiv_E W_{g(e)}$ for all e. Define $h(e) = f(g(e))$, so $h \leq_T A$. Now for all $e \in \text{dom}(\{e\}^C(e))$:

$$W_{h(e)} = W_{f(g(e))} \not\equiv_E W_{g(e)} \equiv_E W_{\{e\}^C(e)};$$

$$W_{h(e)} \not\equiv_E W_{\{e\}^C(e)};$$

so

$$h(e) \neq \{e\}^C(e),$$

as required.   ∎

6.6 Notes.   Historically, results 4.5, 4.6, 4.10, 4.11, 4.14, 4.17, 4.18, and 6.1 through 6.5 were obtained first by the person mentioned but were later combined with a number of other results which will all appear in the joint paper by Jockusch, Lerman, Solovay, and Soare [ta]. Other results in this paper include a density theorem for the ω-classes of r.e. degrees (where the equivalence class of A is $\{W : W \sim_\omega A\}$) and also a generalization of the Arslanov Completeness Criterion V.5.1 where the hypothesis "A r.e." is replaced by "A d.r.e." (By Exercise V.5.9 this cannot be strengthened to "$A \in \Delta_2^0$".) By a more difficult argument the latter result can be strengthened further to prove that if B is n-r.e. in $\emptyset^{(\alpha)}$, for $\alpha \leq \omega$, and $\emptyset^{(\alpha)} \leq_T B$ and $f \leq_T B$ but has no α-fixed points (namely, $(\forall x)\neg[W_x \sim_\alpha W_{f(x)}]$) then $\emptyset^{(\alpha+1)} \leq_T B$.

Part D

Advanced Topics and Current Research Areas in the R.E. Degrees and the Lattice of R.E. Sets

Chapter XIII

Promptly Simple Sets,
Coincidence of Classes of R.E. Degrees,
and an Algebraic Decomposition of the R.E. Degrees

The equivalence of **ENC** and **NC** as well as the algebraic properties of **M** and **NC** mentioned in Chapter IX §5 are best proved by first introducing the notion of a promptly simple set. So far the properties of an r.e. set A we have been studying have been *static* properties such as being creative, simple, hypersimple etc. In studying the algebraic structure of the lattice $\mathcal{E}$ of r.e. sets and particularly their automorphisms, it became clear that a more *dynamic* notion was required that took into account not merely *whether* certain elements are in A, but also *how fast* they are enumerated into A by comparison with elements being enumerated in other r.e. sets under the standard simultaneous enumeration I.3.11 of all r.e. sets.

Such a dynamic notion appeared in the hypotheses of the Extension Theorem of Soare [1974b, p. 91] needed to generate automorphisms of the lattice $\mathcal{E}$ of r.e. sets. Lerman and Soare [1980a] tried to approximate this dynamic property with the static notion of d-simple set which they used to answer a question of D. A. Martin about classes of r.e. degrees. Maass [1982] introduced the elegant and very fruitful dynamic notion of promptly simple set which implied d-simplicity as well as the complicated dynamic property in the hypotheses of the Extension Theorem, and he used it to synthesize and extend results by Soare [1974b] and [1982a] on automorphisms of $\mathcal{E}$. It was quickly recognized that promptly simple sets had important and unexpected connections elsewhere in r.e. sets and degrees.

In §1 we define the promptly simple sets, and relate their properties to the simple sets and to the permitting method that we have studied in Chapter V. In §2 we prove that the class **PS** of degrees containing promptly simple sets surprisingly coincides with both the noncappable degrees **NC**, and also the effectively noncappable degrees **ENC** of IX §5. In §3 we prove that $\mathbf{R} = \mathbf{NC} \cup \mathbf{M}$ gives an algebraic decomposition of $\mathbf{R}$ into the disjoint union of a (strong) filter and an ideal, thereby displaying more uniformity of algebraic structure of the r.e. degrees than had been supposed. In §4 we prove that $\mathbf{PS} = \mathbf{LC}$, the low cuppable degrees. Indeed **PS** coincides with two other apparently unrelated classes as discussed in Notes 2.7. (See Definition II.1.14 for definitions of strong filter, filter, and ideal.)

1. Promptly Simple Sets and Degrees

Recall that we are assuming Convention II.2.6 concerning the enumeration $\{W_{e,s}\}_{e,s \in \omega}$ so that at every stage s at most one set W_e receives at most one new element.

1.1 Definition. If $\{V_{e,s}\}_{e,s \in \omega}$ is any recursive enumeration of a sequence of r.e. sets such that $V_e = \bigcup_s V_{e,s}$ (for example $V_{e,s} = W_{e,s}$) then $x \in V_{e, \text{ at } s}$ denotes that $x \in V_{e,s} - V_{e,s-1}$ if $s > 0$ and $x \in V_{e,0}$ if $s = 0$.

Post defined a coinfinite r.e. set A to be *simple* if $W_e \cap A \neq \emptyset$ for every infinite r.e. set W_e. For A to be promptly simple, some element x entering W_e at stage s must enter A "promptly", namely by the end of stage $p(s)$ in the enumeration of A for some recursive function p.

1.2 Definition. A coinfinite r.e. set A is *promptly simple* if there is a recursive function p and a recursive enumeration $\{A_s\}_{s \in \omega}$ of A such that for every e

$$(1.1) \qquad W_e \text{ infinite} \implies (\exists s)\,(\exists x)\,[x \in W_{e, \text{ at } s} \cap A_{p(s)}].$$

(Note that we may assume that p is nondecreasing by replacing $p(s)$ if necessary by $\max\{p(t) : t \leq s\}$.)

The definition of prompt simplicity is independent of the particular enumeration in the following sense.

1.3 Proposition. *If A is promptly simple and $\{\hat{A}_s\}_{s \in \omega}$ and $\{V_{e,s}\}_{e,s \in \omega}$ are strong arrays of finite sets such that $A = \bigcup_s \hat{A}_s$, $W_e = \bigcup_s V_{e,s}$, $\hat{A}_s \subseteq \hat{A}_{s+1}$, $V_{e,s} \subseteq V_{e,s+1}$, and $\max(\{e : V_{e,s} \neq \emptyset\})$ is recursively bounded, then there is a recursive function q such that for all e,*

$$W_e \text{ infinite} \implies (\exists s)\,(\exists x)\,[x \in V_{e, \text{ at } s} \cap \hat{A}_{q(s)}].$$

Proof. Let A be promptly simple via p with respect to $\{A_s\}_{s \in \omega}$. Given s, compute for each $x \in V_{e, \text{ at } s}$ the least t such that $x \in W_{e,t}$ and let $q(s)$ be the least number u such that $\hat{A}_u \supseteq A_{p(t)}$ for all such t.   ▯

Most simple sets in the literature, such as Post's simple set S in V.1.3, are automatically promptly simple, and often with p the identity function. The following characterization which does not mention enumerations shows that prompt simplicity is recursively invariant. This characterization is similar to the analogous recursion theoretic characterization of nonspeedable sets (see Soare [1977, Theorem 2.4]) which are related to promptly simple sets by Maass's theorem [1982, Theorem 17] that any two sets with both properties are automorphic.

1.4 Theorem (Maass). *The following are equivalent for an r.e. set A:*

(i) *A is promptly simple;*

(ii) *A is coinfinite and there is a recursive function f such that, for all $e \in \omega$,*

(1.2) $$W_{f(e)} \subseteq W_e,$$

(1.3) $$W_{f(e)} \cap \overline{A} = W_e \cap \overline{A},$$

and

(1.4) $$W_e \text{ infinite} \implies W_e - W_{f(e)} \neq \emptyset.$$

(iii) *The same as* (ii) *but with* (1.4) *replaced by*

(1.5) $$W_e \text{ infinite} \implies W_e - W_{f(e)} \text{ infinite.}$$

Proof. (i) $\implies$ (ii). Given an enumeration $\{A_s\}_{s \in \omega}$ of A and a recursive function p which satisfy (1.1), let

$$W_{f(e)} = \{\, x : (\exists s)\, [x \in W_{e,s} - A_{p(s)}]\,\}.$$

Now $W_{f(e)}$ is r.e. uniformly in e, and $W_{f(e)}$ certainly satisfies (1.2) and (1.3). By (1.1), if W_e is infinite, $W_e - W_{f(e)} \neq \emptyset$.

(ii) $\implies$ (iii). Let h be a recursive function so that

$$W_{h(e,x)} = \begin{cases} W_e & \text{if } x = 0, \\ W_e - \{0, 1, \ldots, x-1\} & \text{if } x > 0. \end{cases}$$

If W_e is infinite, so is $W_{h(e,x)}$ for every x. Let f be as in (ii). Let f' be defined by

$$W_{f'(e)} = \{\, x : x \in \bigcap_{y \leq x} \{W_{f(h(e,y))}\}\,\}.$$

Certainly $W_{f(h(e,y))} \subseteq W_{h(e,y)} \subseteq W_e$ so that $W_{f'(e)} \subseteq W_e$. Furthermore, $W_{f'(e)} \cap \overline{A} = W_e \cap \overline{A}$ since if $x \in W_e \cap \overline{A}$, then $x \in W_{h(e,y)} \cap \overline{A}$ for each $y \leq x$. But $W_e - W_{f'(e)}$ is infinite whenever W_e is infinite since $W_{h(e,y)} - W_{f(h(e,y))}$ is nonempty whenever W_e is infinite; i.e., $W_e - W_{f'(e)}$ has an element greater than y for any fixed y.

(iii) $\implies$ (i). Given an enumeration $\{A_s\}_{s\in\omega}$ of A and a function f as in (iii), we define p to satisfy (1.1). Let

$$p(s) = (\mu t)\,(\forall x)\,(\forall e)\,[x \in W_{e,s} \Rightarrow x \in A_t \cup W_{f(e),t}];$$

p is recursive since for each s there are only finitely many pairs x, e such that $x \in W_{e,s}$ and for each of these pairs there is a $t \geq s$ so that $x \in A_t \cup W_{f(e),t}$ by (1.3). That p satisfies (1.1) is a direct consequence of (1.5) and the definition of p. ∎

The following lemma will be essential in several theorems below.

1.5 Lemma (Slowdown Lemma).   *Let $\{U_{e,s}\}_{e,s\in\omega}$ be a strong array of finite sets such that $U_{e,s} \subseteq U_{e,s+1}$ and $U_e = \bigcup_s U_{e,s}$. Then there is a recursive function g such that for all e, $W_{g(e)} = U_e$, and $W_{g(e),s} \cap U_e$, at $s = \emptyset$ (namely any element enumerated in U_e appears strictly later in $W_{g(e)}$).*

Proof. By the Recursion Theorem define the recursive function g by

$$W_{g(e)} = \{\, x : (\exists s)\,[x \in U_{e,s} - W_{g(e),s}]\,\}. ∎$$

We frequently wish to define an r.e. set U_e in stages $\{U_{e,s}\}_{s\in\omega}$. By the Recursion Theorem we may assume that there is a recursive function g such that $W_{g(e)} = U_e$. The crucial point about the Slowdown Lemma 1.5 is that if we enumerate x in U_e at stage s then the least t such that $x \in W_{g(e),t}$ satisfies $t > s$. This is important in Theorem 1.6 below.

A promptly simple set differs from an ordinary simple set in that it must "promptly" intersect any infinite r.e. set. We now wish a broader property which holds for all r.e. sets A in a promptly simple *degree*. Recall from V §3 that any r.e. set A in a nonzero degree must permit any infinite r.e. set W_e in the sense that

$$(\exists x)\,(\exists s)\,[x \in W_{e,\text{ at }s}\ \&\ A_s \restriction x \neq A \restriction x].$$

The next theorem characterizes r.e. sets in a promptly simple degree as those which *promptly* permit.

1.6 Theorem (Promptly Simple Degree Theorem).   *Let A be an r.e. set and $\{A_s\}_{s\in\omega}$ a recursive enumeration of A. Then A has promptly simple degree iff there is a recursive function p such that for all s, $p(s) \geq s$, and for all e,*

(1.6) W_e *infinite* $\implies (\exists x)\,(\exists s)\,[x \in W_{e,\text{ at }s}\ \&\ A_s \restriction x \neq A_{p(s)} \restriction x]$,

namely A "promptly permits" on some element $x \in W_e$.

Proof. ($\implies$). Let $B = \{e\}^A$ where B is promptly simple via $q(s)$ satisfying (1.1), and let $\{B_s\}_{s\in\omega}$ be a recursive enumeration of B. We define p

satisfying (1.6) and simultaneously construct a strong array $\{U_{e,s}\}_{e,s\in\omega}$ to which we apply the Slowdown Lemma 1.5. Set $p(0) = 0$.

Stage $s > 0$. (Define $p(s)$.) Choose the unique e and x (if they exist) such that $x \in W_{e,\text{ at }s}$. If there exists $y \notin U_{e,s-1}$ such that $\{e\}_s^{A_s}(y) \downarrow = B_s(y)$ and $u(A_s; e, y, s) < x$, then enumerate the least such y in $U_{e,s}$. Find the least t such that $y \in W_{g(e),t}$, where $W_{g(e)}$ is obtained from $\{U_{e,s}\}_{e,s\in\omega}$ by Lemma 1.5. Let $p(s)$ be the least $v \ge q(t)$ such that $B_v(y) = \{e\}_v^{A_v}(y)$. If e and x do not exist define $p(s) = p(s-1) + 1$. (This ends the construction.)

Now if W_e is infinite, then U_e is infinite (because $\overline{B}$ is infinite and $B = \{e\}^A$). Hence, by the prompt simplicity of B, there exists

$$y \in W_{g(e), \text{ at } t} \cap B_{q(t)}.$$

But $y \in U_{e, \text{ at } s}$ for some $s < t$ such that $\{e\}_s^{A_s}(y) \downarrow = B_s(y) = 0$. Now $y \in B_{q(t)} - B_s$ implies $A_s \restriction u \ne A_{p(s)} \restriction u$, where $u = u(A_s; e, y, s)$ because $p(s) \ge q(t)$. But y entered U_e only for the sake of some $x \in W_{e, \text{ at } s}$, $x > u$, so (1.6) is satisfied for W_e. Note that only $B \le_T A$ is used in proving this direction. We use this fact in proving Corollary 1.9.

($\Longleftarrow$). Given $p(s)$ satisfying (1.6) we use the usual permitting and coding methods to construct $B \equiv_T A$ such that B is promptly simple via the identity function. We must meet, for all e, the requirement:

$$P_e : W_e \text{ infinite} \implies (\exists x)\,(\exists s)\,[x \in W_{e, \text{ at } s} \cap B_s].$$

Define $B_0 = \emptyset$.

Stage $s > 0$. Let $\overline{B}_{s-1} = \{b_0^{s-1} < b_1^{s-1} < \cdots\}$.

Step 1 (for prompt simplicity). Choose the unique x and e (if they exist) such that $x \in W_{e, \text{ at } s}$. If $x > b_e^{s-1}$ and P_e is not yet satisfied, compute $A_{p(s)}$, and if $A_s \restriction x \ne A_{p(s)} \restriction x$ then enumerate x in B.

Step 2 (to code A into B). For each $x \in A_s - A_{s-1}$, enumerate b_x^{s-1} in B.

This completes the enumeration of B. Now $B \le_T A$ since if $x \in B_s - B_{s-1}$ then $A \restriction x \ne A_{s-1} \restriction x$. But B is promptly simple since if W_e is infinite then the conclusion of P_e is satisfied by the construction, (1.6), and (ii) $\Longrightarrow$ (iii) of Theorem 1.7 below (whose proof does not use Theorem 1.6). Also $A \le_T B$ since if $b_x = \lim_s b_x^s$ and $B_s \restriction b_x = B \restriction b_x$ then $x \in A$ iff $x \in A_s$. ▯

Although not used as often as Theorem 1.6, the following theorem gives some additional characterizations of an r.e. set A having promptly simple degree, and is used in the proof of Theorem 4.2 below.

1.7 Theorem. *Let A be an r.e. set and $\{A_s\}_{s\in\omega}$ a recursive enumeration of A. Then the following are equivalent:*

(i) *A has promptly simple degree.*

(ii) *There is a recursive function p satisfying (1.5).*

(iii) *The same as (ii) but with "$(\exists^\infty x)$" in place of "$(\exists x)$" in (1.5), where $\exists^\infty x$ denotes "there exist infinitely many x," namely*

(1.7) W_e infinite $\implies$ $(\exists^\infty x)\,(\exists s)\,[x \in W_{e,\text{ at }s}\ \&\ A_s \restriction x \neq A_{p(s)} \restriction x]$.

(iv) *Whenever $\{U_{e,s}\}_{e,s\in\omega}$ is a strong array of finite sets such that $W_e = \bigcup_s U_{e,s}$ and $U_{e,s} \subseteq U_{e,s+1}$ there is a recursive function $p(s)$ satisfying (1.6) with "$U_{e,s}$" in place of "$W_{e,s}$".*

(v) *The same as (ii) but with "$W_e = \omega$" in place of "W_e infinite".*

Proof. (i) $\iff$ (ii) was established in Theorem 1.6 and (iii) $\implies$ (ii) is obvious.

(ii) $\implies$ (iii). Given $q(s)$ satisfying (ii), we define $p(s)$ satisfying (iii). Using the Recursion Theorem define

$$W_{g(e,n)} = \{\, x : x > n\ \&\ (\exists s)\,[x \in W_{e,s} - W_{g(e,n),s}]\,\}.$$

For each s, find x and e (if they exist) such that $x \in W_{e,\text{ at }s}$. For each $n < x$ find the least t such that $x \in W_{g(e,n),t}$. Define $p(s)$ to be the maximum of $q(t)$ over all such t if x and e exist and arbitrary otherwise.

Fix n. Now if W_e is infinite then $W_{g(e,n)}$ is infinite so

$$(\exists x)\,(\exists t)\,[x \in W_{g(e,n),\text{ at }t}\ \&\ A_t \restriction x \neq A_{q(t)} \restriction x].$$

But then $x \in W_{e,\text{ at }s}$ for some $s < t$ and $x > n$. However, $q(t) \leq p(s)$ so

$$x \in W_{e,\text{ at }s} \qquad \text{and} \qquad A_s \restriction x \neq A_{p(s)} \restriction x.$$

Since n was arbitrary, there are infinitely many such x, so p satisfies (iii).

(iv) $\implies$ (ii). Define $U_{e,s} = W_{e,s}$.

(iii) $\implies$ (iv). Let $q(s)$ satisfy (iii). We define $p(s)$ to satisfy (iv). Apply Lemma 1.5 to $\{U_{e,s}\}_{e,s\in\omega}$ to obtain $g(e)$. Given s, for all x and $e \leq s$ such that $x \in U_{e,s}$ find t such that $x \in W_{g(e),\text{ at }t}$ and let $p(s)$ be the maximum of $q(t)$ over such t if x and e exist and arbitrary otherwise.

(ii) $\implies$ (v). Immediate.

(v) $\implies$ (ii). Let $q(s)$ satisfy (v). We define $p(s)$ satisfying (ii), by first defining a strong array $\{U_{e,s}\}_{e,s\in\omega}$ and applying Lemma 1.5. If $x \in W_{e,\text{ at }s}$ and $x \notin U_{e,s-1}$ (where $U_{e,-1} = \emptyset$) then enumerate in $U_{e,s}$ all $y \leq x$, $y \notin U_{e,s-1}$. If there is no such x let $U_{e,s} = U_{e,s-1}$. Apply Lemma 1.5 to $\{U_{e,s}\}_{e,s\in\omega}$ to obtain g. Given s, find x and e such that $x \in W_{e,\text{ at }s}$ and then t such that $x \in W_{g(e),\text{ at }t}$ (since $W_{g(e)} \supseteq W_e$) and define $p(s) = q(t)$ if x and e exist and arbitrary otherwise.

Now if W_e is infinite then $W_{g(e)} = \omega$, so

$$(\exists y)\,(\exists t)\,[y \in W_{g(e),\text{ at }t}\ \&\ A_t \restriction y \neq A_{q(t)} \restriction y].$$

Choose s such that $y \in U_{e, \text{ at } s}$. Then $s < t$ by hypothesis on g. Choose the unique $x \in W_{e, \text{ at } s}$. Now $y \leq x$ and $s < t \leq q(t) = p(s)$, so

$$A_s \restriction x \neq A_{p(s)} \restriction x. \quad \Box$$

1.8 Definition. Let **PS** denote the degrees of promptly simple sets.

1.9 Corollary. *If* $\mathbf{b} \in \mathbf{PS}$ *and* $\mathbf{b} \leq \mathbf{a} \in \mathbf{R}$ *then* $\mathbf{a} \in \mathbf{PS}$.

Proof. Suppose B_0 and A are r.e. sets, B_0 is promptly simple, and $B_0 \leq_T A$. Then by the proof of ($\Longrightarrow$) of Theorem 1.6 there exists p satisfying (1.6) for A. Hence by ($\Longleftarrow$) of 1.6 there is a promptly simple set $B \equiv_T A$. $\quad \Box$

1.10–1.14 Exercises

1.10. Prove that Post's simple set S of Theorem V.1.3 is promptly simple.

1.11. Prove directly from Definition 1.2 that if A is promptly simple via a nondecreasing recursive function p' and a recursive enumeration $\{A_s\}_{s \in \omega}$ of A, then there is a recursive function p such that $p'(s) \leq p(s)$ for all s, and for every e

$$(1.8) \qquad W_e \text{ infinite} \implies (\exists^\infty x)\, (\exists s)\, [x \in W_{e, \text{ at } s} \cap A_{p(s)}].$$

Hint. Define h as in the proof of (ii) $\implies$ (iii) in Theorem 1.4. Define p by

$$p(s) = \begin{cases} p'(t) & \text{where } t \text{ is the least stage such that } t > s \text{ and} \\ & (\exists e)\, (\exists x)\, (\exists y)\, [x \in W_{e, \text{ at } s} \,\&\, x \in W_{h(e,y), \text{ at } t}], \\ & \text{if such } t \text{ exists,} \\ p'(s) & \text{otherwise.} \end{cases}$$

1.12. (i) Prove that if $A \subseteq B$, B is r.e. and coinfinite, and A is promptly simple then B is also promptly simple.

(ii) Prove that the promptly simple sets are closed under intersection and hence form a filter in $\mathcal{E}$. *Hint.* Let A and B be promptly simple via $p(s)$ and $q(s)$ respectively satisfying (1.8). Hence, if W_e is infinite then U_e is also where

$$U_{e,s} = \{\, x : (\exists t \leq s)\, [x \in W_{e, \text{ at } t} \cap A_{p(t)}]\,\}.$$

Let $W_{g(e)} = U_e$ by Lemma 1.5. If U_e is infinite then

$$(\exists x)\, [x \in W_{g(e), \text{ at } t} \cap B_{q(t)}].$$

Define a prompt simplicity function for $A \cap B$.

1.13. Let A be an r.e. set of promptly simple degree. Let B be the deficiency set of A (under some enumeration of A) as defined in Exercise V.2.5. Using Theorem 1.6 prove that B is a promptly simple set.

1.14. Construct a low promptly simple set. *Hint.* Combine the promptly simple set requirements with the lowness requirements of Theorem VII.1.1.

1.15 Notes. Definition 1.2 of a promptly simple set was given by Maass [1982] where he proved Theorem 1.4 and Exercise 1.11, introduced the notion of an r.e. generic set, and showed that each r.e. generic set must be both low and promptly simple. Using the automorphism techniques of Soare [1974b and 1982a], Maass proved that any two sets which are both promptly simple and low (indeed which merely both have complements which are semi-low as defined in Exercise IV.4.6(b)) are automorphic. Maass, Shore and Stob [1981] obtained results 1.9 and 1.12. Ambos-Spies, Jockusch, Shore and Soare [1984] obtained results 1.3, 1.5, 1.6, 1.7, and 1.13.

Other notions of genericity for r.e. sets have been studied by Jockusch [ta], and Ingrassia [1981 and ta]. Jockusch introduces a new notion called *e*-genericity and shows that, like the Maass generic r.e. sets, the *e*-generic sets have many properties arising from finite injury priority constructions such as lowness and prompt simplicity. However, Jockusch [ta] also shows that if **a** and **b** are incomparable r.e. degrees with infimum **c**, then none of the degrees **a**, **b**, **c** can contain an *e*-generic set. Hence, *e*-generic degrees are non-branching and strongly noncappable. Harrington and Slaman [ta] have also used promptly simple sets in their proof that the elementary theory of the r.e. degrees $(\mathbf{R}, <)$ has degree $\mathbf{0}^{(\omega)}$. (See Chapter XVI.)

2. Coincidence of the Classes of Promptly Simple Degrees, Noncappable Degrees, and Effectively Noncappable Degrees

We can now show that the promptly simple degrees **PS** of Definition 1.8 coincide with the noncappable degrees **NC** of IX.5.1, and also with the effectively noncappable degrees **ENC** of IX.5.3. The following easy result which shows that **PS** $\subseteq$ **ENC** established the first connection between promptly simple degrees and noncappable degrees.

2.1 Theorem. *If* **a** *is a promptly simple degree then* **a** *is effectively noncappable.*

Proof. Fix a promptly simple set $A \in \mathbf{a}$, a recursive enumeration $\{A_s\}_{s\in\omega}$ of A, and a recursive function $p(s)$ satisfying (1.6) and such that $p(s) \geq s$ for all s. For every nonrecursive r.e. set B, we shall find (uniformly in an index of B) a nonrecursive r.e. set C so that $C \leq_T A$ and $C \leq_T B$, and these reductions are uniform in indices for A and B. To ensure that C is nonrecursive we make $\overline{C}$ infinite and we meet the usual simplicity requirements

$$P_e : W_e \text{ infinite} \implies C \cap W_e \neq \emptyset.$$

To meet P_e while ensuring that $C \leq_T A$ we shall enumerate certain r.e. sets U_e and assume by the Recursion Theorem that $U_e = W_{g(e)}$ for some fixed recursive function g which satisfies the Slowdown Lemma 1.5. To ensure that $C \leq_T B$ we shall require that if $B \upharpoonright x = B_s \upharpoonright x$ then $C \upharpoonright x = C_s \upharpoonright x$. Assume that for each s, $B_{s+1} - B_s$ has exactly one element; denote this element by b_s. Construct C as follows. Let $C_0 = \emptyset$.

Stage $s + 1$. Find the least e (if it exists) such that

$$W_{e,s+1} \cap C_s = \emptyset,$$

and

$$(\exists x)\, [x \in W_{e,s+1} \ \& \ x > 2e \ \& \ x > b_s].$$

Enumerate into $U_{e,s+1}$ the least such x satisfying $x \notin U_{e,s}$. Find t such that $x \in W_{g(e)}$, at t. Enumerate x in C_{s+1} iff $A_{p(t)} \upharpoonright x \neq A_s \upharpoonright x$. (Note that t exists since $W_{g(e)} = U_e$. Furthermore, by the Slowdown Lemma 1.5, for each $x \in U_{e,s+1} - U_{e,s}$, $x \in W_{g(e)}$, at t for some $t > s + 1$, so $p(t) > s$.)

Now $C \leq_T A$ since $A_s \upharpoonright x = A \upharpoonright x$ implies $C_s \upharpoonright x = C \upharpoonright x$. Each requirement P_e is satisfied since otherwise W_e is infinite and because B is nonrecursive $W_{g(e)}$ is infinite and is a counterexample to A and p satisfying (1.6). $\quad \square$

The next theorem is the most difficult in this chapter and yields the main equivalences **ENC** = **NC** = **PS** as well as the fact that these classes all form strong filters. The proof is interesting because it uses a "gap-cogap" construction such as those found in $0'''$-priority constructions discussed in Chapter XIV. (See Chapter XIV §2 for a classification of $0^{(n)}$-priority arguments.) A similar construction using a "gap-cogap" argument is found in Fejer and Soare [1980] in the proof of the plus-cupping theorem. See Notes 2.7 and Theorem 4.2 for other classes equivalent to **PS**.

2.2 Theorem.   **ENC** = **NC** = **PS**.

2.3 Corollary.   **ENC**, **NC**, *and* **PS** *each form strong filters in* **R**.

Proof. By IX.5.5 **ENC** forms a strong filter in **R**. $\square$

Proof of Theorem 2.2. Clearly **ENC** $\subseteq$ **NC**. By Theorem 2.1, **PS** $\subseteq$ **ENC**. It suffices to prove that **NC** $\subseteq$ **PS**.

Fix an r.e. set B and a recursive enumeration $\{ B_s \}_{s \in \omega}$ of B. We shall construct an r.e. set A by a recursive enumeration $\{ A_s \}_{s \in \omega}$ such that either A is nonrecursive and $\deg(A) \cap \deg(B) = \mathbf{0}$, or else B has promptly simple degree.

To attempt to meet the first alternative we use the usual minimal pair method as presented in IX §1. Namely, we shall construct A to satisfy for every $e \in \omega$ the requirements

$$P_e : W_e \text{ infinite} \implies W_e \cap A \neq \emptyset,$$

$$N_e : \{e\}^A = \{e\}^B = f_e \text{ total} \implies f_e \text{ is recursive.}$$

From $\{A_t\}_{t \leq s}$ and $\{B_t\}_{t \leq s}$ define the recursive functions

$$l(e,s) = \max\{x : (\forall y < x) [\{e\}_s^{A_s}(y)\downarrow = \{e\}_s^{B_s}(y)]\}$$

and

$$m(e,s) = \max\{l(e,t) : t < s\}.$$

Call s an *e-expansionary* stage if $l(e,s) > m(e,s)$. (Thus, if $\{e\}^A = \{e\}^B$ and $\{e\}^A$ is total then there are infinitely many e-expansionary stages.)

Simultaneously with the construction of A we define for each e a partial recursive function p_e such that either requirement N_e is met or else p_e is total and witnesses that B has promptly simple degree, namely (using Theorem 1.6) that p_e satisfies for all $i \in \omega$ the requirement

$$R_{e,i} : W_i \text{ infinite} \implies (\exists x)(\exists s)[x \in W_{i,\text{ at }s} \ \& \ B_s \restriction x \neq B_{p_e(s)} \restriction x].$$

Before giving the full construction we sketch the basic module for a single requirement. Fix e. We attempt to satisfy the requirement

$$R_e : N_e \quad \text{or} \quad (\forall i)R_{e,i}.$$

If we satisfy R_e by the second clause then B has promptly simple degree as witnessed by p_e so we need not meet requirements $R_{e'}$, $e' \geq e$. For each i we define a partial recursive function $\psi_{e,i}$ such that if B does not have promptly simple degree, $\{e\}^A = \{e\}^B$ is total, and i is minimal such that p_e fails to satisfy $R_{e,i}$ then $\psi_{e,i} = \{e\}^B$ so that N_e is met. To accomplish this we define a recursive "restraint" function $r(e,i,s)$ which is the restraint imposed by $R_{e,i}$, and which restrains elements from entering A. We define $r(e,s) = \max\{r(e,i,s) : i \leq s\}$, which is the restraint imposed by R_e, and it prevails against the positive requirements of lower priority; namely P_j, $j \geq e$.

At stage $s+1$ we *open* an $R_{e,i}$-gap by choosing the least i (if one exists) such that $R_{e,i}$ is not yet satisfied and such that there exist $x \in W_{i,\text{ at }s}$ and $y \notin \text{dom}(\psi_{e,i,s})$ with $y < l(e,s)$ and $u_y < x$ where

$$u_y = \tilde{u}(B_s; e, y, s) =_{\text{dfn}} \max\{u(B_s; e, y', s) : y' \leq y\}.$$

We define $\psi_{e,i,s+1}(y) = \{e\}_s^{B_s}(y)$ for the least such y, and $r(e,j,s+1) = 0$ for all $j \geq i$. (It follows that $\psi_{e,i,s+1}(z)$ is defined for all $z < y$ by the minimality of y and because $z < y$ implies $z < y < l(e,s)$ and $u_z \leq u_y < x$.)

This gap is later *closed* at stage $t+1$ where $t > s$ is the next e-expansionary stage after s. At stage $t+1$ we define $p_e(s) = t$ and set $r(e,i,t+1) = t$ as A-restraint (since by convention $\tilde{u}(A_s; e, y, t) < t$). Notice that if

$$B_s \upharpoonright u_y \neq B_t \upharpoonright u_y$$

then $B_s \upharpoonright x \neq B_t \upharpoonright x$, so p_e satisfies requirement $R_{e,i}$ via x. Thus, if $R_{e,i}$ is never satisfied then any value $\psi_{e,i}(y) = w$ once defined is protected by the B-side, namely, $\{e\}_v^{B_v}(y) = w$ at all later stages v in this gap, i.e., $s + 1 \leq v \leq t$.

Hence, if $\{e\}^A = \{e\}^B$ is total and B is not of promptly simple degree, choose the least i such that $R_{e,i}$ is never satisfied. In this case infinitely many $R_{e,i}$-gaps are opened and closed. Then $\psi_{e,i} = \{e\}^B$ because once $\psi_{e,i}(y) = w$ is defined at some stage, the B-side holds the computation at all later stages which lie in some $R_{e,i}$-gap, and the A-side holds the computation (because of the A-restraint $r(e,i,s)$) during all the corresponding cogaps (intervals between gaps). Furthermore, $\liminf_s r(e,s) < \infty$ since at each stage s when an $R_{e,i}$-gap is opened,

$$r(e,s) = \max\{r(e,j,s) : j < i\}$$

(since all restraint imposed by $R_{e,j}$, $j \geq i$, is dropped when $R_{e,i}$ opens a gap). Of course if there is no i such that infinitely many $R_{e,i}$-gaps are opened then we may have $\lim_s r(e,i,s) > 0$ for all i and therefore perhaps $\liminf_s r(e,s) = \infty$, but in this case if $\{e\}^A = \{e\}^B$ is total then B is of promptly simple degree and we need not meet the requirements $R_{e'}$, $e' \geq e$. This ends the basic module for a single requirement R_e.

The strategy σ_e just given for meeting a single requirement R_e, say R_0, produces an A-restraint function $r(0,s)$ such that $\liminf_s r(0,s) < \infty$. As in the minimal pair construction in IX §1 or Fejer and Soare [1980] we modify the strategy σ_e for R_e when $e > 0$, so that the various restraint functions $r(j,s)$, $j \leq e$, drop back simultaneously, namely $\liminf_s \tilde{r}(e,s) < \infty$ where $\tilde{r}(e,s) = \max\{r(j,s) : j \leq e\}$. To do this, R_e must guess the value of $k = \liminf_s \tilde{r}(e-1,s)$ (the maximum restraint imposed at stage s by any $R_{e'}$, $e' < e$) and must simultaneously play infinitely many strategies σ_e^k, $k \in \omega$, one for each possible value of k exactly as in IX §1.

In addition, we arrange that σ_e^k is allowed to open an $R_{e,i}^k$-gap (and drop its A-restraint) only at a stage $s \in S_e^k$, where $S_e^k = \{s : \tilde{r}(e-1,s) = k\}$, where we define $r(-1,s) = 0$ for all s. Hence, $S_0^0 = \omega$ and $S_0^k = \emptyset$ for $k > 0$. However, σ_e^k is allowed to *close* that gap (thereby reimposing A-restraint and defining $p_e^k(s)$) at *any* stage t (providing t is an e-expansionary stage). Thus, we have a sufficiently small amount of restraint so that $\liminf_s \tilde{r}(e,s) < \infty$ (if $\deg(B)$ is not promptly simple), and yet we close the gaps often enough so that if $\{e\}^A = \{e\}^B$ then every $R_{e,i}^k$-gap is closed (at the next e-expansionary stage) so that p_e^k is total. In the following proof we use the notation $r(e, s)$ in place of $\tilde{r}(e, s)$ to denote the maximum restraint imposed by all R_j for $j \leq e$.

Construction of A, p_e^k and $\psi_{e,i}^k$.
   Stage $s = 0$. Do nothing.
   Stage $s + 1$. For each $e \leq s$ perform in increasing order of e the following steps.

   Step 1. Let $k = r(e - 1, s + 1)$. For each $j > k$ cancel any gap or restraint previously imposed by $R_{e,i}^j$, for any i.
   Step 2 (closing gaps). If s is not an e-expansionary stage go to Step 3. Otherwise, if there is an $R_{e,i}^j$-gap which was opened at some stage $v+1 < s+1$ and has not been closed or cancelled, then declare the gap to be *closed*, define $p_e^j(t) = s$ for all $t \leq v$ not in domain p_e^j, and let $R_{e,i}^j$ assign s as A-restraint (since $s > u(A_s; e, y, s)$ for all $y < l(e, s)$).
   Step 3 (opening gaps). Let $s' = \max\{t < s : r(e - 1, t) = k\}$ if such t exists and $s' = 0$ otherwise. Choose the least $i \leq s$ such that $R_{e,i}^k$ is not yet satisfied, no $R_{e,i}^k$-gap is open, and

$$(\exists y) \left[(\exists x) \, [x \in W_{i,s} - W_{i,s'}] \ \& \ y \notin \mathrm{dom}\left(\psi_{e,i,s}^k \right) \right.$$
$$\left. \& \ l(e, s) > y \ \& \ \tilde{u}(B_s; e, y, s) < x \right],$$

where

$$\tilde{u}(B_s; e, y, s) = \max\{ u(B_s; e, z, s) : z \leq y \}.$$

Choose the least such y and then the least corresponding x and open an $R_{e,i}^k$-gap by defining $\psi_{e,i,s+1}^k(y) = \{e\}_s^{B_s}(y)$, and cancelling for all $j \geq i$ any A-restraint associated with $R_{e,j}^k$. If i fails to exist do nothing. (Note that some $R_{e,i}^k$-gap may have been closed at Step 2 and a new one opened at Step 3 in which case any A-restraint put on by $R_{e,i}^k$ at Step 2 is cancelled at Step 3.) Let $r(e, s + 1)$ be the maximum of the A-restraint still imposed by $R_{e',i'}^{k'}$, for some $e' \leq e$, $k' \leq s$, $i' \leq s$.
   Step 4 (making A simple). If $W_{e,s+1} \cap A_s = \emptyset$ and

$$(\exists y) \, [y \in W_{e,s+1} \ \& \ y > 2e \ \& \ y > r(e, s + 1)],$$

choose the least such y and enumerate y in A.
   This completes the construction.

Assume that B is not of promptly simple degree. Hence, for all e, there exists i such that $R_{e,i}$ is not satisfied. We must show that for all e requirement N_e is satisfied and $\liminf_s r(e, s) < \infty$, since then it is automatic by Step 4 of the construction that A is simple.

Fix e and assume by induction that for all $e' < e$, $N_{e'}$ is met and that

$$\liminf_s r(e - 1, s) < \infty.$$

Let $k = \liminf_s r(e - 1, s)$. Let $S_e^k = \{ s : r(e - 1, s) = k \}$. Assume that $\{e\}^A = \{e\}^B$ is a total function. Choose the least i such that W_i is infinite but $R_{e,i}^k$ is not satisfied. Since there are infinitely many $R_{e,i}^k$-gaps, $\psi_{e,i}^k$ and p_e^k are total and recursive. Choose s_0 such that for all $s \geq s_0$: (1) no $R_{e,i'}^{k'}$-gap is opened or closed at stage s if $k' < k$ or $i' < i$; (2) $P_{e'}$ does not contribute an element to A at stage s if $e' \leq e$.

Now suppose that $\psi_{e,i}^k(y) = w$ is first defined at some stage $s + 1 > s_0$. We claim by induction on $v > s$ that either

(1) $\{e\}_v^{B_v}(y) = w$, or
(2) $\{e\}_v^{A_v}(y) = w$,

and hence that $f_e(y) = w$. (Thus, $f_e(y) = \psi_{e,i}^k(y)$ for almost all y so N_e is satisfied.)

To prove the claim note that at stage $s + 1$ we open an $R_{e,i}^k$-gap via y and some $x \in W_{i,s} - W_{i,s'}$, where s' is defined as is Step 3 of the construction. Choose v, $s' < v \leq s$, such that $x \in W_i$, at v. Necessarily $p_{e,s}^k(t)$ is undefined for all t, $s' < t$. (Note that $p_e^k(t)$ is defined only when a gap begun at a stage $\geq t$ is closed. However, $R_{e,i'}^k$-gaps are only begun at stages $\in S_e^k$, and s' is the most recent such stage $< s$.) By choice of s_0, no $R_{e,i}^k$-gap is ever cancelled after s_0, so the above gap must be closed at stage $t + 1$, where t is the next e-expansionary stage $> s$. Now $p_e^k(v) = t$, and $v \leq s < t$, so $B_s \upharpoonright x = B_t \upharpoonright x$ because $R_{e,i}$ is never satisfied. But $\tilde{u}(B_s; e, y, s) < x$, so (1) holds for all v, $s + 1 \leq v \leq t$, namely, those stages v in the $R_{e,i}^k$-gap.

Now at stage $t + 1$, this $R_{e,i}^k$-gap for y is closed, and $R_{e,i}^k$ sets $t \geq \tilde{u}(A_t; e, y, t)$ as A-restraint. But by choice of s_0, no such A-restraint is ever injured after s_0. Hence, this A-restraint remains in force until the next stage $s_1 + 1 \geq t + 1$ at which the next $R_{e,i}^k$-gap is opened via $y_1 = y + 1$. Thus, (2) holds for all v, $t + 1 \leq v \leq s_1$, namely, those stages v in the $R_{e,i}^k$-cogap. But since $\tilde{u}(X; e, y_1, v) \geq \tilde{u}(X; e, y, v)$ for all v, the above argument shows that (1) holds for all stages v in the $R_{e,i}^k$-gap opened by y_1. Thus,

(1) holds for all v in the $R_{e,i}^k$-gaps and
(2) holds for all v in the $R_{e,i}^k$-cogaps.

Finally, let $r(e)$ be the maximum of k and the restraint imposed by $R_{e,i'}^{k'}$ for $k' < k$ or $i' < i$. Now $r(e, s) = r(e)$ at every stage s when an $R_{e,i}^k$-gap is opened. Hence, $r(e) = \liminf_s r(e, s) < \infty$. ▯

2.4–2.6 Exercises

2.4.   Prove that there is a recursive function f such that if $\deg(W_e)$ is half of a minimal pair then $\deg(W_e)$ and $\deg(W_{f(e)})$ form a minimal pair. *Hint.* Use the proof of Theorem 2.2.

2.5.   Prove that if A and B are r.e. sets such that $K \leq_{wtt} A \oplus B$ but $K \nleq_{wtt} B$ then $\deg(A)$ is noncappable.

2.6.   Prove that if an r.e. degree $\mathbf{c}$ is not half of a minimal pair, then $\mathbf{c}$ bounds a minimal pair. *Hint.* If $\mathbf{c}$ is not half of a minimal pair, choose any minimal pair $(\mathbf{a}, \mathbf{b})$ and find a minimal pair $(\mathbf{a}_1, \mathbf{b}_1)$ each below $\mathbf{c}$.

2.7 Notes. Maass, Shore and Stob [1981] proved Theorem 2.1. Ambos-Spies, Jockusch, Shore and Soare [1984] proved results 2.2, 2.3, 2.4, and 2.5. They also showed that three other classes of r.e. degrees coincide with **PS**, namely **LC** of Theorem 4.2, the degrees **G** in the orbit of an r.e. generic set under automorphisms of $\mathcal{E}$, and **SPH̄**, the degrees of non-hh-simple sets with a certain splitting property defined by Maass, Shore, and Stob [1981].

Strictly speaking the proof of Theorem 2.2 is a $0'''$-priority argument (using the classification of Chapter XIV) because a $0'''$-oracle is required after the construction to determine exactly how each requirement R_e is finally satisfied. However, the present construction does not require the explicit use of trees as in Chapter XIV but merely the nested strategies method as in the minimal pair construction in IX §1. Furthermore, if we know a priori that B is not promptly simple, then it requires only a $0''$-oracle to tell exactly how each requirement is satisfied.

3. A Decomposition of the R.E. Degrees Into the Disjoint Union of a Definable Ideal and a Definable Filter

We now prove that **NC** and **M** form a definable filter and a definable ideal in **R**, respectively, and hence $\mathbf{R} = \mathbf{M} \cup \mathbf{NC}$ gives a pleasing algebraic decomposition of **R**. Clearly, both **M** and **NC** are definable in the elementary theory of $(\mathbf{R}, \leq)$. By Corollary 2.3 **NC** forms a strong filter in **R**.

3.1 Theorem.   **M** *is an ideal in* **R**.

Proof. By Definition IX.5.1 if $\mathbf{a} \in \mathbf{M}$ and $\mathbf{b} \leq \mathbf{a}$ then $\mathbf{b} \in \mathbf{M}$. Since $\mathbf{M} = \overline{\mathbf{PS}}$ by Theorem 2.2, it suffices to show that $\overline{\mathbf{PS}}$ is closed under join, namely that if $\mathbf{a}, \mathbf{b} \in \mathbf{R}$ and $\mathbf{a} \cup \mathbf{b} \in \mathbf{PS}$ then either $\mathbf{a} \in \mathbf{PS}$ or $\mathbf{b} \in \mathbf{PS}$. Choose r.e. sets A, B, C such that $C = A \cup B$ is of promptly simple degree via $\{C_s\}_{s \in \omega}$ and

$q(s)$ meeting (1.5), $A \subseteq 2\omega$, the even numbers, and $B \subseteq 2\omega + 1$, the odd numbers.

Let $\{A_s\}_{s \in \omega}$ and $\{B_s\}_{s \in \omega}$ be recursive enumerations of A and B such that $C_s = A_s \cup B_s$. We shall define a recursive function $p(s)$ and partial recursive functions $\hat{p}^i(t)$ for all $i \in \omega$, such that either A is of promptly simple degree via p satisfying (1.5) or else any witness W_i to the failure of p guarantees that $\hat{p}^i$ is total and that B is of promptly simple degree via $\hat{p}^i$ satisfying (1.5). Applying Theorem 1.6 we attempt to satisfy, for all i and j, the requirement

$$P_{i,j} : (\exists x)\,(\exists s)\,[x \in W_{i,\text{ at } s} \ \& \ A_s \restriction x \neq A_{p(s)} \restriction x]$$

$$\vee (\exists y)\,(\exists t)\,[x \in W_{j,\text{ at } t} \ \& \ B_t \restriction y \neq B_{\hat{p}^i(t)} \restriction y].$$

During the construction we define r.e. sets $U_{i,j}$ and assume that $g(i,j)$ is the corresponding function satisfying Lemma 1.5. The sets $U_{i,j}$ are used to "force" numbers to enter C (and hence also A or B) promptly.

Construction of p and $\hat{p}^i$.
    Stage $s = 0$. Set $p(0) = 0$.
    Stage $s > 0$. Find the unique x and i (if they exist) such that $x \in W_{i,\text{ at } s}$. For each $j \leq s$ find the least $t < s$ and $y < s$ such that

(3.1) $$y \in W_{j,\text{ at } t} \ \& \ t \notin \text{dom}(\hat{p}_s^i) \ \& \ z_{i,j} < y,$$

where $z_{i,j} = (\mu z)\,[z \notin U_{i,j,s}]$. If t and y exist, enumerate $z_{i,j}$ in $U_{i,j,s+1}$, and let $v_{i,j}$ be the least v such that $z_{i,j} \in W_{g(i,j),v}$, and otherwise let $v_{i,j} = s+1$. (Necessarily $s < v_{i,j}$ by Lemma 1.5.) Define $p(s) = \max\{q(v_{i,j}) : j \leq s\}$. Define $\hat{p}^i(t) = p(s)$ for all $t \leq s$, $t \notin \text{dom}(\hat{p}_s^i)$. If x and i fail to exist define $p(s) = p(s-1) + 1$.
    This ends the construction.

We claim that if A is not of promptly simple degree then B is. Choose i such that W_i is infinite but for all x,

(3.2) $$x \in W_{i,\text{ at } s} \implies A_s \restriction x = A_{p(s)} \restriction x.$$

Since W_i is infinite it follows that $\hat{p}^i$ is total. If W_j is infinite then $U_{i,j}$ is infinite, so $W_{g(i,j)}$ is infinite. Hence, there exist x, s and y such that $x \in W_{i,\text{ at } s}$, y satisfies (3.1), and $C_s \restriction z_{i,j} \neq C_{q(v)} \restriction z_{i,j}$ where $v = v_{i,j}$. But $p(s) \geq q(v)$ and $A_s \restriction x = A_{p(s)} \restriction x$ so $B_t \restriction y \neq B_{\hat{p}^i(t)} \restriction y$ since $z_{i,j} < y$ and $t < s < v < q(v) \leq p(s) = \hat{p}^i(t)$. Hence, B is of promptly simple degree via $\hat{p}^i$. ▯

Theorem 3.1 shows that the join of two cappable degrees is cappable. Also any $\mathbf{a} \in \mathbf{M}$ is equal to $\mathbf{b} \cup \mathbf{c}$ for strictly smaller degrees $\mathbf{b}, \mathbf{c} \in \mathbf{M}$ by the Sacks Splitting Theorem VII.3.2 and the downward closure of $\mathbf{M}$. However, we cannot necessarily choose $\mathbf{b}$ and $\mathbf{c}$ to form a minimal pair because Lachlan [1979] has constructed an r.e. degree $\mathbf{a}$ which bounds no minimal pair (see Chapter XIV §4). It is easy to see that any r.e. degree $\mathbf{d} > \mathbf{0}$ either bounds a minimal pair or is part of one, so Lachlan's degree $\mathbf{a}$ is in $\mathbf{M}$. L. Welch has shown [1981] that there is no r.e. degree $\mathbf{a} < \mathbf{0}'$ such that $\mathbf{a} \geq \mathbf{b}$ for every $\mathbf{b} \in \mathbf{M}$. Thus, $\mathbf{M}$ is not contained in any proper principal ideal of $\mathbf{R}$. Dually, a straightforward cone-avoidance argument shows that $\mathbf{NC}$ is not contained in any proper principal filter of $\mathbf{R}$. Also note that $\mathbf{M}$ is not a maximal ideal since by Theorem 4.2 the ideal generated by $\mathbf{M} \cup \{\mathbf{a}\}$ is proper for every low r.e. degree $\mathbf{a}$. (There are low r.e. degrees $\mathbf{a} \notin \mathbf{M}$ since there are low r.e. degrees with join $\mathbf{0}'$ so low r.e. degrees can be low cuppable and hence noncappable.) However, $\mathbf{M}$ is a prime ideal (i.e., $\mathbf{a} \cap \mathbf{b} \in \mathbf{M}$ implies $\mathbf{a} \in \mathbf{M}$ or $\mathbf{b} \in \mathbf{M}$) because its complement $\mathbf{NC}$ is a filter.

3.2 Exercise

3.2.   Prove that if $\mathbf{a}_1, \mathbf{a}_2, \ldots, \mathbf{a}_n \in \mathbf{M}$ then there exists an r.e. degree $\mathbf{b} > \mathbf{0}$ such that $\mathbf{a}_i \cap \mathbf{b} = \mathbf{0}$ for all $i \leq n$, and furthermore an index of a representative of $\mathbf{b}$ may be found effectively from indices of representatives of $\mathbf{a}_1, \mathbf{a}_2, \ldots, \mathbf{a}_n$. *Hint.* Use Exercise 2.4 and Theorem 3.1.

4. Cuppable Degrees and the Coincidence of Promptly Simple and Low Cuppable Degrees

4.1 Definition. (i) An r.e. degree $\mathbf{a}$ is *cuppable* (i.e., *cups*) if there exists an r.e. degree $\mathbf{b} < \mathbf{0}'$ such that $\mathbf{a} \cup \mathbf{b} = \mathbf{0}'$.

(ii) Furthermore, $\mathbf{a}$ is *low cuppable* (i.e., *low cups*) if $\mathbf{b}$ may be chosen to be low.

(iii) Let $\mathbf{LC}$ denote the class of low cuppable degrees.

Clearly, the notion of cuppable is dual to the notion of cappable of IX.5.1. Cupping properties have played an important role in the study of r.e. degrees and the proof of the undecidability of $\mathbf{R}$ as discussed in Notes 4.4. For example, Harrington proved the pleasing cup or cap theorem (see Fejer and Soare [1981, Corollary 2.4]) which asserts that every r.e. degree either caps or cups and that some degrees both cup *and* cap. We now continue the equivalence of Theorem 2.2 by proving that $\mathbf{NC} = \mathbf{LC}$, namely that every r.e. degree either caps or low cups but none does both, thereby eliminating the overlap in Harrington's theorem.

4.2 Theorem.   NC = LC.

Proof. By the Low Non-Diamond Theorem XI.3.4 no half of a minimal pair is low cuppable so **LC** $\subseteq$ **NC**. Thus it suffices to prove that **NC** $\subseteq$ **LC**. Recall that **NC = PS** by Theorem 2.2.

Choose a set B of promptly simple degree which is contained in the odd numbers, $2\omega + 1$, and a recursive function $p(s)$ satisfying (1.6). We wish to build a low r.e. set $A \subseteq 2\omega$, the even numbers, such that $K \leq_T A \oplus B$. Note that $A \oplus B \equiv_T A \cup B$. Choose j such that $W_j = K$ and let $K_s = W_{j,s}$. We have a list of "coding markers" $\{\Gamma_n\}_{n\in\omega}$, and we let Γ_n^s denote the position of Γ_n at the end of stage s. We arrange that for all n and s, Γ_n^s is even and

(1) $n \in K - K_s \implies (A_s \cup B_s) \restriction (\Gamma_n^s + 1) \neq (A \cup B) \restriction (\Gamma_n^s + 1)$,
(2) $\Gamma_n^s \leq \Gamma_n^{s+1}$ and $\Gamma_n^s < \Gamma_{n+1}^s$,
(3) $\Gamma_n^s < \Gamma_n^{s+1} \implies (A_s \cup B_s) \restriction (\Gamma_n^s + 1) \neq (A \cup B) \restriction (\Gamma_n^s + 1)$, and
(4) $(\forall n)\ [f(n) =_{\mathrm{dfn}} \lim_s \Gamma_n^s < \infty]$.

These conditions clearly guarantee that $K \leq_T A \cup B$ since $f \leq_T A \cup B$ by (3) and (4), and for each n if s is such that $(A \cup B) \restriction (f(n) + 1) = (A_s \cup B_s) \restriction (f(n) + 1)$ then $n \in K$ iff $n \in K_s$ by (1).

To make A low we meet for all e the lowness requirement

$$N_e : (\exists^\infty s)\ [\{e\}_s^{A_s}(e)\ \text{converges}] \implies \{e\}^A(e)\ \text{converges}.$$

We accomplish this by attempting to clear from the A-use, $A \restriction u(A_s; e, e, s)$, of the computation $\{e\}_s^{A_s}(e)$ all markers Γ_n, $n \geq e$, by using the prompt simplicity of B to force $B \restriction \Gamma_n^s$ to change. During the construction we define r.e. sets U_e, $e \in \omega$. Let g be the corresponding recursive function obtained by Lemma 1.5.

Construction of A.
   Stage $s = 0$. Set $\Gamma_n^0 = 2n$ for all $n \in \omega$.
   Stage $s + 1$.
      Step 1. Find the least e such that $\{e\}_s^{A_s}(e)$ converges and

$$\Gamma_e^s \leq u(A_s; e, e, s).$$

(If no such e exists go to Step 2.) Enumerate Γ_e^s in U_e. Find the least stage t such that $\Gamma_e^s \in W_{g(e),t}$. (By Lemma 1.5, $s < t$.)
      Case 1 (Free clear). $B_s \restriction \Gamma_e^s \neq B_{p(t)} \restriction \Gamma_e^s$. Move all markers Γ_i, $i \geq e$, maintaining their order to new even positions in $\overline{A}_s$ and greater than both $u(A_s; e, e, s)$, and their old positions.
      Case 2 (Capricious destruction). $B_s \restriction \Gamma_e^s = B_{p(t)} \restriction \Gamma_e^s$. Enumerate Γ_e^s in A (thereby capriciously destroying the computation $\{e\}_s^{A_s}(e)$), and move all markers Γ_i, $i \geq e$, maintaining their order, to new even positions in $\overline{A}_s$ greater than their old positions.

Step 2. If $n \in K_{s+1} - K_s$ enumerate the current position of Γ_n into A and move all markers Γ_m, $m \geq n$, to new even positions not yet in A.

This ends the construction.

Lemma 1. $(\forall n)\, [f(n) = \lim_s \Gamma_n^s < \infty]$.

Proof. If not find the least n such that Γ_n moves infinitely often, and choose s_0 such that $K_s \upharpoonright (n+1) = K_{s_0} \upharpoonright (n+1)$, and $\Gamma_m^{s_0} = \lim_s \Gamma_m^s$ for all $m < n$ and $s > s_0$. Now since Γ_n moves infinitely often after s_0, $W_{g(n)}$ is infinite but Case 1 never applies after stage s_0 (else the computation would remain cleared forever). At each stage $s + 1 > s_0$ when the construction applies to Γ_n^s, $x = \Gamma_n^s$ is enumerated in $U_{n,s+1}$ and hence in $W_{g(n),t}$ for some $t > s$ but $B_t \upharpoonright x = B_{p(t)} \upharpoonright x$, so $W_{g(n)}$ violates p satisfying (1.6).

Lemma 2. $(\forall e)\, [N_e$ is satisfied$]$ and hence A is low.

Proof. Choose a stage s such that $\{e\}_s^{A_s}(e)$ converges and for all $i \leq e$, $\Gamma_i^s = \lim_t \Gamma_i^t$. Now $\Gamma_e^s > u = u(A_s; e, e, s)$ else some Γ_i, $i \leq e$, moves at stage $s+1$ (no matter which case applies) contrary to the choice of s. Hence $A_s \upharpoonright u = A \upharpoonright u$ so $\{e\}^A(e)$ converges. ▯

4.3 Corollary (Harrington Cup or Cap Theorem).   *Every r.e. degree either cups or caps.* ▯

4.4 Notes. A degree $\mathbf{a} \in \mathbf{R}$ has the *anticupping* (*a.c.*) *property* if there exists an r.e. degree $\mathbf{b} < \mathbf{a}$ such that for no r.e. $\mathbf{c} < \mathbf{a}$ does $\mathbf{a} = \mathbf{b} \cup \mathbf{c}$. Consequence IX (0.2) of Shoenfield's conjecture asserts that no $\mathbf{a} \in \mathbf{R}$ has the a.c. property. Lachlan [1966c] disproved IX (0.2). Ladner and Sasso [1975] proved that every $\mathbf{a} \in \mathbf{R}^+ =_{\mathrm{dfn}} \mathbf{R} - \{\mathbf{0}\}$ has a predecessor $\mathbf{b} \in \mathbf{L}_2 \cap \mathbf{R}^+$ with the a.c. property. Yates, Cooper [1974a], and Harrington [1976] showed that $\mathbf{0}'$ has the a.c. property and Harrington showed that all degrees $\mathbf{a} \in \mathbf{H}_1$ have it, as proved in D. Miller [1981a]. The proof of this is particularly interesting because in it we have restraint functions $r(e, s)$ where $\liminf_s r(e, s) = \infty$. The solution is to use the pinball machine model in VIII §6 but to spread out the restraint imposed by negative requirement N_e over all gates G_i, $i \geq e$ (not merely at G_e as before) so it will still be true that only finitely many residents are permanent residents of each gate G_e. It follows from the next paragraph that not every degree has the a.c. property but the exact classification of such degrees remains open.

At the opposite extreme of a.c. property is the *plus-cupping* property which asserts of a degree $\mathbf{a} \in \mathbf{R}$ that

$$(\forall\, \mathbf{b} \leq \mathbf{a})\, (\forall\, \mathbf{d} \geq \mathbf{a})\, [\mathbf{0} < \mathbf{b} \implies (\exists\, \mathbf{c} < \mathbf{d})\, [\mathbf{b} \cup \mathbf{c} = \mathbf{d}]],$$

where the quantifiers range over degrees in $\mathbf{R}$. This property asserts that every nonzero degree $\mathbf{b} \leq \mathbf{a}$ can be nontrivially cupped to every $\mathbf{d} \geq \mathbf{a}$. Using the $0'''$-priority method of Lachlan [1975a] discussed in Chapter XIV, Harrington [1978] constructed an r.e. degree $\mathbf{a} > \mathbf{0}$ with the plus-cupping property. If we take $\mathbf{d} = \mathbf{0}'$ then the proof requires only a minimal pair type argument as in Theorem 2.2 and Chapter IX §1, rather than a full tree construction as in Chapter XIV. This is shown in Fejer and Soare [1981, Theorem 2.1]. From this version one can immediately deduce the Harrington Cup or Cap Theorem 4.3 and the Harrington Cup and Cap Theorem which asserts that some r.e. degrees both cup and cap. Cupping properties of the r.e. degrees were developed further with the $0'''$-priority method by Harrington and Shelah [1982a and 1982b] to prove the undecidability of the elementary theory of $(\mathbf{R}, \leq)$.

Chapter XIV

The Tree Method and $0'''$-Priority Arguments

In Chapter VII we introduced the finite injury priority method and in Chapters VIII and IX the infinite injury priority method. These methods have also been called the $0'$-priority and $0''$-priority methods, respectively, as explained in §2. Lachlan [1975a] introduced a new more powerful method which may be viewed as a finite injury argument on top of an infinite injury argument. Lachlan used the method to prove his Nonsplitting Theorem which asserts that the Sacks Splitting Theorem VII.3.2 and Density Theorem VIII.4.1 cannot be combined. When first introduced, the method was not very well understood because the proof appeared to be composed of a series of technical devices one on top of the next, each designed to correct the shortcomings of its predecessor. The method was informally referred to during the 1970's as the "monster method" because of its apparent great complexity.

Some of the key ideas were taken out and used elsewhere; for example, the gap-cogap method was used in the Plus-Cupping Theorem of Harrington [1978] (see also Fejer and Soare [1981]) and later in Theorem XIII.2.2. The method gradually became better understood (Harrington [1982]) and was then used by Harrington and Shelah [1982a and 1982b] to introduce enough definability into the structure of the r.e. degrees $(\mathbf{R}, <)$ in order to prove the undecidability of its elementary theory. This new method, now known as the $0'''$-priority method, is immensely powerful. It promises to be as useful during the next decade or two as the $0''$-priority method was during the 1960's and 1970's. It has already had several other important applications recently such as Slaman's theorem [ta] on the density of the branching degrees.

We have chosen to illustrate this method by proving the Nonbounding Theorem of Lachlan [1979], which is probably the easiest example of the method. The method depends heavily upon the use of trees in priority arguments. These trees are very useful in their own right independent of the $0'''$-priority method. We first illustrate the use of trees in $0'$- and $0''$-priority arguments in §1 and §3, with some general discussion about trees and the classification of $0^{(n)}$-priority arguments in §2. (We pronounce "$0'$-priority" as "zero prime priority," and similarly for $0^{(n)}$.) In §4 we prove the Nonbounding Theorem.

1. The Tree Method With 0′-Priority Arguments

In Theorem VII.2.1 we gave the standard proof of the original Friedberg-Muchnik Theorem illustrating the finite injury priority method. We now recast the same proof using trees. In this case there is no particular advantage in using trees, but it will serve to introduce the reader to tree constructions in a very simple setting.

We introduce some general terminology and notation about trees, which we shall use in this chapter. Let Λ be a countable (often finite) set with a linear ordering $<_\Lambda$ and define T to be the tree $\Lambda^{<\omega}$, the finite sequences of elements of Λ, and $[T]$ to be the set of infinite paths through T, where h is a infinite path through T if $h \upharpoonright n \in T$ for all n. Lower case Greek letters $\alpha, \beta, \gamma, \ldots$ range over $\Lambda^{<\omega}$, and f and g range over Λ^ω. Let $|\alpha|$ denote the length of α. Let $\alpha \subseteq \beta$ ($\alpha \subset \beta$) denote that string β extends (properly extends) α. Let λ denote the empty string. Let $\langle a \rangle$ denote the string consisting of element a alone. Let $\alpha^\frown \beta$ denote the concatenation of string α followed by string β. Let $(\mu\beta \subseteq \alpha)\, P(\beta)$ denote the string $\beta \subseteq \alpha$ of least length satisfying predicate P.

1.1 Definition. Let $\alpha, \beta \in T$.

(i) α is to the *left* of β ($\alpha <_\mathrm{L} \beta$) if

$$(\exists a, b \in \Lambda)\, (\exists \gamma \in T)\, [\gamma^\frown \langle a \rangle \subseteq \alpha \ \& \ \gamma^\frown \langle b \rangle \subseteq \beta \ \& \ a <_\Lambda b].$$

(ii) $\alpha \leq \beta$ if $\alpha <_\mathrm{L} \beta$ or $\alpha \subseteq \beta$.
(iii) $\alpha < \beta$ if $\alpha \leq \beta$ and $\alpha \neq \beta$.

Note that $\alpha \leq \beta$ is a kind of modified Kleene-Brouwer ordering. If $\alpha \subset \beta$ then α is a *predecessor* of β and β is a *successor* of α. (Thus, we view the tree T as growing downward with λ as the top node.)

We assume the reader is familiar with the notation of Theorem VII.2.1, which we now restate.

1.2 Theorem (Friedberg-Muchnik).   *There exist r.e. sets A and B such that $A \not\leq_\mathrm{T} B$ and $B \not\leq_\mathrm{T} A$.*

Recall that it suffices to meet for all e the requirements of VII.2.1,

$$R_{2e} : A \neq \{ e \}^B,$$

and

$$R_{2e+1} : B \neq \{ e \}^A.$$

For this section we fix $\Lambda = \{ 0, 1 \}$ with $<_\Lambda$ the usual order, and $T = \Lambda^{<\omega}$. For each $\alpha \in T$ we have an α-strategy which is a version of the basic strategy in Theorem VII.2.1, and which is designed to attempt to satisfy requirement R_i, where $i = |\alpha|$. Namely α has a candidate (potential witness) $x(\alpha)$ and a restraint $r(\alpha)$ whose values at the end of stage s are denoted by $x(\alpha, s)$ and $r(\alpha, s)$, respectively. Now α *requires attention at stage* $s + 1$ if $|\alpha| = 2e$ and

$$(1.1) \qquad \{e\}_s^{B_s}(x(\alpha, s))\downarrow = 0 \ \& \ r(\alpha, s) = -1,$$

or $|\alpha| = 2e + 1$ and (1.1) holds with A_s in place of B_s. At some stage $s + 1$ a certain α which requires attention and whose guess "seems correct" will *receive attention* (i.e., *act*), at which time we enumerate $x(\alpha, s)$ in A if $|\alpha| = 2e$ (in B if $|\alpha| = 2e + 1$, respectively), set $r(\alpha, s + 1) = s$ to preserve the computation, and *initialize* (and therefore possibly injure) all nodes $\beta > \alpha$ (since these are the nodes of T of weaker priority than α) by defining $r(\beta, s + 1) = -1$ and redefining $x(\beta, s + 1)$.

The difference between this and the previous construction is that previously some R_e may act as many as 2^e times, since R_e begins anew after each injury. Here, each of the 2^e strings α of length e acts at most once, since α is equipped with a "guess" about the outcomes of attempts to satisfy R_i, for all $i < e$, and α will only act when this guess "seems correct." Requirement R_e will ultimately be met by the unique α of length e with the correct guess, namely, $\alpha = f \restriction e$, where $f \in 2^\omega$ is the *true path* defined as follows.

1.3 Definition. The *true path* $f \in 2^\omega$ through T (for the construction below) is defined by induction on n. Let $\alpha = f \restriction n$. Define

$$f(n) = \begin{cases} 0 & \text{if } (\exists s) \, [R_\alpha \text{ acts at stage } s], \\ 1 & \text{otherwise.} \end{cases}$$

Although f is not recursive, clearly $f \leq_T \emptyset'$, and indeed, $f(n) = \lim_s \delta_s(n)$ where we define the recursive sequence $\{\delta_s\}_{s \in \omega}$ of strings in T (such that $|\delta_s| = s$) as follows by induction on $n < s$. Given $\alpha = \delta_s \restriction n$. If $n < s$ define

$$\delta_s(n) = \begin{cases} 0 & \text{if } (\exists t \leq s) \, [R_\alpha \text{ acts at stage } t], \\ 1 & \text{otherwise.} \end{cases}$$

Notice that for each n, $\delta_{s+1} \restriction n \leq \delta_s \restriction n$ (i.e., as s increases, δ_s never moves rightward), so $\lim_s \delta_s(n)$ exists and equals $f(n)$. (In the more general $0''$ or $0'''$-tree constructions we can only guarantee that $f(n) = \liminf_s \delta_s(n)$ in the sense of §2, namely, that f is the leftmost path visited infinitely often by the recursive sequence $\{\delta_s\}_{s \in \omega}$.)

Thus each $\alpha \in T$ may be identified with a guess about which $\beta \subset \alpha$ will eventually act, namely, α guesses that $\beta = \alpha \restriction k$ will act iff $\alpha(k) = 0$. If $\alpha \subseteq \delta_s$ then at stage $s + 1$ α's guess *seems correct* and α is eligible to act. (Note that at such a stage α is guessing that the only strings $\beta \subset \alpha$ which will ever act have already done so.)

Tree Proof of Theorem 1.2.

Construction of A and B.

For each $\alpha \in T$ we have parameters $x(\alpha)$ and $r(\alpha)$ whose values at the end of stage s are denoted by $x(\alpha, s)$ and $r(\alpha, s)$. When a parameter is assigned a value it retains that value until a new value is assigned. Let $\omega^{[\alpha]}$ denote $\omega^{[n]}$ where n is the code number for α under some effective coding of the members of T. To *initialize* α at stage s means to define $r(\alpha, s) = -1$, and $x(\alpha, s)$ to be the least $y \in \omega^{[\alpha]}$, $y \notin A_s \cup B_s$, $y > s$, and $y >$ all previous values of the parameter $x(\alpha)$.

Stage $s = 0$. Set $A_0 = B_0 = \emptyset$ and initialize all $\alpha \in T$.

Stage $s + 1$. Let α be the $\subset$-minimal $\gamma \subseteq \delta_s$ which requires attention. Set $r(\alpha, s + 1) = s$, enumerate $x(\alpha, s)$ in A if $|\alpha|$ is even and in B if $|\alpha|$ is odd, and initialize all γ, $\alpha < \gamma$. We say that α *acts.* (If no such α exists do nothing.)

Lemma. *For every i, requirement R_i is satisfied by $\alpha = f \restriction i$.*

Proof. Fix i, let $\alpha = f \restriction i$, and assume the lemma for all $j < i$. Choose s minimal such that $\alpha \subseteq \delta_t$ for all $t \geq s$, using the fact that $\delta_{v+1} \restriction i \leq \delta_v \restriction i$ for all v. Now $r(\alpha, s) = -1$ and α is never initialized after stage s. Let $x = \lim_t x(\alpha, t) = x(\alpha, s)$. Suppose $i = 2e$. If α never acts then $\{e\}^B(x) \uparrow$ or $\{e\}^B(x) \downarrow \neq 0$. If α acts at some stage $t > s$, then $\{e\}_t^{B_t \restriction t}(x) \downarrow = 0$ and $A_t(x) = 1$, but $x(\beta, t) > t$ for all $\beta > \alpha$ so β cannot later injure α by contributing some $y \leq t$ to $A \cup B$. Furthermore, if $\beta < \alpha$ then β will never act after stage t by hypothesis on s. In either case $\{e\}^B(x) \neq A(x)$. $\square$

Notice that there is no injury in the construction along the true path. (We say that node β is *injured* at stage t if β acts at some stage $s < t$ and then β is initialized at stage t.) Namely, if $\alpha \subset \beta \subset f$ then α never injures β, because β has a correct guess about whether α will ever act and β only acts after its guess seems correct. Note that even stronger we have here that no $\alpha \subset f$ is *ever* injured by *any* $\beta \in T$. This is not true for $\emptyset''$-constructions.

1.4–1.5 Exercises

1.4. Give a tree proof of Theorem VII.1.1.

1.5. Give a tree proof of Theorem VII.3.1.

2. The Tree Method in Priority Arguments and the Classification of $0'$, $0''$ and $0'''$-Priority Arguments

With the simple tree proof of §1 in mind, we now make some general remarks on the use of trees in the priority method before we study some more difficult examples. Some of these remarks were suggested by Harrington [1982] and in private conversations, although the use of trees was introduced by Lachlan [1975a] and used in Lachlan [1979]. We adopt some of the notation, terminology and conventions of Lachlan [1979], which we find particularly convenient, although our proof in §4 of the Nonbounding Theorem will be different from Lachlan's.

To prove a theorem in recursion theory we first write down a list of requirements, R_i, $i \in \omega$, sufficient to establish it. We then formulate a strategy called the *basic module* for satisfying a single such requirement in isolation. The basic module may require infinitely many actions and may have several possible final outcomes (perhaps infinitely many).

Let Λ be a set of symbols for the possible outcomes, and $<_\Lambda$ an appropriate ordering of Λ, generally chosen so that if $a <_\Lambda b$ and a "appears correct" infinitely often, then b is not the correct final outcome. (For example, in §1, $\Lambda = \{0, 1\}$ where outcome 1 denotes the outcome that $\{e\}^B(x)\uparrow$ or $\{e\}^B(x)\downarrow \neq 0$; and outcome 0 denotes that $\{e\}^B(x)\downarrow = 0$ and that we enumerate x in A.)

If there are several different types of requirements (e.g., positive and negative) we may need a different type of basic module for each, and the outcome set Λ will vary according to the type of the requirement. (For simplicity, we first consider the case where there is only one type of basic module and a single outcome set Λ. The other types can easily be combined with this method in the obvious way as in §4.)

Next, we define the priority tree $T = \Lambda^{<\omega}$ and use the notation and terminology about T defined in §1, especially Definition 1.1. For each $\alpha \in T$ we define an α-strategy for meeting requirement R_i, where $i = |\alpha|$. The α-strategy is merely the natural adaptation of the basic module but takes into account that for each $k < |\alpha|$, $\alpha(k) = a \in \Lambda$ means that α "guesses" that the outcome of the β-strategy for $\beta = \alpha \upharpoonright k$ is a. (This gives α a considerable advantage over the standard linear version of the same argument where R_i has no information about the outcomes of strategies for higher priority requirements R_j, $j < i$.) We defer for the moment the specification of the whole construction and exactly when the α-strategy is allowed to act.

At the end of the construction, we define the *true path* $f \in \Lambda^\omega$ by induction on n as in §1. Namely, if $\alpha = f \upharpoonright n$, let $f(n) \in \Lambda$ be the final outcome of the α-strategy. In presenting the full recursive construction, $\mathcal{C} = \bigcup\{\mathcal{C}_\alpha : \alpha \in T\}$, where $\mathcal{C}_\alpha$ is that portion of the construction performed by the α-strategy, our main objective is to ensure that what Harrington [1982] calls the *construction along the true path,* namely, $\hat{C} = \bigcup \{ C_\alpha : \alpha \subset f \}$, succeeds because at the end we shall verify

(2.1) $(\forall i) [\alpha = f \restriction i \implies \alpha$ satisfies requirement $R_i]$.

Of course, $\hat{C}$ is not really a "construction" in any effective sense, since f is not recursive, but $\hat{C}$ is merely a part of the whole recursive construction C. Nevertheless, $\hat{C}$ is the only essential part of C because by (2.1) the only α's which will matter at the end are those $\alpha \subset f$. However, since we cannot recursively identify *during* the construction those $\alpha \subset f$, we must specify possible action C_α for *every* $\alpha \in T$ in such a way that if $\alpha \subset f$ then α is given at least the minimal environment required by the original basic module to succeed.

How well can we identify f during the construction? In our example of §1, the outcomes $\Lambda = \{0, 1\}$ were each finitary in nature and we defined a recursive sequence $\{\delta_s\}_{s \in \omega}$ such that $f(n) = \lim_s \delta_s(n)$. In an infinite injury (namely, $0''$) construction the outcomes will usually include infinitary as well as finitary outcomes. For example, in the Thickness Lemma VIII.1.1 we have positive requirements of the form $P_e : B^{[e]} =^* A^{[e]}$ where we are constructing an r.e. set A, and B is a given r.e. set such that $B^{[e]}$ is either $\omega^{[e]}$ or is finite. If $|\alpha| = i$ and $R_i = P_e$ then the final outcomes for α are $\Lambda = \{0, 1\}$ where 1 denotes that R_i contributes at most *finitely* many elements to A, while 0 denotes that R_i contributes *infinitely* many. In cases such as the minimal pair construction IX.1.2, or the Thickness Lemma VIII.1.1, R_i may be a negative requirement N_e with an associated recursive restraint function $r(e, s)$ such that $\liminf_s r(e, s)$ exists. The set of outcomes for N_e will be $\Lambda = \omega$ where outcome k denotes that $\liminf_s r(e, s) = k$, as in §3.1.

We shall define a recursive approximation to f, $\{\delta_s\}_{s \in \omega}$, where $\delta_s \in \Lambda^{<\omega}$ and $|\delta_s| = s$. Fix s and define $\delta_s(n)$ by induction on n for $n < s$. Suppose $\alpha = \delta_s \restriction n$. Using the Thickness Lemma example above, if R_n is N_e, define $\delta_s(n) = r(e, s)$. If R_n is P_e then define $\delta_s(n) = 0$ if $A_s^{[e]} - A_t^{[e]} \neq \emptyset$, where t is the greatest stage $< s$ such that $\alpha \subseteq \delta_t$, and $\delta_s(n) = 1$ otherwise. The crucial point is that f is the leftmost path visited infinitely often by $\{\delta_s\}_{s \in \omega}$, namely, that $f \restriction n = \liminf_s \delta_s \restriction n$, for all $n \in \omega$, in the sense that if $\alpha = f \restriction n$ then

(2.2) $(\exists^{<\infty} s) [\delta_s <_{\mathrm{L}} \alpha]$,

and

(2.3) $(\exists^{\infty} s) [\alpha \subseteq \delta_s]$.

Furthermore, we shall arrange that

(2.4) $(\forall \beta) [\delta_s <_{\mathrm{L}} \beta \implies \beta$ is initialized at the end of stage $s]$, where β is *initialized at stage* s as in §1 by resetting all its parameters and cancelling any pending β-action. At stage $s + 1$, those $\alpha \subseteq \delta_s$ are said to *appear correct*, and s is called an α-*stage* for such α. Furthermore, α is only eligible to act when α appears correct, namely,

$$(2.5) \qquad (\forall \alpha)\,(\forall s)\,[\alpha \text{ acts at stage } s + 1 \implies \alpha \subseteq \delta_s].$$

(The picture is that at the end of stage s, those β to the right of δ_s, i.e., $\delta_s <_L \beta$, are initialized and any pending β-action or β-restraint is cancelled. During stage $s+1$, those $\alpha \subseteq \delta_s$ are eligible to act because their "guess" is in agreement with δ_s. Those $\beta <_L \delta_s$ are regarded as "asleep" during stage $s+1$ while waiting for some later stage t such that $\beta \subseteq \delta_t$. While β is asleep, any pending β-action or β-restraint is preserved but no new β-action is taken. If $\alpha = f \restriction n$ then like the princess who awakens only once every hundred years, α will awaken at each of the infinitely many α-stages, possibly take some action, and will go to sleep until the next α-stage, during which interval α's action is preserved. During the recursive construction $\mathcal{C}$ none of all the β, $|\beta| = i$, working on R_i know which β will be chosen for royalty (namely, which β will be $\alpha =_{\text{dfn}} f \restriction i$) and will succeed in satisfying R_i. Thus each β simply "does his duty to logic" by doing his best to satisfy R_i. Those $\beta <_L \alpha$ will act finitely often and will possibly not satisfy R_i, while those β, $\alpha <_L \beta$, will be initialized infinitely often, thus also possibly failing to satisfy R_i, while only α is certain to succeed.)

Fix $\alpha = f \restriction n$ and consider the construction from the point of view of α. As in Harrington [1982] decompose $T - \{\alpha\}$ into the disjoint sets of nodes to the left, right, above, and below α:

$$(2.6) \qquad \begin{aligned} L &= \{\beta : \beta <_L \alpha\}, \\ R &= \{\beta : \alpha <_L \beta\}, \\ A &= \{\beta : \beta \subset \alpha\}, \\ B &= \{\beta : \alpha \subset \beta\}, \end{aligned}$$

and let $\mathcal{C}_X = \{\mathcal{C}_\beta : \beta \in X\}$ for $X \subseteq T$. By (2.2) and (2.5) there are only finitely many stages when any $\beta <_L \alpha$ acts, say none after s_0, so $\mathcal{C}_L$ has only a finite effect upon α. Those $\beta \in R$ will be initialized just before any α-stage, and so $\mathcal{C}_R$ will never give any interference when α wants to act. Finally, choose $s_1 \geq s_0$ so that for all $\beta \subset \alpha$ if β acts only finitely often then β acts at no stage $s \geq s_1$. Any action performed by α at an α-stage $> s_1$ will remain in force until the next α-stage. In most $0''$-constructions we can show that the basic module can be adapted to succeed if played on an infinite recursive set of stages (the α-stages $> s_1$) in place of *all* stages $s \in \omega$, and so the α-strategy will succeed. (This modification is the main contribution of the version of the minimal pair proof in Theorem IX.1.2 and is an essential feature of any tree argument at level $0''$ or $0'''$.) For $0'''$-constructions, the

α-strategy requires not only this but also the active "cooperation" by $\mathcal{C}_A$ as we describe later in §4.4. Roughly, the guiding principle is what Harrington [1982] calls the *golden rule*, namely, that α must behave towards $\beta \in R$ ($\beta \in B$, respectively) as α would have L (A, respectively) behave towards α, providing that β's guess appears correct often enough. Thus, α achieves the cooperation of $\mathcal{C}_A$ and actively cooperates with certain $\beta \supset \alpha$ when β seems correct.

The tree method suggests a classification of priority arguments into either $0'$, $0''$ or $0'''$-arguments as follows. We define a construction $\mathcal{C}$ to be of level $0^{(n)}$ if it requires a $\emptyset^{(n)}$-oracle to compute (at the end of the construction) exactly how each requirement R_i was satisfied. In all tree arguments (even $0'''$) the approximation $\{\delta_s(n)\}_{s \in \omega}$ will be recursive as a function of s and n. For the finite injury constructions, as in §1, we have $f(n) = \lim_s \delta_s(n)$, so $f \leq_T \emptyset'$, while in the infinite injury constructions we merely have $f \upharpoonright n = \liminf_s \delta_s \upharpoonright n$ as in (2.2) and (2.3), so $f \leq_T \emptyset''$. But if $\alpha = f \upharpoonright n$, then $f(n)$ is the true outcome of the α-strategy, and α satisfies R_n by (2.1). Hence the finite injury arguments have level $0'$ and the infinite injury arguments level $0''$.

In most tree arguments, the true path f is recursive in $\emptyset''$ because $f \upharpoonright n = \liminf_s \delta_s \upharpoonright n$ as above and $\{\delta_s\}_{s \in \omega}$ is a recursive sequence. Thus, there is a mystery as to how any construction $\mathcal{C}$ can be classified as having level $0'''$. The answer is that (2.1) no longer holds. Namely, in a typical $0'''$-construction (for example, in §4) the requirements R_i are more complicated than before, being in our example of the logical form

$$(2.7) \qquad R_i \equiv P_{i,1} \lor P_{i,2} \lor \ldots \lor P_{i,m} \lor (\exists X_i) \, [N_i(X_i) \ \& \ (\forall j) \, R_{i,j}(X_i)],$$

where X_i is some object being built during the construction such as an r.e. set (as in (2.8)) or a Turing reduction as in Lachlan [1975a] or Slaman [ta].

For example, in the Lachlan Nonbounding Theorem of §4, to construct a nonrecursive r.e. set C whose degree does not bound a minimal pair we must meet for all i the requirement

$$(2.8) \qquad \begin{aligned} R_i : \ &\Phi_i^C \neq A_i \lor \Psi_i^C \neq B_i \lor A_i \text{ is recursive} \lor B_i \text{ is recursive} \\ &\lor (\exists D_i) \, [D_i \leq_T A_i \ \& \ D_i \leq_T B_i \ \& \ (\forall j) \, [D_i \neq \overline{W}_j]]. \end{aligned}$$

where $\{(A_i, B_i, \Phi_i, \Psi_i)\}_{i \in \omega}$ is a standard enumeration of all 4-tuples (A, B, Φ, Ψ) such that A, B are r.e. sets and Φ, Ψ are p.r. functionals, and where D_i is an r.e. set. (Note that (2.8) is of the form of (2.7) where $m = 4$, $P_{i,k}$, $k \leq 4$, is the kth clause of (2.8), N_i asserts "$D_i \leq_T A_i$ and $D_i \leq_T B_i$" and the subrequirement $R_{i,j}$ asserts "$D_i \neq \overline{W}_j$.")

Since the overall requirement R_i is too complicated to attack all at once, we begin to construct D_i and describe a basic module to satisfy $R_{i,j}$ for a fixed i and j. The set $\Lambda = \{s, g2, g1, w, 1, 0\}$ of all possible outcomes of the two modules in §4 will contain a special subset $\Lambda_1 = \{ g2, g1, 1 \}$. We shall define in §4 the priority tree $T \subseteq \Lambda^{<\omega}$, certain recursive functions $i, j : T \to \omega$, and the true path $f \in [T]$ where $[T]$ denotes the set of infinite paths through T as in Definition 4.3. Each node $\alpha \in T$ of even length will be assigned to test the hypotheses $\Phi_i^C = A_i$ and $\Psi_i^C = B_i$ for $i = i(\alpha)$. Each $\alpha \in T$ of odd length will be associated with a variant of the basic module which tries to satisfy the subrequirement $R_{i(\alpha), j(\alpha)}$. The significance of $\Lambda_1 = \{ g1, g2, 1 \}$ is that if $\alpha \subset f$ and $f(|\alpha|) \in \Lambda_1$ then one of the first four disjuncts of (2.8) holds for $i = i(\alpha)$ and α witnesses that R_i is immediately satisfied. Fix i. If no $\alpha \subset f$ witnesses that R_i is satisfied in this way then for every j there will be some odd $\beta \subset f$, with $i(\beta) = i$ and $j(\beta) = j$, such that β witnesses that subrequirement $R_{i(\beta), j(\beta)}$ is satisfied, in which case $f(|\beta|) \in \Lambda - \Lambda_1$. In this case R_i is satisfied by the fifth disjunct of (2.8).

Now a $\emptyset'''$-oracle suffices to determine exactly how each requirement R_i is satisfied as follows. As usual $f \leq_T \emptyset''$. For R_0 we ask the $\emptyset'''$-oracle whether there exists $\alpha \subset f$ such that $i(\alpha) = 0$ and $f(|\alpha|) \in \Lambda_1$. If so then we let α_0 be the least such α, and we can determine exactly which of the first four clauses of (2.8) holds. If not we let α_0 be the longest $\alpha \subset f$ of even length such that $i(\alpha) = 0$, and the construction will associate with α_0 the r.e. set D_0 satisfying the fifth clause of (2.8). Given α_i we can similarly decide how requirement R_{i+1} is satisfied by searching for α_{i+1} from among the nodes α, $\alpha_i \subset \alpha \subset f$.

An interesting feature of such a construction is that if R_i is satisfied in the former way as witnessed by α_i then we regard this satisfaction of R_i as an "injury" to all requirements $R_{i'}$, for $i' > i$, and we force the strategies for these requirements to begin anew on nodes $\beta \supset \alpha_i$. This is called "finite injury along the true path" and is explained further in §4.3.

3. The Tree Method With $0''$-Priority Arguments

3.1. Trees Applied to an Ordinary $0''$-Priority Argument

In Theorem IX.1.2 we constructed a minimal pair of r.e. degrees. We now give a tree version of that construction that helps to understand it better. We assume the definitions and terminology used in IX.1.2.

We define the restraint function $r(e, s)$ using the tree $T = \Lambda^{<\omega}$ where $\Lambda = \omega$ with the usual order. We shall define a recursive sequence of strings $\{ \delta_t \}_{t \in \omega}$ which approximate the true path f as in §2. For $\alpha \in T$ a stage s is an α-stage if $\alpha \subseteq \delta_s$ or $s = 0$. Let S^α be the set of α-stages. The α-strategy is the same as the basic module but with S^α in place of ω. Namely, define $l(\alpha, s) = l(e, s)$, where $e = |\alpha|$, and

$$m(\alpha, s) = \max\{ l(\alpha, t) : t \leq s \ \& \ t \in S^\alpha \}.$$

A stage s is *α-expansionary* if $s = 0$ or s is an α-stage and $l(\alpha, s) > m(\alpha, s-1)$. The restraint function for the α-strategy alone is

$$(3.1) \quad r(\alpha, s) = \begin{cases} 0 & \text{if } s \text{ is } \alpha\text{-expansionary,} \\ \text{the greatest } \alpha\text{-expansionary} & \text{otherwise.} \\ \quad \text{stage } t < s \end{cases}$$

For $s \geq 0$ define $\delta_s \in T$, $|\delta_s| = s$, as follows. Given $\alpha = \delta_s \restriction e$, $e < s$, define

$$\delta_s(e) = r(e, s) =_{\text{dfn}} \max\{ r(\beta, s) : \beta \leq \alpha \}.$$

Note that for $\beta <_L \alpha$, $r(\beta, s)$ is defined by the second clause of (3.1) so $\delta_s \restriction e$ suffices to define $r(\beta, s)$ for all $\beta \leq \alpha$.) Now the true path $f \in \Lambda^\omega$ is defined by $f(e) = r(e) =_{\text{dfn}} \liminf_s r(e, s)$, and has the property that $f \restriction e = \liminf_s \delta_s \restriction e$ in the sense of (2.2) and (2.3). The rest of the proof is the same as in Theorem IX.1.2.

This definition produces essentially the same restraint function $r(e, s)$ as the definition in Theorem IX.1.2 but it more clearly illustrates how the α-strategies are combined. The reason that the use of trees can be eliminated in the proof as given in Chapter IX, is that the argument there resembles a Markov process in that the only necessary bit of information for the α-strategy to know is $\alpha(e-1) = r(e-1, s)$, where $e = |\alpha|$, because we have ensured that $r(e-1, s) > r(j, s)$ for all $j < e-1$. In the present argument the α-strategy needs to know only $\max\{ r(\beta, s) : \beta < \alpha \}$. This applies to many other constructions also, such as the plus-cupping theorem (see Fejer and Soare [1981]). However, certain other constructions are best done on a tree if they have both a minimal pair type restraint function $r(e, s)$ and infinitary positive requirements, as we now see in the following theorem which is perhaps the most typical example of a $0''$-priority argument using trees.

3.2. A $0''$-Priority Argument Which Requires the Tree Method

Lachlan [1966b] constructed a pair of maximal sets whose degrees form a minimal pair. By Martin's result XI.2.5 the degrees of maximal sets are precisely the high r.e. degrees. Hence, Lachlan's result is equivalent to the existence of a minimal pair of high r.e. degrees.

We are interested in the latter result because in the proof we shall construct r.e. sets A and B where we must guess not only about the value of $\liminf_s r(e, s)$ as in §3.1, but also about which rows $A^{[e]}$ and $B^{[e]}$ are infinite. We construct A and B as thick subsets of a certain r.e. set C as described below so that $A^{[e]} =^* B^{[e]} =^* C^{[e]}$ for all e, and therefore we need merely guess about whether $C^{[e]} = \omega$ or $C^{[e]} =^* \emptyset$. For this section, we shall use the priority tree,

$$(3.2) \qquad T = \{ \alpha : \alpha \in \omega^{<\omega} \ \& \ (\forall e) \, [\alpha(2e+1) \in \{0, 1\}] \}.$$

The true path f through T will be such that

$$(3.3) \qquad\qquad f(2e) = r(e) =_{\text{dfn}} \liminf_s r(e, s),$$

and

$$(3.4) \qquad\qquad f(2e + 1) = \begin{cases} 0 & \text{if } C^{[e]} \text{ is infinite,} \\ 1 & \text{if } C^{[e]} \text{ is finite.} \end{cases}$$

The method used here can be adapted to give tree proofs for virtually all the standard infinite injury constructions, for example, those in Chapters VIII or IX. More importantly, the tree method also applies to all those 0″-arguments requiring both infinitary positive requirements (as in the usual infinite injury) and also a minimal pair type of restraint function $r(e, s)$ as in Theorem IX.1.2 rather than the type of restraint function for avoiding an upper cone as in a typical infinite injury argument such as the Sacks Density Theorem VIII.4.1 or the Sacks Jump Theorem VIII.3.1. In the latter two cases the restraint functions can be made to drop back simultaneously using the "true stages" method of Chapter VIII §1. The minimal pair type restraint function requires the nested strategies method of Theorem IX.1.2 to make the restraint functions drop back simultaneously and hence requires the tree method when combined with infinitary positive requirements. Thus, this section should be viewed as a more powerful method than that of Chapter VIII for handling infinite injury arguments. For example, the tree method can be used to prove Exercise XII.2.13 where the "true stages" infinite injury method of Chapter VIII §1–§4 did not suffice and we needed the pinball machine approach to infinite injury as described in Chapter VIII §5. The tree method also gives a more perspicuous proof of Theorem XII.5.1 (see Exercise 3.8).

3.1 Theorem (Lachlan [1966b]). *There exists a minimal pair of high r.e. degrees.*

Proof. We shall construct nonrecursive r.e. sets A and B whose degrees form a minimal pair by satisfying for every e the following negative requirement as in Theorem IX.1.2,

$$N_e : \{e\}^A = \{e\}^B = g \text{ total} \implies g \text{ is recursive.}$$

Now, analogously to the set B of Corollary VIII.1.2, fix an r.e. set C such that for all e,

$$(3.5) \qquad\qquad e \in \emptyset'' \implies C^{[e]} \text{ is finite,}$$

and

$$(3.6) \qquad\qquad e \notin \emptyset'' \implies C^{[e]} = \omega^{[e]}.$$

To ensure that A and B have high degree we construct them as *thick subsets* of C, namely, $A, B \subseteq C$, and meeting for every e the positive ("thickness") requirements,

$$P_e^A : A^{[e]} =^* C^{[e]},$$

and

$$P_e^B : B^{[e]} =^* C^{[e]}.$$

Note that $\{ P_e^A \}_{e \in \omega}$ and $\{ P_e^B \}_{e \in \omega}$ ensure that $\emptyset'' \leq_T A'$ and $\emptyset'' \leq_T B'$ by the Limit Lemma III.3.3 because (3.5) and (3.6) guarantee that

$$\lim_x A(\langle x, e \rangle) = \lim_x (B\langle x, e \rangle) = 1 - \emptyset''(e).$$

Let T be as in (3.2). For each s we shall define $\delta_s \in T$, $|\delta_s| = 2s$, such that for all n, $f \restriction n = \lim \inf_s \delta_s \restriction n$ as in §2.

3.2 Definition. An A-computation $\{ e \}_s^{A_s}(x) = y$ is *α-correct* if

$$(3.7) \qquad \begin{aligned} &(\forall i < e) \, [\alpha(2i + 1) = 0 \\ &\qquad \implies (\forall z) \, [[\alpha(2i) < z \leq u(A_s; e, x, s) \ \& \ z \in \omega^{[i]}] \implies z \in A_s^{[i]}]], \end{aligned}$$

and similarly for a B-computation (with B_s in place of A_s).

(The intuition is that by (3.3) α guesses that

$$\alpha(2i) = r(i) =_{\mathrm{dfn}} \lim \inf_s r(i, s).$$

Also by (3.4), (3.5), (3.6), and P_i^A, if $\alpha(2i + 1) = 0$, then α guesses that $C^{[i]} = \omega^{[i]}$ and hence $A^{[i]} =^* \omega^{[i]}$, and indeed, α believes that all numbers $z > r(i)$, $z \in \omega^{[i]}$, will eventually enter $A^{[i]}$. Thus, α does not "believe in" a computation until these elements have entered A.)

We now adapt the basic module to $\alpha \in T$, $|\alpha| = 2e$, by defining the recursive functions

$$l(\alpha, s) = \max\{ x : (\forall y < x) \, [\{ e \}_s^{A_s}(y) \! \downarrow \, = \{ e \}_s^{B_s}(y) \! \downarrow$$
$$\text{and these computations are } \alpha\text{-correct}] \},$$

$$m(\alpha, s) = \max\{ l(\alpha, t) : t \leq s \ \& \ t \in S^\alpha \}.$$

A stage s is *α-expansionary*: if $s = 0$; or if $s > 0$, s is an α-stage, and $l(\alpha, s) > m(\alpha, s - 1)$. (If $|\alpha|$ is odd, say $2e + 1$, these three definitions have no meaning, since α is assigned to P_e^A and P_e^B.)

Construction of A and B.

Stage $s = 0$. Let $A_0 = B_0 = \emptyset$, and $r(\alpha, 0) = 0$ for all $\alpha \in T$, and $r(\alpha, s) = 0$ for all α such that $|\alpha|$ is odd and for all s.

Stage $s + 1$. First define $\delta_s \in T$, $|\delta_s| = 2s$, by performing the steps $e < s$ below in increasing order of e. Next, choose the least integer $x = \langle y, e \rangle < s$

such that $x \in C_s^{[e]} - A_s^{[e]}$ or $x \in C_s^{[e]} - B_s^{[e]}$ and $x > r(e, s) = \delta_s(2e)$. Enumerate x in A if $x \notin A_s$ and in B otherwise. If no such x exists, do nothing. Notice that at most one element enters either A or B at any stage, and no element enters both.

  Step e, $0 \le e < s$. Given $\alpha = \delta_s \upharpoonright 2e$, define $\delta_s(2e)$ and $\delta_s(2e + 1)$ as follows. For $\beta \le \alpha$, $|\beta|$ even, define

$$(3.8) \quad r(\beta, s) = \begin{cases} 0 & \text{if } \beta \subseteq \alpha \text{ and} \\ & s \text{ is } \beta\text{-expansionary,} \\ \text{the greatest } \beta\text{-expansionary} \\ \quad \text{stage } t < s & \text{otherwise.} \end{cases}$$

Define

$$\delta_s(2e) = r(e, s) =_{\mathrm{dfn}} \max\{\, r(\beta, s) : \beta \le \alpha \,\}.$$

Now let $\gamma = \delta_s \upharpoonright (2e + 1)$. Define

$$\delta_s(2e + 1) = \begin{cases} 0 & \text{if } |C_s^{[e]}| > |C_t^{[e]}| \text{ where } t \text{ is the greatest } \gamma\text{-stage} < s, \\ 1 & \text{otherwise.} \end{cases}$$

(The intuition is that γ is assigned to P_e^A and P_e^B and γ "wakes up" only at γ-stages. If the cardinality of $C^{[e]}$ has increased since γ was last awake, then our current guess is that $C^{[e]}$ will be infinite, so we set $\delta_s(2e + 1) = 0$ according to (3.4), P_e^A, and P_e^B; and $\delta_s(2e + 1) = 1$ otherwise.)

  If $e < s - 1$, go to step $e + 1$. Otherwise, define $r(\beta, s) = r(\beta, s - 1)$ for all β such that $r(\beta, s)$ has not yet been defined and complete stage $s + 1$.

Lemma 1 (True Path Lemma).   *There exists $f \in \Lambda^\omega$, called the* true path, *such that*

  (i) $(\forall n)$ $[f \upharpoonright n = \liminf_s \delta_s \upharpoonright n$ as defined in (2.2) and (2.3)$]$,
  (ii) $(\forall e)$ $[f(2e) = r(e) =_{\mathrm{dfn}} \liminf_s r(e, s) < \infty]$,
  (iii) $(\forall e)$ $[[f(2e + 1) = 0 \iff |C^{[e]}| = \infty]$
        $\&$ $[f(2e + 1) = 1 \iff |C^{[e]}| < \infty]]$.

Proof. We prove (i), (ii) and (iii) simultaneously by induction. Fix $e \ge 0$, let $\alpha = f \upharpoonright 2e$, and assume (i) for $n = 2e$, and (ii) and (iii) for all $e' < e$. Choose t_0 minimal such that there is no $s > t_0$, such that $\delta_s <_L \alpha$. Choose $t_1 > t_0$ such that $\alpha \subseteq \delta_{t_1}$ and let $k = \max\{\, r(\beta, t_1) : \beta < \alpha \,\}$. Now, as in Lemma 1 of Theorem IX.1.2, if there are infinitely many α-expansionary stages s, then $r(\alpha, s) = 0$ at each such s and $f(2e) = r(e) = k$. Otherwise, there is a largest such stage v, $r(\alpha, s) = v$ for almost all s, and $r(e) = \max\{\, k, v \,\}$. In either case, we have (i) for $n = 2e + 1$ and (ii) for e. Now let $\gamma = f \upharpoonright (2e + 1)$. If $C^{[e]}$ is finite then $\delta_s(2e + 1) = 1$ for almost all s. If $|C^{[e]}| = \infty$ then there are infinitely many s such that $\delta_s \supset \gamma^\frown\langle 0\rangle$ because there are infinitely many γ-stages. In either case, we have (i) for $n = 2e + 2$ and (iii) for e.

Lemma 2. $(\forall e)\, [A^{[e]} =^* C^{[e]} =^* B^{[e]}]$. *(More precisely, for all e, $A^{[e]} \subseteq C^{[e]}$ and $C^{[e]} \subseteq \{0, 1, \ldots, r(e)\} \cup A^{[e]}$, and similarly for $B^{[e]}$ in place of $A^{[e]}$.)*

Proof. Choose any $x \in C^{[e]}$, such that $x > r(e)$. Choose s_0 such that $x \in C_{s_0}$, and for all $y < x$, $y \in A \cup B$ iff $y \in A_{s_0} \cup B_{s_0}$. At the next stage $s+1 > s_0$ such that $r(e, s) = r(e)$, x is enumerated in A_{s+1} if $x \notin A_s$, and x is enumerated in B_{s+1} if $x \in A_s - B_s$. In the former case choose the next stage $t+1 > s+1$ such that $r(e, t) = r(e)$. Thus, x is enumerated in B_{t+1}.

Lemma 3. $(\forall e)\, [\{e\}^A = \{e\}^B = g$ total $\implies g$ is recursive].

Proof. Assume that $\{e\}^A = \{e\}^B = g$ is a total function. Let $\alpha = f \restriction 2e$. We shall show that the α-strategy satisfies N_e. By Lemma 1(iii), α has the correct guess about $|C^{[i]}|$, for all $i < e$, namely, $\alpha(2i + 1) = 0$ iff $|C^{[i]}| = \infty$. By Lemma 1(i) choose an α-stage t such that there is no $s \geq t$ with $\delta_s <_L \alpha$. Note that by Lemma 1(ii) we have

$$(3.9) \qquad (\forall i < e)\, (\forall s \geq t)\, [r(i) \leq r(i, s)],$$

and hence

$$(3.10) \qquad (\forall i < e)\, (\forall x \leq r(i))\, [x \in A^{[i]} \cup B^{[i]} \iff x \in A_t^{[i]} \cup B_t^{[i]}].$$

By Lemma 1(iii), we also have

$$(3.11) \qquad (\forall i < e)\, [C^{[i]} \text{ finite} \implies C_t^{[i]} = C^{[i]}],$$

because if $C^{[i]}$ is finite, then $\alpha(2i + 1) = 1$, but if some x enters $C^{[i]}$ at a stage $v > t$ then at some stage $s \geq v > t$ we must have $\delta_s \supseteq (\alpha \restriction 2i)^\frown \langle 0 \rangle$, so $\delta_s <_L \alpha$ contrary to the choice of t. Now we shall also assume t has been chosen sufficiently large so that

$$(\forall i < e)\, [C^{[i]} \text{ finite} \implies [A_t^{[i]} = A^{[i]} \ \& \ B_t^{[i]} = B^{[i]}]].$$

To recursively compute $g(p)$ find the least α-expansionary stage $v > t$ such that $l(\alpha, v) > p$. Let $g(p) = \{e\}_v^{A_v}(p)$. Note that v exists because if $\{e\}^A(p) = q$ with use u, then the computation $\{e\}_s^{A_s}(p) = q$ will be α-correct cofinitely often, namely, it will be correct for all $s > t$ such that $z \in C_s$ for all $z \in \omega^{[i]}$ such that $i < e$, $\alpha(2i + 1) = 0$, and $r(i) < z \leq u$. The same is true for B-computations. As in Theorem IX.1.2 Lemma 3, it now follows by induction on $s \geq v$ that for all $s \geq v$

$$(3.12) \qquad \{e\}_s^{A_s}(p) = q \text{ is an } \alpha\text{-correct computation,}$$

or

$$(3.13) \qquad \{e\}_s^{B_s}(p) = q \text{ is an } \alpha\text{-correct computation.}$$

The extra point here, beyond those in Theorem IX.1.2 Lemma 3, is that by (3.10), (3.11) and the definition of α-correct computation, we cannot have $z \in A^{[i]}_{s+1} - A^{[i]}_s$ or $z \in B^{[i]}_{s+1} - B^{[i]}_s$ if $i < e$, $s > t$, and z is used in an α-correct computation $\{e\}^{A_s}_s(p) = q$ or $\{e\}^{B_s}_s(p) = q$. As usual, the restraint function $r(e, s)$ prevents any $z \in \omega^{[i]}$ for $i \geq e$ from entering $A \cup B$ and destroying a computation (3.12) or (3.13), unless s is an α-expansionary stage, in which case *both* (3.12) and (3.13) hold for s.   ▯

3.3 Remark. In the proof of Theorem 3.1 and in essentially all $0''$-tree constructions, we could have modified the tree so that there is no injury in the "construction along the true path" $\hat{C} = \bigcup \{ C_\alpha : \alpha \subset f \}$. Namely, if $\beta \subset \alpha \subset f$ then the α-strategy is never injured by β in the following sense.

Let $\alpha = f \upharpoonright 2e$. Define a slightly modified restraint function $\tilde{r}(\beta, s)$ as in (3.8) but in the second clause of (3.8) replace t by u, the greatest element used in an A- or B-computation at step t on any argument $x < l(\alpha, t)$. (Note that $\tilde{r}(\alpha, s) \leq r(\alpha, s)$ and is sufficient restraint to show that the α-strategy succeeds.) We say that α is *injured at stage* $s + 1$ if $z \in (A_{s+1} - A_s) \cup (B_{s+1} - B_s)$ for some $z \leq \tilde{r}(\alpha, s)$. Furthermore, if $z \in \omega^{[i]}$ and $\beta = \delta_s \upharpoonright (2i + 1)$, we say β *injures* α.

Now α can be injured at most finitely often by those $\beta <_L \alpha$. In the proof of Theorem 3.1, α can be also injured finitely often by those $\beta \subset \alpha$ if $|\beta| = 2i + 1$ and $\alpha(2i + 1) = 1$ because α knows that $C^{[i]}$ is finite but does not know its exact members. This can be remedied by replacing the tree T of (5.1) by $T = \omega^{<\omega}$ and (3.4) by

$$f(2i + 1) = \begin{cases} 0 & \text{if } |C^{[i]}| = \infty, \\ k + 1 & \text{if } C^{[i]} = D_k, \end{cases}$$

where $\{D_k\}_{k \in \omega}$ is the usual canonical indexing of finite sets. Now if $\alpha = f \upharpoonright 2e$ then α has sufficient information about $C^{[i]}$ for all $i < e$, and defines α-correct computations accordingly, so that α is never injured by any $\beta \subset \alpha$.

3.4 Remark. In the above construction we could have let the priority tree be simply $T = 2^{<\omega}$. Now $f(2e + 1)$ is defined by (3.4) as before. If $\alpha = f \upharpoonright 2e$, then $f(2e) = 0$ if there are infinitely many α-expansionary stages and $f(2e) = 1$ otherwise.

The construction and proof are carried out as above, except that in the definition of α-correct computation in (3.7) we replace "$\alpha(2i) < z$" by "$r(i, s) < z$," define $r(e, s)$ as above, but define $\delta_s(2e)$ as follows. If $\alpha = \delta_s \upharpoonright 2e$, let $\delta_s(2e) = 0$ if s is α-expansionary, and $\delta_s(2e) = 1$ otherwise. Lemma 1(ii) is modified by letting $r(e)$ be as previously, but $f(2e)$ be as defined here. In the proofs of the lemmas no information is lost, because if $\delta_s(2e) = 1$ and $r(e, s) = k$, this is equivalent to the former case where

$\delta_s(2e) = k$. In particular, Lemma 3, (3.9), (3.10), and (3.11) all hold as before.

This approach has some advantages because from now on, particularly in the $0'''$-constructions, it is convenient to use only finitely branching priority trees T. (However, α does not have complete information about all $\beta \subset \alpha$ as in Remark 3.3, and so α may be injured by such β.)

3.5–3.8 Exercises

3.5 (Lachlan).   Use the method of Theorem 3.1 to construct directly two maximal sets whose degrees form a minimal pair.

3.6.   Give a tree proof of the Sacks Jump Theorem VIII.3.1.

3.7 (Shore).   Combine the tree method of Theorem 3.1 with the ideas of Chapter XIII §2 to prove that for every r.e. degree $\mathbf{b}$,

$$\mathbf{b} \in \mathbf{PS} \vee (\exists \, \text{r.e.} \ \mathbf{a}) \, [\mathbf{a}' = \mathbf{0}'' \ \& \ \mathbf{a} \cap \mathbf{b} = \mathbf{0}].$$

3.8 (Slaman).   Give a proof of Theorem XII.5.1 using the tree method in Theorem 3.1.

4. The Tree Method With a $0'''$-Priority Argument: The Lachlan Nonbounding Theorem

4.1 Preliminaries

It is an open question to decide exactly which degrees $\mathbf{a} \in \mathbf{R}^+$ bound minimal pairs.   Cooper (Exercise XI.2.15) showed that any high degree $\mathbf{a} \in \mathbf{H}_1$ will suffice, but in the following theorem, Lachlan shows that not every $\mathbf{a} \in \mathbf{R}^+$ will suffice. (See also the remarks following Theorem XIII.3.1.)

4.1 Nonbounding Theorem (Lachlan [1979]).   *There exists an r.e. degree* $\mathbf{c} > \mathbf{0}$ *with no r.e. minimal pair below it.*

For this section we now adopt some notation of Lachlan [1979] which is convenient for this and other $0'''$-arguments.   Upper case Greek letters $\Phi^A, \Psi^A, \ldots$ denote A-partial recursive functions previously denoted by $\{e\}^A$, $e \in \omega$. When we define an r.e. set A or p.r. functional Φ during a recursive construction, then we use $A[s], \Phi_e[s]$ to denote the result by the end of stage s of the construction (formerly denoted by $A_s, \{e\}_s$, etc.). If m is a parameter then $m[s]$ denotes the value at the end of stage s. We also allow whole expressions to be qualified by "$[s]$." For example, $\Phi^A(x)[s]$ denotes

$\Phi[s]^{A[s]}(x)$. The advantage is that during the construction we regard A, Φ, m and so on as in a state of formation. During a given stage (which may involve many substages) we can allow the notation A to denote the set of elements so far enumerated in A, m to denote the current value of the parameter m, and so on. When necessary to avoid confusion we append s and write $A[s]$, $m[s]$ or $\Phi[s]$ to denote the result by the *end* of stage s.

Proof of Theorem 4.1. We must construct an r.e. set C satisfying for all i the requirements:

$$P_i : C \neq \overline{W}_i,$$

(4.1)

$$R_i : [\Phi_i^C = A_i \;\&\; \Psi_i^C = B_i \;\&\; A_i \text{ is not recursive} \;\&\; B_i \text{ is not recursive}]$$

$$\implies (\exists \text{ r.e. } D_i) [D_i \leq_T A_i \;\&\; D_i \leq_T B_i \;\&\; D_i \text{ nonrecursive}],$$

where $\{ (A_i, B_i, \Phi_i, \Psi_i) \}_{i \in \omega}$ is a standard indexing of all 4-tuples (A, B, Φ, Ψ) such that A and B are r.e. sets and Φ and Ψ are p.r. functionals. Suppose that A_i, B_i, Φ_i and Ψ_i are supplied by a recursive enumeration in a uniformly effective fashion. Without loss of generality, we may assume for all x and s,

(4.2) $$x \in A_i[s] - A_i[s-1] \implies \Phi_i^C(x)[s] = 1,$$

and

(4.3) $$x \in B_i[s] - B_i[s-1] \implies \Psi_i^C(x)[s] = 1.$$

This is achieved by withholding a number x from A_i and B_i until $\Phi_i^C(x)$ and $\Psi_i^C(x)$ take the value 1. (This does no harm since we are only concerned with $(A_i, B_i, \Phi_i, \Psi_i)$ if $A_i = \Phi_i^C$ and $B_i = \Psi_i^C$.) Define

(4.4) $$l^{\Phi_i}[s] = \max\{ x : (\forall y < x) [A_i(y)[s] = \Phi_i^C(y)[s]] \},$$

and similarly $l^{\Psi_i}[s]$ with B_i and Ψ_i in place of A_i and Φ_i. The advantage of (4.2) and (4.3) is that at some stage if $x < l^{\Phi_i}$ we can preserve $A_i \restriction x$ by preserving $C \restriction u$ where u is the greatest element used in the computations $\Phi_{i,s}^{C_s}(y)$, $y < x$, and likewise for $B_i \restriction x$.

4.2 The Basic Module for Meeting a Subrequirement (Designing the Computer Chip)

To understand our strategy, it is convenient to rewrite requirement R_i in the following form:

$$R_i : \Phi_i^C = A_i \;\&\; \Psi_i^C = B_i \implies$$
(4.5)
$$[A_i \text{ recursive} \vee B_i \text{ recursive}$$
$$\vee (\exists \text{r.e. } D_i) [D_i \leq_T A_i \;\&\; D_i \leq_T B_i \;\&\; (\forall j) [D_i \neq \overline{W}_j]].$$

Define the subrequirement

(4.6) $$R_{i,j} : D_i \neq \overline{W}_j.$$

Our strategy is to attempt to satisfy R_i by constructing an r.e. set D_i to satisfy the last disjunct of R_i. The basic module below attempts to meet only a single subrequirement $R_{i,j}$ (consistent with making $D_i \leq_T A_i, B_i$ uniformly). It either achieves that objective, or else it guarantees that either A_i or B_i is recursive, in which case R_i is satisfied at once and no further subrequirements $R_{i,j'}$, $j' \neq j$, need to be examined. (The following basic module is not the same as in Lachlan [1979], but more similar to that hinted at by Harrington [1982], although the latter gives no details. This basic module gives rise to a simpler tree of outcomes than in Lachlan, although the initial module is slightly more complicated, having two types of gaps instead of one.)

Fix i and j and drop the subscripts from A_i, B_i, Φ_i and Ψ_i. We have parameters x, r_1, r_2 and r, which should be viewed as functions of the stage s. A parameter once given a value retains that value until a new value is assigned. Now x is the current candidate to satisfy $R_{i,j}$, r_1 and r_2 are restraint functions which prevent elements from entering C in order to preserve $A \upharpoonright x$ and $B \upharpoonright x$, respectively, and $r = \max\{r_1, r_2\}$. To *reset* the candidate x at stage $s + 1$ means to cancel the old candidate $x[s]$ and to define $x[s+1]$ to be the least $y \in \omega^{[\langle i,j \rangle]}$, $y > s+1$, and y is greater than any previous candidates for any module. To *initialize* (the basic module) means to reset x and to set $r_1 = r_2 = 0$. We initialize at stage $s = 0$ and at any later stage when some $z \leq r$ enters C (due to the action for some requirement P_k), or when the module is "injured" by the action of one of higher priority, in which case the basic module begins all over on a new candidate x.

From now on we assume as in (4.5) that $\Phi^C = A$ and $\Psi^C = B$. The basic module consists of the following steps. (See Diagram 4.1.)

Step 1. Wait for s such that $x \in W_j[s]$. At stage $s + 1$ *open an A-gap* by setting $r_1[s + 1] = 0$ and go to step 2.

Step 2. Wait for the least $t \geq s+1$ such that $l^\Phi[t] > x$. At stage $t+1$ *close the A-gap* by performing step 2(a) or 2(b) according to which case applies.

Step 2(a). (Successful close). Suppose $A[s] \upharpoonright x \neq A[t] \upharpoonright x$. Open a B-gap by defining $r_2[t + 1] = 0$. (Note that r_1 remains 0 so $r[t + 1] = 0$.) Go to step 3.

Step 2(b). (Unsuccessful close). Suppose $A[s] \upharpoonright x = A[t] \upharpoonright x$. Define $r_1[t + 1] = t$ (to preserve $A[t] \upharpoonright x$), reset x, and go to step 1.

Step 3. Wait for the least $v \geq t+1$ such that $l^\Psi[v] > x$. At stage $v + 1$ *close the B-gap* by performing step 3(a) or 3(b) according to which case applies.

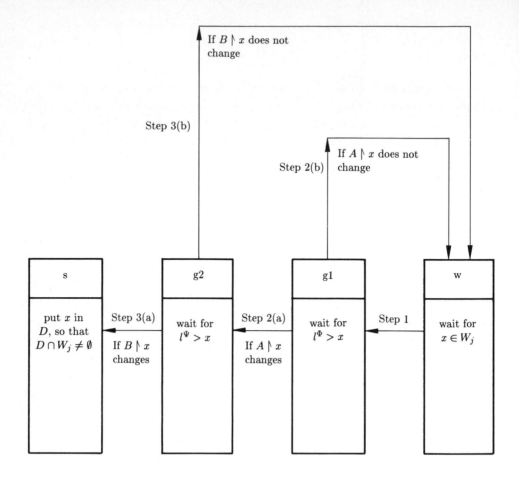

Diagram 4.1. Diagram of the Basic Module as a Finite Automaton

Step 3(a). (Successful close). Suppose $B[v] \restriction x \neq B[t] \restriction x$. Enumerate x in D, and stop.

Step 3(b). (Unsuccessful close). Suppose $B[v] \restriction x = B[t] \restriction x$. Define $r_2[v+1] = v$ (to preserve $B[v] \restriction x$), reset x, and go to step 1.

(Note that at the opening of an A-gap (B-gap), the dropping of the restraint r_1 (r_2) will allow other positive requirements to enumerate elements into C, thereby causing Φ^C (Ψ^C) to change and thus l^Φ (l^Ψ, respectively) to drop down.)

First note that if the hypotheses of (4.5) hold, then $D \leq_T A$ and $D \leq_T B$ uniformly because we require both A- and B-permission (namely, changes in $A \restriction x$ and $B \restriction x$) before x enters D. Now $R_{i,j}$ is satisfied if we either wait forever for s such that $x \in W_j[s]$, so $D \cup W_j \neq \omega$, or if some B-gap is closed successfully, in which case x enters D and $D \cap W_j \neq \emptyset$. Every A-gap (B-gap) must be eventually closed because $\Phi^C = A$ ($\Psi^C = B$).

Suppose that there are infinitely many B-gaps, say, begun at stages $\{t_n\}_{n \in \omega}$ and closed (necessarily unsuccessfully) at stages $\{v_n\}_{n \in \omega}$, where $t_0 < v_0 < t_1 < v_1 < \cdots$. The intervals of stages $\{s : t_n \leq s < v_n\}$ are called B-*gaps*, since $r_2 = 0$ in these intervals and we are free to enumerate any number into C, while the intervals $\{s : v_n \leq s < t_{n+1}\}$ are called B-*cogaps*, during which time we have $r_2 = v_n$.

Let us suppose that the basic module is *injured* only finitely often, namely, that there exists s_0 such that

$$(4.7) \qquad (\forall s > s_0)\,(\forall z)\,[z \in C[s+1] - C[s] \implies z > r[s]].$$

Then we can show that B is recursive as follows. To compute $B(p)$ choose $t > s_0$ such that $x = x[t] > p$, and a B-gap is begun at stage $t+1$. Since the B-gap is closed unsuccessfully at $v+1 > t+1$, $B[t] \restriction x = B[v] \restriction x$. If the next B-gap is begun at stage $t'+1 > v+1$, then the C-restraint r_2 ensures that $B[v] \restriction x = B[t'] \restriction x$. Now the argument continues with x replaced by $x' = x[t'+1] > x$, and establishes that $B[t] \restriction x = B \restriction x$. Hence $p \in B$ iff $p \in B[t]$. Note that $\lim_t x[t] = \infty$ since each new candidate is chosen to be greater than any previous candidates.

If there are only finitely many B-gaps but infinitely many A-gaps then A is recursive by the same argument with r_1, A and Φ in place of r_2, B and Ψ, respectively, and with $s_1 > s_0$ large enough to satisfy (4.7) and greater than all the stages at which the B-gaps were closed (so that after s_1 every A-gap is closed unsuccessfully).

Finally, note that $\liminf_s r[s] < \infty$. If there are only finitely many A- and B-gaps then $\lim_s r[s]$ exists. If there are infinitely many B-gaps then $\liminf_s r[s] = 0$, since $r[s] = 0$ if a B-gap is begun at stage s. If there are finitely many B-gaps and infinitely many A-gaps, then $k = \lim_s r_2[s]$ exists, and so $r[s] = k$ for all $s > s_1$ such that an A-gap is begun at stage s, because $r_1[s] = 0$.

It is useful to view the basic module as a kind of finite automaton M, whose states are $\{ s, g2, g1, w \}$ arranged in order from left to right as listed, and whose transitions between states are those given by steps 1, 2(a), 2(b), 3(a) and 3(b) of the basic module. Namely, M begins in the starting state w, attempting to move leftward to the final state s. While M is in state w ($g1$, $g2$), M waits for the exit condition associated with the corresponding step 1 (2, 3) to be satisfied. Then M moves left or right to a new state, according to which transition step applies. Note that the desired leftward progress towards the final state s may be interrupted by an unsuccessful close of an A-gap (B-gap) and therefore by a return to the starting state w. This may cause M to perform infinite cycles returning rightward infinitely often from state $g1$ or $g2$ or both.

Let $F[s]$ denote the state of M at the end of stage s. The possible *outcomes* of the basic module are

$$S = \{ s, g2, g1, w \},$$

with ordering $<_S$ as listed. These are divided into the finitary ($\mathbf{0'}$) outcomes $S^{\mathrm{Fin}} = \{ s, w \}$ and the infinitary ($\mathbf{0''}$) outcomes $S^{\mathrm{Inf}} = \{ g1, g2 \}$.

We define the outcome as follows. If $\lim_s F[s] = a$, then $a \in S^{\mathrm{Fin}}$ and the outcome is a. If $\lim_s F[s]$ does not exist, then the outcome is $g2$ if there are infinitely many B-gaps (infinitely many right exits from $g2$), and the outcome is $g1$ otherwise (in which case there are infinitely many right exits from $g1$).

The states of M are shown in Diagram 4.1 as rectangles. (Thus the outcome of the basic module is the $<_S$-least (i.e., leftmost) $a \in S$ such that M either is in, or passes through, rectangle a at infinitely many stages.)

The table shown in Diagram 4.2 summarizes the progress made on R_i or $R_{i,j}$ by the basic module for each outcome $a \in S$.

Fixing i and repeating the basic module for all $j \in \omega$, it is now clear that there is a nonrecursive r.e. set C which satisfies a single requirement R_i of (4.5) and such that D_i can be found uniformly in each of A_i and B_i. It then follows, using the Recursion Theorem, that there is no uniform effective procedure for passing from the index of a nonrecursive r.e. set C to indices of r.e. sets $A, B \leq_T C$, whose degrees form a minimal pair, together with indices of reductions Φ and Ψ from C to A and B.

4.3 The Priority Tree (The Computer Architecture of the Machine)

If we view the basic module in §4.2 as a kind of atomic "computer chip" for performing a small fixed task, then this section deals with how these chips are "hard wired together" to produce the computer architecture of

Outcome	s	g2	g1	w
Effect on $R_{i,j}$	win	none	none	win
Effect on R_i	none	B_i is recursive; R_i is satisfied	A_i is recursive; R_i is satisfied	none

Diagram 4.2. Analysis of Outcomes.

the entire machine. The architecture involves arranging the chips into a certain tree (the *priority tree*) and then assigning a task (requirement) to each chip according to its position on the tree. A great deal of information can be obtained by simply studying the priority tree before proceeding to the construction. In §4.5 we give the construction which corresponds to the "operating system" and tells us what to do at each stage, namely, how the flow of control passes from node to node as we descend along a certain path on the tree *during* a stage of the construction.

Look at the requirement R_i defined in (4.1). Note that the first two hypotheses,

$$(4.8) \qquad \Phi_i^C = A_i \quad \text{and} \quad \Psi_i^C = B_i,$$

are Π_2^C-conditions. We shall approximate these conditions by measuring whether the lim sup of the minimum of $l^{\Phi_i}[s]$ and of $l^{\Psi_i}[s]$ is infinite on a certain infinite set of stages. These approximations suffice and they are simply Π_2-conditions. Hence, like the Π_2-conditions in §3, each can be conveniently added as a node β on the tree above those nodes α which are assigned to work on a subrequirement $R_{i,j}$ for some j. Thus, α may assume that these hypotheses (4.8) are satisfied, and therefore α needs to act only when this guess "seems correct". In order to measure when (4.8) "seems correct" we define the following recursive functions.

4.2 Definition. (i) Let $l^{\Phi_i}[s]$ and $l^{\Psi_i}[s]$ be as in (4.4). Define

$$l_i[s] = \min\{\, l^{\Phi_i}[s], l^{\Psi_i}[s]\,\},$$

and

$$m_\alpha^i[s] = \max\{\, l_i[t] : t \leq s \ \& \ t+1 \in S^\alpha\,\},$$

where S^α will be defined in Definition 4.9.

(ii) A stage v is α-*expansionary* if $v = 0$; or if $v = s+1$, $s+1 \in S^\alpha$, and $l_i[s] > m_\alpha^i[s-1]$, where $i = i(\alpha)$ and $i(\alpha)$ is defined as in Definition 4.4.

4.3 Definition. (i) Let $\Lambda = \{\, s, g2, g1, w, 0, 1\,\}$ with the ordering $<_\Lambda$ in the order as listed. (This is the set of possible outcomes for the strategies on any node.)

(ii) Define the *priority tree*

$$T = \{\, \alpha \in \Lambda^{<\omega} : (\forall n)\, [\alpha(2n) \in \{0,1\} \ \& \ \alpha(2n+1) \in \{\, s, g2, g1, w\,\}]\,\}.$$

(iii) $[T] = \{\, h : h \in \Lambda^\omega \ \& \ (\forall n)\, [h \restriction n \in T]\,\}$. We call these $h \in [T]$ the (infinite) *paths* through T.

(iv) A node $\alpha \in T$ is *even* if $|\alpha|$ is even and *odd* otherwise.

We shall define functions $i, j : T \to \omega$. Every *even* node $\alpha \in T$ will be assigned to the (Φ_i, Ψ_i)-hypothesis (4.8) for $i = i(\alpha)$. The outcome 1 for α's strategy denotes roughly that there are at most finitely many α-expansionary stages so (4.8) fails for i, and hence α believes that R_i is satisfied. The outcome 0 denotes that there are infinitely many α-expansionary stages, and hence that further action to satisfy R_i must be taken by nodes $\beta \supset \alpha$.

Every *odd* node $\alpha \in T$ will be assigned to some subrequirement $R_{i,j}$, for $i = i(\alpha)$ and $j = j(\alpha)$, and the α-strategy will be a variant of the basic module having as possible outcomes $\{\, s, g2, g1, w\,\}$. The outcome $g2$ $(g1)$ denotes that there are infinitely many B_α-gaps (infinitely many A_α-gaps but finitely many B_α-gaps) and hence that B_α $(A_\alpha$, respectively) is recursive. The outcome s (w) denotes that subrequirement $R_{i,j}$ is satisfied because $D_{i(\alpha)} \cap W_{j(\alpha)} \neq \emptyset$ $(D_{i(\alpha)} \cup W_{j(\alpha)} \neq \omega$, respectively).

4.4 Definition. We define partial functions $i, j : T \to \omega$ and functions $L_0, L_1 : T \to P(\omega) = \{\, X : X \subseteq \omega\,\}$, called *lists*, by induction on $n = |\alpha|$ as follows. The function i will be total, and j will be defined exactly on the odd nodes.

For $n = 0$. Define $i(\lambda) = 0$, and $L_0(\lambda) = L_1(\lambda) = \omega$.

For $n > 0$. Let $\alpha = \beta^\frown\langle a \rangle$ for $a \in \Lambda$. Assume that $i(\beta)$, $L_0(\beta)$, and $L_1(\beta)$ are defined and also that $j(\beta)$ is defined if β is odd. First define $L_0(\alpha)$ and $L_1(\alpha)$ as in the next paragraph. Next if α is even define $i(\alpha) = (\mu i)\,[i \in L_0(\alpha)]$, and let $j(\alpha)$ be undefined. If α is odd define $i(\alpha) = i_0$ and $j(\alpha) = j_0$ where $\langle i_0, j_0 \rangle = (\mu n)\,[n \in L_1(\alpha)]$.

Define $L_0(\alpha)$ and $L_1(\alpha)$ as follows. Let $i = i(\beta)$ and if β is odd let $j = j(\beta)$.

Case 1. β is even.

Case 1A. $a = 1$. Define

$$L_0(\alpha) = (L_0(\beta) - \{\,i\,\}) \cup \{\,i' : i' > i\,\},$$

and

$$L_1(\alpha) = (L_1(\beta) - \{\,\langle i, k \rangle : k \in \omega\,\}) \cup \{\,\langle i', k \rangle : i' > i \ \& \ k \in \omega\,\}.$$

Case 1B. $a = 0$. Define

$$L_0(\alpha) = L_0(\beta) - \{\,i\,\},$$

and

$$L_1(\alpha) = L_1(\beta).$$

Case 2. β is odd.

Case 2A. $a \in \{\,g2, g1\,\}$. Define $L_0(\alpha)$ and $L_1(\alpha)$ as in Case 1A.

Case 2B. $a \in \{\,s, w\,\}$. Define $L_0(\alpha) = L_0(\beta)$, and $L_1(\alpha) = L_1(\beta) - \{\,\langle i, j \rangle\,\}$.

The intuition is that as we descend the tree T along any path $h \in [T]$, L_0 represents a "list" of indices i corresponding to the (Φ_i, Ψ_i)-hypotheses of (4.8) and which will be assigned in decreasing order of priority to even nodes $\alpha \in T$. The "list" L_1 represents indices $\langle i, j \rangle$ corresponding to the subrequirements $R_{i,j}$ and which will be assigned to odd nodes. If $\alpha = \beta^\frown\langle a \rangle$, and $a \in \{\,g1, g2, 1\,\}$ then α believes that R_i is satisfied so in constructing $L_0(\alpha)$ $(L_1(\alpha))$ from $L_0(\beta)$ $(L_1(\beta))$ we can remove i $(\{\,\langle i, k \rangle : k \in \omega\,\}$, respectively). However, we regard this "action" to satisfy R_i as "injuring" $R_{i'}$ for all $i' > i$, so these i' $(\{\,\langle i', k \rangle : k \in \omega\,\}$, respectively) must be placed back in $L_0(\alpha)$ $(L_1(\alpha))$ so that all such requirements $R_{i'}$, $i' > i$, begin afresh on nodes $\gamma \supseteq \alpha$.

(There is an obvious analogy between descending along any path $h \in [T]$ and passing through the stages of a finite injury construction such as Theorem VII.2.1. In the latter we begin at stage 0 with a list $L_0 = \omega$ of indices of requirements R_i, which must be satisfied. At each stage $s + 1$ we choose the least $i \in L_s$ such that R_i requires attention. We allow R_i to act and we define

$$L_{s+1} = (L_s - \{\,i\,\}) \cup \{\,i' : i' > i\,\},$$

since all $R_{i'}$, $i' > i$, are injured at stage $s+1$ and must be allowed to act at later stages $v > s+1$.)

4.5 Definition. Fix $\alpha \in T$.

(i) Define

$$\tau(\alpha) = (\mu\beta \subseteq \alpha) \, [\beta \text{ is even } \& \, i(\beta) = i(\alpha)$$
$$\& \, \neg(\exists\gamma)_{\beta \subseteq \gamma \subset \alpha}[i(\gamma) < i(\alpha) \, \& \, \alpha(|\gamma|) \in \{\, g2, g1, 1 \,\}]].$$

If no such β exists let $\tau(\alpha)$ be undefined.

(ii) If $\tau(\alpha)$ is defined, then for $i = i(\alpha)$ define the *i-region* containing α to be

$$E(\alpha, i) = \{\, \beta : \beta \in T \, \& \, \tau(\alpha) \subseteq \beta \, \& \, \tau(\beta) = \tau(\alpha) \, \& \, i(\beta) = i(\alpha) \,\}.$$

(iii) We call $\tau(\alpha)$ the *top* of the *i*-region $E(\alpha, i)$ because it is the $\subset$-least node there. It is also the only even node in $E(\alpha, i)$.

(iv) We call γ an *i-boundary* if $\gamma = \beta^\smallfrown\langle a \rangle$, $i = i(\beta)$, and $a \in \{\, g2, g1, 1 \,\}$.

4.6 Lemma. *Suppose E is an i-region with top τ and $i = i(\tau)$. Then*

(i) $\neg(\exists\beta \in E) \, [\tau \subset \beta = \alpha^\smallfrown\langle a \rangle \, \& \, i(\alpha) < i \, \& \, a \in \{\, g2, g1, 1 \,\}]$,

(ii) $\neg(\exists\beta \in E) \, [\tau \subset \beta \, \& \, \beta \text{ is even } \& \, i(\beta) = i]$.

Proof. The proof of (i) is immediate from the definitions of $\tau(\alpha)$ and $E(\alpha, i)$. To prove (ii) assume to the contrary that β is a $\subset$-minimal node in E satisfying the negation of (ii). Now $i \in L_0(\beta)$ but $i \notin L_0(\tau^+)$ where $\tau^+ = \tau^\smallfrown\langle a \rangle$ for $a = \beta(|\tau|)$. Hence, by Definition 4.4 there is some γ, $\tau \subset \gamma \subset \beta$ with $i(\gamma) < i$ and $\beta(|\gamma|) \in \{\, g2, g1, 1 \,\}$ contrary to (i).   ▯

The intuition is that for any $i \in \omega$ as we descend along any path $h \in [T]$ we eventually come to some even $\tau \subset h$ for which we define $i(\tau) = i$, thereby starting a new i-region E with top τ. However, if we meet a node γ, $\tau \subset \gamma \subset h$, such that $\gamma = \beta^\smallfrown\langle a \rangle$ is an i'-boundary for some $i' < i$, then β believes that $R_{i'}$ has been satisfied (and this will be true if $h = f$, the *true path*), so β *injures* our attempt to meet R_i, and we begin a new i-region with top some τ', $\gamma \subseteq \tau' \subset h$. After finitely many injuries this process stabilizes at some $\tau(h, i)$ defined below which is the top of the final i-region begun in descending along h. The following lemma asserts that for each i as we descend along any path $h \in [T]$ we cross finitely many i'-boundaries, for $i' < i$, so $E(h, i)$ exists as defined in Definition 4.8.

4.7 Finite Injury Along Any Path Lemma. *For every path $h \in [T]$ and every $i \in \omega$,*

(i) $(\exists^{<\infty}\alpha \subset h) \, [i(\alpha) = i \, \& \, h(|\alpha|) \in \{\, g2, g1, 1 \,\}]$,

(ii) $(\exists^{<\infty}\alpha \subset h) \, [i(\alpha) = i \ \& \ \alpha$ is even$]$.

Proof. Fix h and i and assume the lemma for all $i' < i$. Choose $\beta \subset h$ such that for all α, $\beta \subseteq \alpha \subset h$, if $i(\alpha) < i$ then α is odd and $h(|\alpha|) \notin \{\,g2, g1, 1\,\}$. To prove (i) assume that there is some γ, $\beta \subseteq \gamma \subset h$, $i(\gamma) = i$, and $a \in \{\,g2, g1, 1\,\}$, where $a = h(|\gamma|)$. Let $\gamma^+ = \gamma^\frown\langle a\rangle$. Then it follows by induction on the length of α that

$$(\forall\alpha)_{\gamma^+ \subseteq \alpha \subset h}[i \notin L_0(\alpha) \ \& \ (\forall k) \, [\langle i, k\rangle \notin L_1(\alpha)]],$$

and hence

$$(\forall\alpha)_{\gamma^+ \subseteq \alpha \subset h}[i(\alpha) \le i \implies h(|\alpha|) \notin \{\,g2, g1, 1\,\}].$$

To prove (ii) choose the least even γ, $\beta \subseteq \gamma \subset h$, such that $i(\gamma) = i$. Let $\gamma^+ = \gamma^\frown\langle h(|\gamma|)\rangle$. Then by choice of β and Definition 4.4, we have

$$(\forall\alpha)_{\gamma^+ \subseteq \alpha \subset h}[i \notin L_0(\alpha)],$$

and hence

$$(\forall\alpha)_{\gamma^+ \subseteq \alpha \subset h}[\alpha \text{ even} \implies i(\alpha) > i]. \quad \square$$

Lemma 4.7 allows us to define the final i-region for $h \in [T]$ as follows.

4.8 Definition. Fix any $i \in \omega$ and any path $h \in [T]$. Let α be the $\subset$-maximal even node $\beta \subset h$ such that $i(\beta) = i$. (Note that α exists by Lemma 4.7 and Definition 4.4.)
(i) Define $E(h, i) = E(\alpha, i)$, the *final i-region* of h.
(ii) Define $\tau(h, i) = \tau(\alpha) = \alpha$, the *top* of the final i-region $E(h, i)$.

Notice that

$$(4.9) \qquad (\forall\gamma) \, [[\tau(h, i) \subseteq \gamma \subset h \ \& \ i(\gamma) = i] \implies \gamma \in E(h, i)].$$

(One should think of the final i-region $E = E(h, i)$ as a kind of downward cone of nodes with top $\tau = \tau(h, i)$. Of course, E contains many more nodes besides those γ, $\tau \subset \gamma \subset h$, $i(\gamma) = i$, in which we are really interested. However, E is not a true cone (i.e., consisting of all $\gamma \supseteq \tau$, $i(\gamma) = i$), because there may exist β, $\tau \subset \beta \not\subset h$ such that $i(\beta) < i$ and $\gamma \supseteq \beta^\frown\langle a\rangle$ for $a \in \{\,g2, g1, 1\,\}$. In this case $\gamma \notin E$ even if $i(\gamma) = i$.)

4.4 Intuition for the Priority Tree and the Proof

In the next section we shall define the *true path* $f \in [T]$ for the construction by induction on n where $\alpha = f \restriction n$, and $f(n) = a$ indicates that a is the true outcome to the strategy of α. If α is odd then the true outcome $a \in \{\,s, g2, g1, w\,\}$ for α is defined as in the basic module in §4.2. If α is even then $a = 0$ iff there are infinitely many α-expansionary stages, and $a = 1$ otherwise.

The true path f will enable us to verify that each requirement R_i defined in (4.1) has been satisfied. Fix i and consider the final i-region $E = E(f, i)$ with top $\tau = \tau(f, i)$. If $f(|\tau|) = 1$ then $\limsup_{s \in S^\tau} m^i_\tau[s] < \infty$ so either $\Phi^C_i \neq A_i$ or $\Psi^C_i \neq B_i$ and R_i is immediately satisfied. Otherwise, $f(|\tau|) = 0$ and $\limsup_{s \in S^\tau} m^i_\tau[s] = \infty$ so any odd $\alpha \in E$ with $i(\alpha) = i$ which opens a gap eventually closes that gap.

We think of τ as the place where we originate our strategy to define the nonrecursive set D_i and recursive functionals $\Theta^{A_i}_i = D_i$ and $\widehat{\Theta}^{B_i}_i = D_i$ to satisfy the conclusion of R_i in (4.1). To ensure that D_i is not recursive we must be sure that for almost every j there is some odd node α, $\tau \subset \alpha \subset f$, such that $i(\alpha) = i$ and $j(\alpha) = j$. If there exists an odd node α, $\tau \subset \alpha \subset f$, such that $i = i(\alpha)$ and $f(|\alpha|) \in \{g2, g1\}$ then A_i or B_i is recursive and R_i is immediately satisfied. Thus, if the hypotheses of (4.1) are satisfied, α satisfies $\tau \subset \alpha \subset f$, and $i = i(\alpha)$ then $f(|\alpha|) \in \{s, w\}$ indicating that subrequirement $R_{i(\alpha), j(\alpha)}$ is satisfied.

Suppose that all the hypotheses of R_i in (4.1) are satisfied. Let $\tau = \tau(f, i)$. Then we must construct a nonrecursive r.e. set $D_\tau \leq_T A_i, B_i$. Define

$$D_\tau = \{ x(\alpha) : \alpha \in E(f, i) \ \& \ i(\alpha) = i \text{ and } x(\alpha) \text{ is}$$
$$\text{enumerated in } D_\tau \text{ under (4.14) to meet } R_{i(\alpha), j(\alpha)} \}.$$

Then D_τ is r.e. because $E(f, i)$ is a recursive set of nodes. Secondly, D_τ is nonrecursive because for almost every $j \in \omega$ there is some α, $\tau \subset \alpha \subset f$, with $i(\alpha) = i$, $j(\alpha) = j$, and $f(|\alpha|) \in \{s, w\}$, and hence $R_{i,j}$ is met. Finally, $D_\tau \leq_T A_i, B_i$ since for any odd $\alpha \in E(f, i)$ with $i = i(\alpha)$, any α-gap once opened is eventually closed so for any α-candidate $x = x(\alpha)$ there is a stage v at which x is either cancelled or enumerated in D_τ at which time the functionals $\Theta^{A_i}_i(x)[v]$ and $\widehat{\Theta}^{B_i}_i(x)[v]$ may be redefined to equal $D_\tau(x)[v]$. (See Lemma 4.17 for details.)

The key point is that since (4.8) holds we know that $\tau^\frown\langle 0 \rangle \subset f$ and all $\alpha \in E(f, i)$ have $\tau^\frown\langle 0 \rangle \subset \alpha$. Hence, every $\alpha \in E(f, i)$ with $i = i(\alpha)$ will eventually close any open gap so the fate of $x(\alpha)$ can be decided. This will be false for an arbitrary even α with $i = i(\alpha)$ so we must use the set D_τ above to satisfy the conclusion of R_i in (4.1).

Thus, the real *injury* to R_i as we descend along $f \in [T]$ is the following. When we open a new i-region at some even $\tau \subset f$ by defining $i(\tau) = i$ we specify a candidate $(D_\tau, \Theta_\tau, \widehat{\Theta}_\tau)$ for the triple $(D_i, \Theta_i, \widehat{\Theta}_i)$ to satisfy the conclusion of (4.1). If we later pass an i'-boundary β for $\tau \subset \beta \subset f$ and $i' < i$, we must begin a new i-region at some τ', $\beta \subseteq \tau' \subset f$, and also build at τ' a new candidate $(D_{\tau'}, \Theta_{\tau'}, \widehat{\Theta}_{\tau'})$ for the final triple. The significance of Lemma 4.7 and Definition 4.8 is that as we descend along *any* path $h \in [T]$

this process stabilizes with a final value $(D_\tau, \Theta_\tau, \widehat{\Theta}_\tau)$ for $\tau = \tau(h, i)$ and hence a final value $(D_\tau, \Theta_\tau, \widehat{\Theta}_\tau)$ for the triple $(D_i, \Theta_i, \widehat{\Theta}_i)$. (Explicit mention of Θ_i and $\widehat{\Theta}_i$ during the construction may be omitted since these are defined in Lemma 4.17 using delayed A-permitting and delayed B-permitting.)

4.5 The Construction (The Operating System of the Machine)

We shall denote the sets $A_{i(\alpha)}$ and $B_{i(\alpha)}$ by A_α and B_α, and the functionals $\Phi_{i(\alpha)}$ and $\Psi_{i(\alpha)}$ by Φ_α and Ψ_α. For each $\alpha \in T$ we have a strategy which attempts to satisfy requirement

$$P_\alpha : C \neq \overline{W}_k,$$

where $k = |\alpha|$. In addition, for each odd $\alpha \in T$ we have a strategy whose aim is to satisfy the subrequirement

$$R_\alpha : D_{i(\alpha)} \neq \overline{W}_{j(\alpha)}.$$

(Thus, there are two *different* strategies associated with every odd node α.) To achieve the latter we have a variant of the basic module called the α-module which attempts to satisfy subrequirement $R_{i(\alpha), j(\alpha)}$.

During the construction we regard $A_\alpha, B_\alpha, \Phi_\alpha, \Psi_\alpha, C, W_j$ and so on as being in a state of formation. At a given point during the construction if we write A_α we mean the finite set of elements so far enumerated in A_α, and similarly for Φ_α. When necessary to avoid confusion, we append $[s]$ and write $A_\alpha[s]$ or $\Phi_\alpha[s]$ to denote the result by the end of stage s. However, in the verification after the construction, we also use the notations A_α, Φ_α and so on to denote the completed sets and functionals.

To state the construction, it is useful to have for odd nodes $\alpha \in T$ the following parameters (as in the basic module) which have the following meanings:

$x(\alpha) = $ the current candidate for the α-module,
$r(\alpha, 1) = $ the restraint imposed on C to preserve $A_\alpha \restriction x(\alpha)$,
$r(\alpha, 2) = $ the restraint imposed on C to preserve $B_\alpha \restriction x(\alpha)$,
$r(\alpha) = \max\{\, r(\alpha, 1), r(\alpha, 2) \,\}$,
$F(\alpha) = $ the current state $a \in \{\, s, g2, g1, w \,\}$ of the α-module.

The values of the parameters may change during the construction. A parameter once given a value retains that value until redefined. Let $w^{[\alpha]}$ denote $w^{[n]}$ where n is the code number for α in some effective coding of T. Assume that α is odd. To *reset the candidate* $x(\alpha)$ at stage s means to cancel the old candidate $x(\alpha)[s-1]$ (if $s > 0$) and to define $x(\alpha)[s]$ to be the least $y \in w^{[\alpha]}$, $y > s$ and y greater than all previous candidates. To *initialize node* α means: to reset $x(\alpha)$ and to set $F(\alpha) = w$ unless $F(\alpha) = s$; to set $r(\alpha, 1) = r(\alpha, 2) = r(\alpha) = 0$; to *cancel* any A_α- or B_α-gaps; and to remove any links which have bottom equal to α. (Links will be defined later.) If α is even the phrase "initialize node α" means to remove any links which have top equal to α. Furthermore, if α is even we define $r(\alpha, 1)[s] = r(\alpha, 2)[s] = r(\alpha)[s] = 0$ for all s.

At the end of the construction we shall define the *true path* $f \in [T]$ of the construction. We shall approximate f by defining during each stage s of the construction a string $\delta[s] \in T$ in such a way that we can prove

$$f = \liminf_s \ \delta[s],$$

namely $f \restriction n = \liminf_s \ \delta[s] \restriction n$, for all n, as defined in (2.2) and (2.3). During stage $s + 1$ we shall define $\delta[s + 1]$ using a series of substages $t \le s$ at which we define a string δ_t such that $\delta_t \subseteq \delta_{t+1}$ and $\delta[s + 1]$ is the last δ_t defined. Only string δ_t will be allowed to act at substage t.

4.9 Definition. Fix any $\alpha \in T$.

(i) We say that $s + 1$ is an α-*stage* if $\alpha \subseteq \delta[s + 1]$. In addition 0 is an α-stage. Let V^α be the set of α-stages.

(ii) We say $s+1$ is a *genuine α-stage* if $\delta_t = \alpha$ at some substage t of stage $s + 1$, where this δ_t will be defined during the construction. Let S^α denote the set of genuine α-stages.

(iii) Suppose further that α is even. Let $i = i(\alpha)$. We say that v is an α-*expansionary stage* if $v = 0$ or $v = s + 1$, $s + 1$ is a genuine α-stage, and $l_i[s] > m_\alpha^i[s - 1]$.

At stage s if (an odd node) α opens an A_α-gap we shall construct a *link* (τ, α) with bottom α and top (the even node) $\tau = \tau(\alpha)$. In this case we shall have $\tau^\smallfrown\langle 0 \rangle \subseteq \alpha$ and $F(\alpha)[s] = g1$. (Notice that α needs precisely a τ-expansionary stage in order to close its gap according to (4.13) and that τ is associated with the (Φ_i, Ψ_i)-hypothesis (4.8) which if true will ensure that there are infinitely many τ-expansionary stages.) The link (τ, α) remains in force until either: α closes the A_α-gap unsuccessfully (and $F(\alpha) = w$ again); or α closes a B_α-gap (either successfully or unsuccessfully) at which time the link (τ, α) is *removed*; or node α is initialized and the link (τ, α) is removed. (A link (τ, α) if removed may later be recreated.)

4.10 Definition. If at stage $s+1$ there is a link (τ, α) and there is a substage t of stage $s + 1$ such that $\delta_t = \tau$ and $\delta_{t+1} = \alpha$ then we say that the link (τ, α) is *travelled at stage $s + 1$*.

The link (τ, α) should be viewed physically as a wire or short circuit starting at τ and leading directly from τ down to α. During stage $s + 1$ if we have $\delta_t = \tau$ and $s + 1$ is a τ-expansionary stage then we always travel this link (τ, α) directly to α, because α is now in a position to close its A_α-gap or B_α-gap. The purpose of the link is to allow the flow of control of the construction to pass directly from node τ where the (Φ_i, Ψ_i)-hypothesis (4.8) has just been further verified (by $s + 1$ being a τ-expansionary stage) down to some node $\alpha \supseteq \tau^\frown\langle 0 \rangle$ with $i(\alpha) = i(\tau) = i$ where α is in a gap and needs this hypothesis to close the gap.

Travelling this link may cause us to pass over other nodes γ, $\tau \subset \gamma \subset \alpha$, of lower global priority, $i(\gamma) > i(\alpha)$, which would otherwise receive attention before α if we simply descended the tree without using links. For such γ, $s + 1$ is a γ-stage but not a genuine γ-stage since there is no substage t such that $\delta_t[s + 1] = \gamma$ at which γ may act. Thus, to consider whether a given α-strategy succeeds we must examine the *genuine* α-stages rather than merely the α-stages.

The construction is as follows.

Stage $s = 0$. Initialize all nodes $\alpha \in T$. Define $\delta[0] = \lambda$.

Stage $s + 1$. The construction will proceed by substages $t \leq s + 1$. We refer to substage t of stage $s + 1$ as *stage* $(s + 1, t)$. The value of parameter p at the end of substage t will be denoted by p_t, which abbreviates its complete specification $p_t[s + 1]$. At the end of substage t we shall have $\delta_t \in T$ such that $\delta_t \subseteq \delta_{t+1} \subseteq \delta[s + 1]$, $t \leq s$, and only δ_t may act during substage t.

Substage $t = 0$. Define $\delta_0 = \lambda$. Go to substage 1.

Substage $t + 1 \leq s + 1$. We are given δ_t and, for all odd $\alpha \in T$, $F_t(\alpha) \in \{s, g2, g1, w\}$, the current state of the α-module. We first define δ_{t+1} as follows.

Case 1. δ_t is odd. Define $\delta_{t+1} = \delta_t^\frown\langle F_t(\delta_t)\rangle$.

Case 2. δ_t is even.

Case 2A. Stage $s + 1$ is not δ_t-expansionary. Define $\delta_{t+1} = \delta_t^\frown\langle 1 \rangle$.

Case 2B. Stage $s + 1$ is δ_t-expansionary, and there now exists no link with top δ_t. Define $\delta_{t+1} = \delta_t^\frown\langle 0 \rangle$.

Case 2C. Stage $s + 1$ is δ_t-expansionary and there now exists a link with top δ_t and bottom some $\beta \supseteq \delta_t^\frown\langle 0 \rangle$. (We have ensured that β will be unique as shown in Lemma 4.11.) Define $\delta_{t+1} = \beta$.

Let $\alpha = \delta_{t+1}$. We say that α *requires attention* at stage $(s + 1, t + 1)$ if one of the following conditions holds (as defined precisely later):

(4.10)
$$\begin{cases} \text{ready to satisfy } P_\alpha, \\ \text{ready to open an } A_\alpha\text{-gap}, \\ \text{ready to close an } A_\alpha\text{-gap}, \\ \text{ready to close a } B_\alpha\text{-gap}. \end{cases}$$

If α does not require attention and $t < s$ go to substage $t + 2$. If α requires attention, choose the first of clauses (4.11)–(4.14) which applies and perform the action indicated below. If $t < s$ and α opens an A_α-gap, or α

closes an A_α-gap successfully and opens a B_α-gap, then go to substage $t + 2$. Otherwise, go to stage $s + 2$, and define $\delta[s + 1] = \delta_{t+1}$.

(4.11)   Ready to satisfy P_α:
   (a) $C \cap W_k = \emptyset$, where $k = |\alpha|$,
   (b) $(\exists y) [y \in W_k \cap \omega^{[\alpha]}$ & $y > \tilde{r}(\alpha)]$, where $\tilde{r}(\alpha) = \max\{r(\beta) : \beta \leq \alpha\}$.

   Action. Enumerate the least such y in C. Initialize all $\beta \geq \alpha$.

(4.12)   Ready to open an A_α-gap:
   (a) α is odd,
   (b) $\tau(\alpha)$ is defined,
   (c) $F_t(\alpha) = w$,
   (d) $x(\alpha) \in W_{j(\alpha)}$,
   (e) $x(\alpha) < m_\alpha^{i(\alpha)}[s]$.

   Action. (This action is called *Step 1.*)   Open an A_α-gap by defining $F(\alpha) = g1$, and $r(\alpha, 1) = 0$. Initialize all γ such that $\alpha^\frown \langle w \rangle \leq \gamma$. Create a *link* (τ, α) with *bottom* α and *top* $\tau = \tau(\alpha)$.

(4.13)   Ready to close an A_α-gap:
   (a) α is odd,
   (b) $F_t(\alpha) = g1$,
   (c) $s + 1$ is a τ-expansionary stage where $\tau = \tau(\alpha)$.

   Action. (This action is called *Step 2.*) Let $v + 1 < s + 1$ be the stage when the current A_α-gap was opened, and let $x = x(\alpha)[s]$. Close the A_α-gap as follows.
   Step 2A. (Successful close.) If $A_\alpha[s] \upharpoonright x \neq A_\alpha[v] \upharpoonright x$ then open a B_α-gap by defining $F(\alpha) = g2$ and $r(\alpha, 2) = 0$. (Hence, $r(\alpha) = 0$ also.) Initialize all γ such that $\alpha^\frown \langle g1 \rangle \leq \gamma$.
   Step 2B. (Unsuccessful close.) If $A_\alpha[s] \upharpoonright x = A_\alpha[v] \upharpoonright x$, define $F(\alpha) = w$ and $r(\alpha, 1) = s$ (to preserve $A_\alpha \upharpoonright x$), reset $x(\alpha)$, and initialize all γ such that $\alpha^\frown \langle w \rangle \leq \gamma$. Remove the link (τ, α).

(4.14)   Ready to close a B_α-gap:
   (a) α is odd,
   (b) $F_t(\alpha) = g2$,
   (c) $s + 1$ is a τ-expansionary stage where $\tau = \tau(\alpha)$.

   Action. (This action is called *Step 3.*) Let $v + 1 < s + 1$ be the stage when the current B_α-gap was opened, and let $x = x(\alpha)[s]$. Remove the link (τ, α) and close the B_α-gap as follows.
   Step 3A. (Successful close.) If $B_\alpha[s] \upharpoonright x \neq B_\alpha[v] \upharpoonright x$, then enumerate x in $D_{\tau(\alpha)}$, and define $F(\alpha) = s$. Initialize all γ, $\alpha^\frown \langle g2 \rangle \leq \gamma$. Set $r(\alpha, 1) = r(\alpha, 2) = 0$.

Step 3B. (Unsuccessful close.) If $B_\alpha[s] \upharpoonright x = B_\alpha[v] \upharpoonright x$, then define $F(\alpha) = w$ and $r(\alpha, 2) = s$ (to preserve $B_\alpha \upharpoonright x$), reset $x(\alpha)$, and initialize all γ, $\alpha^\smallfrown \langle g1 \rangle \leq \gamma$.

This completes the construction.

4.6 The Verification

4.11 The Link Lemma. (i) *If (ρ, γ) is a link then $\gamma \supseteq \rho^\smallfrown \langle 0 \rangle$. Furthermore, ρ is even, γ is odd, $\rho = \tau(\gamma)$, and $i(\rho) = i(\gamma)$.*

(ii) *Any link (ρ, γ) once created may be travelled at most twice before it is removed. (However, after removal a new link (ρ, γ) with the same top and bottom may be created at a later stage.)*

(iii) *If (ρ_1, γ_1) is a link created at some stage s, and before it is removed, another link (ρ_2, γ_2) is later created (either at a later substage of stage s or at a later stage) such that $\rho_1^\smallfrown \langle 0 \rangle \subseteq \gamma_2$ then $\gamma_1 \subset \gamma_2$ and either: $\rho_2 \subset \rho_1$ in case $i(\gamma_2) < i(\gamma_1)$ and there is no i'-boundary β such that $\gamma_1 \subset \beta \subseteq \gamma_2$ and $i' < i(\gamma_2)$; or $\gamma_1 \subset \rho_2$ otherwise. (Hence, distinct links existing simultaneously may be nested but never crossing, and if link (ρ_1, γ_1) is nested within link (ρ_2, γ_2) then $i(\gamma_2) < i(\gamma_1)$ so the outer link (ρ_2, γ_2) has higher priority, and it is created later than the inner link (ρ_1, γ_1).)*

(iv) *If the link (ρ, γ) is removed at some stage then at the end of that stage there is no link (ρ_1, γ_1) with $\rho_1 \subseteq \rho$ and $\rho \subseteq \gamma_1$.*

Proof. (i). The properties following the word "furthermore" are immediately obvious from the creation of links which can only be done according to (4.12) and only when γ opens an A_γ-gap. To see that $\gamma \supseteq \rho^\smallfrown \langle 0 \rangle$, suppose for a contradiction that $\gamma \supseteq \rho^\smallfrown \langle 1 \rangle$. Let $i = i(\gamma)$. Thus, by definition $\rho = \tau(\gamma)$, $i = i(\rho)$, and ρ is even. Let $\rho^+ = \rho^\smallfrown \langle 1 \rangle$. By Definition 4.4, $i \notin L_0(\rho^+)$ and for all j, $\langle i, j \rangle \notin L_1(\rho^+)$. But since $i = i(\gamma)$ and γ is odd we must have $\langle i, j \rangle \in L_1(\gamma)$ for some j. By Definition 4.4 this can only happen if there exists ξ, $\rho^+ \subseteq \xi \subset \gamma$ such that $i(\xi) < i$ and $\xi^\smallfrown \langle a \rangle \subseteq \gamma$ for some $a \in \{ g2, g1, 1 \}$. But then $\tau(\gamma) \neq \rho$ by Definition 4.5.

(ii). Suppose that link (ρ, γ) is created at stage $(s + 1, t + 1)$. Then $\delta_{t+1} = \gamma$, and γ opens an A_γ-gap at stage $(s + 1, t + 1)$. Hence, at the end of stage $s + 1$, $F(\gamma) = g1$ and this remains true either forever; or until γ is initialized (in which case the link (ρ, γ) is removed); or until the link is next travelled at some stage $s_1 + 1 > s + 1$. In the latter case $\delta_{t_1+1} = \gamma$ for some substage $t_1 + 1$ and γ closes the A_γ-gap. (This is because the fact that the link is travelled implies that $s_1 + 1$ is a ρ-expansionary stage and $\rho = \tau(\gamma)$ so all the conditions of (4.13) are met.) If the A_γ-gap is closed unsuccessfully at $(s_1 + 1, t_1 + 1)$ then the link (ρ, γ) is removed. Otherwise, γ opens a B_γ-gap at stage $(s_1 + 1, t_1 + 1)$, $F(\gamma)[s_1 + 1] = g2$, and the link (ρ, γ) remains until removed or until it is next travelled at some stage $s_2 + 1 > s_1 + 1$. At this stage γ closes the B_γ-gap and the link (ρ, γ) is removed regardless of whether the closure was successful or unsuccessful.

(iii). Suppose that link (ρ_1, γ_1) is created at stage (s, t_1), and that (iii) is true up to the time the second link (ρ_2, γ_2) is created. Then $\delta_{t_1} = \gamma_1$, and γ_1 opens an A_{γ_1}-gap at stage (s, t_1). Suppose another link (ρ_2, γ_2) is created at some substage $t_2 > t_1$ during stage s. Then $\gamma_1 \subset \gamma_2$ because $\delta_{t_1} \subset \delta_{t_2} = \gamma_2$. Now since γ_1 opened an A_{γ_1}-gap at stage (s, t_1) we know that $\gamma_1 \,\hat{}\, \langle g1 \rangle \subset \gamma_2$, and $\gamma_1 \,\hat{}\, \langle g1 \rangle$ is therefore an $i(\gamma_1)$-boundary by Definition 4.5. Thus, if $i(\gamma_1) < i(\gamma_2)$ then $\gamma_1 \subset \rho_2$ by Definition 4.5. If $i(\gamma_2) = i(\gamma_1)$ then by Definition 4.4 there must have been an i'-boundary β for some β, $\gamma_1 \subset \beta \subset \gamma_2$, with $i' < i(\gamma_1)$, so again $\gamma_1 \subset \rho_2$ by Definition 4.5 because γ_1 and γ_2 lie in different $i(\gamma_1)$-regions. If $i(\gamma_2) < i(\gamma_1)$ then $\rho_2 \subset \rho_1$ by Definition 4.5 unless there is an i'-boundary β such that $\gamma_1 \subset \beta \subset \gamma_2$ and $i' < i(\gamma_2)$, in which case $\gamma_1 \subset \rho_2$.

Now suppose that a link (ρ_2, γ_2) with $\rho_1 \,\hat{}\, \langle 0 \rangle \subseteq \gamma_2$ is created at some stage (v, t) for $v > s$ and that (iii) holds at all stages prior to (v, t). Now v is a ρ_1-stage. If it is a genuine ρ_1-stage then the link (ρ_1, γ_1) is travelled at stage (v, t') for some $t' < t$ since v is a ρ_1-expansionary stage because it is a genuine γ_2-stage and $\rho_1 \,\hat{}\, \langle 0 \rangle \subset \gamma_2$. Since the link (ρ_1, γ_1) was not removed at stage (v, t), γ_1 remains in an A_{γ_1}- or B_{γ_1}-gap so $\gamma_1 \,\hat{}\, \langle gk \rangle \subset \gamma_2$ for some $k \in \{1, 2\}$. Hence, by the same analysis as in the preceding paragraph $\gamma_1 \subset \gamma_2$ and either $\gamma_1 \subset \rho_2$ or $\rho_2 \subset \rho_1$ under the same conditions as before.

Suppose v is not a genuine ρ_1-stage. However, by hypothesis v is a ρ_1-stage. Hence, there must exist a link (ρ_3, γ_3) such that $\rho_3 \subset \rho_1$ and $\rho_1 \,\hat{}\, \langle 0 \rangle \subseteq \gamma_3$ and such that (ρ_3, γ_3) is travelled at stage v. Since (iii) holds prior to stage (v, t) we know $\gamma_1 \subset \gamma_3$, namely, link (ρ_1, γ_1) is nested within (ρ_3, γ_3). Hence, there is some substage $t' < t$ such that $\delta_{t'}[v] = \rho_3$, $\delta_{t'+1}[v] = \gamma_3$, and $\gamma_3 \subseteq \gamma_2$. Since γ_1 remains in a gap during this time we have $\gamma_1 \,\hat{}\, \langle gk \rangle \subset \gamma_2$ for some $k \in \{1, 2\}$. Hence, the same analysis as above applies so $\gamma_1 \subset \gamma_2$ and either $\gamma_1 \subset \rho_2$ or $\rho_2 \subset \rho_1$ exactly as before.

(iv). Assume that link (ρ, γ) is removed at stage $(s + 1, t + 1)$. If γ is initialized at stage $s + 1$ because of the action of some $\alpha < \gamma$ then $\alpha < \rho$ so every node $\gamma_1 \supseteq \rho$ is initialized at stage $s + 1$ and any link with bottom γ_1 is removed. Otherwise, at stage $s + 1$, $\delta_t = \rho$, $\delta_{t+1} = \gamma$, the link (ρ, γ) is travelled, and γ either closes an A_γ-gap unsuccessfully or closes a B_γ-gap. Since link (ρ, γ) is travelled, there is no other link (ρ_1, γ_1) at $(s + 1, t)$ such that $\rho_1 \subset \rho$ and $\rho \subseteq \gamma_1$. By (i) and (iii) there is no other link (ρ_1, γ_1) at stage $(s + 1, t)$ with $\rho_1 = \rho$ and $\gamma_1 \neq \gamma$. However, since the link (ρ, γ) is removed at $(s + 1, t + 1)$ under the action of either (4.13) or (4.14) we pass immediately to stage $s + 2$ with no further substages $t' > t + 1$, according to the construction as described between lines (4.10) and (4.11). Hence, no new links are created during stage $s + 1$ after substage $t + 1$. This establishes (iv) and Lemma 4.11. ▯

We now make formal the informal definition in §4.4 of the true path f.

4.12 Definition. The true path $f \in [T]$ for the construction is defined by induction on n as follows. Let $\alpha = f \restriction n$.
Case 1. α is even. Define

$$f(n) = \begin{cases} 0 & \text{if } (\exists^\infty s) \ [s \text{ is } \alpha\text{-expansionary}], \\ 1 & \text{otherwise.} \end{cases}$$

Case 2. α is odd. Define

$$f(n) = \begin{cases} a & \text{if } \lim_s \ F(\alpha) \ [s] \text{ exists and equals } a \in \{s, w\}, \\ gk & \text{if } \liminf_s \ F(\alpha)[s] = gk, \ \ k \in \{1, 2\}, \end{cases}$$

(Note that $\lim_s \ F(\alpha)[s] = gk, \ \ k \in \{1, 2\}$ is impossible, but in the second case we have $\liminf_s \ F(\alpha)[s] = g2$ if α opens infinitely many B_α-gaps and $g1$ if α opens finitely many B_α-gaps and infinitely many A_α-gaps.)

4.13 Leftmost Path Lemma. *Fix any n and let $\alpha = f \restriction n$.*

(i) $(\exists^{<\infty} s) \ [\delta[s] <_L \alpha]$,
(ii) S^α *is infinite where* S^α *is the set of genuine α-stages defined in Definition 4.9(ii).*

Proof. Clearly (i) and (ii) hold for $n = 0$. Fix $n \geq 0$, and assume (i) and (ii) for n. Let $\beta = f \restriction n$, $a = f(n)$ and $\alpha = \beta^\smallfrown \langle a \rangle$. We prove (i) and (ii) for α.
Fix s_1 such that for all stages $s \geq s_1$:

$$(4.15) \qquad\qquad\qquad \delta[s] \not<_L \beta;$$

$$(4.16) \qquad\qquad \text{if } \beta \text{ is odd then } F(\beta) \ [s] \not<_\Lambda a;$$

$$(4.17) \qquad \text{if } \beta \text{ is even and } a = 1, \text{ then } s \text{ is not a } \beta\text{-expansionary stage;}$$

$$(4.18) \qquad \neg(\exists \gamma) \ [[\gamma \leq \beta \vee \gamma = \alpha] \ \& \ P_\gamma \text{ acts at stage } s].$$

(i) Suppose to the contrary that $\delta[s] <_L \alpha$ for infinitely many s. Choose $b <_\Lambda a$ such that $\beta^\smallfrown \langle b \rangle \subset \delta[s]$ for infinitely many s. Let $\beta^+ = \beta^\smallfrown \langle b \rangle$. Now by (4.15) and (4.16), there is no genuine β^+-stage $s \geq s_1$. Hence, at infinitely many stages s there is some link (ρ, γ) of the form

$$(4.19) \qquad\qquad\qquad \rho \subseteq \beta \ \& \ \beta^+ \subseteq \gamma$$

such that (ρ, γ) is travelled at s.
By inductive hypothesis the set S^β of genuine β-stages is infinite. Consider any stage $s_2 + 1 > s_1$, $s_2 + 1 \in S^\beta$. Choose t such that $\delta_t[s_2 + 1] = \beta$.

At stage $(s_2 + 1, t)$ there can be at most one link (ρ, γ) satisfying (4.19) and we must have $\rho = \beta$. Furthermore, this link must be travelled at some stage $s_3 + 1 \geq s_2 + 1$ before any new link satisfying (4.19) can be created. Now by Lemma 4.11(ii) this specific link (ρ, γ) can be travelled at most twice after which it is removed. By Lemma 4.11(iv) when it is removed at some stage $s_4 + 1 \geq s_2 + 1$ there are no links satisfying (4.19). But since there is no genuine β^+-stage $\geq s_1$, it follows simultaneously by induction on $s > s_4 + 1$ that at every stage $s > s_4 + 1$:

$$(4.20) \qquad \text{there is no link } (\rho, \gamma) \text{ satisfying (4.19)};$$

and

$$(4.21) \qquad \beta^+ \not\subseteq \delta[s].$$

(ii). We must prove that S^α is infinite. Fix $k \in \omega$. It suffices to find $s_3 \in S^\alpha$ such that $k \leq s_3$.

Case 1. β is odd.

Case 1A. $\lim_s F(\beta)\,[s] = a \in \{s, w\}$. Choose $s_2 \geq \max\{k, s_1\}$ such that $F(\beta)\,[s] = a$ for all $s \geq s_2$. Now for all $s \geq s_2$, if $s \in S^\beta$ then $s \in S^\alpha$ because if $\delta_t = \beta$ at stage s then $\delta_{t+1} = \alpha$ by Case 1 of the definition of δ_{t+1}.

Case 1B. $\alpha = \beta^\frown\langle gk \rangle$ for $k \in \{1, 2\}$. Suppose $k = 2$. (The case $k = 1$ is similar.) Then β opens infinitely many B_β-gaps. But if β opens a B_β-gap at stage (s, t) then $s \in S^\beta$, $\delta_t[s] = \beta$, $F_t(\beta) = g2$, and $\delta_{t+1}[s] = \alpha$ so $s \in S^\alpha$.

Case 2. β is even.

Case 2A. $a = 1$. Using (i) choose $s_2 \geq s_1$ such that $\delta[s] \not<_L \alpha$ at any $s \geq s_2$. Choose the least $s_3 \in S^\beta$, $s_3 \geq \max\{k, s_2\}$. Choose t such that $\delta_t[s_3] = \beta$. Then s_3 is not β-expansionary and hence $\delta_{t+1} = \alpha$ by Case 2A in the definition of δ_{t+1}. Thus, $s_3 \in S^\alpha$.

Case 2B. $a = 0$. Since $\beta^\frown\langle 0 \rangle = \alpha \subset f$ there are infinitely many β-expansionary stages $s' > k$ and by Definition 4.9(iii) each such s' is a *genuine* β-stage so $\delta_t = \beta$ for some substage t of stage s'. Hence, $\delta_{t+1} = \alpha$ unless there is a link (ρ, γ) such that $\rho = \beta$ and $\alpha \subseteq \gamma$. If so then by Lemma 4.11(ii) this link can be travelled at most twice. Since there are infinitely many β-expansionary stages, the link (ρ, γ) will be travelled until it is removed at some stage $s_2 > s'$. By Lemma 4.11(iv) at the end of stage s_2 there is no link (ρ', γ') with $\rho' \subseteq \beta$ and $\alpha \subseteq \gamma'$, and this condition persists until the next β-expansionary stage $s_3 > s_2$. Choose t such that $\delta_t[s_3] = \beta$. Then $\delta_{t+1}[s_3] = \alpha$ by Case 2B in the definition of δ_{t+1}. ☐

4.14 Restraint Lemma. *If* $\alpha = f \upharpoonright n$ *and* $\alpha^+ = \alpha^\frown\langle f(n) \rangle$ *then*

$$\tilde{r}(\alpha) =_{\mathrm{dfn}} \lim \{\tilde{r}(\alpha)[s] : s \in V^{\alpha^+}\} \text{ exists}$$

where V^γ is the set of γ-stages as in Definition 4.9(i) and

$$\tilde{r}(\alpha)[s] =_{\mathrm{dfn}} \max\{\, r(\beta)[s] : \beta \le \alpha \,\}.$$

Proof. Fix $\alpha = f \upharpoonright n$, let α^- be the predecessor (if any) of α, and assume inductively that $\tilde{r}(\alpha^-)$ exists where we define

$$\tilde{r}(\alpha^-) = \lim \{\, \tilde{r}(\alpha^-)[s] : s \in V^\alpha \,\}.$$

If α is even then $r(\alpha)[s] = 0$ so the lemma holds. Assume α is odd. By Lemma 4.13(i) choose v such that for all $s \ge v$, $\delta[s] \not\prec_L \alpha$, and if $s \in V^\alpha$ then $\tilde{r}(\alpha^-)[s] = \tilde{r}(\alpha^-)$. Now as in the basic module in §4.2, $r(\alpha) =_{\mathrm{dfn}} \liminf_s r(\alpha)[s]$ exists and $r(\alpha)[s] = r(\alpha)$ at almost every s such that $\alpha^+ \subseteq \delta[s]$ (because if $f(n) = gk$ for $k \in \{1, 2\}$ then $r(\alpha, k)[s] = 0$ at such an s). Hence, there exists $z \ge v$ and an integer $\tilde{r}(\alpha)$ such that $\tilde{r}(\alpha)[s] = \tilde{r}(\alpha)$ for all $s \in V^{\alpha^+}$, $s \ge z$. ▯

4.15 Nonrecursiveness Lemma. *C is nonrecursive.*

Proof. Assume for a contradiction that $\overline{C}$ is r.e., say $\overline{C} = W_k$, for $k > 0$. (We choose $k > 0$ because $\alpha = \lambda$ never acts.) Choose α such that $\alpha \subset f$ and $|\alpha| = k$. Note that $C^{[\alpha]} = \emptyset$ because elements of $\omega^{[\alpha]}$ enter C only to meet requirement $P_k = P_{|\alpha|}$. It follows that $W_k^{[\alpha]} = \omega^{[\alpha]}$. By Lemma 4.13(ii) choose $y \in \omega^{[\alpha]}$, $y \in W_k[s]$, $y > \tilde{r}(\alpha)[s]$, where $s+1$ is a genuine α-stage such that no $\beta < \alpha$ acts to satisfy P_β according to (4.11) at any stage $\ge s+1$. Since $C[s] \cap W_k[s] = \emptyset$, α acts at stage $s+1$ according to (4.11) to satisfy P_α, so $C[s+1] \cap W_k[s+1] \ne \emptyset$, a contradiction. ▯

4.16 Truth of Outcome Lemma. *Let $\alpha = f \upharpoonright n$.*
    (i) *If $f(n) = 1$, then $\neg[\Phi_\alpha^C = A_\alpha$ & $\Psi_\alpha^C = B_\alpha]$.*
    (ii) *If $f(n) = 0$, then there are infinitely many α-expansionary stages.*
(Hence, $\limsup_{s \in S^\alpha} m_\alpha^{i(\alpha)}[s] = \infty$.)
    (iii) *If $f(n) = g2$ then B_α is recursive.*
    (iv) *If $f(n) = g1$ then A_α is recursive.*
    (v) *If $f(n) = w$ and $\tau(\alpha)$ is defined then $D_{\tau(\alpha)} \cup W_{j(\alpha)} \ne \omega$.*
    (vi) *If $f(n) = s$ then $D_{\tau(\alpha)} \cap W_{j(\alpha)} \ne \emptyset$, and indeed $x(\alpha) \in D_{\tau(\alpha)} \cap W_{j(\alpha)}$, where $x(\alpha)$ is the (unique) element enumerated into $D_{\tau(\alpha)}$ under the action of (4.14).*

Proof. (i). By Lemma 4.13(ii) there are infinitely many genuine α-stages. By the Definition 4.12 of f there are only finitely many α-expansionary stages. Hence, by Definition 4.9(iii) there are only finitely many i-expansionary stages for $i = i(\alpha)$. Therefore, $\Phi_i^C \ne A_i$ or $\Psi_i^C \ne B_i$.
    (ii). This is immediate by the Definition 4.12 of f.

(iii). Suppose $f(n) = g2$. Let $\alpha^+ = \alpha^\frown\langle g2 \rangle$. Choose s_0 such that $\alpha^+ \leq \delta[s]$ for all $s \geq s_0$, and for all $\gamma \subseteq \alpha^+$, P_γ is satisfied before stage s_0 if at all. Thus, α is never initialized at any stage $s \geq s_0$. Now (4.7) holds for s_0 with $r(\alpha)[s]$ in place of $r[s]$ so the same proof as in the basic module shows that B_α is recursive.

(iv). The proof is similar but with $g1$ and A_α in place of $g2$ and B_α.

(v). Let $f(n) = w$. Then $\lim_s F(\alpha)[s] = w$ and $\lim_s x(\alpha)[s] = x(\alpha)$ both exist. Choose s_0 so that these values are constant for all $s \geq s_0$. Hence, by the construction $x(\alpha) \notin D_{\tau(\alpha)}$. Now suppose $\tau(\alpha)$ is defined, and $x(\alpha) \in W_{j(\alpha)}$. Then for almost every stage s, all the conditions of (4.12) for α to open an A_α-gap are satisfied. But by Lemma 4.13(ii) there are infinitely many genuine α-stages, so α opens an A_α-gap at some stage $s > s_0$ contrary to hypothesis.

(vi). If $f(n) = s$, then at some stage s, (4.14) applied, α closed a B_α-gap successfully and enumerated $x = x(\alpha)[s]$ in $D_{\tau(\alpha)}$. But necessarily $x \in W_{j(\alpha),s}$ since at some stage $s' < s$, α opened an A_α-gap with candidate

$$x = x(\alpha)[s'] \in W_{j(\alpha),s'}. \quad \Box$$

4.17 The i-Region Lemma. *Fix i such that $\Phi_i^C = A_i$, $\Psi_i^C = B_i$, A_i is nonrecursive, and B_i is nonrecursive. Then there exists a nonrecursive r.e. set D such that $D \leq_T A_i$ and $D \leq_T B_i$.*

Proof. By Lemma 4.7 and Definition 4.8 fix $\tau = \tau(f, i)$. Let $\tau = f \restriction m$. Then by 4.7, 4.8, Lemma 4.16(i), and the hypotheses of Lemma 4.17 we have that $f(m) = 0$, and there is no $\gamma \supset \tau^\frown\langle 0 \rangle$ such that: $i(\gamma) \leq i$, $\gamma \subset f$, and either γ is even, or γ is odd and $f(|\gamma|) \in \{ g2, g1 \}$.

Define the r.e. set

$$D = D_\tau = \{ x(\alpha) : \alpha \in E(\tau, i)$$
$$\& \; \alpha \text{ enumerates } x(\alpha) \text{ in } D_\tau \text{ according to (4.14)} \}.$$

Clearly, D is r.e. since $E(\alpha, i)$ is a recursive set of nodes. To see that D is nonrecursive, suppose for a contradiction that $\overline{D}$ is r.e. Find some α, $\tau \subset \alpha \subset f$ such that $i(\alpha) = i$ and $W_{j(\alpha)} = \overline{D}$. This is possible since by Lemma 4.16 and the hypotheses of Lemma 4.17 there is no $\beta \subset f$ with $i(\beta) = i$ and $f(|\beta|) \in \{ g2, g1, 1 \}$. Hence, by Definition 4.4, for almost every $j \in \omega$ there exists an odd α, $\tau \subset \alpha \subset f$ with $i(\alpha) = i$ and $j(\alpha) = j$. Let $\alpha = f \restriction n$. We must have $f(n) \in \{ s, w \}$. By Lemma 4.16, if $f(n) = s$ when $x(\alpha) \in D \cap W_{j(\alpha)}$ and if $f(n) = w$ then $x(\alpha) \notin D \cup W_{j(\alpha)}$. Hence, $\overline{D} \neq W_{j(\alpha)}$.

Finally, to prove that $D \leq_T A_i$ and $D \leq_T B_i$ it suffices to prove that for all odd $\alpha \in E(\tau, i)$ any α-gap once opened is later closed or cancelled. Then to A_i-recursively test whether $x \in D$ find a stage s such that $A_i[s] \restriction x = A_i \restriction x$. Find a stage $v > s$ such that for any $\alpha \in E(\tau, i)$ having $x(\alpha)[s] = x$, by the end of stage v we have either that the A_α-gap is closed unsuccessfully, or the B_α-gap is closed, or node α is initialized. Now $x \in D$ iff $x \in D[v]$. (The proof for B_i in place of A_i is the same.)

To prove that any such α-gap is eventually closed or cancelled, fix any odd $\alpha \in E(\tau, i)$ and suppose that α opens an A_α-gap at stage s. Then a link (τ, α) is created at stage s. But $\tau\hat{\ }\langle 0 \rangle \subseteq \alpha$ and there are infinitely many τ-expansionary stages by Lemma 4.16(ii). Thus, the link (τ, α) is eventually travelled and the A_α-gap is closed at some stage $v > s$ (if it is not cancelled first). At stage v either $x = x(\alpha)$ is cancelled or else a B_α-gap is opened which must later be closed by the same argument at some stage $z > v$.   ▯

This lemma completes the proof of Theorem 4.1.   ▯

4.18 Exercise

4.18° (Downey).   Construct a high r.e. set C such that if A and B are r.e. sets such that $A \leq_{wtt} C$ and $B \leq_{wtt} C$ then $\deg(A)$ and $\deg(B)$ do not form a minimal pair. *Hint.* Combine the method of §4 with that of Theorem 3.1 so that the priority tree has three types of nodes, those of §4 plus two-branching nodes as in Theorem 3.1 for the thickness requirements. Notice that this theorem is in contrast to Cooper's result for $A \leq_T C$ and $B \leq_T C$ in Exercise XI.2.15.

4.19 Notes. An earlier version of this proof of Theorem 4.1 appeared in Soare [1985]. The present version incorporates some suggestions of Slaman including the use of links and of the even nodes to test the Φ_i- and Ψ_i-hypotheses. This enables us to eliminate some of the purely technical features of the earlier version, thereby giving a more intuitive construction and proof. S. Lempp and V. Harizanov also made many corrections and suggestions.

Theorem 4.1 stands in contrast to the other result by Lachlan [1979] that there exists a nonzero r.e. degree $\mathbf{c}$ such that, if $\mathbf{d}$ is r.e. and $\mathbf{0} < \mathbf{d} \leq \mathbf{c}$, then there is a minimal pair of r.e. degrees below $\mathbf{d}$. This is proved by a straightforward variation of the minimal pair construction. Cooper showed that this degree $\mathbf{c}$ can be made high so that the Nonbounding Theorem 4.1 cannot be extended to make the nonbounding r.e. degree $\mathbf{c}$ high. (This contrasts with Exercise 4.18 where $\leq_T$ is replaced by $\leq_{wtt}$.)

Chapter XV

Automorphisms of the Lattice of R.E. Sets

In this chapter we study the group Aut $\mathcal{E}$ (Aut $\mathcal{E}^*$) of automorphisms of the lattice of r.e. sets $\mathcal{E}$ ($\mathcal{E}^*$). In §1 of this chapter we continue in the spirit of Chapter X by proving that creativeness is elementary lattice theoretic and hence invariant under Aut $\mathcal{E}$, while in §3 we introduce an easy finite injury method for generating automorphisms to prove that hypersimplicity is not invariant under Aut $\mathcal{E}$. In §2 we give some basic facts about automorphisms and show that any $\Phi \in$ Aut $\mathcal{E}^*$ is induced by some permutation p of ω, and hence Φ is induced by some $\Psi \in$ Aut $\mathcal{E}$. The main objective for the rest of the chapter is to prove that if A and B are any two maximal sets then there is an automorphism Φ of $\mathcal{E}$ such that $\Phi(A) = B$. This requires a new and much more difficult method for generating automorphisms that is contained in the Extension Theorem 4.5 which we motivate and state in §4 and prove in §6. In §5 we verify the hypotheses of the Extension Theorem for the case of maximal sets. The Extension Theorem is the principal tool for generating automorphisms of $\mathcal{E}$ in the sense that all known difficult results on automorphisms of $\mathcal{E}$ either appeal directly to it or else are modifications of it. In Chapter XVI §1 we state some further results and open questions about automorphisms.

1. Invariant Properties

It was observed in Definition II.1.15 that the r.e. sets under inclusion form a distributive lattice, which we denoted by $\mathcal{E}$ and studied in Chapter X. In Definition X.1.2 we defined $\mathcal{E}^*$ to be the quotient lattice of $\mathcal{E}$ modulo finite sets. Recall from Definition X.1.3 that a property P of r.e. sets is *invariant* or *lattice theoretic* in $\mathcal{E}$ ($\mathcal{E}^*$) if P is invariant under Aut $\mathcal{E}$ (Aut $\mathcal{E}^*$) and P is *elementary lattice theoretic* (*e.l.t.*) if P is first order definable over $\mathcal{E}$ ($\mathcal{E}^*$). We saw in Chapter X §1 that a property P of r.e. sets which is closed under finite differences is e.l.t. in $\mathcal{E}$ iff P is e.l.t. in $\mathcal{E}^*$. The same will follow later for "invariant" in place of "e.l.t." by Corollary 2.8.

If a property is e.l.t. then in some cases the first order formula defining the property is obvious as in the case of the properties: recursive, simple, finite, and maximal (see Chapter X §1). In other cases it is not at all obvious that such a defining property exists and some clever proof is required, for example, in the case of the property hh-simple (see Corollary X.2.8). In this section we present an equally clever and unexpected proof that the property of being creative is e.l.t.

1.1 Theorem (Harrington).   *An r.e. set A is creative iff*

$$(1.1) \qquad (\exists C \supset A)\, (\forall B \subseteq C)\, (\exists R)\, [R \text{ is recursive}$$
$$\& \ R \cap C \text{ is nonrecursive} \ \& \ R \cap A = R \cap B],$$

where all variables range over $\mathcal{E}$.

1.2 Corollary (Harrington).   *The property of being creative is elementary lattice theoretic.*   ⬚

1.3 Definition. (i) Let Aut $\mathcal{E}$ (Aut $\mathcal{E}^*$) denote the group of automorphisms of $\mathcal{E}$ ($\mathcal{E}^*$). The symbols Φ, Ψ will denote automorphisms of $\mathcal{E}$ or $\mathcal{E}^*$.
   (ii) For A and $B \in \mathcal{E}$ we say A is *automorphic to* B and write $A \cong_{\mathcal{E}} B$ ($A^* \cong_{\mathcal{E}*} B^*$) if there exists $\Phi \in$ Aut $\mathcal{E}$ (Aut $\mathcal{E}^*$) such that $\Phi(A) = B$ ($\Phi(A^*) = B^*$).
   (iii) The *orbit* of $A \in \mathcal{E}$, written orbit(A), is the class $\{\, B : A \cong_{\mathcal{E}} B \,\}$.
   (iv) The *orbit* of $A^* \in \mathcal{E}^*$ is the class $\{\, B^* : A^* \cong_{\mathcal{E}*} B^* \,\}$.

1.4 Corollary (Harrington).   *The creative sets constitute an orbit.*

Proof. By Corollary 1.2 and the fact that for any two creative sets A and B there is a recursive isomorphism, and hence an automorphism of $\mathcal{E}$, which maps A to B (see Theorems I.5.4 and II.4.8).   ⬚

Proof of Theorem 1.1. ($\Longrightarrow$). Since all creative sets are recursively isomorphic we can choose A to be *any* creative set rather than some given one. Define

$$A = \{\, \langle x, y \rangle : x \in K \ \& \ \langle x, y \rangle \in W_y \,\}.$$

(Clearly A is creative by Corollary II.4.8 since $K \leq_1 A$ because if we choose $W_{y_0} = \omega$ then $x \in K$ iff $\langle x, y_0 \rangle \in A$.) Next define $C = \{\, \langle x, y \rangle : x \in K \,\}$. Given B r.e. and $B \subseteq C$ choose y_0 such that $W_{y_0} = B$. Define $R = \{\, \langle x, y_0 \rangle : x \in \omega \,\}$. Now $R \cap C = \{\, \langle x, y_0 \rangle : x \in K \,\}$ which is not recursive, and

$$R \cap A = R \cap B = \{\, \langle x, y_0 \rangle : x \in K \ \& \ \langle x, y_0 \rangle \in W_{y_0} = B \,\},$$

because $B \subseteq C$.
   ($\Longleftarrow$). (The following method slightly resembles that of Exercise II.4.16.) Fix A and C satisfying (1.1) so that $A \subset C$. Fix r.e. indices a, c, and k for A, C, and K, respectively, and define $A_s = W_{a,s}$, $C_s = W_{c,s}$,

and $K_s = W_{k,s}$, noting that Convention II.2.6 holds. We wish to construct a recursive function f such that $K \leq_m A$ via f. To do so we first build an r.e. set $B \subseteq C$ by an enumeration $\{B_s\}_{s \in \omega}$ such that $B \smallsetminus C = \emptyset$ and hence $B = C \smallsetminus B$ (see Definition II.2.9). Now since $B \subseteq C$ there must exist a recursive set $R = W_e$ satisfying the matrix of (1.1). For each e we shall define a p.r. function ψ_e. We shall have that $f = \psi_{e_0}$ (or at least $f = \psi_i$ for some $i \leq e_0$) where e_0 is the least e such that $R = W_e$ satisfies the matrix of (1.1).

The basic strategy for a single e is the following. Suppose e is minimal such that $R = W_e$ satisfies the matrix of (1.1). Now $W_e \cap C$ is nonrecursive so $W_e - C$ is non-r.e. and hence $|W_e \smallsetminus C| = \infty$ (as in (2.1) of Chapter X). Furthermore,

$$(1.2) \qquad\qquad W_e \cap A = W_e \cap B.$$

Given e define $\psi_e = f$ as follows. Let ψ be a 1:1 recursive function with range $W_e \smallsetminus C$ which is r.e. by Exercise II.2.10 and is infinite. For each x, when x enters K enumerate $\psi(x)$ in B, which forces $\psi(x)$ to be later enumerated in A by (1.2). Hence, $x \in K$ iff $\psi(x) \in A$ by (1.2).

This completes the basic strategy for a single e. Since we cannot effectively identify the minimal e above we must play the i-strategy for all i simultaneously. To prevent some $i < e$ from infinitely often interfering with the minimal e we define a new value $\psi_{i,s+1}(x)$ only if

$$(1.3) \qquad (\forall y < x) \, [\psi_{i,s}(y)\downarrow \ \& \ [y \in K_s \iff \psi_{i,s}(y) \in A_s]].$$

Thus, except for finitely many elements taken by $i < e$, the strategy for the minimal e succeeds exactly as above.

Construction of B and $\{\psi_i\}_{i \in \omega}$.
    Stage $s = 0$. Set $B_0 = \emptyset = \psi_{i,0}$.
    Stage $s + 1$.
        Step 1. Find the unique z (if it exists) such that $z \in C_{s+1} - C_s$. Find the least i (if it exists) such that $z \in W_{i,s}$ and (1.3) holds for i where $x = (\mu v) \, [\psi_{i,s}(v)\uparrow]$. Define $\psi_{i,s+1}(x) = z$. If z or i fails to exist do nothing.
        Step 2. For all i, if $x \in K_{s+1}$ and $\psi_{i,s+1}(x) \downarrow$ enumerate $\psi_{i,s+1}(x)$ in B.
End of construction.

If i is minimal such that ψ_i is total, then $K \leq_m A$ via ψ_i. (Notice that $i \leq e$, where e is minimal such that $R = W_e$ satisfies the matrix of (1.1).) Hence, A is creative by Corollary II.4.8.

Note that the above proof shows that if A and C satisfy (1.1) then $A \subset_\infty C$, and indeed this may be seen simply by choosing $B = \emptyset$ and noticing that the corresponding set R is such that $R \cap A = \emptyset$ and $R \cap C$ is nonrecursive. Hence, $R \cap C \subseteq C - A$ and $R \cap C$ is infinite. Moreover, $C - A$ is not r.e.

because otherwise if we choose $B = C - A$ then the corresponding set R is such that $R \cap C = \emptyset$ (since $R \cap C = (R \cap A) \cup (R \cap B)$, $R \cap A = R \cap B$, and $A \cap B = \emptyset$), contradicting the fact that $R \cap C$ is nonrecursive.

1.5 Notes. Harrington (unpublished) has claimed that the orbit of creative sets is the only orbit consisting of a single nonrecursive m-degree. Namely,

$$(\forall A \in \mathcal{E}) \, [A \text{ nonrecursive and noncreative}$$
$$\Longleftrightarrow (\exists \Phi \in \text{Aut } \mathcal{E}) \, [\Phi(A) \not\equiv_m A]],$$

although this result uses the full automorphism machinery. Harrington also has claimed that there is a noncreative set $A \in \mathcal{E}$ such that

$$\text{orbit}(A) \subseteq \{ W_e : W_e \equiv_T K \},$$

namely, an orbit containing only Turing complete but not creative sets.

2. Some Basic Properties of Automorphisms of $\mathcal{E}$

2.1 Definition. A permutation p of ω *induces* an automorphism Φ of $\mathcal{E}$ ($\mathcal{E}^*$) if for all n, $p(W_n) = \Phi(W_n)$ $((p(W_n))^* = \Phi(W_n^*))$, respectively).

It is obvious that every recursive permutation trivially induces an automorphism of $\mathcal{E}$, and that every automorphism of $\mathcal{E}$ induces an automorphism of $\mathcal{E}^*$. It is natural to ask: whether there are any other automorphisms of $\mathcal{E}$ or of $\mathcal{E}^*$; if so how many there are; whether every one is induced by a permutation of ω; and what the relationship is between automorphisms of $\mathcal{E}$ and automorphisms of $\mathcal{E}^*$. In this section we answer these questions, and we develop some definitions and notation to be used later.

Kent (Exercise 2.11) used cohesive sets to show that there are $2^{\aleph_0}$ automorphisms of $\mathcal{E}$, but the corresponding induced automorphisms of $\mathcal{E}^*$ are all trivial. Lachlan used the following more subtle but still easy argument to obtain the same result for $\mathcal{E}^*$ in place of $\mathcal{E}$. However, in Lachlan's method, $\Phi \in \text{Aut } \mathcal{E}^*$ is induced by a permutation p which is obtained by "piecing together" recursive permutations in a nonrecursive fashion, in such a way that for any A and B, if $p(A) = B$ then $A \equiv B$. Thus, the method produces no new members in the $\mathcal{E}^*$-orbit of A^* beyond those obtained by mere recursive permutations. Therefore, to prove the noninvariance of hypersimplicity in §3 or the maximal set theorem in §4–§6, a new device for generating automorphisms had to be invented.

2.2 Theorem (Lachlan). *There are $2^{\aleph_0}$ automorphisms of $\mathcal{E}^*$.*

Proof. For each n, we define a recursive set R_n such that $R_n \subseteq R_{n-1}$, R_n is infinite, and $R_{n-1} - R_n$ is infinite. For convenience, let $R_{-1} = \omega$. Choose $R_n \subseteq R_{n-1} - W_n$ if $R_{n-1} \cap W_n$ is finite, and $R_n \subseteq R_{n-1} \cap W_n$ otherwise. In either case, we have R_n recursive, R_n infinite, and $R_{n-1} - R_n$ infinite. Note that $R_n \subseteq W_n$ or $R_n \subseteq \overline{W}_n$, for all n.

For all $n \in \omega$ and $j \in \{0, 1\}$, choose partial recursive functions ψ_j^n which are permutations of $R_{n-1} - R_n$, such that ψ_0^n is the identity, and ψ_1^n carries S_n into some $T \neq^* S_n$ for some infinite r.e. set $S_n \subseteq R_{n-1} - R_n$. For any function $f : \omega \to \{0, 1\}$, define the permutation p_f of ω, such that p_f restricted to $R_{n-1} - R_n$ is $\psi_{f(n)}^n$. Clearly p_f is total and thus is a permutation of ω because for every m there exists n such that $W_n = \{m\}$ so that $m \in \overline{R}_n$ (and hence $m \in \mathrm{dom}\ p_f$). Since $\psi_0^n(S_n) \neq^* \psi_1^n(S_n)$, p_f and p_g cannot induce the same automorphism on $\mathcal{E}^*$ if $f \neq g$.

It suffices to show that $p_f(W_n)$ and $p_f^{-1}(W_n)$ are r.e. for each n. Clearly, $\bigcup\{\psi_{f(i)}^i : i \leq n\}$ is a partial recursive function with domain $\overline{R}_n$. Therefore, for all n, $p_f(\overline{R}_n \cap W_n) = \bigcup\{\psi_{f(i)}^i(\overline{R}_n \cap W_n) : i \leq n\}$ which is r.e. Furthermore, $p_f(R_n \cap W_n) = R_n \cap W_n$ since either $R_n \subseteq W_n$, or R_n and W_n are disjoint. Thus,

$$p_f(W_n) = p_f(R_n \cap W_n) \cup p_f(\overline{R}_n \cap W_n)$$

is r.e. Likewise, $p_f^{-1}(W_n)$ is r.e.   ▯

2.3 Corollary. *Not every automorphism of $\mathcal{E}^*$ is induced by a recursive permutation of ω.*   ▯

On the other hand, we shall show that every automorphism of $\mathcal{E}^*$ is induced by *some* permutation of ω. First we need a convenient way to describe an arbitrary $\Phi \in \mathrm{Aut}\ \mathcal{E}^*$ relative to our fixed indexing $\{W_n\}_{n \in \omega}$.

2.4 Definition. An automorphism $\Phi \in \mathrm{Aut}\ \mathcal{E}^*$ is *presented by h*, a permutation of ω, if $\Phi(W_n^*) = W_{h(n)}^*$ for all n.

2.5 Theorem. *Every $\Phi \in \mathrm{Aut}\ \mathcal{E}^*$ is induced by some permutation p of ω. Furthermore, if Φ is presented by h, p can be chosen recursive in $(h \oplus \emptyset')'$.*

Proof. Fix $\Phi \in \mathrm{Aut}\ \mathcal{E}^*$, and let Φ be presented by h. For any $x \in \omega$, define e-states,

$$\sigma(e, x) = \{i : i \leq e\ \&\ x \in W_i\},$$
$$\hat{\sigma}(e, x) = \{i : i \leq e\ \&\ x \in W_{h(i)}\}.$$

Note that the sets $\sigma(e, x)$ are uniformly recursive in $\emptyset'$, and the sets $\hat{\sigma}(e, x)$ in $h \oplus \emptyset'$. Let $p = \bigcup_n p_n$, where finite partial functions p_n are defined as follows.

For convenience, let $p_{-1} = \emptyset$, and $\sigma(-1, x) = \hat{\sigma}(-1, x) = \emptyset$ for all x. For n even, let

$x_n = (\mu x) \, [x \notin \text{dom } p_{n-1}]$,

$e_n = \max\{ e : -1 \le e \le n \ \& \ (\exists y) \, [y \notin \text{ran } p_{n-1} \ \& \ \sigma(e, x_n) = \hat{\sigma}(e, y)] \}$,

$y_n = (\mu y) \, [y \notin \text{ran } p_{n-1} \ \& \ \sigma(e_n, x_n) = \hat{\sigma}(e_n, y)]$.

Define $p_n = p_{n-1} \cup \{ (x_n, y_n) \}$. Note that e_n and hence y_n may be found recursively in $(h \oplus \emptyset')'$. For n odd, we let $y_n = (\mu y) \, [y \notin \text{ran } p_{n-1}]$, and proceed as above with x_n and y_n interchanged, and σ and $\hat{\sigma}$ interchanged. Clearly, $p = \bigcup_n p_n$ is a permutation and $p \le_T (h \oplus \emptyset')'$. Since $\Phi \in \text{Aut } \mathcal{E}^*$, $\lim_n e_n = \infty$, and it easily follows by induction on n that $p(W_n) =^* W_{h(n)}$ for all n. ▯

2.6 Corollary. *Every $\Phi \in \text{Aut } \mathcal{E}^*$ is induced by some $\Psi \in \text{Aut } \mathcal{E}$.*

Proof. The above p induces $\Psi \in \text{Aut } \mathcal{E}$, which induces Φ. ▯

2.7 Corollary. *If $A, B \in \mathcal{E}$ are infinite and coinfinite then $A \equiv_{\mathcal{E}} B$ if and only if $A^* \equiv_{\mathcal{E}^*} B^*$.*

Proof. If $A \equiv_{\mathcal{E}} B$ then clearly $A^* \equiv_{\mathcal{E}^*} B^*$. Conversely, if $A^* \equiv_{\mathcal{E}^*} B^*$ via Φ, then in the proof of Theorem 2.5, set $W_0 = A$ and $W_{h(0)} = B$, so that the permutation p obtained satisfies $p(A) = B$. ▯

By Corollary 2.7 to show that $A \equiv_{\mathcal{E}} B$ it is immaterial whether we construct $\Phi \in \text{Aut } \mathcal{E}$ or $\Phi \in \text{Aut } \mathcal{E}^*$. We normally do the latter, which is usually easier.

2.8 Corollary. *Any property of r.e. sets which is well-defined on $\mathcal{E}^*$ is invariant under $\text{Aut } \mathcal{E}$ just if it is invariant under $\text{Aut } \mathcal{E}^*$.* ▯

Note that the proof of Theorem 2.5 establishes the following fact, which will be useful later.

2.9 Corollary. *Given any permutation h of ω, define $\sigma(e, x)$ and $\hat{\sigma}(e, x)$ as in the proof of Theorem 2.5. Then h presents some $\Phi \in \text{Aut } \mathcal{E}^*$ if and only if for every e and every e-state σ_0,*

$$\{ x : \sigma(e, x) = \sigma_0 \} \text{ is infinite} \iff \{ x : \hat{\sigma}(e, x) = \sigma_0 \} \text{ is infinite}. ▯$$

Let $\mathcal{L}$ be any countable sublattice of the lattice $\mathcal{N} = \langle 2^\omega; \subseteq, \cup, \cap \rangle$ such that $\mathcal{L}$ is closed under finite differences and contains $\emptyset$ and ω. The above theorem and corollaries clearly hold with $\mathcal{L}$ and $\mathcal{L}^* (= \mathcal{L}/\mathcal{F})$ in place of $\mathcal{E}$ and $\mathcal{E}^*$ if $\emptyset'$ is replaced by $\oplus\{ A : A \in \mathcal{L} \}$, namely $\{ \langle m, n \rangle : m \in A_n \}$ where $\{ A_n \}_{n \in \omega}$ are the members of $\mathcal{L}$. The method of Theorem 2.5 can thus be used to give an alternate (but similar) proof of Lachlan's result (Exercise 2.12) that for such sublattices $\mathcal{L}_1$ and $\mathcal{L}_2$, $\mathcal{L}_1 \cong \mathcal{L}_2$ just if $\mathcal{L}_1^* \cong \mathcal{L}_2^*$.

To present some $\Phi \in \operatorname{Aut} \mathcal{E}^*$ it is clearly equivalent (and usually easier) to give functions f and g satisfying

$$(2.1) \qquad (\forall n)\, [\Phi(W_n^*) = W_{f(n)}^* \ \& \ \Phi^{-1}(W_n^*) = W_{g(n)}^*],$$

in which case we say that Φ is *presented by the pair* (f, g). (Given f and g, one constructs a corresponding presentation h recursive in $f \oplus g$ by the Padding Lemma I.3.2, and conversely given h, let $f = h$ and $g = h^{-1}$.)

Construction of an automorphism Φ consists of building a permutation p of ω which induces the proposed Φ, and simultaneously giving functions f and g (not necessarily recursive) satisfying (2.1). The permutation p guarantees that the proposed Φ preserves the inclusion ordering (and is thus a 1:1 homomorphism from $\mathcal{E}$ into $\mathcal{N}$), while the functions f and g ensure that p and p^{-1} map $\mathcal{E}$ into $\mathcal{E}$. Often the permutation p is merely implicit and is achieved by the e-state condition of Corollary 2.9.

2.10 Definition.   An automorphism Φ of $\mathcal{E}^*$ is *effective* if Φ is presented by some recursive permutation h (or equivalently by a pair of recursive functions $\langle f, g \rangle$).

Clearly, any recursive permutation p induces an effective automorphism, but not conversely. For example, Martin's method in §3 always produces effective automorphisms and yet suffices to prove noninvariance of hypersimplicity. Nevertheless, since effective automorphisms are not sufficient for the maximal set result Theorem 4.6, one must examine nonrecursive presentations h. For example, one might expect to get noneffective automorphisms corresponding to presentations $h \leq_T \emptyset'$ but this is false, by Exercise 2.13. Therefore, one must consider presentations $h \leq_T \emptyset''$ as we do in §4.

2.11–2.13 Exercises

2.11 (Kent [1962]).   Without using Theorem 2.2 prove directly that there are $2^{\aleph_0}$ automorphisms of $\mathcal{E}$. *Hint.* Fix any cohesive set C. Consider the class

$$\mathcal{C} = \{\, p : p \text{ is a permutation of } \omega \ \& \ (\forall x)\, [x \notin C \implies f(x) = x]\,\}.$$

Show that each $p \in \mathcal{C}$ induces an automorphism of $\mathcal{E}$, and that $\mathcal{C}$ contains $2^{\aleph_0}$ members. What is the induced automorphism on $\mathcal{E}^*$ for each such $p \in \mathcal{C}$?

2.12 (Lachlan [1968c, Lemma 14]).   Let $\mathcal{L}_1$ and $\mathcal{L}_2$ be countable sublattices of $\mathcal{N}$ closed under $=^*$. Prove that $\mathcal{L}_1 \cong \mathcal{L}_2$ iff $\mathcal{L}_1^* \cong \mathcal{L}_2^*$, where $\mathcal{L}^*$ denotes $\mathcal{L}/\mathcal{F}$.

2.13 (Jockusch).   Prove that if $\Phi \in \text{Aut } \mathcal{E}^*$ is presented by some permutation $h \leq_T \emptyset'$ then Φ is presented by some recursive permutation h_1. *Hint.* Use the Limit Lemma.

3. Noninvariant Properties

In Chapter X §1 we saw that the following properties are elementary lattice theoretic and hence invariant: recursive, simple, and maximal. In Corollary X.2.8 and Theorem 1.1 more subtle definitions showed that hh-simplicity and creativeness are also invariant. In this section we show that hypersimplicity is not invariant. The proof is an easy finite injury argument, and it helps to prepare us for the more difficult method later for generating automorphisms of $\mathcal{E}^*$.

3.1 Theorem (D. A. Martin).   *There exists a hypersimple set H and $\Phi \in \text{Aut } \mathcal{E}$ such that $\Phi(H)$ is not hypersimple.*

Proof. We define effectively a sequence of recursive permutations $\{ p_s \}_{s \in \omega}$ and an increasing sequence of finite sets $\{ H_s \}_{s \in \omega}$ such that $H = \bigcup_s H_s$ is hypersimple, $p = \lim_s p_s$ is a permutation of ω, $S = p(H)$ is not hypersimple, and p induces an automorphism of $\mathcal{E}$. Let S_s denote $p_s (H_s)$. We shall have $S_s \subseteq S_{s+1}$ for all s so that S is r.e. Let $\{ F_n^e \}_{n \in \omega}$ be the eth (candidate for a) strong array, arranged so that F_n^e is a finite set, and either F_n^e is defined for all n or else is undefined for all $n >$ some m.

An index e is *satisfied at stage s* if either some F_n^e and F_m^e are defined by stage s and $F_n^e \cap F_m^e \neq \emptyset$, with $n \neq m$, or else F_n^e is defined and is a subset of H_s. The *complete e-state of x at stage s* is the pair of e-states $\langle \sigma, \tau \rangle$ where

$$\sigma = \{ i : i \leq e \ \& \ x \in W_{i,s} \},$$

and

$$\tau = \{ i : i \leq e \ \& \ p_s(x) \in W_{i,s} \}.$$

Construction of p and H.
   Stage $s = 0$. Let p_0 be the identity, and $H_0 = \emptyset$.
   Stage $s + 1$. Look for an index e such that e is not satisfied at stage s, and such that there is a member F_n^e of the eth strong array and a number $x \in F_n^e - H_s$ such that for some $y \notin H_s$:

   (i) x and y are in the same complete e-state at stage s;
   (ii) $\min(x, y, p_s(x), p_s(y)) \geq e$;
   (iii) $p_s(y) > 2x$; and
   (iv) $y > x$.

If e does not exist, set $p_{s+1} = p_s$ and $H_{s+1} = H_s$. Otherwise, let e_s be the least such e, n_s minimal for e_s, and x_s and y_s minimal for e_s and n_s. Define

$$p_{s+1}(z) = \begin{cases} p_s(y_s) & \text{if } z = x_s, \\ p_s(x_s) & \text{if } z = y_s, \\ p_s(z) & \text{otherwise.} \end{cases}$$

Let $H_{s+1} = H_s \cup \{x_s\}$. This ends the construction. Notice that $S_{s+1} = S_s \cup \{p_s(y_s)\}$.

Lemma 1. $S = \bigcup_s S_s$ *is coinfinite and not hypersimple.*

Proof. If $z \in S$, then $z = p_{s+1}(x_s) \neq p_s(x_s)$ for some s. For each s there is at most one such z and $z = p_s(y_s)$. Condition (iii) ensures that for every n there are $\leq n$ members of S smaller than $2n$, because if $z = p_s(y_s)$ enters S at stage $s+1$ then x_s enters H and $2x_s < z$. Hence, S is not hypersimple by Theorem V.2.3.

Lemma 2. $(\forall z) [p(z) = \lim_s p_s(z) \text{ exists}]$.

Proof. Fix z and choose t such that $H_t \upharpoonright (z+1) = H \upharpoonright (z+1)$. Let $s \geq t$ and $p_{s+1}(z) \neq p_s(z)$. Then $z = x_s$ or $z = y_s$. If $z = x_s$ then $z \in H_{s+1} - H_s$, a contradiction. If $z = y_s$ then by condition (iv) some $x_s < y_s$ is in $H_{s+1} - H_s$ which cannot happen at $s \geq t$. Thus, $p(z) = p_t(z)$.

Lemma 3. *If* $H = \bigcup_s H_s$ *is coinfinite then H is hypersimple. Furthermore,* $\lim_s e_s = \infty$.

Proof. Assume to the contrary that $\{F_n^e\}_{n \in \omega}$ witnesses the nonhypersimplicity of H with e minimal. If $i < e$, then by the minimality of e, the ith candidate for a strong array is finite or else i is eventually satisfied. Choose s_0 such that $e_s \geq e$ for all $s \geq s_0$.

A complete e-state is *well-resided* if infinitely many members of $\overline{H}$ finally settle in it. All but finitely many members of $\overline{H}$ lie in well-resided complete e-states. Hence, there is an n such that either $F_n^e \subseteq H$, or else $F_n^e - H \neq \emptyset$ and all $x \in F_n^e - H$ have $x \geq e$ and $p(x) \geq e$ and lie in well-resided complete e-states. Choose the least such n. If the latter case holds, choose the least such x. Then there are infinitely many stages s at which there exists some $y > x$, with $y \geq e$, $p(y) = p_s(y) > 2x$, $p(y) > e$, $y \notin H_s$, and y is in the same complete e-state as x. Hence, x is eventually enumerated in H, a contradiction.

To see that $\lim_s e_s = \infty$, fix e and by induction choose s_0 such that $e \leq e_s$ for all $s \geq s_0$. Now if $e_s = e$ for infinitely many $s \geq s_0$ then there are infinitely many n such that $F_n^e \cap \overline{H}$ contains some element x_n^e waiting forever to be interchanged with some y_n in the same complete e-state. But infinitely many of these x_n^e lie in the same complete e-state, and so one is interchanged with some y and x_n^e enters H.

Lemma 4. *For all z, $p^{-1}(z) = \lim_s p_s^{-1}(z)$ exists. Hence, p is a permutation of ω by Lemma 2.*

Proof. If $p_{s+1}^{-1}(z) \neq p_s^{-1}(z)$ then $z = p_s(x_s)$ or $z = p_s(y_s)$ and $e_s \leq z$ by condition (ii). However, $e_s \leq z$ for at most finitely many s by Lemma 3.

Lemma 5. *H is hypersimple.*

Proof. $H = p^{-1}(S)$, so H is coinfinite. Apply Lemma 3.

Lemma 6. *p induces an automorphism of $\mathcal{E}$ and in fact an effective automorphism of $\mathcal{E}^*$.*

Proof. Recursively in $\emptyset'$ define the function $k(e)$ to be the least stage t such that $e_s \geq e$ for all stages $s \geq t$. Define functions f and g by

$$W_{f(n)} = \bigcup \{ p_s(W_{n,s}) : s \geq k(n) \},$$

and

$$W_{g(n)} = \bigcup \{ p_s^{-1}(W_{n,s}) : s \geq k(n) \}.$$

Clearly, $p(W_n) = W_{f(n)}$ and $p^{-1}(W_n) = W_{g(n)}$ for all n, because after stage $k(n)$, p only interchanges elements in the same complete e-state by condition (i). Both f and g are recursive in $\emptyset'$. To see that p induces an *effective* automorphism $\Phi \in \mathrm{Aut}\, \mathcal{E}^*$, apply Exercise 2.13 or replace $k(n)$ above by 0 to produce *recursive* functions f and g such that $p(W_n) =^* W_{f(n)}$ and $p^{-1}(W_n) =^* W_{g(n)}$. ▯

Note that in the above proof, the eth positive requirement, which asserts that $F_j^e \subset H$ for some j, once met, is never injured although it may require attention finitely often (to put all of F_j^e into H) before being fully met. The eth negative requirement, which asserts that only elements in the same complete e-state may be interchanged, may be injured by the ith positive requirement only if $i < e$.

3.2 Exercise

3.2. Prove that for every nonzero r.e. degree $\mathbf{a}$ there exist a hypersimple set H and non-hypersimple set S both in $\mathbf{a}$ and $\Phi \in \mathrm{Aut}\,\mathcal{E}$ such that $\Phi(H) = S$. *Hint.* Fix an r.e. set $A \in \mathbf{a}$. When the conditions (i) through (iv) of Theorem 3.1 are satisfied define p_{s+1} as above, and assign x_s and $p_s(y_s)$ as followers of requirement R_e but do not enumerate x_s in H (and $p_s(y_s)$ in S) until A permits on x_s. Clearly $S \leq_T H \leq_T A$ because $x_s < p_s(y_s)$. To achieve equality code A into S.

4. The Statement of the Extension Theorem and Its Motivation

Our main objective for the remainder of this chapter is to prove that if A and B are any two maximal sets then there exists $\Phi \in \operatorname{Aut} \mathcal{E}$ such that $\Phi(A) = B$. The principal tool for accomplishing this is the Extension Theorem, which we state and motivate in this section. In §5 we prove that given maximal sets A and B, the necessary enumerations can be given satisfying the hypotheses of the Extension Theorem. In §6 we prove the Extension Theorem.

Fix two maximal sets A and B and consider some naive (and unsuccessful) attempts to construct $\Phi \in \operatorname{Aut} \mathcal{E}$ such that $\Phi(A) = B$. First, with Exercise 2.11 in mind, we might try to find a permutation p such that: (i) p is the identity on $A \cap B$; (ii) $p(\overline{A}) =^* \overline{B}$; and (iii) p induces an automorphism of $\mathcal{E}^*$. However, by Exercise 4.11, any permutation satisfying (i) and (ii) for A and B nonrecursive and $\overline{A} \cap \overline{B} =^* \emptyset$ cannot satisfy (iii). Next, we might try to extend Martin's method of Theorem 3.1 to construct Φ. However, this method always yields *effective* automorphisms and we now prove that there are maximal sets which are not *effectively* automorphic.

4.1 Theorem (Soare [1974b]). *There exist maximal sets A and B such that for every effective $\Phi \in \operatorname{Aut} \mathcal{E}^*$, $\Phi(A^*) \neq B^*$.*

Proof. Recall the Yates construction in Theorem X.3.3 of a maximal set A where markers $\{\Gamma_e\}_{e \in \omega}$ move to maximize their e-states with respect to a standard enumeration $\{U_{n,s}\}_{n,s \in \omega}$ (as defined in Definition II.2.8). Note that in this construction $\overline{A} \subseteq U_0$ iff $U_0 \setminus A$ is infinite.

First we show how to handle a single requirement R_e, which asserts that $A^* \not\equiv_{\mathcal{E}^*} B^*$ via Φ with presentation φ_e. We specify standard enumerations $\{U_{n,s}\}_{n,s \in \omega}$ and $\{V_{n,s}\}_{n,s \in \omega}$. Maximal sets A and B are constructed from these by the method of Theorem X.3.3. We let $U_{n+2} = V_{n+2} = W_n$ for all n, and define U_0, U_1, V_0, and V_1 so as to meet R_e.

Let $U_1 = W_{i_1} =$ the even numbers, enumerated so that $U_1 \setminus A$ is infinite. Let $V_0 = W_{\varphi_e(i_1)}$ if $\varphi_e(i_1)$ is defined, and $= \emptyset$ otherwise. Define

$$
U_0 = \begin{cases} \text{odd numbers} & \text{if } V_0 \setminus B \text{ is infinite,} \\ \text{a finite subset of the odd numbers} & \text{otherwise.} \end{cases}
$$

Assume that U_0 is enumerated so that $U_0 \setminus A$ is infinite iff $V_0 \setminus B$ is infinite. Let $W_{i_0} = U_0$, and define $V_1 = W_{\varphi_e(i_0)}$ if $\varphi_e(i_0)$ is defined and $= \emptyset$ otherwise.

Clearly, A and B will be maximal sets. Now suppose that $A^* \equiv_{\mathcal{E}^*} B^*$ via Φ with presentation φ_e. Then $\varphi_e(i_0)$ and $\varphi_e(i_1)$ are defined, $\Phi(U_0^*) = V_1^*$, and $\Phi(U_1^*) = V_0^*$. Now $V_0 \setminus B$ is infinite because otherwise $U_0 \setminus A$ is finite (by the enumeration of U_0) so $\overline{A} \subseteq^* U_1$ (by the construction of A since $U_1 \setminus A$ is infinite), and hence $\overline{B} \subseteq^* V_0$ (because $\Phi(U_1^*) = V_0^*$ and $\Phi(A^*) = B^*$) which forces $V_0 - B$ and thus $V_0 \setminus B$ to be infinite. Now $V_0 \setminus B$ infinite implies that $U_0 \setminus A$ is infinite by the definition of U_0.

But $U_0 \setminus A$ and $V_0 \setminus B$ infinite imply that $\overline{A} \subseteq U_0$ and $\overline{B} \subseteq V_0$ by the construction in Theorem X.3.3. However, $\overline{A} \subseteq U_0$ and $\Phi(U_0^*) = V_1^*$ imply that $\overline{B} \subseteq^* V_1$. Thus,

$$\overline{B} \subseteq^* V_0 \cap V_1 =^* \Phi(U_0 \cap U_1) =^* \emptyset,$$

contradicting the maximality of B.

This method may be modified as follows to simultaneously handle all requirements R_e, $e \in \omega$. The u.r.e. sequences $\{U_n\}_{n \in \omega}$, $\{V_n\}_{n \in \omega}$ are grouped in collections of finite sequences called "e-blocks". The 0-blocks are U_0, U_1, $U_2 = W_0$, and V_0, V_1, $V_2 = W_0$, respectively, where U_0, U_1, V_0, and V_1 are defined as above with $e = 0$. Assume that for some $e \geq 0$ we have defined both sequences through the e-blocks, and let $\{U_n\}_{n \leq j}$ and $\{V_n\}_{n \leq j}$, for some j, be the resulting sequences. Let $\sigma(j, x, s)$ ($\hat{\sigma}(j, x, s)$) denote the j-state of x with respect to $\{U_{n,s}\}_{n \leq j}$ ($\{V_{n,s}\}_{n \leq j}$, respectively). The $(e+1)$-blocks of $\{U_n\}_{n \in \omega}$ and $\{V_n\}_{n \in \omega}$ will be $X_1, Y_1, X_2, Y_2, \ldots, X_m, Y_m, W_{e+1}$ and $\hat{X}_1, \hat{Y}_1$, $\ldots, \hat{X}_m, \hat{Y}_m, W_{e+1}$, respectively. The pairs $\langle X_k, Y_k \rangle$ and $\langle \hat{X}_k, \hat{Y}_k \rangle$ play the role of $\langle U_0, U_1 \rangle$ and $\langle V_0, V_1 \rangle$ of the earlier proof, respectively, correspond to a fixed j-state $\sigma_k = \hat{\sigma}_k$, and are arranged in decreasing order according to the order of σ_k in the usual ranking of j-states. For a fixed j-state σ_k, $0 \leq k \leq 2^{j+1}$, define the r.e. set

Let $\hat{C}_k = W_{\varphi_e(i')}$, for some i' such that $W_{i'} = C_k$, if $\varphi_e(i')$ is defined, and $\hat{C}_k = \emptyset$ otherwise. Note that if σ_k is the true j-state of (cofinitely many members of) $\overline{A}$ then $\overline{A} \subseteq^* C_k$. Thus, if $A^* \equiv_{\mathcal{E}^*} B^*$ with presentation φ_e, then $\overline{B} \subseteq^* \hat{C}_k$, so $\hat{C}_k \setminus B$ is infinite. Define $X_k = C_k^{\text{odd}}$, $Y_k = C_k^{\text{even}}$, $\hat{X}_k = W_{\varphi_e(i_1)} \cap \hat{C}_k$, and $\hat{Y}_k \subseteq W_{\varphi_e(i_0)} \cap \hat{C}_k$ for $W_{i_0} = X_k$, and $W_{i_1} = Y_k$ analogously as before, such that $X_k \setminus A$ is infinite iff $\hat{X}_k \setminus B$ is infinite.

If $\sigma_{k'}$ is the true j-state of $\overline{A}$ then for all $k < k'$, C_k and hence X_k and Y_k will be finite and will cease interfering with $X_{k'}$ and $Y_{k'}$, which will witness that $A^* \not\equiv_{\mathcal{E}^*} B^*$ with presentation φ_e as above. ∎

4.2 Definition. A uniformly r.e. (u.r.e.) sequence of r.e. sets $\{U_n\}_{n \in \omega}$ is a *skeleton* if

$$(\forall e)\,(\exists n)\,[W_e =^* U_n].$$

Fix any two maximal sets A and B. Let us return to the problem of proving that $A \equiv_{\mathcal{E}} B$. Rather than working directly with $\{W_n\}_{n \in \omega}$ it will be sufficient (and will avoid the obstacles of Theorem 4.1) if we work instead with skeletons $\{U_n\}_{n \in \omega}$ and $\{V_n\}_{n \in \omega}$ defined for A and B respectively. The automorphism Φ will be effective with respect to these skeletons in the sense that we shall specify a permutation p of ω and recursive functions f and g such that $p(A) = B$ and

(4.1) $$(\forall n)\,[p(U_n) =^* W_{f(n)} \ \& \ p^{-1}(V_n) =^* W_{g(n)}].$$

Since $\{\,U_n\,\}_{n\in\omega}$ and $\{\,V_n\,\}_{n\in\omega}$ are skeletons there are (nonrecursive) functions F and G such that

(4.2) $$(\forall n)\,[W_n =^* U_{F(n)} \ \& \ W_n =^* V_{G(n)}].$$

Thus, the automorphism Φ will be presented (in the sense of (2.1)) by the pair $\langle f \circ F, g \circ G \rangle$ (with respect to $\{W_n\}_{n\in\omega}$). The noneffectiveness of the presentation comes from the fact that F and G are not recursive but are only Σ_3 total functions (because the relation $=^*$ in (4.2) is Σ_3) and hence are recursive in $\emptyset''$.

In Theorem 5.1 we shall define the skeletons $\{\,U_n\,\}_{n\in\omega}$ and $\{\,V_n\,\}_{n\in\omega}$ for A and B, respectively, and give a simultaneous enumeration of all these sets. Let us suppose that this has been done. How do we define f and g? From now on we let $\hat{U}_n$ denote $W_{f(n)}$ and $\hat{V}_n$ denote $W_{g(n)}$ so that our intended correspondence is:

(4.3) $$\Phi : A \mapsto B$$
$$U_n \mapsto \hat{U}_n$$
$$\hat{V}_n \mapsto V_n$$

It is convenient to picture two copies of ω with A, U_n, $\hat{V}_n$ on the "left hand side" and B, $\hat{U}_n$, V_n on the "right hand side" as suggested by (4.3). The problem of constructing $\hat{U}_n$, $\hat{V}_n$, and p may be split into two parts, the "$\overline{A}$ to $\overline{B}$" part, (4.4), corresponding to $p(\overline{A}) = \overline{B}$, and the "$A$ to B" part, (4.5), corresponding to $p(A) = B$. These parts assert that for all n,

(4.4) $$p(\overline{A} \cap U_n) =^* (\overline{B} \cap \hat{U}_n) \ \& \ p(\overline{A} \cap \hat{V}_n) =^* (\overline{B} \cap V_n),$$

and

(4.5) $$p(A \cap U_n) =^* (B \cap \hat{U}_n) \ \& \ p(A \cap \hat{V}_n) =^* (B \cap V_n).$$

Using Corollary 2.9 we may completely suppress the role of p by considering full e-states.

4.3 Definition. (i) Let $\{\,X_e\,\}_{e\in\omega}$ and $\{\,Y_e\,\}_{e\in\omega}$ be u.r.e. sequences of r.e. sets. The *full e-state* $\nu(e, x)$ of x with respect to (w.r.t.) $\{\,X_e\,\}_{e\in\omega}$ and $\{\,Y_e\,\}_{e\in\omega}$ is the triple $\langle e, \sigma, \tau \rangle$ where σ is the e-state of x with respect to $\{\,X_e\,\}_{e\in\omega}$ and τ is the e-state of x with respect to $\{\,Y_e\,\}_{e\in\omega}$.

(ii) Given simultaneous enumerations $\{X_{e,s}\}_{e,s\in\omega}$ and $\{Y_{e,s}\}_{e,s\in\omega}$ of $\{X_e\}_{e\in\omega}$ and $\{Y_e\}_{e\in\omega}$ we define the approximation to $\nu(e,x)$ at the end of stage s to be

$$\nu(e,x,s) = \langle e, \sigma(e,x,s), \tau(e,x,s)\rangle,$$

where

$$\sigma(e,x,s) = \{i : i \le e \ \& \ x \in X_{i,s}\},$$

and

$$\tau(e,x,s) = \{i : i \le e \ \& \ x \in Y_{i,s}\}.$$

By the method of Corollary 2.9 the $\overline{A}$ to $\overline{B}$ part (4.4) is equivalent to the condition

(4.6) $(\forall\nu)\,[(\exists^\infty x \in \overline{A})\,[\nu(e,x) = \nu \text{ w.r.t. } \{U_n\}_{n\in\omega} \text{ and } \{\hat{V}_n\}_{n\in\omega}]$
$\Longleftrightarrow (\exists^\infty x \in \overline{B})\,[\nu(e,x) = \nu \text{ w.r.t. } \{\hat{U}_n\}_{n\in\omega} \text{ and } \{V_n\}_{n\in\omega}]].$

The A to B part (4.5) is equivalent to the condition

(4.7) $(\forall\nu)\,[(\exists^\infty x \in A)\,[\nu(e,x) = \nu \text{ w.r.t. } \{U_n\}_{n\in\omega} \text{ and } \{\hat{V}_n\}_{n\in\omega}]$
$\Longleftrightarrow (\exists^\infty x \in B)\,[\nu(e,x) = \nu \text{ w.r.t. } \{\hat{U}_n\}_{n\in\omega} \text{ and } \{V_n\}_{n\in\omega}]].$

In the proof of the Extension Theorem in §6 we shall give a simultaneous enumeration of all the sets A, B, $\{U_n\}_{n\in\omega}$, $\{V_n\}_{n\in\omega}$, $\{\hat{U}_n\}_{n\in\omega}$, and $\{\hat{V}_n\}_{n\in\omega}$. The notations $X \setminus Y$ and $X \diagdown Y$ of Definition II.2.9 are understood as being taken with respect to this simultaneous enumeration which will respect the order of the simultaneous enumeration in Theorem 5.2 of some elements in these sets.

We may think of this as a game between two teams: team I, which is trying to build the automorphism Φ, and team II, which is trying to defeat Φ. Team I has two team members: player Ia, who is responsible for satisfying (4.6), and player Ib, who is responsible for satisfying (4.7). Player Ia must cause *enough* enumeration of elements into $\hat{U}_n$ and $\hat{V}_n$ to satisfy (4.6). However, player Ia must do this with great caution because elements enumerated by player Ia into $\hat{U}_n$ ($\hat{V}_n$) may later be enumerated (by player II) into B (A, respectively) thereby causing difficulty for player Ib in satisfying (4.7). For example, if

(4.8) $|\hat{U}_0 \diagdown B| = \infty \quad \text{and} \quad U_0 \cap A =^* \emptyset,$

then player Ib has an immediate loss before he has begun and cannot satisfy (4.7).

Let $\hat{U}_n^+$ ($\hat{V}_n^+$) denote the set of elements enumerated into $\hat{U}_n$ ($\hat{V}_n$) by player Ia and $\hat{U}_n^-$ ($\hat{V}_n^-$) those enumerated by player Ib. Note that $\hat{U}_n^+ =$

$\hat{U}_n \setminus B$ and $\hat{U}_n^- = B \setminus \hat{U}_n$ so that $\hat{U}_n = \hat{U}_n^- \cup \hat{U}_n^+$, and likewise for $\hat{V}_n$ and A. Player Ia must guarantee certain minimal hypotheses, such as

(4.9) $$|\hat{U}_0^+ \setminus B| = \infty \implies |U_0 \setminus A| = \infty,$$

and its dual condition

(4.10) $$|\hat{V}_0^+ \setminus A| = \infty \implies |V_0 \setminus B| = \infty,$$

in order to avoid an immediate loss for player Ib like that in (4.8). The point of the Extension Theorem 4.5 is that if player Ia is sufficiently restrained in enumerating $\hat{U}_n^+$ and $\hat{V}_n^+$ so as to guarantee these minimal hypotheses then player Ib can extend $\hat{U}_n^+$ and $\hat{V}_n^+$ to $\hat{U}_n$ and $\hat{V}_n$ (by enumerating the extra elements in $\hat{U}_n^-$, $\hat{V}_n^-$) satisfying (4.7) so that team I succeeds in constructing Φ. This is somewhat similar to the proof that open games are determined because if the player for the closed set can avoid a loss at every finite situation in the game, then he wins the game in the end.

Of course, the conditions (4.9) and (4.10) must be replaced by ones taking into account full e-states instead of merely single r.e. sets as is done in (4.11) and (4.12). Replacing single r.e. sets of (4.9) by full e-states yields an obvious necessary condition (4.11) below, which we refer to as the assertion that "A covers B". The dual condition (4.12) asserts that "B co-covers A." (This terminology will be further explained in §6.)

4.4 Definition.   Given full e-states $\nu = \langle e, \sigma, \tau \rangle$ and $\nu' = \langle e, \sigma', \tau' \rangle$ we let $\nu \leq \nu'$ denote that $\sigma \subseteq \sigma'$ and $\tau \supseteq \tau'$. (This relation $\leq$ is pronounced "is covered by.") (Recall that $\sigma \subseteq \sigma'$ denotes inclusion as finite subsets of ω and not extension as initial segments in $2^{<\omega}$.)

4.5 Extension Theorem (Soare [1974a]).   *Assume that A and B are infinite r.e. sets and that $\{U_n\}_{n\in\omega}$, $\{V_n\}_{n\in\omega}$, $\{\hat{U}_n^+\}_{n\in\omega}$, and $\{\hat{V}_n^+\}_{n\in\omega}$ are u.r.e. sequences of r.e. sets. Assume that there is a simultaneous enumeration of all the sets above such that for all n,*

$$A \setminus \hat{V}_n^+ = \emptyset = B \setminus \hat{U}_n^+.$$

For each full e-state $\nu = \langle e, \sigma, \tau \rangle$, define the r.e. sets

$$D_\nu^A =_{\text{dfn}} \{ x : (\exists s) [x \in A_{s+1} - A_s \ \& \ \sigma = \{ i : i \leq e \ \& \ x \in U_{i,s} \}$$
$$\& \ \tau = \{ i : i \leq e \ \& \ x \in \hat{V}_{i,s}^+ \} \},$$

and

$$D_\nu^B =_{\text{dfn}} \{ x : (\exists s) [x \in B_{s+1} - B_s \ \& \ \sigma = \{ i : i \leq e \ \& \ x \in \hat{U}_{i,s}^+ \}$$
$$\& \ \tau = \{ i : i \leq e \ \& \ x \in V_{i,s} \} \}.$$

Furthermore, assume that

(4.11) $(\forall \nu) \, [D_\nu^B \text{ infinite } \implies (\exists \nu' \geq \nu) \, [D_{\nu'}^A \text{ infinite}]],$

and

(4.12) $(\forall \nu) \, [D_\nu^A \text{ infinite } \implies (\exists \nu' \leq \nu) \, [D_{\nu'}^B \text{ infinite}]].$

Then one can extend the r.e. sets $\hat{U}_n^+$ to $\hat{U}_n$ and $\hat{V}_n^+$ to $\hat{V}_n$ such that (4.7) is satisfied.

The Extension Theorem will be proved in §6.

4.6 Theorem (Soare [1974]).   *Given any two maximal sets A and B there is an automorphism Φ of $\mathcal{E}$ such that $\Phi(A) = B$.*

Proof. Fix maximal sets A and B. In Theorem 5.1 we shall define skeletons $\{U_n\}_{n \in \omega}$ and $\{V_n\}_{n \in \omega}$ which depend upon A and B, respectively. In Theorem 5.2 we shall then define u.r.e. sequences $\{\hat{U}_n^+\}_{n \in \omega}$ and $\{\hat{V}_n^+\}_{n \in \omega}$ satisfying (4.6) and give a simultaneous enumeration of all the above r.e. sets which satisfies the hypotheses (4.11) and (4.12) of the Extension Theorem. By the conclusion of the Extension Theorem and Corollary 2.9 there exists a 1:1 map p from A to B satisfying (4.7) so $A^* \equiv_{\mathcal{E}^*} B^*$. By Corollary 2.7 $A \equiv_{\mathcal{E}} B$.   ▯

4.7 Corollary (Martin).   *Turing degree is not lattice invariant.*

Proof. By Theorem XI.2.3 there are maximal sets A and B such that $A \equiv_{\mathrm{T}} \emptyset'$ and $B <_{\mathrm{T}} \emptyset'$.   ▯

Corollary 4.7 was originally proved by Martin using an argument like that in the proof of Theorem 3.1.

4.8 Corollary (Soare [1974a]).   *For any $k > 0$, let $A_1, A_2, \ldots, A_k$ and $B_1, B_2, \ldots, B_k$ be two sequences each containing k maximal sets whose complements are pairwise disjoint. Then there exists an automorphism Φ of $\mathcal{E}$ such that $\Phi(A_i) = B_i$ for all $i \leq k$.*

Proof. Find recursive sets R_i, $1 \leq i \leq k$, such that $\omega = \bigcup_{i \leq k} R_i$, $R_i \cap R_j = \emptyset$ for $i \neq j$, and $R_i \supseteq \overline{A}_i$ for all $i \leq k$. Likewise, find recursive sets $\hat{R}_i$, $i \leq k$, with B_i in place of A_i. Apply Theorem 4.6 to find a 1:1 partial function p_i with domain R_i and which maps R_i onto $\hat{R}_i$ such that $p_i(\overline{A}_i) = \overline{B}_i$ and p_i induces an isomorphism from $\{W_n \cap R_i\}_{n \in \omega}$ to $\{W_n \cap \hat{R}_i\}_{n \in \omega}$. Then $p = \bigcup_{i \leq k} p_i$ is a permutation of ω which induces the desired Φ.   ▯

A set A is *quasimaximal of rank n* if A is the intersection of n maximal sets with pairwise disjoint complements.

4.9 Corollary. *Let A and B be any two quasimaximal sets of the same rank. Then there exists $\Phi \in \text{Aut } \mathcal{E}$ such that $\Phi(A) = B$.* ▯

4.10 Corollary. *For every $k \geq 1$, the group $\text{Aut } \mathcal{E}^*$ is k-ply transitive on the coatoms of $\mathcal{E}^*$.*

Proof. Apply Corollary 4.9 and the fact that if $A_1^*, A_2^*, \ldots, A_k^*$ are distinct coatoms in $\mathcal{E}^*$ then there exist maximal sets $B_1, B_2, \ldots, B_k$ whose complements are pairwise disjoint and such that $B_i \in A_i^*$ for all $i \leq k$. ▯

4.11 Exercise

4.11. Let A and B be r.e. and nonrecursive such that $\overline{A} \cap \overline{B} \neq^* \emptyset$. Prove that there is no permutation p of ω such that: (i) p is the identity on $A \cap B$; (ii) $p(\overline{A}) =^* \overline{B}$; and (iii) p induces an automorphism of $\mathcal{E}^*$. *Hint.* Let C be a major subset of A. Prove that $p(C) =^* \overline{A} \cup (C \cap B)$ and that $p(C)$ cannot be r.e.

5. Satisfying the Hypotheses of the Extension Theorem for Maximal Sets

Fix maximal sets A and B from now on. In this section we first specify in Theorem 5.1 skeletons $\{ U_n \}_{n \in \omega}$ and $\{ V_n \}_{n \in \omega}$ for A and B, respectively. In Theorem 5.2 we define sequences $\{ \hat{U}_n^+ \}_{n \in \omega}$ and $\{ \hat{V}_n^+ \}_{n \in \omega}$ satisfying (4.6), and we give a simultaneous enumeration of all the above sets which satisfies the hypotheses of the Extension Theorem 4.5.

For any r.e. set C define

$$\mathcal{C}(C) = \{ W_e : \overline{C} \subseteq W_e \text{ or } W_e \subseteq^* C \}.$$

Note that if C is coinfinite then $\mathcal{C}^*(C) = \mathcal{E}^*$ (namely $\mathcal{C}(C)$ is a skeleton) iff C is maximal. To facilitate the proof of Theorem 5.2 we need to give a simultaneous enumeration of a sequence $\{ Z_n \}_{n \in \omega}$ comprising all sets of $\mathcal{C}(C)$ which is order preserving in the sense that for all n and m with $n < m$, almost every x which while in $\overline{C}$ is enumerated in *both* Z_n and Z_m appears in Z_n *first*. This very strong property can be achieved because we need to consider only those Z_i such that $Z_i \supseteq \overline{C}$.

5.1 Order-Preserving Enumeration Theorem. *Given any coinfinite r.e. set C, there is a sequence $\{ Z_n \}_{n \in \omega}$, with $Z_0 = C$, such that $\{ Z_n^* \}_{n \in \omega} = \mathcal{C}^*(C)$, and there is a simultaneous enumeration of $\{ Z_n \}_{n \in \omega}$ such that:*

$$(5.1) \qquad (\forall n) \, [Z_n \setminus C \text{ infinite} \iff Z_n - C \text{ infinite} \iff Z_n \supseteq \overline{C}],$$

and

(5.2) $(\forall n)(\forall m > n)$ (a.e. x) $[x \in (Z_n \setminus C) \cap (Z_m \setminus C) \implies x \in (Z_n \setminus Z_m)]$.

Proof. Fix C. Define a sequence $\{Y_n\}_{n \in \omega}$ such that $\{Y_n^*\}_{n \in \omega} = C^*(C)$ by $Y_0 = C$,

$$Y_{2n+1} = W_n \cap C,$$

and

$$Y_{2n+2} = \{x : x \in W_n \ \& \ (\forall y \le x)[y \in C \cup W_n]\}.$$

(The device for Y_{2n+2} is similar to the definition of V_e in Theorem X.4.6 about major subsets.) Fix a simultaneous enumeration of $\{Y_n\}_{n \in \omega}$ such that $Y_{2n+1} = C \searrow Y_{2n+1}$ and

(5.3) $(\forall n)[Y_n \setminus C$ infinite $\iff Y_n - C$ infinite $\iff Y_n \supseteq \overline{C}]$,

where $C_s = Y_{0,s}$.

We now define $\{Z_n\}_{n \in \omega}$ as follows. (It will be obvious that (5.3) holds with Z_n in place of Y_n.) Define $Z_{0,s} = Y_{0,s}$ for all s.

We shall specify a function $F(e)$ such that $F(0) = 0$, $Z_{F(e)} = Y_e$ for each e, and Z_n is finite for $n \notin \operatorname{ran} F$. Our choice of $F(e)$ will depend upon Y_e as well as the $(e-1)$-state of $\overline{C}$ with respect to $\{Y_n\}_{n \in \omega}$, which is well-defined by (5.3). For each $e > 0$ and $\sigma \subseteq \{0, 1, \ldots, e-1\}$, we have a movable marker, Γ_e^σ, which comes to rest on an index $F(e)$ such that $Z_{F(e)} = Y_e$ in case the $(e-1)$-state of $\overline{C}$ with respect to $\{Y_n\}_{n \in \omega}$ is σ. An integer n is *fresh* at some stage if n has never been occupied by a marker. Define $Z_{e,s}$ for all $e > 0$ in stages s as follows.

Stage $s = 0$. Let $Z_{e,0} = \emptyset$ for all $e > 0$.

Stage $s + 1$. Enumerate one new pair $\langle x_0, y_0 \rangle$ where $x_0 \in Y_{y_0}$. If $y_0 = 0$ go to stage $s + 2$. Otherwise, perform the following steps.

Step 1. If $x_0 \in C_{s+1}$ leave all markers fixed. Otherwise, move those markers Γ_e^σ, for which $y_0 < e$ and $y_0 \notin \sigma$, to the first fresh integers maintaining the present order of the markers being moved.

Step 2. For each $\sigma \subseteq \{0, 1, \ldots, s\}$ assign a marker Γ_{s+1}^σ to the first fresh integer in order of the s-state ranking of σ with greater s-states preceding lesser ones.

Step 3. For each e, $0 < e \le s + 1$, and each $\sigma \subseteq \{0, 1, \ldots, e-1\}$ let $F(e, \sigma, s+1)$ denote the present position of marker Γ_e^σ. Let (x', e', σ') be the first triple (x, e, σ) with the following properties P1 and P2. If there exists such (x', e', σ') enumerate x' in $Z_{F(e', \sigma', s+1), s+1}$. Otherwise, go to stage $s + 2$.

(P1) $x \in Y_{e, s+1} \ \& \ x \notin Z_{F(e, \sigma, s+1), s}$.

(P2)   Either $x \in C_{s+1}$, or $\sigma = \emptyset$, or the elements of σ are $a_1 < a_2 < \cdots < a_k$, and $(\forall j \le k)[x \in Z_{F(a_j, \sigma_j, s+1), s}]$, where $\sigma_j = \{a_1, a_2, \ldots, a_{j-1}\}$.

Lemma 1.   *For all e and σ, marker Γ_e^σ moves infinitely often iff $\overline{C} \subseteq Y_y$ for some $y < e$, $y \notin \sigma$.*

Proof. By Step 1 of the construction and (5.3).

Lemma 2.   *For all n, if $Z_n \setminus C$ is infinite then some marker Γ_e^σ comes to rest on n such that*

$$(\forall y) \; [y \in \sigma \cup \{e\} \implies \overline{C} \subseteq Y_y].$$

Proof. By Step 3 of the construction and (5.3).

Now let $e_1 < e_2 < \cdots$ be a listing of all e such that $\overline{C} \subseteq Y_e$. For all e let $\rho_e = \{e_j : e_j < e\}$. By Lemma 1, $\Gamma_e^{\rho_e}$ comes to rest on some integer which we denote by $F(e)$, and $Z_{F(e)} = Y_e$ by Step 3. The markers $\Gamma_e^{\rho_e}$, $e \in \omega$, once appointed, remain in order of e (by Steps 1 and 2), and hence, for all i and e, $i < e$ iff $F(i) < F(e)$. By Lemmas 1 and 2, for all $i \notin \{F(e_j) : j \in \omega\}$, $Z_i \setminus C$ is finite. The last clause (P2) of Step 3 ensures that for a.e. x if $x \in (Z_{F(e_i)} \setminus C) \cap (Z_{F(e_j)} \setminus C)$, where $i < j$, then $x \in Z_{F(e_i)} \setminus Z_{F(e_j)}$.

Note that (5.3) guarantees (5.1). Namely, if $n = F(e)$ and $x \in Z_n \setminus C$ then $x \in Y_e \setminus C$. Hence, if $Z_n \setminus C$ is infinite then $Y_e \setminus C$ is infinite so $Y_e \supseteq \overline{C}$ by (5.3) and $Z_n \supseteq \overline{C}$ since $Y_e = Z_n$. If $n \notin \operatorname{ran} F$ then $Z_n \setminus C$ is finite by Lemma 2.   ∎

5.2 Theorem.   *Given maximal sets A and B, let $\{U_n\}_{n\in\omega}$ and $\{V_n\}_{n\in\omega}$ be the skeletons with simultaneous enumerations given by Theorem 5.1 for A and B, respectively. Then there exist u.r.e. sequences $\{\hat{U}_n^+\}_{n\in\omega}$ and $\{\hat{V}_n^+\}_{n\in\omega}$ and a simultaneous enumeration of all the sets $A = U_0$, $B = V_0$, $\{U_n\}_{n\in\omega}$, $\{V_n\}_{n\in\omega}$, $\{\hat{U}_n^+\}_{n\in\omega}$, and $\{\hat{V}_n^+\}_{n\in\omega}$ which satisfy the hypotheses of the Extension Theorem and (4.6) which (because A and B are maximal, $U_n \in \mathcal{C}(A)$, and $V_n \in \mathcal{C}(B)$) here asserts that*

$$(5.4) \qquad (\forall n) \; [[\overline{A} \subseteq U_n \iff \overline{B} \subseteq^* \hat{U}_n^+] \;\&\; [\overline{B} \subseteq V_n \iff \overline{A} \subseteq^* \hat{V}_n^+]].$$

Proof. We shall change the *given* enumeration to produce a *new* enumeration specified in the construction below. We let X_s denote the finite set of elements enumerated in set X by the end of stage s in the *new* enumeration. At the end of stage s below, we shall have defined $U_{n,s}$, $V_{n,s}$, $\hat{U}_{n,s}^+$, and $\hat{V}_{n,s}^+$. Given $A_s = U_{0,s}$ and $B_s = V_{0,s}$, define ψ_s to be the 1:1 partial function with domain $\overline{A}_s$ which maps $\overline{A}_s$ onto $\overline{B}_s$ in increasing order. Let ψ_{-1} be the identity function on ω. Let $\psi = \lim_s \psi_s$.

We shall use ψ to govern our enumeration of the sets $\hat{U}_n^+$ and $\hat{V}_n^+$. We must enumerate $\hat{U}_n^+$ and $\hat{V}_n^+$ to satisfy (5.4). We shall arrange that for almost all $x \in \overline{A}$, $x \in U_n$ iff $\psi(x) \in \hat{U}_n^+$ and similarly for $\overline{B}$, V_n, and $\hat{V}_n^+$. The main problem is that we may enumerate so many elements in $\hat{U}_n^+$ that we violate

(4.11). The solution is to avoid enumerating an element y in $\hat{U}_n^+$ until the state ν to which it would be raised is successfully covered by an element entering A. We can accomplish this because of the extreme simplicity of the sets $\{U_n\}_{n\in\omega}$, namely, $|U_n\cap\overline{A}| = \infty$ iff $\overline{A}\subseteq U_n$ (and similarly for V_n and B), and because of the order preserving enumeration property of Theorem 5.1.

Construction.

Stage $s = 4t$. Let x be the unique element enumerated (by Player II) in some set U_e at stage t under the given simultaneous enumeration of $\{U_n\}_{n\in\omega}$. Enumerate x in $U_{e,s}$. Let $A_s = U_{0,s}$.

Stage $s = 4t+1$. Do as in stage $4t$ but with V_n and B in place of U_n and A, respectively.

Stage $s = 4t+2$. Let (x', e') be the first pair (x, e) with the following properties P1–P4. If there exists such a pair (x', e'), enumerate $\psi_{s-1}(x')$ in $\hat{U}_{e',s}^+$. Otherwise, do nothing at this stage.

(P1) $x \in U_{e,s-1} - A_{s-1}$ and $\psi_{s-1}(x) \notin \hat{U}_{e,s-1}^+$.

(P2) $(\forall i < e)\ [x \in U_{i,s-1} \iff \psi_{s-1}(x) \in \hat{U}_{i,s-1}^+]$.

(P3) $(\forall i < e)\ [x \in \hat{V}_{i,s-1}^+ \iff \psi_{s-1}(x) \in V_{i,s-1}]$.

Define

$$\sigma_0 = \{i : i \le e\ \&\ \psi_{s-1}(x) \in \hat{U}_{i,s-1}\},$$
$$\tau_0 = \{i : i \le x\ \&\ \psi_{s-1}(x) \in V_{i,s-1}\},$$
$$t_x = (\text{the unique } t < s-1)\ [x \in U_{e,t+1} - U_{e,t}].$$

(P4) $(\exists v)\ (\exists y)\ (\exists \sigma_1)\ (\exists \tau_1)\ [t_x < v < s - 1\ \&\ y \in A_{i,v+1} - A_{i,v}$
$\&\ \sigma_1 \supseteq \{e\}\cup\sigma_0\ \&\ \tau_1 \subseteq \tau_0$
$\&\ \sigma_1 = \{i : i \le e\ \&\ y \in U_{i,v}\}$
$\&\ \tau_1 = \{i : i \le x\ \&\ y \in \hat{V}_{i,v}^+\}].$

(Property P4 asserts that *after* x appeared in U_e there appeared in A some element y whose pair of states $\langle\sigma_1,\tau_1\rangle$ at that time covers the pair $\langle\sigma_0\cup\{e\},\tau_0\rangle$ now proposed for $\psi_{s-1}(x)$, in the sense that $\sigma_1 \supseteq \sigma_0\cup\{e\}$ and $\tau_1 \subseteq \tau_0$. The crucial point is that we measure σ_0 only with respect to $i \le e$ not $i \le x$ as for τ_0, which would impose more restraint on enumerating $\psi_{s-1}(x)$ in $\hat{U}_e^+$ because in general $x > e$. By the order preserving enumeration property this weaker assumption will suffice to prove Lemma 1. We can afford to impose the stronger assumption "$i \le x$" on τ_1, however, since either $\psi_{s-1}(x)$ is eventually enumerated in V_i, so that this imposes no restraint, or else $\psi_{s-1}(x) \in \overline{V}_i$ forever, in which case $\hat{V}_i^+$ is finite so the condition $x \in \hat{V}_i^+$ affects at most finitely many elements $x \in \overline{A}$.)

Stage $s = 4t + 3$. Similarly, attempt to enumerate some element $\psi_{s-1}^{-1}(x')$ in $\hat{V}_{e',s}^+$ where A, U_n, $\hat{U}_n^+$, V_n, $\hat{V}_n^+$, and $\psi_{s-1}(x)$ in stage $4t + 2$ are replaced by B, V_n, $\hat{V}_n^+$, U_n, $\hat{U}_n^+$, and $\psi_{s-1}^{-1}(x)$, respectively. Also in (P3) (which now refers to U_i and $\hat{U}_i^+$) replace "$(\forall i < e)$" by "$(\forall i \leq e)$" to reflect the higher priority of the matching $U_e \leftrightarrow \hat{U}_e^+$ than $\hat{V}_e^+ \leftrightarrow V_e$.

Lemma 1. *The hypotheses of the Extension Theorem are met by the above u.r.e. sequence and simultaneous enumeration.*

Proof. It is obvious that $B \setminus \hat{U}_n^+ = \emptyset = A \setminus \hat{V}_n^+$ for all n because we enumerate only elements $\psi(x)$ in $\hat{U}_n^+$ ($\psi^{-1}(x)$ in $\hat{V}_n^+$). By "speeding up" the enumeration of A if necessary we may assume that infinitely many x appear in A before appearing in any U_n, $n > 0$, or $\hat{V}_n^+$, $n \geq 0$, and similarly for B, V_n, and $\hat{U}_n^+$. Thus, we need to verify (4.11) and (4.12) only for states $\sigma \neq \emptyset$. We verify (4.11); the verification of (4.12) is dual.

Fix e and full e-state $\nu_0 = \langle e, \sigma_0, \tau_0 \rangle$ with $\sigma_0 \neq \emptyset$. Assume that $D_{\nu_1}^A$ is finite for all $\nu_1 \geq \nu_0$. We must show that for every $\tau_1 \subseteq \tau_0$ only finitely many x whose e-state with respect to $\{V_n\}_{n \in \omega}$ is τ_1 are allowed to enter $\hat{U}_{\sigma_0}^+ =_{\mathrm{dfn}} \bigcap \{\hat{U}_n^+ : n \in \sigma_0\}$. Since e-states increase with time it follows that $D_{\nu_0}^B$ must be finite also.

Let $e_0 = \max(\sigma_0)$. It follows by P2 of the construction and the order enumeration property (5.2) of Theorem 5.1 for $\{U_n\}_{n \in \omega}$ that $\hat{U}_{e_0}^+ \setminus \hat{U}_n^+$ is finite for all $n < e_0$. Hence, if there are infinitely many $x \in \hat{U}_{e_0}^+$, almost all such x enter $\hat{U}_{e_0}^+$ only after $x \in \hat{U}_n^+$ for all $n \in \sigma_0 - \{e_0\}$. But by P4 of the construction and our assumption that $D_{\nu_1}^A$ is finite for all $\nu_1 \geq \nu_0$, at most finitely many x already in $\bigcap \{\hat{U}_n^+ : n \in \sigma_0 - \{e_0\}\}$ will be allowed to enter $\hat{U}_{e_0}^+$ while the e-state of x with respect to $\{V_n\}_{n \in \omega}$ is some $\tau_1 \subseteq \tau_0$.

Lemma 2. *(5.4) is satisfied.*

Proof. We prove (5.4) by induction on e. Fix e and assume that (5.4) holds for all $n < e$. If $U_e \not\supseteq \overline{A}$ then $U_e \setminus A$ is finite by (5.1) and hence $\hat{U}_e^+ \setminus B$ and $\hat{U}_e^+ - B$ are finite. Assume that $U_e \supseteq \overline{A}$. Choose y_0 such that for all $x \in \overline{A}$, $x \geq y_0$ implies

$$(\forall i < e) \, [[x \in U_i \iff \psi(x) \in \hat{U}_i^+] \ \& \ [x \in \hat{V}_i^+ \iff \psi(x) \in V_i]].$$

We claim that for any $x_0 \in \overline{A}$, $x_0 \geq y_0$ implies that $\psi(x_0) \in \hat{U}_e^+$. Let $m = \max\{e, x_0\}$. Choose s_0 such that for all $s \geq s_0$, $\psi_s(x) = \psi_{s_0}(x)$, and x_0 and $\psi(x_0)$ have reached by stage s_0 their final m-states with respect to each of the sequences $\{U_n\}_{n \in \omega}$, $\{\hat{U}_n^+\}_{n \in \omega}$, $\{V_n\}_{n \in \omega}$, and $\{\hat{V}_n^+\}_{n \in \omega}$. To see that $\psi(x_0)$ must be in $\hat{U}_e^+$ define

$$\tau_0 = \{i : i \leq x_0 \ \& \ \psi(x_0) \in V_i\}.$$

Now if $i \leq x_0$ and $i \notin \tau_0$, then $V_i - B$ is finite and hence $V_i \setminus B$ is finite by (5.1). But for any i, $V_i \setminus B$ finite implies $\hat{V}_i^+ \setminus A$ finite. Hence, there exist $\sigma_1 \supseteq \sigma_0$ and $\tau_1 \subseteq \tau_0$ such that

$$(\forall t)\, (\exists v > t)\, (\exists y)\, [y \in A_{v+1} - A_v$$
$$\&\ \sigma_1 = \{\, i : i \leq e\ \&\ y \in U_{i,v} \,\}$$
$$\&\ \tau_1 = \{\, i : i \leq x_0\ \&\ y \in \hat{V}_{i,v}^+ \,\}.$$

But then by P4, $\psi(x)$ is eventually enumerated in $\hat{U}_e^+$. The case of $\psi^{-1}(V_e) =^* \hat{V}_e^+$ is handled similarly. Theorem 5.2 follows from Lemmas 1 and 2. ∎

6. The Proof of the Extension Theorem 4.5

6.1 The Machines

We fix two copies of the natural numbers, ω and $\hat{\omega}$. We let variables $x, y, \ldots$ ($\hat{x}, \hat{y}, \ldots$) range over ω ($\hat{\omega}$). In accordance with (4.3) we regard A, U_n, and $\hat{V}_n$ as subsets of ω, and B, $\hat{U}_n$, and V_n as subsets of $\hat{\omega}$. From now on we measure the full e-state of $x \in \omega$ with respect to $\{U_n\}_{n \in \omega}$ and $\{\hat{V}_n\}_{n \in \omega}$ and of $\hat{x} \in \hat{\omega}$ with respect to $\{\hat{U}_n\}_{n \in \omega}$ and $\{V_n\}_{n \in \omega}$ so that

$$\nu(e, x, s) = \langle e, \{\, i \leq e : x \in U_{i,s} \,\}, \{\, i \leq e : x \in \hat{V}_{i,s} \,\} \rangle,$$

and

$$\nu(e, \hat{x}, s) = \langle e, \{\, i \leq e : \hat{x} \in \hat{U}_{i,s} \,\}, \{\, i \leq e : \hat{x} \in V_{i,s} \,\} \rangle,$$

where $U_{i,s}$, $\hat{V}_{i,s}$, $\hat{U}_{i,s}$ and $V_{i,s}$ are specified in the construction in §6.2. Hence, $\nu(e, x) = \lim_s \nu(e, x, s)$ and $\nu(e, \hat{x}) = \lim_s \nu(e, \hat{x}, s)$.

In addition to Definition 4.4 of $\nu \leq \nu'$ we need the following definitions about states ν.

6.1 Definition. Let $\nu = \langle e, \sigma, \tau \rangle$ and $\nu' = \langle e', \sigma', \tau' \rangle$.

  (i) ν σ-*exactly covers* ν' ($\nu \geq_\sigma \nu'$) iff $e = e'$, $\sigma = \sigma'$, and $\tau' \supseteq \tau$.

  (ii) ν τ-*exactly covers* ν' ($\nu \geq_\tau \nu'$) iff $e = e'$, $\tau = \tau'$, and $\sigma' \subseteq \sigma$.

  (iii) ν is an *initial segment* of ν' ($\nu \preceq \nu'$) iff $e \leq e'$, $\sigma = \sigma' \cap \{0, 1, \ldots, e\}$, and $\tau = \tau' \cap \{0, 1, \ldots, e\}$.

  (iv) The *length* of ν, written $|\nu|$, is e.

We say that x ($\hat{x}$) has *state* ν *at the end of stage* s if $\nu \preceq \nu(x, x, s)$ ($\nu \preceq \nu(x, \hat{x}, s)$) and that x has *final state* ν if $\nu \preceq \nu(x, x)$ ($\nu \preceq \nu(x, \hat{x})$).

The relations in Definitions 4.4 and 6.1 are essential in the proof for the following reason. To achieve (4.7) we wish to consider elements $x \in \omega$ and $\hat{y} \in \hat{\omega}$ which can be brought to the same final state ν, to bring both x and $\hat{y}$ to this state, and then to match them. This is difficult because we can enumerate only the sets $\{\hat{U}_n\}_{n\in\omega}$ and $\{\hat{V}_n\}_{n\in\omega}$ while the "opponent" enumerates the others. Suppose x is in state $\nu_1 = \langle e, \sigma_1, \tau_1 \rangle$ and $\hat{y}$ is in state $\nu_0 = \langle e, \sigma_0, \tau_0 \rangle$. Then it is within *our* power to bring x and $\hat{y}$ to the same state iff $\nu_1 \geq \nu_0$. In this case we first raise x to state $\nu_2 = \langle e, \sigma_1, \tau_0 \rangle$ by enumerating x in $\hat{V}_n$ for all $n \in \tau_0 - \tau_1$. Next we raise $\hat{y}$ to state $\nu_2 \geq_\tau \nu_0$ by enumerating $\hat{y}$ in $\hat{U}_n$ for all $n \in \sigma_1 - \sigma_0$. Now x and $\hat{y}$ can be successfully matched together. (Note that for $\hat{y} \in \hat{M}$ we write $\nu(y, \hat{y}, s)$, not $\nu(\hat{y}, \hat{y}, s)$.)

The actual enumeration of x in $\hat{V}_n$ and $\hat{y}$ in $\hat{U}_n$ is more complicated than this and is best described using two identical *pinball machines* M and $\hat{M}$. The elements $x \in \omega$ ($\hat{x} \in \hat{\omega}$) are played on machine M ($\hat{M}$). As soon as x is enumerated in A (B), element x ($\hat{x}$) is placed above hole H ($\hat{H}$) in machine M ($\hat{M}$). Machine M is shown in Diagram 6.1. Machine $\hat{M}$ is identical except that everything in $\hat{M}$ wears a hat (e.g., track X and rule R_n become track $\hat{X}$ and rule $\hat{R}_n$), and sets U_n and V_n of M are replaced by V_n and $\hat{U}_n$, respectively, in $\hat{M}$ for reference in rules $\hat{R}_n$.

The *surface of the machine* M is that portion of the diagram covered by arrows and is divided into tracks C, C_1, C_2, and D. An element x originally *enters* M from hole H when x is enumerated in A, and we say that x is *in machine* M thereafter. Next x proceeds during a sequence of consecutive *moves* in the direction of the arrows, called *downwards*, until x enters either pocket P or Q. Pocket P is subdivided into *boxes* B_ν such that each element $x \in B_\nu$ satisfies $\nu \preceq \nu(x, x, s)$, and B_ν contains at most finitely many permanent residents. The exact motion of x and its possible enumeration in sets $\hat{V}_n$, for certain $n \in \omega$, are controlled by rules R_1, R_2, R_3, and R_4, which will be described later.

After reaching a pocket, x may later be removed and placed again above hole H if x is enumerated in U_n for some $n \leq x$. In this case x later moves from hole H to track C, and the above process is repeated. Since x can only be placed above hole H after a change in $\sigma(x, x, s)$ it is clear that x moves to hole H at most finitely often and thus eventually comes to rest permanently in either pocket P or Q.

Intuitively, the state raising and matching of elements x and $\hat{y}$ mentioned above is intended to be accomplished by machines M and $\hat{M}$ very roughly as follows. The matching is established by a back and forth procedure where the "back" part is the matching $P \leftrightarrow \hat{Q}$ of elements $x \in P$ with elements $\hat{y} \in \hat{Q}$, and the "forth" part is the analogous matching $Q \leftrightarrow \hat{P}$. (We describe only the former since the latter is entirely dual with the usual changes, namely, with

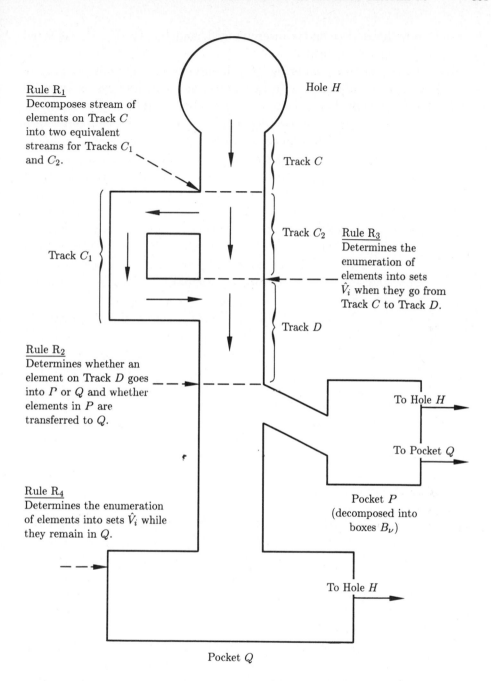

Rule R_1
Decomposes stream of
elements on Track C
into two equivalent
streams for Tracks C_1
and C_2.

Hole H

Track C

Track C_2

Rule R_3
Determines the
enumeration of
elements into sets
$\hat{V}_i$ when they go from
Track C to Track D.

Track C_1

Track D

Rule R_2
Determines whether an
element on Track D goes
into P or Q and whether
elements in P are
transferred to Q.

To Hole H

To Pocket Q

Rule R_4
Determines the enumeration
of elements into sets $\hat{V}_i$ while
they remain in Q.

Pocket P
(decomposed into
boxes B_ν)

To Hole H

Pocket Q

Diagram 6.1. Machine M

A and B interchanged, x and $\hat{x}$ interchanged, and U_n, $\hat{V}_n$, V_n, $\hat{U}_n$ replaced by V_n, $\hat{U}_n$, U_n, $\hat{V}_n$, respectively.)

Most of the elements $\hat{y}$ entering $\hat{M}$ will enter pocket $\hat{Q}$ from which they will seek an appropriate mate x from among the elements passing track D of M. While $\hat{y}$ lies in pocket $\hat{Q}$ the state of $\hat{y}$ will be raised from $\nu_0 = \langle e, \sigma_0, \tau_0 \rangle$ to some $\nu_2 = \langle e, \sigma_1, \tau_0 \rangle$, $\nu_2 \geq_r \nu_0$, according to rule $\hat{R}_4$, by enumerating $\hat{y}$ in $\hat{U}_n$ for $n \in \sigma_1 - \sigma_0$. (The choice of which ν_2 to use for $\hat{y}$ will be based upon the frequency of ν_2 occurring among the states of elements x which previously passed track D.) Secondly, various elements x on track C_2 of M will be raised from some state $\nu_1 = \langle e, \sigma_1, \tau_1 \rangle$, $\tau_1 \subseteq \tau_0$, to ν_2 according to rule R_3, by enumerating x in $\hat{V}_n$ for $n \in \tau_0 - \tau_1$, and x will be placed on track D. (This enumeration of x in $\hat{V}_n$ by rule R_3 is likewise governed by the states of previous elements $\hat{y}$ in $\hat{M}$.) Finally, rule R_2 may remove x from track D and place x in box B_{ν_2} of pocket P providing that $\nu_2 \preceq \nu(x, x, s)$ and that some element $\hat{y}$ in $\hat{Q}$ in state ν_2 "requests" x as a mate. Later, a change in membership of elements $\hat{z} \in \hat{Q}$ or a change in their state may cause this request to be cancelled. In this case x is removed from pocket P, and usually x then enters pocket Q from which x seeks a mate $\hat{z}$ in pocket $\hat{P}$, thereby changing the role of x from "mate sought" to "mate seeker."

Furthermore, $\hat{y}$ may not be content in having merely *one* mate x in B_{ν_2}, but fearing the loss of x (due to the enumeration by the opponent of x in U_n for some $n \leq |\nu_2|$) may request *finitely many* mates x in state ν_2 to be placed in box B_{ν_2}. However, we guarantee that there are at most finitely many *permanent* residents of B_{ν_2} so that the final correspondence is finite to one but not necessarily 1:1. This suffices for satisfying (4.7).

6.2 The Construction

We now give the formal description of the construction although the exact action will also depend on the rules R_1, R_2, R_3, and R_4 ($\hat{R}_1$, $\hat{R}_2$, $\hat{R}_3$, and $\hat{R}_4$), which will be given in §6.4. We fix the enumeration, called the *given enumeration* of U_n, V_n, $\hat{U}_n^+$, $\hat{V}_n^+$, $n \in \omega$, which was given by the hypotheses of the Extension Theorem 4.5. As usual we shall define a *new* enumeration in the construction below and let X_s denote the set of elements enumerated in X by the end of stage s in this new enumeration. For convenience in stabilizing the final position of element x within M we withhold from U_n (V_n, $\hat{U}_n$, $\hat{V}_n$), under Case 3 Step 1 below, the finitely many elements $x \leq n$ so that each x is enumerated in only finitely many sets. This has no effect on (4.7) which is concerned with infinitary behavior.

Construction.

   Stage $s = 0$. Do nothing.

   Stage $s + 1$. Adopt the first case which holds.

Case 1. Some element x is on track C, C_1, C_2, or D ($\hat{C}$, $\hat{C}_1$, $\hat{C}_2$, or $\hat{D}$) at the end of stage s. (The element x will be unique.) Apply rule R_1 if it is on track C ($\hat{C}$), rule R_3 ($\hat{R}_3$) if it is on track C_2 ($\hat{C}_2$), and rule R_2 ($\hat{R}_2$) if it is on track D ($\hat{D}$). If x is on track C_1 remove x and place x on track D at stage $s + 1$.

Case 2. Some element is above hole H ($\hat{H}$). Take the least such element (choosing the element above H if there are equally small elements above both H and $\hat{H}$) and put it on track C ($\hat{C}$) if it was above hole H ($\hat{H}$).

Case 3. Otherwise.

Step 1. Let the given enumeration enumerate one new element x ($\hat{x}$) into one r.e. set $Y = U_n$ or $\hat{V}_n^+$ ($= V_n$ or $\hat{U}_n^+$). If $n \leq x$ ($\hat{x}$) enumerate x ($\hat{x}$) in U_n (V_n) if $Y = U_n$ (V_n) and in $\hat{U}_n$ ($\hat{V}_n$) if $Y = \hat{U}_n^+$ ($\hat{V}_n^+$).

Step 2. If $Y = U_0 = A$ ($V_0 = B$) then put x ($\hat{x}$) above hole H ($\hat{H}$). We say that element x ($\hat{x}$) is *in machine* M ($\hat{M}$) at all stages $\geq s + 1$.

Step 3. If x ($\hat{x}$) is now in box B_ν ($\hat{B}_\nu$) of pocket P ($\hat{P}$), $Y = U_n$ (V_n), and $n \leq |\nu|$, remove x ($\hat{x}$) from pocket P ($\hat{P}$) and place x ($\hat{x}$) above hole H ($\hat{H}$). If x ($\hat{x}$) is now in pocket Q ($\hat{Q}$) remove x ($\hat{x}$) and place x ($\hat{x}$) above hole H ($\hat{H}$). At the end of stage $s + 1$ we apply rule R_4 ($\hat{R}_4$) to every element now in pocket Q ($\hat{Q}$). (Notice that each element x comes to rest because x can only be placed again above hole H if there has been a change in its x-state since the last time x was above hole H.)

This ends the construction.

This construction produces a simultaneous enumeration of all the sets $A = U_0$, $B = V_0$, $\{U_n\}_{n>0}$, $\{V_n\}_{n>0}$, $\{\hat{U}_n\}_{n\in\omega}$, and $\{\hat{V}_n\}_{n\in\omega}$. If X is a set (pocket or box), X_s refers to the elements which have been enumerated in X by the end of stage s (which reside in X at the end of stage s). If X is a pocket or a box let X_∞ denote the permanent residents of X.

6.3 The Requirements and the Motivation for the Rules

Before giving the precise description of the rules we give some motivation. Our goal is to achieve (4.7). To do so it suffices to satisfy for all ν the following requirement.

S_ν: If infinitely many elements of $B(A)$ have final state ν, then there is at least one element $\geq |\nu|$ of $A(B)$ which has final state ν.

The reason is that if infinitely many elements of B have final state ν, then for every $k > |\nu|$ there is a state $\nu_k \succeq \nu$ of length k such that infinitely many elements of B have final state ν_k. Thus, by requirement S_{ν_k} there is an element $\geq k$ in A which has final state ν_k.

To satisfy requirement S_ν we attempt to match every element $\hat{y} \in \hat{Q}$ with at least one element $x \in P$ in the same state. We position the elements

$\hat{y} \in \hat{Q}_s$ on a tree of states such that for every node ν of the tree and every stage s at most one element, denoted by $\hat{q}(\nu, s)$, rests on node ν, and $\hat{q}(\nu, s)$ is in state ν.

6.2 Definition.   Define by induction on $|\nu|$ the function

$$\hat{q}(\nu, s) = (\mu \hat{y}) \, [\hat{y} \in \hat{Q}_s \ \& \ \nu \preceq \nu(y, \hat{y}, s)$$
$$\& \ (\forall \nu' \prec \nu) \, [\hat{y} > \hat{q}(s, \nu')]].$$

If $\hat{y}$ fails to exist then $\hat{q}(\nu, s)$ is undefined ($\hat{q}(\nu, s) \uparrow$), and otherwise $\hat{q}(\nu, s) \downarrow$.

Notice that for every $\hat{y} \in \hat{Q}_s$ there is a unique state ν such that $\hat{y} = \hat{q}(\nu, s)$ (because there are $y + 1$ states ν of $\hat{y}$ and at most y of them can be such that $\hat{q}(\nu, s) = \hat{z}$ for some $\hat{z} < \hat{y}$). During the action of the machines we try to find a partner x for $\hat{q}(\nu, s)$, and we place x in box B_ν of pocket P according to rule R_2. For this correspondence to succeed we need to guarantee for every ν the two conditions,

(6.1)
$$\hat{q}(\nu, s) \uparrow \implies (\forall x \in B_{\nu, s}) \, (\exists t \geq s) \, [x \text{ is removed from box } B_\nu \text{ at stage } t],$$

and

(6.2) $|B_{\nu, \infty}| < \infty.$

(Property (6.1) will be guaranteed by Rule 2 Step 1, and (6.2) by the construction Case 3 Step 3 and rule R_2 Step 2 and Step 3.) Because of (6.1) and (6.2), to meet requirements S_ν for every ν it suffices to meet for every ν the following key requirement,

R_ν: If pocket $\hat{Q}$ (Q) has infinitely many permanent residents in final state ν then pocket P $(\hat{P})$ has at least one permanent resident in final state ν.

There are clearly two strategies for meeting requirement R_ν. The *first strategy* is to attempt to find a permanent resident for box B_ν. (This is done by rule R_2). Since the "opponent" controls the enumeration of the sets $\{U_n\}_{n \in \omega}$ this may be impossible since every x in A in state ν may later be raised by the opponent to some final state $\nu' \geq_\tau \nu$. The *second strategy* is to raise the state of almost every element $\hat{y}$ in state ν in $\hat{Q}$ to some state $\nu' \geq_\tau \nu$ for which we *can* play the first strategy. (This is done by rule $\hat{R}_4$.) The decision as to which strategy to play and, in the second case, into which state $\nu' \geq_\tau \nu$ to raise elements $\hat{y}$ in state ν in $\hat{Q}$ is the main problem of the whole construction. The intuition for its solution is roughly the following. Based upon the states of those elements which previously have passed track D (and therefore have been candidates to enter box B_ν) we specify a certain list M_s (defined more exactly in Definition 6.6) of those states ν' for which it seems possible at stage s to satisfy requirement $R_{\nu'}$ via the first strategy. Given M_s we formally define P_s as follows.

6.3 Definition.   Define

$$P_s = \{\, \nu : (\exists \nu' \in M_s)\, [\nu' \geq_\tau \nu]\,\}.$$

Intuitively, P_s consists of those ν for which it appears at stage s that we can successfully apply the second strategy and then the first.

6.4 Definition.   Define

$$d(\hat{y}, s) = \max(\{\,-1\,\} \cup \{\, e \geq 0 : \nu(e, \hat{y}, s) \in P_s$$
$$\&\ [d(\hat{y}, t) \geq e \text{ at all stages } t < s$$
$$\text{at which } \hat{y} \text{ was already in machine } \hat{M}]\,\}).$$

Intuitively, $d(\hat{y}, s)$ is the maximum length $|\nu|$ for which $\nu \preceq \nu(y, \hat{y}, s)$ and at stage s it seems that we can play at least *one* of the two strategies to satisfy R_ν. The last clause in Definition 6.4 ensures that $\lambda s[d(\hat{y}, s)]$ is nonincreasing in s after the stage at which $\hat{y}$ enters machine $\hat{M}$ so that $d(\hat{y}) =_{\text{dfn}} \lim_s d(\hat{y}, s)$ exists.

If at some stage s there is an element $\hat{y}$ in machine $\hat{M}$ in some state $\nu \notin P_s$ then it seems that neither of the two strategies will satisfy R_ν so we say requirement R_ν *requires attention at stage s*, and we perform the following drastic action. We cancel pending action (such as lists) to meet requirements $R_{\nu'}$ for $|\nu'| > |\nu|$, and thus all $R_{\nu'}$ are *injured at stage $s + 1$*. This is formally accomplished by *excluding* from M_{s+1} (under Condition (a) in Definition 6.6) all ν' such that $|\nu'| > |\nu|$. This forces $d(\hat{y}, s + 1) \leq |\nu|$ for all $\hat{y} \in \hat{Q}_{s+1}$ so that $\hat{y}$ is only concerned with requirements $R_{\nu''}$ such that $|\nu''| \leq |\nu|$.

It is the responsibility of rule R_3 to see that requirement R_ν requires attention at most finitely often. Whenever requirement R_ν requires attention we also add to a list $\mathcal{H}$ all states $\nu_1 \geq_\tau \nu$. Later elements x on track C_2 may be raised to state ν_1 and may be placed on track D by rule R_3. This increases the probability that $\nu_1 \in M_s$ and hence that $\nu \in P_s$ so that requirement R_ν will not require attention again. The key lemma is Lemma 5 of §6.5 which shows that if R_ν threatens to require attention infinitely often then this action by rule R_3 will force $\nu \in P_s$ for cofinitely many s so that R_ν cannot require attention more than finitely often.

6.4 The Rules

We now give a precise description of the rules. First we need another definition.

6.5 Definition. (i) Let X (= C, C_1, C_2, or D) be a track of M. If x is on track X at the end of stage s define $S_s(X)$ as the sequence of all states $\nu \preceq \nu(x, x, s)$, and we say that x *causes* ν in $S_s(X)$. Define $S_s(X)$ to be empty if there is no such x.

(ii) Let X (= P or Q) be a pocket of M. Define

$$S_s(X) = \{\, \nu : (\exists x)\, [\nu \preceq \nu(x, x, s)$$
$$\&\ [x \in X_s - X_{s-1} \vee \nu(x, x, s) \neq \nu(x, x, s-1)]]\,\}.$$

We say that x *causes* ν in $S_s(X)$. (Recall by Case 1 of the construction that if x is on track X at stage s then x and X are unique.)

(iii) Let $S(X)$ be the concatenation of the sequences $S_s(X)$ for all $s \in \omega$. We say that x *causes* ν in $S(X)$ if x causes ν in $S_s(X)$ for some s.

(iv) For $\hat{X}$ a track or pocket of $\hat{M}$ define $S_s(\hat{X})$ and $S(\hat{X})$ similarly using $\nu(x, \hat{x}, s)$. For the rest of this chapter we let X be a variable which stands for track C, C_1, C_2, or D or for pocket P or Q, and likewise for $\hat{X}$ and tracks or pockets of $\hat{M}$.

Rule R_1 is very similar to the Friedberg Splitting Theorem X.2.1 and is intended to ensure that if ν occurs infinitely often in $S(C)$ then ν occurs infinitely often in both $S(C_1)$ and $S(C_2)$ as we prove in Lemma 1 in §6.5. For rule R_1 we fix an infinite sequence called *list* R such that for every ν and every $i \in \{1, 2\}$ the ordered pair $\langle \nu, i \rangle$ occurs on R infinitely often. Each occurrence of $\langle \nu, i \rangle$ on R is initially *unmarked*, but later may become *marked*.

Rule R_1. Suppose x is on track C at the end of stage s. Let $\langle \nu, i \rangle$ be the first unmarked member on list R such that $\nu \preceq \nu(x, x, s)$. Mark $\langle \nu, i \rangle$ and place x on track C_i at stage $s + 1$.

During applications of rule R_2 we define for every box B_ν of pocket P a certain list $\mathcal{L}(B_\nu)$ of states ν'. Let $\mathcal{L}_s(B_\nu)$ denote the states ν' added to $\mathcal{L}(B_\nu)$ by the end of stage s. When first added the occurrence of ν' is *unmarked* although it may later become *marked*. As usual the same state ν' may occur many times in $\mathcal{L}(B_\nu)$. (Intuitively, unmarked states $\nu' \in \mathcal{L}_s(B_\nu)$ are those for which we are willing to add a new element x in state ν' to box B_ν if such an x should appear on track D and if $\hat{q}(\nu, s)$ needs it.)

Rule R_2. Suppose x is on track D at the end of stage s. Let $s' < s$ be the last stage before s such that some element was on track D at stage s'. (If no such stage exists let $s' = 0$.)

Step 1. For each ν such that

$$(\exists t)_{s' \leq t \leq s}\, [\hat{q}(\nu, t) \neq \hat{q}(\nu, s)]$$

remove from pocket P every $y \in B_{\nu,s}$. For each such y, if $\nu(y,y,t') \neq \nu(y,y,s)$ where y last entered pocket P at stage t' then place y above hole H. Otherwise, place y in pocket Q.

Step 2. For each ν such that $B_{\nu,s} = \emptyset$ add ν' to list $\mathcal{L}_{s+1}(B_\nu)$ for all ν' such that $\nu \preceq \nu'$, $|\nu'| \leq s$, and ν' does not already appear unmarked on the list $\mathcal{L}_s(B_\nu)$.

Step 3. Check whether there exist some states ν and ν' such that $\nu \preceq \nu' \preceq \nu(x,x,s)$, $\nu' \in \mathcal{L}_{s+1}(B_\nu)$ and ν' is currently unmarked on $\mathcal{L}_{s+1}(B_\nu)$. If so choose the ν of least length, then the first corresponding unmarked ν' on $\mathcal{L}_{s+1}(B_\nu)$. Put x in box B_ν of pocket P at stage $s+1$ and mark that occurrence of ν' on list $\mathcal{L}_{s+1}(B_\nu)$. If no such ν and ν' exist, put x in pocket Q.

This concludes the description of rule R_2.

6.6 Definition. For each stage s define a set $\mathcal{M}_s$ of states by induction on s. Define $\mathcal{M}_0 = \emptyset$, and

$$\mathcal{M}_{s+1} = \{ \nu : [\nu \in \mathcal{M}_s \ \& \ \nu \text{ is not excluded from } \mathcal{M}_{s+1}]$$
$$\vee [\nu \notin \mathcal{M}_s \ \& \ \nu \in S_{s+1}(D)] \}.$$

We say that ν is *excluded from* $\mathcal{M}_{s+1}$ if $\nu \in \mathcal{M}_s$ and one of the following two conditions holds.

Condition (a). $(\exists \nu') (\exists \hat{X}) [|\nu'| < |\nu| \ \& \ \nu' \in S_s(\hat{X}) - P_s]$.

Condition (b). $(\exists \nu' \preceq \nu) [B_{\nu',s} = \emptyset$
$\& \ (\forall t)_{|\nu| \leq t \leq s} (\forall \nu'' \preceq \nu') [\hat{q}(\nu'',t) \downarrow = \hat{q}(\nu'',s)]]$.

Using $\mathcal{M}_s$ we define P_s and $d(\hat{y},s)$ as in Definitions 6.3 and 6.4. Also define

$$\mathcal{M}_\omega = \{ \nu : (\text{a.e. } s) [\nu \in \mathcal{M}_s] \},$$

and

$$P_\omega = \{ \nu : (\text{a.e. } s) [\nu \in P_s] \}.$$

(We define sets of states $\hat{\mathcal{M}}_s$, $\hat{P}_s$, and a function $\hat{d}(x,s)$ for $x \in Q_s$ in an analogous manner with the usual changes.) This concludes Definition 6.6.

For rule R_3 we define a list $\mathcal{H}$ of states ν as follows. Let $\mathcal{H}_s$ consist of all $\nu(e,\hat{y},t)$ such that $e \leq \hat{y}$, $t \leq s$, $\hat{y}$ is in machine $\hat{M}$ at stage t and for some $\hat{X}$ we have $\nu(e,\hat{y},s) \in S_s(\hat{X}) - P_s$, where $\hat{X} = \hat{C}_1, \hat{C}_2, \hat{C}, \hat{D}, \hat{P}$, or $\hat{Q}$. We write $\mathcal{H}$ for the concatenation of $\mathcal{H}_s$, $s \in \omega$, and $\mathcal{H}_{\leq s}$ for the concatenation of $\mathcal{H}_t$, $t \leq s$.

Rule R_3. Suppose x is on track C_2 at the end of stage s. Let $\nu = \langle e, \sigma, \tau \rangle$ be the first unmarked state in $\mathcal{H}_{\leq s}$ such that $\nu(e,x,s) \geq \nu$. Enumerate x in the set $\hat{V}_i$ for all $i \in \tau - \tau(e,x,s)$ and mark this occurrence of ν in $\mathcal{H}_{\leq s}$. (If no such ν exists do no enumeration.) Put x on track D at stage $s+1$.

Rule $\hat{R}_4$. Suppose $\hat{y}$ is in pocket $\hat{Q}$ at the end of stage $s+1$, $e = d(\hat{y}, s)$ and $\nu(e, \hat{y}, s) = \langle e, \sigma, \tau \rangle \notin M_s$. Choose from among all states $\nu' = \langle e, \sigma', \tau \rangle \in M_s$ such that $\sigma' \supseteq \sigma$, that ν' which has occurred most often in the sequence $S_s(D)$. For that ν' enumerate $\hat{y}$ in $\hat{U}_i$ for all $i \in \sigma' - \sigma$.

We have given the rules R_1, R_2, R_3, and $\hat{R}_4$ necessary for the correspondence $P \leftrightarrow \hat{Q}$. The rules $\hat{R}_1$, $\hat{R}_2$, $\hat{R}_3$, and R_4 for the correspondence $Q \leftrightarrow \hat{P}$ are exactly dual with the usual changes. (Notice that this enumeration does not satisfy condition (2) of Definition II.2.8 because rule R_3 ($\hat{R}_4$) may enumerate element x (several elements $\hat{y}$) in several sets $\hat{V}_i$ ($\hat{U}_i$) rather than just one element in one r.e. set. This could be avoided by adding more stages but it does no harm to do the enumeration as above.)

6.5 The Verification

We now give an exact analysis of the construction using a series of lemmas which establish (4.7). A trivial proof by induction on the enumeration shows the following. Every $x \in A$ ($\hat{x} \in B$) is placed above hole H ($\hat{H}$) at some stage and x ($\hat{x}$) is in machine M ($\hat{M}$) thereafter. No number remains forever above hole H ($\hat{H}$). At every stage there is at most one element on any one of the tracks C, C_1, C_2, D, $\hat{C}$, $\hat{C}_1$, $\hat{C}_2$, or $\hat{D}$. This number is moved downwards at the next stage. Furthermore, an element $x \in A$ can move upwards in machine M (i.e., from pocket P or Q to hole H) only if x is enumerated in some new set U_i, for $i \leq x$. No element x moves directly from one box B_ν in P to another box $B_{\nu'}$, although x may be recycled to hole H and be put into a different box $B_{\nu'}$ when it enters P the next time. Therefore, every $x \in A$ moves only finitely often in machine M and eventually comes to rest forever either in pocket Q or in some box B_ν of pocket P. The analogous facts hold for elements $\hat{y} \in B$ in machine $\hat{M}$. The following lemmas deal only with the first half of requirement R_ν for the correspondence $P \leftrightarrow \hat{Q}$. The analogous lemmas for the correspondence $Q \leftrightarrow \hat{P}$ are entirely dual with the usual changes.

Lemma 1. *Every state which occurs infinitely often in $S(C)$ also occurs infinitely often in $S(C_1)$, $S(C_2)$, and $S(D)$.*

Proof. Assume ν occurs in $S(C)$ infinitely often. Then rule R_1 ensures that ν occurs in $S(C_i)$ infinitely often for $i \in \{1, 2\}$ because each of the infinitely many occurrences of $\langle \nu, i \rangle$ on list $\mathcal{R}$ is eventually marked when an element x in state ν on track C is placed on track C_i. But each element on track C_1 enters track D at the next stage so ν occurs in $S(D)$ infinitely often.

Lemma 2. *For every permanent resident $\hat{y}$ of $\hat{Q}$ there exists a unique state ν such that $\hat{y} = \lim_s \hat{q}(\nu, s)$. This state ν satisfies $\nu \preceq \lim_s \nu(y, \hat{y}, s)$. Furthermore, for every ν, if $\lim_s \hat{q}(\nu, s)$ exists then $\lim_s \hat{q}(\nu', s)$ exists for every $\nu' \preceq \nu$.*

Proof. Fix $\hat{y}$ and assume the first part of the lemma for all $\hat{z} < \hat{y}$. Choose stage s_0 such that

$$(\forall s \geq s_0)\ (\forall \hat{z} < \hat{y})\ (\forall \nu)\ [[\hat{z} \in \hat{Q}_s \iff \hat{z} \in \hat{Q}_{s_0}]$$
$$\&\ [\hat{q}(\nu, s_0) = \hat{z} \implies \hat{q}(\nu, s) = \hat{z}]$$
$$\&\ \nu(y, \hat{y}, s_0) = \nu(y, \hat{y}, s)\ \&\ \hat{y} \in \hat{Q}_s].$$

Then for some ν, $\hat{y} = \hat{q}(\nu, s_0)$, and for all $\nu' \prec \nu$, $\hat{q}(\nu', s_0) < \hat{y}$. Thus, for all $s > s_0$, $\hat{y} = \hat{q}(\nu, s)$ and $\hat{q}(\nu', s) = \hat{q}(\nu', s_0)$ for all $\nu' \preceq \nu$.

For the second part choose s_0 such that $\hat{y} = \hat{q}(\nu, s)$ for all $s \geq s_0$. If $\nu' \prec \nu$ and $\lim_s \hat{q}(\nu', s)$ does not exist then by Definition 6.2 there exists $s_1 > s_0$ such that $\hat{y} = \hat{q}(\nu', s_1)$, a contradiction.

Lemma 3. *Assume that ν is a state such that only finitely many states $\nu' >_\tau \nu$ occur in $S(D)$. Then there are only finitely many stages s at which ν is excluded from M_{s+1} according to Condition (b).*

Proof. Assume that the lemma is false for ν. Then by the definition of M_{s+1}, there are infinitely many stages s such that $\nu \in M_{s+1} - M_s$ and ν is later excluded from M_{t+1} for some $t > s$. Therefore, ν occurs infinitely often in $S(D)$. This also implies by the definition of Condition (b) that there is some $\nu_0 \preceq \nu$ of minimal length such that box B_{ν_0} has no permanent resident and $\hat{q}(\nu') = \lim_s \hat{q}(\nu', s)$ exists for all $\nu' \preceq \nu_0$. Therefore, there are infinitely many stages when box $B_{\nu_0} = \emptyset$ by rule R_2 Step 3. Thus, ν occurs infinitely often on the list $\mathcal{L}(B_{\nu_0})$ according to rule R_2 Step 2. Hence, infinitely many elements x in state ν enter box B_{ν_0} according to rule R_2 Step 3 (because ν appears in $S(D)$ infinitely often and because ν occurs at most finitely often on the list $\mathcal{L}(B_{\nu'})$ for every $\nu' \prec \nu_0$ by the minimality of ν_0). But all these elements x later leave box B_{ν_0}. Since $\lim_s \hat{q}(\nu_0, s)$ exists, finitely many x leave B_{ν_0} because of rule R_2 Step 1 so almost every x leaves because of the construction Case 3 Step 3 when x is enumerated in U_n, for some $n \leq |\nu_0|$. Every such x leaves box B_{ν_0} in some state $\nu_1 >_\tau \nu_0$, is placed above hole H, and later enters track C still in state ν_1. Hence, there is some state $\nu' >_\tau \nu$, $\nu' \succeq \nu_1$, such that ν' appears in $S(C)$ infinitely often and therefore ν' appears in $S(D)$ infinitely often by Lemma 1 contradicting the hypothesis about ν.

Lemma 4. *Fix e. Assume that there are at most finitely many stages s such that some state ν of length $\leq e$ is excluded from M_{s+1} via Condition (a). Then*

$$(\forall \nu)_{|\nu|=e} \left[(\exists^{\infty} s) \left[\nu \in P_s \right] \implies \nu \in P_\omega \right].$$

Proof. Assume that there are at most finitely many stages s such that some state of length $\leq e$ is excluded from M_{s+1} via Condition (a). Fix ν of length e such that $\nu \in P_s$ for infinitely many s. Note that by Definition 6.3,

(6.3) $\nu \in P_s \implies (\exists \nu' \geq_\tau \nu) \left[\nu' \in M_s \right].$

If $\nu' \in M_\omega$ then $\nu \in P_\omega$. Otherwise, if ν' satisfies (6.3) for infinitely many s then ν' is infinitely often added to M_{s+1} and later excluded, in which case ν' occurs in $S(D)$ infinitely often. Choose $\nu' \geq_\tau \nu$ such that ν' occurs in $S(D)$ infinitely often and no $\nu'' >_\tau \nu'$ occurs in $S(D)$ infinitely often. Now by Lemma 3, ν' is excluded from M_{s+1} at most finitely often via Condition (b). Hence, $\nu' \in M_\omega$ and therefore $\nu \in P_\omega$.

Lemma 5. *Let $\hat{X}$ be a track or pocket of machine $\hat{M}$ and X a track or pocket of machine M.*

(i) $(\forall \nu) (\forall \hat{X}) \left[\nu \text{ in } S(\hat{X}) \text{ infinitely often} \implies \nu \in P_\omega \right],$

(ii) $(\forall \nu) (\forall X) \left[\nu \text{ in } S(X) \text{ infinitely often} \implies \nu \in \hat{P}_\omega \right].$

Proof. We prove (i) and (ii) simultaneously by induction on $|\nu|$. Assume that (i) and (ii) hold for all ν such that $|\nu| < e$. Then every state of length $\leq e$ is excluded from M_{s+1} or $\hat{M}_{s+1}$ via Condition (a) for at most finitely many s. It suffices to prove (i) for every ν of length e since (ii) is dual.

Assume for a contradiction that (i) fails for some state ν_1 of length e. Fix some $\hat{X}$ such that ν_1 occurs in $S(\hat{X})$ infinitely often. Hence, $\nu_1 \in S_s(\hat{X}) - P_s$ for infinitely many s by Lemma 4. Therefore,

(6.4) $(\forall \nu) \left[\nu \in M_\omega \implies |\nu| \leq e \right],$

by Condition (a) exclusion from M_{s+1}.

Fix infinitely many different elements $\hat{y}_j$, and stages t_j, $j \in \omega$, such that for all $j \in \omega$, $\hat{y}_j$ causes $\nu_1 \in S_{t_j}(\hat{X}) - P_{t_j}$. Let $\mathcal{T}_j$ be the finite sequence of states ν such that

$$(\exists s \leq t_j) \left[\hat{y}_j \text{ is in machine } \hat{M} \text{ at stage } s \text{ and } \nu(e, \hat{y}_j, s) = \nu \right].$$

Let $\mathcal{T}$ be the concatenation of the sequences $\mathcal{T}_j$, $j \in \omega$. Since ν_1 occurs in $\mathcal{T}$ infinitely often, the following two claims (proved separately below) imply that $\nu_1 \in P_\omega$, a contradiction.

Claim 1. *If ν occurs infinitely often in T and some $\nu' \geq \nu$ occurs infinitely often in $S(C)$, then some $\nu'' \geq_\tau \nu$ occurs infinitely often in $S(D)$ and $\nu \in P_\omega$.*

Proof of Claim 1. State ν occurs in list $\mathcal{H}$ for rule R_3 infinitely often because ν occurs in T infinitely often. Assume that $\nu' \geq \nu$ and ν' occurs in $S(C)$ infinitely often. Let $\nu = \langle e, \sigma, \tau \rangle$ and $\nu' = \langle e, \sigma', \tau' \rangle$, with $\sigma \subseteq \sigma'$ and $\tau \supseteq \tau'$. Then ν' occurs in $S(C_2)$ infinitely often by Lemma 1. Let $\nu'' = \langle e, \sigma', \tau \rangle$ so that $\nu'' \geq_\tau \nu$. Now by rule R_3 infinitely many elements x on track C_2 in state ν' are raised to state ν'' and then placed on track D. Thus, for infinitely many s, $\nu'' \in M_s$ and therefore $\nu \in P_s$. Hence, $\nu \in P_\omega$ by Lemma 4.

Claim 2. *If ν occurs infinitely often in T then some $\nu' \geq \nu$ occurs infinitely often in $S(C)$.*

Proof of Claim 2. Assume for a contradiction that Claim 2 in false. Fix $\nu_2 = \langle e, \sigma_2, \tau_2 \rangle$ so that σ_2 is minimal and τ_2 is minimal for σ_2 such that the claim fails for ν_2. Because of the "covering property" hypothesis (4.11) of the Extension Theorem 4.5 it cannot be that infinitely many $\hat{y}_j$ were already in state ν_2 when they entered machine $\hat{M}$. Therefore, there exists a state $\nu_3 = \langle e, \sigma_3, \tau_3 \rangle \neq \nu_2$, an infinite set $J \subseteq \omega$, and stages $s_j \leq t_j$, $j \in J$, such that for every $j \in J$, $\hat{y}_j$ is in machine $\hat{M}$ at stage $s_j - 1$, $\nu(s_j - 1, e, \hat{y}_j) = \nu_3$ and $\nu(s_j, e, \hat{y}_j) = \nu_2$.

By the minimality of σ_2 and τ_2 we have that Claim 2 holds for ν_3. Thus, we cannot have $\sigma_3 = \sigma_2$ and $\tau_3 \subset \tau_2$ because then $\nu_3 \geq \nu_2$ so Claim 2 holds for ν_2 because it holds for ν_3. Thus, we may assume that $\sigma_3 \subset \sigma_2$. Hence, there is an infinite set $J' \subseteq J$ such that for every $j \in J'$, rule $\hat{R}_3$ applies to $\hat{y}_j$ at stage s_j (Case 1 below), or for every $j \in J'$, rule $\hat{R}_4$ applies to $\hat{y}_j$ at stage s_j (Case 2 below).

Case 1. For every $j \in J'$ rule $\hat{R}_3$ applies to $\hat{y}_j$ at stage s_j. By the induction hypothesis there are at most finitely many ν on list $\hat{\mathcal{H}}$ for rule $\hat{R}_3$ such that $|\nu| < e$. Hence, for almost every $j \in J'$ during the application of rule $\hat{R}_3$ at stage s_j we mark some $\nu' = \langle e', \sigma', \tau' \rangle$ in $\hat{\mathcal{H}}$ such that $e' \geq e$ and $\langle e, \sigma' \restriction (e + 1), \tau' \restriction (e + 1) \rangle \geq \nu_2$. By the definition of $\hat{\mathcal{H}}$ we know that the element x in M which caused the occurrence of ν' on $\hat{\mathcal{H}}$ earlier passed track C in some state $\nu'' \geq \nu'$ (namely the first time it passed track C after achieving e-state σ'). Hence, some $\nu_2' \geq \nu_2$ occurs on $S(C)$ infinitely often.

Case 2. For all $j \in J'$, rule $\hat{R}_4$ applies to $\hat{y}_j$ at stage s_j. Hence, for almost every $j \in J'$, $d(\hat{y}_j, s_j - 1) \geq e$ by the minimality of σ_2 and Claim 1. First suppose that $d(\hat{y}_j, s_j - 1) > e$ for infinitely many $j \in J'$. Then there is for each of these j some $\nu(j) \succ \nu_2$ in $M_{s_j - 1}$ with $|\nu(j)| > e$. By (6.4) no such $\nu(j)$ is in M_ω. Hence, for infinitely many s some $\nu(j) \succ \nu_2$ is added to M_{s+1} and so $\nu(j)$ occurs on $S_s(D)$. Thus, some $\nu' \geq \nu_2$ occurs infinitely often on $S(C)$. Next suppose that $d(\hat{y}_j, s_j - 1) = e$ for almost every $j \in J'$. Since Claim 2 holds for ν_3, some $\nu' \geq_\tau \nu_3$ occurs infinitely often on $S(D)$ by Claim 1. Choose ν'' to be the maximal ν' with respect to $\leq_\tau$ such that ν' has this property. Hence, ν'' occurs in $S(D)$ infinitely often and $\nu'' \in M_\omega$ by Lemma 3 and the induction hypothesis. But in applying rule $\hat{R}_4$ to $\hat{y}_j$ at stage $s_j - 1$ we raise $\hat{y}_j$ to that state $\nu \geq_\tau \nu_3$, $\nu \in M_{s_j - 1}$, which has occurred most frequently in $S(D)$, so ν_2 occurs infinitely often in $S(D)$ because ν'' does. Thus, some $\nu_2' \geq_\sigma \nu_2$ occurs in $S(C)$ infinitely often since any x on track D was previously on track C in the same σ-state. This proves Claim 2 and therefore Lemma 5.

Lemma 6. (i) *For every $e \in \omega$ there are only finitely many elements $\hat{y} \in B$ such that $d(\hat{y}) =_{\mathrm{dfn}} \lim_s d(\hat{y}, s) < e$.*

(ii) *If $\hat{y}$ remains forever in pocket $\hat{Q}$ and $d(\hat{y}) \geq 0$ then*

$$\lim_s \nu(d(\hat{y}), \hat{y}, s) \in M_\omega.$$

Proof. (i). Any $\hat{y} \in B$ eventually comes to rest in some final state $\nu(y, \hat{y})$ and in pocket $\hat{P}$ or $\hat{Q}$ where it causes all $\nu \preceq \nu(y, \hat{y})$ to be added to $S(\hat{P})$ or $S(\hat{Q})$. Hence, by Lemma 5(i) and Definition 6.4 for every $e \in \omega$, $d(\hat{y}) < e$ for at most finitely many $\hat{y} \in B$.

(ii). Choose s_0 such that after stage s_0, $\hat{y}$ remains in pocket $\hat{Q}$, $d(\hat{y}, s)$ and $\nu(y, \hat{y}, s)$ remain constant, and $d(\hat{y}) \geq 0$. Then for all $s \geq s_0$,

$$\lim_s \nu(d(\hat{y}), \hat{y}, s) = \nu(d(\hat{y}), \hat{y}, s + 1) \in M_s$$

because of the action of rule $\hat{R}_4$ at stage $s + 1$.

Lemma 7. *For every ν if pocket $\hat{Q}$ has infinitely many permanent residents in final state ν then so does pocket P. Furthermore, box B_ν of pocket P has finitely many permanent residents and each has final state ν.*

Proof. Assume that S is an infinite set of permanent residents of pocket $\hat{Q}$ in final state ν. It suffices to show that box B_ν of pocket P has at least one permanent resident. (This suffices because for every $k \geq |\nu|$ there is some $\nu_k \succeq \nu$, $|\nu_k| = k$, such that $\hat{Q}$ has infinitely many permanent residents in final state ν_k, so by the above claim box B_{ν_k} has a permanent resident.)

By Lemma 2 for every $\hat{y} \in S$ there is some state $\nu_{\hat{y}}$ such that $\hat{y} = \lim_s \hat{q}(\nu_{\hat{y}}, s)$ and $\nu_{\hat{y}} \preceq \lim_s \nu(y, \hat{y}, s)$. Since no two elements $\hat{y} \in S$ have the same $\nu_{\hat{y}}$ we have $\nu \preceq \nu_{\hat{y}}$ for almost every $\hat{y} \in S$. Hence, by Lemma 2 we can choose s_0 such that

$$(\forall \nu' \preceq \nu)\,(\forall s \geq s_0)\,[\hat{q}(\nu', s) = \hat{q}(\nu', s_0)].$$

Let $\mathcal{T} = \{\nu : \nu = \lim_s \nu(d(\hat{y}), \hat{y}, s) \ \& \ \hat{y} \in S\}$. By Lemma 6(ii), $\mathcal{T}$ is an infinite subset of $\mathcal{M}_\omega$. Choose any $\nu_1 \in \mathcal{T}$, $|\nu_1| \geq s_0$, and $\nu \preceq \nu_1$. Define

$$t_0 = (\mu t)\,(\forall t' \geq t)\,[\nu_1 \in \mathcal{M}_{t'}].$$

Then $B_{\nu,s} \neq \emptyset$ for all $s \geq t_0$ because otherwise ν_1 is excluded from $\mathcal{M}_{s+1}$ by Condition (b) exclusion. Hence, the list $\mathcal{L}(B_\nu)$ is finite by rule R_2 Step 2. Therefore, at most finitely many elements enter box B_ν by rule R_2 Step 3 because each entering element marks an occurrence of some ν' on list $\mathcal{L}(B_\nu)$. Hence, some element remains forever in box B_ν. By rule R_2 Step 3 and the construction Case 3 Step 3 every element x which enters box B_ν enters in state ν and while in B_ν remains in state ν (although x may be enumerated in U_n for some $n > |\nu|$). Thus, box B_ν has at most finitely many permanent residents and each is in final state ν.

Lemma 8. *For every ν if pocket P has infinitely many residents in final state ν then so does pocket $\hat{Q}$.*

Proof. If pocket P has infinitely many permanent residents x in final state ν then for almost every such x there exists some $\nu' \succeq \nu$ such that x is a permanent resident of box $B_{\nu'}$ and each box $B_{\nu'}$ has finitely many permanent residents by Lemma 7. For each such ν' rule R_2 Step 1 implies that $\hat{y}_{\nu'} = \lim_s \hat{q}(\nu', s)$ exists, and $\hat{y}_{\nu'}$ is in final state ν' by Lemma 2.

   Lemmas 7 and 8 establish (4.7) for the correspondence $P \leftrightarrow \hat{Q}$. The proof for the correspondence $\hat{P} \leftrightarrow Q$ is exactly analogous. This completes the proof of the Extension Theorem 4.5.   ▯

6.7 Notes.   The definitions, notation and fundamental ideas of the proofs presented in sections 4, 5, and 6 were taken from Soare [1974]. Various simplifications and improvements for the presentation were taken from Maass [1983], and still further modifications to that presentation were made here. Extensive corrections and suggestions on the current version of Chapter XV were made by V. Harizanov, S. Lempp, and M. Stob. Further results and open questions on automorphisms are presented in Chapter XVI §1.

Further Results and Open Questions About R.E. Sets and Degrees

The purpose of this chapter is to give a brief overview without proofs of some further results and current open questions about r.e. sets and r.e. degrees which would have been covered in detail in this book if time and space had permitted. The reader may recognize how the diverse results and methods studied in Chapters VII through XV have been combined and extended in these later theorems. No attempt has been made to be comprehensive, and numerous important and current topics in recursion theory have of necessity been omitted, such as the Turing degrees in general, recursive model theory, effective mathematics, computational complexity, and others.

1. Automorphisms and Isomorphisms of the Lattice of R.E. Sets

The automorphism method of Chapter XV has been extended to give a lot of information about the structure of r.e. sets and their algebraic properties in $\mathcal{E}$. Most of these results show surprising and very pleasing uniformity of structure of $\mathcal{E}$ rather than pathology of structure as is the case with some other results about r.e. sets and degrees. Recall Definition X.2.6 of $\mathcal{L}(A)$, the lattice of supersets of $A \in \mathcal{E}$. Robinson (Exercise XI.3.5) used the oracle method for approximating a low r.e. set A to prove that if A is coinfinite then A has a maximal superset. It was observed that Robinson's method only used the weaker hypothesis "$\overline{A}$ is semi-low" (see Exercises IV.4.7 and XI.3.6(a)) in place of "A is low." Soare then combined this lowness method with the automorphism method of Chapter XV to prove the following characterization of $\mathcal{L}(A)$ for such sets.

1.1 Theorem (Soare [1982a]). *If A is r.e. and coinfinite and $\overline{A}$ is semi-low (as defined in Exercise IV.4.6) then $\mathcal{L}(A) \cong \mathcal{E}$, and indeed $\mathcal{L}^*(A)$ is effectively isomorphic to $\mathcal{E}^*$ (written $\mathcal{L}^*(A) \cong^{\mathrm{eff}} \mathcal{E}^*$) in the sense of Definition XV.2.10.*

This shows a very strong resemblance of low r.e. sets to recursive sets because it is obvious that $\mathcal{L}^*(R) \cong^{\mathrm{eff}} \mathcal{E}^*$ if R is recursive and coinfinite. By analogy with the definition of semi-low (= semi-low$_1$) in Exercise IV.4.6 we now define a set S to be *semi-low$_{1.5}$* if

$$\{\, x : W_x \cap S \text{ finite} \,\} \leq_1 \emptyset''.$$

If S is semi-low then S is semi-low$_{1.5}$ but not conversely. Bennison and Soare [1978] studied coinfinite r.e. sets A such that $\overline{A}$ is semi-low or semi-low$_{1.5}$, and they related these index set properties to computational complexity properties of A in terms of Meyer-Fischer type complexity sequences (see the overview in Soare [1982b]). They also proved that every coinfinite r.e. set A such that $\overline{A}$ is semi-low$_{1.5}$ has a maximal superset. Maass combined this method of exploiting the hypothesis "$\overline{A}$ semi-low$_{1.5}$" with an extension of the automorphism machinery used in Theorem 1.1 to obtain the following very pleasing characterization of those r.e. sets A such that $\mathcal{L}^*(A) \cong^{\text{eff}} \mathcal{E}^*$.

1.2 Theorem (Maass [1983]). *If A is r.e. and coinfinite then $\mathcal{L}^*(A) \cong^{\text{eff}} \mathcal{E}^*$ iff $\overline{A}$ is semi-low$_{1.5}$.*

One direction is immediate, while the other requires a further extension of the automorphism machinery of Chapter XV. In the case of both Theorem 1.1 and Theorem 1.2 the isomorphism $\mathcal{L}(A) \cong \mathcal{E}$ has been preceded by the weaker result that A has a maximal superset if $\overline{A}$ is infinite and semi-low (semi-low$_{1.5}$). By analogy therefore, Theorem XI.5.1 suggests that the following open question may have a positive answer.

1.3 Open Question. If A is r.e., coinfinite and low$_2$, is $\mathcal{L}(A) \cong \mathcal{E}$?

By Theorem 1.2 we could not hope to show in Question 1.3 that $\mathcal{L}^*(A) \cong^{\text{eff}} \mathcal{E}^*$ since there exist coinfinite r.e. sets A such that $\overline{A}$ is low$_2$ but not semi-low$_{1.5}$. However, the isomorphism might be effective on some skeleton. A positive answer to Question 1.3 would be a first step in attacking some of the open questions in Questions XI.4.5 and in particular in proving line XI (4.8). Notice that we cannot hope to prove 1.3 for low$_3$ in place of low$_2$ because Theorem XI.4.1 shows that non-low$_2$ r.e. degrees contain atomless r.e. sets, i.e., the low$_n$ degrees for $n \geq 3$ fail the maximal superset test.

From Definition X.2.2 recall the definition of the lattice $\mathcal{E}(S) = \{W \cap S : W \in \mathcal{E}\}$ for $S \subseteq \omega$ and recall that $\mathcal{E}(\overline{A}) \cong \mathcal{L}(A)$ for $A \in \mathcal{E}$. In Chapter X §4 we studied properties of the major subsets B (written $B \subset_m A$) of an r.e. set A. By extending the automorphism machinery of Chapter XV Maass and Stob were able to show that there is a single isomorphism type for the intervals determined by major subsets.

1.4 Theorem (Maass and Stob [1983]).   *If $B_1 \subset_m A_1$ and $B_2 \subset_m A_2$ then $\mathcal{E}(A_1 - B_1) \cong \mathcal{E}(A_2 - B_2)$.*

1.5 Corollary.   *If A and B are r-maximal sets with maximal supersets, then $\mathcal{E}(\overline{A}) \cong \mathcal{E}(\overline{B})$.*

Corollary 1.5 does not classify the automorphism type of such r-maximal sets, however, because given a maximal set M we can find major subsets A and B such that $A \subset_{sm} M$ but not $B \subset_s M$ (see Definition X.4.10 and Exercise X.4.12), so A and B cannot be automorphic. Another proof of this fact together with more information about major subsets and automorphisms can be found in Maass [1985b].

1.6 Open Question.   Classify the automorphism types of r-maximal sets.

A very important algebraic question which is largely open is the following.

1.7 Open Question.   If $A, B \in \mathcal{E}$ and $\mathcal{L}(A) \cong \mathcal{L}(B)$ under what further conditions is A automorphic to B ($A \cong_{\text{aut}} B$)?

We know that $\mathcal{L}(A) \cong \mathcal{L}(B)$ does not *always* imply that A is automorphic to B by the remark following Corollary 1.5. (Alternatively we could choose A to be low and simple and B to be recursive and coinfinite so that $\mathcal{L}(A) \cong \mathcal{L}(B)$ by Theorem 1.1 but not $A \cong_{\mathcal{E}} B$.) By Corollary XV.4.9, $\mathcal{L}(A) \cong \mathcal{L}(B)$ does imply $A \cong_{\text{aut}} B$ if $\mathcal{L}^*(A)$ is finite (and therefore necessarily is a finite Boolean algebra). This is no longer true if $\mathcal{L}^*(A)$ is an *infinite* Boolean algebra even if $\mathcal{L}^*(A)$ is so well behaved that it is the countable atomless Boolean algebra.

1.8 Theorem (Lerman, Shore, and Soare [1978]).   *There exist nonautomorphic r.e. sets A and B such that $\mathcal{L}(A) \cong \mathcal{L}(B)$ and $\mathcal{L}^*(A)$ is the countable atomless Boolean algebra.*

We write $C \subset_{rm} D$ if $C \subset_m D$ and simultaneously C is r-maximal in D. Lerman, Shore, and Soare proved that an r.e. set A has an rm-subset C iff A has a "Δ_3 preference function" (which is a certain property of $\overline{A}$), and also proved that there exist sets A and B satisfying Theorem 1.8 such that A possesses an rm-subset but B does not. In the positive direction Maass has shown that 1.7 holds for the case of Boolean algebras if we add the following stronger hypothesis about the isomorphism $\mathcal{L}(A) \cong \mathcal{L}(B)$.

1.9 Theorem (Maass [1984]).   *If $A, B \in \mathcal{E}$, $\mathcal{L}^*(A)$ is a Boolean algebra, and $\mathcal{L}^*(A) \cong \mathcal{L}^*(B)$ via a Σ_3^0-isomorphism then $A \cong_{\mathcal{E}} B$.*

Theorem 1.9 cannot be improved with Σ_4^0 in place of Σ_3^0 because of Theorem 1.8 and the fact that if A and B are atomless and hh-simple then the standard back and forth construction yields a Σ_4^0 isomorphism between $\mathcal{L}^*(A)$ and $\mathcal{L}^*(B)$.

For an r.e. set A we say that $\overline{A}$ is *semi-low$_2$* if

$$\{\, e : W_e \cap \overline{A} \text{ infinite} \,\} \leq_T \emptyset''.$$

Many classes of r.e. sets (e.g., atomless hyperhypersimple sets) contain particularly well-behaved representatives A such that $\overline{A}$ is semi-low$_2$. It follows from Theorem 1.9 that if A and B have $\mathcal{L}(A) \cong \mathcal{L}(B)$, $\mathcal{L}^*(A)$ an atomless Boolean algebra, and $\overline{A}$ and $\overline{B}$ semi-low$_2$, then $A \cong_{\mathcal{E}} B$. Herrmann [ta] gave a further classification of the orbits of hh-simple sets A with the same $\mathcal{L}(A)$, and showed that they may lie in infinitely many different orbits.

Maass [1982] gave an alternative solution to Post's problem by constructing r.e. sets which are generic in the usual set theoretic sense but where the notion of forcing is taken with respect to tiny universes. Maass showed that all r.e. generic sets A are both promptly simple and low (and hence $\overline{A}$ is semi-low), and therefore of Turing degree strictly between $\mathbf{0}$ and $\mathbf{0}'$.

1.10 Theorem (Maass [1982]). *If A and B are promptly simple, coinfinite, and have semi-low complements then $A \cong_{\mathcal{E}} B$.*

The proof of Theorem 1.10 is not difficult but uses Theorem 1.1 together with a proof that simultaneously the hypotheses of the Extension Theorem XV.4.5 can be satisfied if A and B are both promptly simple. Let **G** be the class of degrees of sets which are automorphic in $\mathcal{E}$ to r.e. generic sets. Ambos-Spies, Jockusch, Shore, and Soare [1984] showed that

$$\mathbf{G} = \mathbf{ENC} = \mathbf{NC} = \mathbf{PS} = \mathbf{LC},$$

where the last three equalities were established in Theorems XIII.2.2 and XIII.4.2. Other results on promptly simple sets are contained in Maass [1985a] and Maass, Shore, and Stob [1981]. A new notion of genericity for r.e. sets, called *e-genericity*, was introduced by Jockusch [1985] and generalized in Nerode and Remmel [1985a] and [ta]. The property of *e*-genericity is easier to work with than r.e. genericity, and much more is known about *e*-generic sets than r.e. generic sets.

Theorem 1.10 is useful for proving properties noninvariant. In Theorem XV.3.1 and Corollary XV.4.7 we saw that h-simplicity and Turing degree are not invariant in $\mathcal{E}$. Stob [1982a] generalized the Extension Theorem method to prove that dense simplicity is not invariant in $\mathcal{E}$. Maass and Stob later observed that this could also be obtained from Theorem 1.10 by first giving a straightforward construction of a dense simple, promptly simple set A such that $\overline{A}$ is semi-low, since it is easy to build a promptly simple B which is low and therefore not dense simple (by Exercise XI.1.11). Thus, the orbit of the r.e. generic sets contains a surprisingly large variety of r.e. sets. It is an open question whether this orbit is definable, and if not whether it can be characterized in some other way. The property of prompt simplicity is known not to be invariant under Aut $\mathcal{E}$ because we can construct maximal sets A and B such that A is promptly simple and B is not. However, Maass, Shore, and Stob [1981] have shown that prompt simplicity implies a certain splitting property which is invariant.

1.11 Open Question.   Does there exist $A \in \mathcal{E}$ such that

$$\deg\{\,W : W \cong_{\mathcal{E}} A\,\} = \mathbf{R}^+?$$

Note that Theorem 1.10 cannot be used to answer this question because **PS** splits **R** by Theorem XIII.2.2.

One of the most interesting algebraic questions about r.e. sets is to classify their orbits. In Chapter XV we saw that creative sets and maximal sets each constitute an orbit. Downey and Stob [ta] define a nonrecursive r.e. set A to be *hemimaximal* if there is a maximal set M and a nonrecursive r.e. set B such that $M = A \cup B$ and $A \cap B = \emptyset$ (for example, if A and B are produced from M by the Friedberg Splitting Theorem X.2.1). They show surprisingly that,

1.12 Theorem (Downey and Stob [ta]).   *Any two hemimaximal sets A and B are automorphic.*

The idea is to first modify Theorem XV.5.2 to obtain $\mathcal{L}(A) \cong \mathcal{L}(B)$ and simultaneously to use the nonrecursiveness of A and B to obtain the covering property hypotheses of the Extension Theorem XV.4.5. Let **HM** be the class of degrees of hemimaximal sets. Downey and Stob show that $\mathbf{H}_1 \subseteq \mathbf{HM}$, $\mathbf{R}^+ \not\subseteq \mathbf{HM}$, and that for every $\mathbf{a} \in \mathbf{R}^+$ there exists $\mathbf{b} \in \mathbf{HM}$, $\mathbf{b} \le \mathbf{a}$. The jumps of degrees of hemimaximal sets are unknown. The hemimaximal sets are the first example of a definable class in $\mathcal{E}$ which forms an orbit and contains nonzero low degrees. Note that **HM** also shows that it is false that the class of degrees of sets in an orbit of nonrecursive sets is closed upwards in **R**.

1.13 Open Question.   Let $\mathcal{C}$ be a class of simple sets which forms an orbit. Does

$$\{\,A : A \text{ is half of a Friedberg splitting of a set in } \mathcal{C}\,\}$$

form an orbit, where A_0 and A_1 form a *Friedberg splitting* of B if conditions (i), (ii), and (iii) of Theorem X.2.1 are satisfied?

This question has been raised by Downey and Stob [ta] who suggested starting by considering Friedberg splittings of low promptly simple sets. A major open question is whether every orbit (of a nonrecursive set) contains $\mathbf{0}'$, namely

1.14 Open Question. For every nonrecursive r.e. set A does there exist $\Phi \in \text{Aut } \mathcal{E}$ such that $\Phi(A) \equiv_T \emptyset'$?

This question appears to be rather difficult. (Some partial results were obtained by Downey and Stob [ta, Theorems 9 and 12] which imply that every low$_2$ simple set, every simple set A with $\overline{A}$ semi-low$_{1.5}$, and every d-simple set with a maximal superset is automorphic to a complete set.) The dual question is whether every orbit contains some degree $\mathbf{a} < \mathbf{0}'$. By Theorem XV.1.1 this cannot be exactly true and indeed Harrington claims further that there exists an orbit consisting of only Turing complete but not creative sets. However, Harrington also claims that a revised version of the question has a positive answer. Namely, he claims that if A is r.e., nonrecursive and not Turing complete, then there exists a set B in the orbit of A such that $B \not\geq_T A$. A weaker open question than 1.14 is whether every nonrecursive r.e. set contains *some* high r.e. set in its orbit.

Martin's Theorem XI.2.3 on degrees of maximal sets leads us to ask whether Question 1.14 can be strengthened by showing that every orbit containing a nonrecursive r.e. set also contains a set of *every* high r.e. degree. Maass, Shore, and Stob [1981] have refuted this by showing that the degrees of non-hh-simple sets with a certain splitting property (definable in $\mathcal{E}$) are exactly the promptly simple degrees **PS** of Theorem XIII.2.2 which are known to nontrivially split $\mathbf{H}_1$. However, the question is still open for $\mathcal{L}(A)$ in place of orbit (A) and is very interesting since it would be the strongest possible generalization of Martin's theorem.

1.15 Open Question. Suppose $A \in \mathcal{E}$ and $\mathbf{d}$ is high. Does there exist an r.e. set $B \in \mathbf{d}$ such that $\mathcal{L}(A) \cong \mathcal{L}(B)$?

Martin's Theorem XI.2.3 established this if $\mathcal{L}^*(A)$ is the two element Boolean algebra, and Lachlan [1968c, p. 27] established this in the case where $\mathcal{L}^*(A)$ is *any* Boolean algebra. Harrington has claimed that for any coinfinite $A \in \mathcal{E}$ there exists $B \in \mathcal{E}$ such that $\mathbf{0} < \deg(B) < \mathbf{0}'$ and $\mathcal{L}(A) \cong \mathcal{L}(B)$ which gives some positive evidence for 1.15. (See also XI.2.20.)

2. The Elementary Theory of $\mathcal{E}$

Let $\text{Th}(\mathcal{E})$ $(\text{Th}(\mathcal{E}^*))$ be the elementary theory of $\mathcal{E}$ $(\mathcal{E}^*)$ in the language $L(\subseteq, \cup, \cap, 0, 1)$ where 0 and 1 are interpreted as the least and greatest elements of the lattice. Lachlan showed [1968c, Theorem 1] that the decision problem for $\text{Th}(\mathcal{E})$ is reducible to that for $\text{Th}(\mathcal{E}^*)$ and conversely. (Indeed he proved the first reduction for any sublattice $\mathcal{L}$ of $\mathcal{N}$ closed under symmetric difference with finite sets where $\mathcal{L}^* = \mathcal{L}/\mathcal{F}$. The first reduction relies on elimination of quantifiers while the second is an easy consequence of the fact that the property of being finite is definable in $\mathcal{E}$ as we have seen in line X (1.2).)

The first major progress about the decision problem for $\mathrm{Th}(\mathcal{E})$ was made by Lachlan [1968d]. Recall that $\mathcal{B}$ denotes the Boolean algebra generated by $\mathcal{E}$. Lachlan added to the language a unary predicate $E(x)$ which is to be interpreted over $(\mathcal{E}^*, \mathcal{B}^*)$ as "$x \in \mathcal{E}^*$". An $\overrightarrow{\forall}\overrightarrow{\exists}$-sentence in this language is one of the form $(\forall x_1)\cdots(\forall x_n)(\exists y_1)\cdots(\exists y_m)\, P(\overrightarrow{x}, \overrightarrow{y})$, where P is quantifier free.

2.1 Theorem (Lachlan [1968d]). *There is a decision procedure for the $\overrightarrow{\forall}\overrightarrow{\exists}$-sentences true in $\mathcal{B}^*$ where quantifiers range over $\mathcal{E}^*$.*

Lachlan began by proving that all consistent existential sentences are true in $\mathcal{E}^*$. A lattice $\mathcal{L}$ is *separated* if for any pair x, y of elements of $\mathcal{L}$ there exists a disjoint pair x_1, y_1 of elements of $\mathcal{L}$ such that $x_1 \leq x$, $y_1 \leq y$, and $x_1 \cup y_1 = x \cup y$. Clearly, $\mathcal{E}^*$ is separated. Lachlan reduced the $\overrightarrow{\forall}\overrightarrow{\exists}$-decision problem for $\mathcal{E}^*$ to the following. Given finite separated lattices $L, L_1, L_2, \ldots, L_k$, such that each L_i is a refinement of L, when is it true that for all sublattices $\mathcal{L}$ of $\mathcal{E}^*$ such that $\mathcal{L} \cong L$, there exists a sublattice $\mathcal{L}'$ of $\mathcal{E}^*$ such that one of the following diagrams commutes for $1 \leq i \leq k$?

$$
\begin{array}{ccc}
L & \overset{\subseteq}{\Rightarrow} & L_i \\[4pt]
\downarrow \cong & & \downarrow \cong \\[4pt]
\mathcal{L} & \overset{\subseteq}{\Rightarrow} & \mathcal{L}'
\end{array}
$$

Diagram 2.1

For each finite separated lattice L, Lachlan produced a "canonical realization" $\mathcal{L} \subseteq \mathcal{E}^*$, $\mathcal{L} \cong L$. These canonical realizations provide necessary conditions for the diagram to commute. To construct the canonical realizations Lachlan needs only the following facts about $\mathcal{E}$:

(1) the Reduction Principle (Corollary II.1.10);
(2) for any infinite $A \in \mathcal{E}$, $\mathcal{E}(A) \cong \mathcal{E}$ (trivial);
(3) there is an infinite coinfinite recursive set;
(4) there is a maximal set (Theorem X.3.3);
(5) the Friedberg Splitting Theorem (Theorem X.2.1) with condition (iii);

(6) the small major subset theorem (Exercise X.4.12).

For example, if the separated lattice L consists of $0, x_1, x_2, x_3, 1$ such that: $x_1 \subset 1$; $x_2, x_3 \subset x_1$; and $x_2 \cap x_3 = 0$; then Lachlan would construct $\mathcal{L} \cong L$ consisting of r.e. sets A_0, A_1, A_2, and A_3 where $A_0 = \omega$, A_1 is a maximal set, $B \subset_{sm} A_1$ and A_2 and A_3 form a Friedberg Splitting of B. These canonical realizations rule out as false a large class of $\overrightarrow{\forall}\overrightarrow{\exists}$-sentences. For example, there is no $W \in \mathcal{E}$ such that W splits $\overline{A}_1$, and there is no $V \in \mathcal{E}$ such that $\overline{A}_1 \subseteq V$ but $\overline{B} \not\subseteq^* V$. Lachlan's refinement theorem [1968d, Theorem 4] then shows that these necessary conditions are also sufficient. He defines a notion of characteristic for finite separated lattices which well-orders them. His refinement theorem either constructs $\mathcal{L}' \subseteq \mathcal{E}^*$ such that Diagram 2.1 commutes, or else shows that a counterexample has already been found among the canonical realizations of a certain finite effectively determined set of lattices of smaller characteristic than that of L. This is the essence of his decision procedure. Using a similar method Stob [1979] gave a decision procedure for the $\overrightarrow{\forall}\overrightarrow{\exists}$-sentences true in $\mathcal{E}(B - A)$ where $A \subset_m B$. By Theorem 1.4 there is only one isomorphism type of such intervals. Stob's result can be derived from Theorems 1.4 and 2.1.

Lerman and Soare [1980b] extended Lachlan's method to give a decision procedure for the $\overrightarrow{\forall}\overrightarrow{\exists}$-sentences of an extended language which are true in $\mathcal{E}^*$. The extended language contains new predicates such as $\text{Max}(x)$ and $\text{Hhs}(x)$ which are to be interpreted as "x is maximal" and "x is hh-simple". The idea is to construct canonical realizations and to prove a refinement theorem as in Lachlan's proof but each step is more complicated now, and requires new structural information about $\mathcal{E}$, because the $\overrightarrow{\forall}\overrightarrow{\exists}$-sentences of the extended language include such statements as "there exists an atomless hh-simple set with an r-maximal subset," or "there exists an r-maximal set".

The next major development was the proof of undecidability of $\text{Th}(\mathcal{E})$ given by Herrmann and independently by Harrington. (Harrington's proof appeared slightly later and exists only in the form of unpublished notes.)

2.2 Theorem (Herrmann [1984b], Harrington [1983]).   *The first order theory of $\mathcal{E}$ is undecidable.*

Both proofs depend upon the undecidability of the theory of a Boolean algebra with a distinguished subalgebra (see M. Rubin [1976] and Burris and McKenzie [1981]). A *Boolean pair* is a pair $(\mathcal{A}_1, \mathcal{A}_2)$ where $\mathcal{A}_1$ is a Boolean algebra and $\mathcal{A}_2$ is a subalgebra defined over $\mathcal{A}_1$ by some unary relation $R(x)$. A Boolean pair is *recursive* if the usual Boolean operations and the relation $R(x)$ are all recursive. Burris and McKenzie [1981] constructed a class of Boolean pairs (isomorphic to recursive Boolean pairs) such that every larger class has an undecidable theory. Thus, if C is a class of Boolean pairs including all isomorphism types of recursive Boolean pairs then C has an undecidable theory. Herrmann proved Theorem 2.2 by finding a class C of Boolean pairs such that: C contains isomorphism types of all recursive Boolean pairs; and C is elementarily definable with parameters (e.d.p.) in $\mathcal{E}$. Herrmann achieved this after a series of preliminary results about definability in $\mathcal{E}$ contained in various papers such as Herrmann [1981], [1983], [1984b], and [ta]. For example, he first proved that (Σ_4, Σ_3) is e.d.p. in $\mathcal{E}$ where Σ_4 is the lattice of Σ_4 sets together with a unary predicate defining exactly the Σ_3 sets. The proof uses ideas from Lachlan's method of Theorem X.7.2 for embedding Σ_3 relations in $\mathcal{E}$. Harrington proved Theorem 2.2 by showing that any Boolean pair $(\mathcal{A}_1, \mathcal{A}_2)$ of Δ_2^0 Boolean algebras is e.d.p. in $\mathcal{E}$. Given $\mathcal{A}_1$ and $\mathcal{A}_2$ Harrington used a $0'''$-construction to construct an hh-simple set A and an r.e. set $B \subseteq A$ defining $(\mathcal{A}_1, \mathcal{A}_2)$ as follows. Let the ideal determined by A and B be

$$ I = \{ X \in \mathcal{E} : (\exists \text{ rec. } C \subseteq A) (\exists D \subset_m A) [X \subseteq C \cup B \cup (A - D)] \}. $$

For $X \in \mathcal{E}(\overline{A})$ choose any recursive set Y_x such that $Y_x - A = X$. He showed that if $\hat{Y}_x$ is another such set then $Y_x \triangle \hat{Y}_x \in I$. Hence, the map $f(X) = Y_x$ produces a lattice homomorphism from $\mathcal{E}(\overline{A})$ to $\mathcal{E}(A)/I$. He constructed A and B such that $(\mathcal{A}_1, \mathcal{A}_2) \cong (\mathcal{E}(A)/I, f(\mathcal{E}(\overline{A})))$. The major remaining open question about $\mathrm{Th}(\mathcal{E})$ is whether it has the greatest degree possible (namely $\mathbf{0}^{(\omega)}$).

2.3 Open Question. What is the degree of $\mathrm{Th}(\mathcal{E})$? In particular, is its degree equal to $\mathbf{0}^{(\omega)}$?

It would also be interesting to classify the quantifier level where $\mathrm{Th}(\mathcal{E})$ becomes undecidable.

2.4 Open Question. Is there a decision procedure for the $\vec{\exists}\vec{\forall}\vec{\exists}$-sentences true in $\mathcal{E}$?

The exact quantifier level from Theorem 2.2 where $\mathrm{Th}(\mathcal{E})$ becomes undecidable has not yet been determined although it is apparently greater than three alternating blocks of quantifiers. On the other hand pushing Theorem 2.1 and the results by Lerman and Soare to the $\vec{\exists}\vec{\forall}\vec{\exists}$ case seems quite hard. Any advance on this problem is likely to lead to a considerable amount of new structural information about $\mathcal{E}$. There are also interesting open questions concerning definability in $\mathcal{E}$. Since every r.e. set A, $A \neq \emptyset$ and $A \neq \omega$, has infinitely many sets in its orbit, it is not possible to define a single nontrivial r.e. set. However, we have seen that various *classes* of r.e. sets are e.l.t., such as maximal, hh-simple, creative and others. A cardinality argument shows that there are properties which are lattice theoretic (l.t.) but not e.l.t.

More specific examples can be derived from results by Lempp [1986] and [ta]. He classified the index sets of hh-simple sets A whose $\mathcal{L}^*(A)$ has a certain rank with respect to the Cantor Bendixson derivative, and he showed that the degrees of these index sets go through levels $\mathbf{0}^{(\alpha)}$ for all recursive successor ordinals α. Any of these classes of r.e. sets whose index set has degree $\mathbf{0}^{(\alpha)}$ for $\alpha \geq \omega$ is l.t. but not e.l.t. At the lower end Lempp showed that $\{\, x : W_x \text{ quasi-maximal} \,\}$ is Σ_5-complete, but it is unknown whether the class of quasi-maximal sets is e.l.t. in $\mathcal{E}$. A compactness argument shows that this class cannot be definable in the pure theory of lattices, but it may be definable in $\mathcal{E}$. The same question is open for other classes near the bottom of the Lempp hierarchy such as the r.e. sets A whose $\mathcal{L}^*(A)$ is the atomic Boolean algebra (i.e., isomorphic to the Boolean algebra of finite and cofinite sets), or has some other relatively simple isomorphism type.

Since the algebraic structure of $\mathcal{E}$ is very complicated it has been suggested that one should first study various quotient lattices. Lachlan [1968c, p. 36] suggested studying the quotient lattice of $\mathcal{E}$ modulo the filter generated by the simple sets. Lerman [1976] proposed majoricity and hh-simplicity as congruence relations for factoring $\mathcal{E}$ (see Exercise X.4.9). Degtev [1978c] has studied some quotient lattices. Much more work remains to be done.

3. The Elementary Theory of the R.E. Degrees

Let $\mathrm{Th}(\mathbf{R})$ denote the elementary theory of the upper semi-lattice of r.e. degrees $(\mathbf{R}, \leq, \cup)$ in the language $L(\leq, \cup)$. The existential theory of $\mathbf{R}$ is decidable because by Exercise VII.2.2 any consistent existential statement is true in $\mathbf{R}$. Attempts have been made to produce a decision procedure for certain subsets of all $\overrightarrow{\forall}\,\overrightarrow{\exists}$-sentences true in $\mathbf{R}$ using results and methods of Chapters VIII through XIV, but these have not yet proved fully successful. The next major development was the proof of undecidability of $\mathrm{Th}(\mathbf{R})$.

3.1 Theorem (Harrington and Shelah [1982a] and [1982b]). $\mathrm{Th}(\mathbf{R})$ *is undecidable.*

This result was extended by Harrington and Slaman who showed that the degree of $\mathrm{Th}(\mathbf{R})$ is as great as possible, namely, $\mathbf{0}^{(\omega)}$.

3.2 Theorem (Harrington and Slaman [ta]). $\mathrm{Th}(\mathbf{R})$ *has degree* $\mathbf{0}^{(\omega)}$.

A simplified proof of Theorem 3.1 will appear in Harrington and Slaman [ta]. The proof depends upon the fact that the theory of finite partial orderings is undecidable, and it uses a $0'''$-priority argument to code an arbitrary Δ_2^0 partial ordering into $\mathbf{R}$ using four parameters. Specifically, given a Δ_2^0

partial ordering $(P, \leq_P)$ one can construct r.e. degrees **a**, **b**, **c**, and **d** coding $(P, \leq_P)$ as follows. Let **S** be the class of degrees $\mathbf{x} \in \mathbf{R}$ satisfying:

(1) $\mathbf{x} \leq \mathbf{a}$,

(2) $\mathbf{x} \cup \mathbf{c} \not\geq \mathbf{b}$, and

(3) **x** is maximal with respect to those degrees $\mathbf{y} \in \mathbf{R}$ satisfying (1) and (2).

Let $\hat{\mathbf{x}} = \mathbf{x} \cup \mathbf{d}$, and $\hat{\mathbf{S}} = \{\hat{\mathbf{x}} : \mathbf{x} \in \mathbf{S}\}$. Then $(\hat{\mathbf{S}}, \leq) \cong (P, \leq_P)$.

3.3 Open Question.   For every pair of r.e. degrees $\mathbf{a} < \mathbf{b}$, is the theory of the r.e. degrees **x**, $\mathbf{a} \leq \mathbf{x} \leq \mathbf{b}$, undecidable? Does it have degree $\mathbf{0}^{(\omega)}$?

3.4 Open Question.   What are the decidable fragments of $\mathrm{Th}(\mathbf{R})$? In particular, is the $\overrightarrow{\forall}\overrightarrow{\exists}$-theory of **R** decidable?

Closely related are questions about definability of r.e. degrees. A degree $\mathbf{d} \in \mathbf{R}$ is *definable* if there is a first order formula $F(x)$ in $L(\leq)$ such that **d** is the unique element which satisfies $F(x)$ in $(\mathbf{R}, \leq)$. Slaman and Harrington have raised the following questions.

3.5 Open Question.   Is there a definable r.e. degree **a** such that $\mathbf{0} < \mathbf{a} < \mathbf{0}'$? Is every r.e. degree definable?

This question is related to the following long-standing question of Sacks [1967].

3.6 Open Question.   Does there exist a degree invariant solution to Post's problem, namely a Gödel number e such that for all sets X and $Y \subseteq \omega$,

$$X <_{\mathrm{T}} W_e^X <_{\mathrm{T}} X' \ \& \ [X \equiv_{\mathrm{T}} Y \implies W_e^X \equiv_{\mathrm{T}} W_e^Y]?$$

Lachlan [1975] showed that if such a solution exists it cannot be uniform in the sense that Gödel numbers for the second $\equiv_{\mathrm{T}}$ are obtained uniformly recursively from indices for the first $\equiv_{\mathrm{T}}$. A strong negative solution to Sacks' question in the context of set theory has been conjectured by Martin. For more information on Martin's conjecture and partial results, see Slaman and Steel [ta].

3.7 Open Question.   Which *classes* of r.e. degrees are definable in $(\mathbf{R}, \leq)$? In particular, is $\mathbf{H}_n$ or $\mathbf{L}_n$ definable for any $n > 0$? (See also XI.4.5 for other open questions about $\mathbf{H}_n$ and $\mathbf{L}_n$.)

Also interesting is the relationship between the r.e. degrees and the Δ_2^0 degrees. Slaman and Woodin [ta] have shown that the class of r.e. degrees is definable with parameters in the Δ_2^0 degrees. It is an open question whether the parameters can be eliminated. Slaman [ta] has also shown that the r.e. degrees do not form a Σ_1 substructure of the Δ_2^0 degrees in the language $L(\leq)$. Slaman proves that the following Δ_0-formula $F(x, y, z)$ with parameters $\mathbf{a}, \mathbf{b}, \mathbf{c} \in \mathbf{R}$ is satisfiable in the Δ_2^0 degrees but not in $(\mathbf{R}, \leq)$,

$$0 < x \leq a \ \& \ x \leq y \ \& \ b \leq y \ \& \ c \not\leq y.$$

3.8 Open Question. Are there any automorphisms of $(\mathbf{R}, \leq)$ other than the identity?

Harrington has raised the question of the interdefinability of $(\omega, +, \cdot)$ and $(\mathbf{R}, \leq)$. An *interpretation* I of ω in $\mathbf{R}$ consists of a set $\mathbf{S} \subseteq \mathbf{R}$, an equivalence relation on $\mathbf{S}$ yielding equivalence classes $[\mathbf{d}]_I$, $\mathbf{d} \in \mathbf{S}$, and definable operations $\oplus, \odot$ on these equivalence classes so that

$$(\omega, +, \cdot) \cong (\{ [\mathbf{d}]_I : \mathbf{d} \in \mathbf{S} \}, \oplus, \odot).$$

We say that $\mathbf{d}$ is a *code* in I for $n \in \omega$ if this isomorphism takes n to $[\mathbf{d}]$. Harrington and Slaman have proved Theorem 3.2 by building an interpretation of ω in $\mathbf{R}$.

3.9 Open Question. Is there an interpretation I of ω in $\mathbf{R}$ such that the relation $R(\mathbf{a}, \mathbf{b})$ is definable, where

$$R(\mathbf{a}, \mathbf{b}) \iff \mathbf{a} \text{ is a code in } I \text{ for } \mathbf{n} \text{ and } W_n \in \mathbf{b}?$$

If so then $(\mathbf{R}, \leq)$ is rigid, in which case the answer to Question 3.8 is no.

If $\mathbf{a}_1, \ldots, \mathbf{a}_n \in \mathbf{R}$, the *n-type* of $\langle \mathbf{a}_1, \ldots, \mathbf{a}_n \rangle$ is the set of all first order formulas $F(x_1, \ldots, x_n)$ in $L(\leq)$ with n free variables satisfied by $\langle \mathbf{a}_1, \ldots, \mathbf{a}_n \rangle$ in $(\mathbf{R}, \leq)$. $\mathrm{Th}(\mathbf{R})$ is $\aleph_0$-categorical iff for every n there are finitely many n-types. Lerman, Shore and Soare [1984] showed that there are infinitely many 3-types in $(\mathbf{R}, \leq)$ and hence that $\mathrm{Th}(\mathbf{R})$ is not $\aleph_0$-categorical. (The latter fact also follows from the proof of Theorem 3.1.) Ambos-Spies and Soare [ta] have used the method in Chapter XIV to show that there exist r.e. degrees $\{ \mathbf{c}_n \}_{n \in \omega}$ which pairwise form minimal pairs, and each of which is nonbounding. From this it can be shown that there exist infinitely many 1-types in $(\mathbf{R}, \leq)$, which was a question left open in Lerman, Shore and Soare [1984].

4. The Algebraic Structure of **R**

Probably the most important open question about the structure of $\mathbf{R}$ is the following.

4.1 Embeddability Question. Characterize those finite lattices which can be embedded in $(\mathbf{R}, \leq, \cup, \cap)$ by a map preserving suprema and infima.

The known results about this question were summarized in Notes IX.2.7. A more detailed discussion appears in Lerman [1985a], and the most recent results appear in Ambos-Spies and Lerman [1986]. Both embeddability and non-embeddability techniques have been pushed quite far, but no clear embeddability criterion has yet emerged. Notice that a solution to Question 4.1 is necessary before we can classify even those *existential* sentences in $L(\leq, \cup, \cap)$ which are true in $\mathbf{R}(\leq, \cup, \cap)$. It would also be interesting to discover which finite lattices are *dense* in $\mathbf{R}(\leq, \cup, \cap)$.

One of the original pieces of evidence for Shoenfield's conjecture (see Chapter IX §1) was the Sacks Density Theorem VIII.4.1. Since then when a new property P has been discovered, such as the branching or nonbranching degrees (Definition IX.4.1), researchers have tried to determine whether that property is *dense*, namely whether for all r.e. degrees $\mathbf{a} < \mathbf{b}$ there exists an r.e. degree $\mathbf{d}$ with property P, $\mathbf{a} < \mathbf{d} < \mathbf{b}$. For example, Fejer [1983] proved the density of the nonbranching degrees, while Slaman [ta] proved the density of the branching degrees. This existence of various different types in each interval gives some positive evidence for Question 3.3 because it may be possible to code an interpretation of ω into the interval $(\mathbf{a}, \mathbf{b})$.

Another consequence of Shoenfield's conjecture is the strong cupping property IX (0.2). This was refuted by Cooper and Yates. Given r.e. degrees $\mathbf{0} < \mathbf{b} < \mathbf{a}$ we say that $\mathbf{b}$ *cups to* $\mathbf{a}$ if there exists an r.e. degree $\mathbf{c} < \mathbf{a}$ such that $\mathbf{b} \cup \mathbf{c} = \mathbf{a}$; if no such $\mathbf{c}$ exists then $\mathbf{b}$ is an *anti-cupping witness* for $\mathbf{a}$. The r.e. degree $\mathbf{a}$ has the *anti-cupping (a.c.) property* if it has an anti-cupping witness. (The conjecture IX (0.2) asserts that no nonrecursive r.e. degree has the a.c. property.) Ladner and Sasso [1975] proved that every degree $\mathbf{a} \in \mathbf{R}^+$ has a predecessor $\mathbf{b} \in \mathbf{L}_2 \cap \mathbf{R}^+$ with the a.c. property.

4.2 Theorem (Cooper [1974a] and Yates). $\mathbf{0}'$ *has the a.c. property.*

Cooper circulated a proof of Theorem 4.2, and Harrington modified the proof so that the witness $\mathbf{b}$ is low. Then Harrington strengthened the result by proving that the witness $\mathbf{b}$ could be high. These proofs relied on the fact that $\mathbf{0}'$ contains a creative set. Harrington then combined this technique with the high permitting method of Cooper (Theorem XI.2.15) to obtain the following generalization.

4.3 Theorem (Harrington). *Every high r.e. degree* $\mathbf{a}$ *has the a.c. property via a high r.e. witness* $\mathbf{b}$.

Harrington's notes [1976] on the proof were never published, but a proof of Theorem 4.3 appears in Miller [1981a]. The proofs of Theorems 4.2 and

4.3 are interesting because they use a restraint function $r(e, s)$ such that $\liminf_s r(e, s) = \infty$ unlike the restraint functions we have seen in Chapters VIII, IX, and XIV.

In contrast Harrington also proved that not *every* degree $\mathbf{a} \in \mathbf{R}$ has the a.c. property.

4.4 Plus Cupping Theorem (Harrington). *The following statement holds where all quantifiers range through* **R**.

$$(\exists \mathbf{a} > \mathbf{0}) \ (\forall \mathbf{b})_{\mathbf{0} < \mathbf{b} < \mathbf{a}} \ (\forall \mathbf{d})_{\mathbf{a} \leq \mathbf{d}} \ (\exists \mathbf{c})_{\mathbf{c} < \mathbf{d}} \ [\mathbf{b} \cup \mathbf{c} = \mathbf{d}].$$

Harrington's notes [1978] on the proof were circulated but were never published. Shoenfield also circulated notes on the proof which were not published. The proof uses a $0'''$-priority argument like that in Chapter XIV. A special case of the theorem for $\mathbf{d} = \mathbf{0}'$ appears in Fejer and Soare [1981] and the proof uses only a $0''$-priority argument similar to the proof of Theorem XIII.2.2. The methods used to achieve cupping played a crucial role in the proof of Theorem 3.1.

Another group of results of current interest concerns relative enumerability and the structure of the r.e. degrees with the jump operator added. We say that $\mathbf{a}$ is *r.e. in and above* (r.e.a. in) $\mathbf{b}$ if $\mathbf{a}$ is r.e. in $\mathbf{b}$ and $\mathbf{a} \geq \mathbf{b}$. Cooper [1972d] conjectured that if $\mathbf{a}$ is r.e.a. in $\mathbf{0}'$ then $\mathbf{a}$ is r.e. in every high degree $\mathbf{d} \leq \mathbf{0}'$. This was refuted by Soare and Stob [1982] who showed that for any $\mathbf{c} \in \mathbf{R}^+$ there exists $\mathbf{a}$ r.e.a. in $\mathbf{c}$ such that $\mathbf{a}$ is not an r.e. degree. (This answers Cooper's question because we relativize the result with $\mathbf{d}$ in place of $\mathbf{0}$ and with $\mathbf{0}'$ in place of $\mathbf{c}$.) The proof cannot be uniform but produces two candidates $\mathbf{a}_1$ and $\mathbf{a}_2$ for $\mathbf{c}$, one of which succeeds. Shore combines this method with the $0'''$-priority method to prove a non-inversion theorem which stands in contrast to the jump inversion theorems of Shoenfield (Exercise VI.4.11) and Sacks (Theorem VIII.3.1).

4.5 Non-Inversion Theorem (Shore [ta]). *There are degrees* $\mathbf{a}_0$ *and* $\mathbf{a}_1$ *r.e.a. in* $\mathbf{0}'$ *such that* $\mathbf{a}_0 \cup \mathbf{a}_1 < \mathbf{0}''$ *and if* $\mathbf{u} < \mathbf{0}'$ *then not both* $\mathbf{a}_0$ *and* $\mathbf{a}_1$ *are r.e. in* $\mathbf{u}$.

A corollary is that not every degree $\mathbf{a}$ r.e.a. in $\mathbf{0}'$ is the jump of a degree which is half of a minimal pair of r.e. degrees. This corollary was also proved by Cooper [ta] using a $0'''$-priority method. The proof of Theorem 4.5 uses a new $0'''$-priority argument where the tree of strategies is $\omega + 1$ branching so that determining the true path requires a $0'''$-oracle.

In a similar vein, Lempp and Slaman [ta] proved that there is no nonzero r.e. degree $\mathbf{a}$ which is *deep* in the sense that for all $\mathbf{w} \in \mathbf{R}$, $\mathbf{w}' = (\mathbf{a} \cup \mathbf{w})'$. The proof also uses a $0'''$-priority argument. Both this result and the Soare-Stob result above refute the existence of nonrecursive r.e. degrees with certain "strongly low" properties.

The Robinson Splitting Theorem XI.3.2 showed that the Sacks Splitting Theorem VII.3.2 and the Sacks Density Theorem VIII.4.1 can be combined for an interval (c, b) of r.e. degrees if the bottom degree c is low. Lachlan proved that the hypothesis "c is low" is necessary.

4.6 Lachlan Nonsplitting Theorem (Lachlan [1975a]).   *There exist r.e. degrees* b *and* c *such that* $c < b$ *and for all r.e. degrees* a_0 *and* a_1

$$[c \leq a_0, \; a_1 \leq b \; \& \; b \leq a_0 \cup a_1] \implies [b \leq a_0 \vee b \leq a_1].$$

The proof of Theorem 4.6 is particularly interesting because it is the first example of a $0'''$-priority argument. It also revealed the complexity of structure of $(\mathbf{R}, <)$ which led to the coding necessary to prove Theorems 3.1 and 3.2.

References

O. Aberth
[1971]   The failure in computable analysis of a classical existence theorem for ordinary differential equations, Proc. Amer. Math. Soc. *30* (1971), 151–156.
[1980]   Computable Analysis, McGraw-Hill, New York, 1980.

U. Abraham and R. A. Shore
[1986]   Initial segments of the Turing degrees of size $\aleph_1$, Israel J. Math., *53* (1986), 1–51.

J. W. Addison
[1958]   Separation principles in the hierarchies of classical and effective descriptive set theory, Fund. Math. *56* (1958), 123–135.

J. W. Addison, L. Henkin, and A. Tarski
[1965]   Symposium on the Theory of Models (editors), North-Holland, Amsterdam, 1965.

D. A. Alton
[1970]   Uniformities in Recursively Enumerable Sets, Ph.D. Dissertation, Cornell University, Ithaca, N.Y., 1970.
[1971]   Recursively enumerable sets which are uniform for finite extensions, J. Symbolic Logic *36* (1971), 271–287.
[1974]   Iterated quotients of the lattice of recursively enumerable sets, Proc. London Math. Soc. *28* (1974), 1–12.
[1975]   A characterization of r-maximal sets, Arch. Math. Logik Grundlag. *17* (1975), 35–36.

K. Ambos-Spies
[1980]   On the Structure of the Recursively Enumerable Degrees, Doctoral Dissertation, University of Munich, 1980.
[1984a] An extension of the non-diamond theorem in classical and α-recursion theory, J. Symbolic Logic *49* (1984), 586–607.
[1984b] On pairs of recursively enumerable degrees, Trans. Amer. Math. Soc. *283* (1984), 507–531.
[1984c] Contiguous r.e. degrees, In: Richter et al. [1984], 1–37.
[1984d] On the Structure of the Polynomial Time Degrees of Recursive Sets, Habilitationsschrift, University of Dortmund, 1984.
[1984e] On the structure of polynomial degrees, In: STACS '84, Symposium on Theoretical Aspects on Computer Science (M. Fontet and K. Mehlhorn, editors), Lecture Notes in Computer Science No. 166, Springer-Verlag, Berlin, Heidelberg, New York, Tokyo, 1984, 198–208.
[1985a] Anti-mitotic recursively enumerable sets, Z. Math. Logik Grundlag. Math. *31* (1985), 461–477.
[1985b] Cupping and noncapping in the r.e. weak truth table and Turing degrees, Arch. Math. Logik Grundlag. *25* (1985), 109–126.
[1985c] Generators of the recursively enumerable degrees, In: Ebbinghaus, Müller, and Sacks [1985], 1–28.
[ta]   Automorphism bases for the recursively enumerable degrees, to appear.

K. Ambos-Spies and P. A. Fejer
[ta]   Degree theoretic splitting properties of recursively enumerable sets, to appear.

K. Ambos-Spies, C. G. Jockusch, Jr., R. A. Shore and R. I. Soare
[1984]   An algebraic decomposition of the recursively enumerable degrees and the coin-
          cidence of several degree classes with the promptly simple degrees, Trans. Amer.
          Math. Soc. *281* (1984), 109–128.

K. Ambos-Spies and M. Lerman
[1986]   Lattice embeddings into the recursively enumerable degrees, J. Symbolic Logic *51*
          (1986), 257–272.

K. Ambos-Spies and R. I. Soare
[ta]   The recursively enumerable degrees have infinitely many one types, to appear.

M. M. Arslanov
[1968]   Two theorems on recursively enumerable sets, Algebra i Logika *7* (1968), 4–8 (Rus-
          sian); Algebra and Logic *7* (1968), 132–134 (English translation).
[1969]   On effectively hypersimple sets, Algebra i Logika *8* (1969), 143–153 (Russian); Al-
          gebra and Logic *8* (1969), 79–85 (English translation).
[1970]   On complete hypersimple sets, Izv. Vyssh. Uchebn. Zaved. Mat. *95* No. 4 (1970),
          30–35 (Russian); Sov. Math. (Iz. VUZ) *95* (1970), 30–35 (English translation).
[1979]   Weakly recursively enumerable sets and limiting computability, Veroyatn. Metod. i
          Kibern. (Probabilistic Meth. and Cybern.) *15* (1979), 3–9 (Russian).
[1981]   On some generalizations of the theorem on fixed points, Izv. Vyssh. Uchebn. Zaved.
          Mat. *228* No. 5 (1981), 9–16 (Russian); Sov. Math. (Iz. VUZ) *25* No. 5 (1981), 1–1
          (English translation).
[1982]   On a hierarchy of the degrees of unsolvability, Veroyatn. Metod. i Kibern. (Proba-
          bilistic Meth. and Cybern.) *18* (1982), 10–17 (Russian).
[1985a]   Families of recursively enumerable sets and their degrees of unsolvability, Izv. Vyssh.
          Uchebn. Zaved. Mat. *275* No. 4 (1985), 13–19 (Russian); Sov. Math. (Iz. VUZ) *29*
          (1985),13–21 (English translation).
[1985b] m-reducibility and fixed points, In: Complexity problems of mathematical logic,
          Kalininskii Gos. Univ., Kalinin, 1985, 11–18 (Russian).
[1985c]   Structural properties of the degrees below $\mathbf{0}'$, Dokl. Akad. Nauk SSSR, N.S. *283*
          No. 2 (1985), 270–273 (Russian).
[1985d]   Effectively hyperimmune sets and dominants, Mat. Zametki *38* No.2 (1985), 302–309
          (Russian); Math. Notes *38* (1985), 677–680 (English translation).
[1985e]   On a class of hypersimple incomplete sets, Mat. Zametki *38* (1985), 872–874, 958
          (Russian); Math. Notes *38* (1985), 984–985 (English translation).

M. M. Arslanov, R. F. Nadirov, and V. D. Solov'ev
[1977]   Completeness criteria for recursively enumerable sets and some general theorems on
          fixed points, Izv. Vyssh. Uchebn. Zaved. Mat. *179* No. 4 (1977), 3–7 (Russian); Sov.
          Math. (Iz. VUZ) *21* No. 4 (1977), 1–4 (English translation).

C. J. Ash
[1986]   Stability of recursive structures in arithmetical degrees, Ann. Pure Appl. Logic *32*
          (1986), 113–135.
[ta]   Recursive labelling systems and stability of recursive structures in hyperarithmetical
          degrees, Trans. Amer. Math. Soc., to appear.
[ta]   Categoricity in hyperarithmetical degrees, to appear.

C. J. Ash and R. G. Downey
[1984]   Decidable subspaces and recursively enumerable subspaces, J. Symbolic Logic *49*
          (1984), 1137–1145.

C. J. Ash and S. S. Goncharov
[ta]   Strong Δ_2^0 categoricity, to appear.

C. J. Ash and A. Nerode
[1981]   Intrinsically recursive relations, In: Crossley [1981], 26 41.

Y. Bar-Hillel
[1970]   Mathematical Logic and Foundations of Set Theory (editor), Proceedings of an
         International Colloquium, Jerusalem, November 11-14, 1968, North-Holland, Ams-
         terdam, London, 1970.

K. J. Barwise
[1977]   Handbook of Mathematical Logic (editor), North-Holland, Amsterdam, New York,
         Oxford, 1977.

K. J. Barwise, H. J. Keisler, and K. Kunen
[1980]   The Kleene Symposium (editors), Proceedings of the Symposium held June 18 24,
         1979 at Madison, Wisconsin, U.S.A., North-Holland, Amsterdam, New York, Ox-
         ford, 1980.

O. V. Belegradek
[1978] m-degrees of the word problem, Sibirsk. Mat. Zh. 19 (1978), 1232 1236 (Russian);
         Siberian Math. J. 19 (1978), 867 870 (English translation).

V. L. Bennison
[1976]   On the Computational Complexity of Recursively Enumerable Sets, Ph.D. Disser-
         tation, University of Chicago, 1976.
[1979]   Information content characterizations of complexity theoretic properties, In: The-
         oretical Computer Science: Fourth G.I. Conference, Aachen, March 26 28, 1979,
         edited by K. Weihrauch, Lecture Notes in Computer Science No. 67, Springer-Verlag,
         Berlin, Heidelberg, New York, 1979, 58 66.
[1980]   Recursively enumerable complexity sequences and measure independence, J. Sym-
         bolic Logic 45 (1980), 417 438.

V. L. Bennison and R. I. Soare
[1978]   Some lowness properties and computational complexity sequences, Theoret. Com-
         put. Sci. 6 (1978), 233 254.

M. Bickford
[1983]   The Jump Operator in Strong Reducibilities, Ph.D. Dissertation, University of Wis-
         consin, 1983.

M. Bickford and C. F. Mills
[ta]   Lowness properties of r.e. sets, J. Symbolic Logic, to appear.

G. Birkhoff
[1933]   On the combination of subalgebras, Proc. Cambridge Philos. Soc. 29 (1933),
         441 464.
[1967]   Lattice theory, 3rd ed., Amer. Math. Soc. Colloquium Publications XXV, Amer.
         Math. Soc., Providence, R.I., 1967.

E. Bishop
[1967]   Foundations of Constructive Analysis, McGraw-Hill, New York, 1967.

M. Blum and I. Marques
[1973]   On complexity properties of recursively enumerable sets, J. Symbolic Logic 38
         (1973), 579 593.

G. Boolos and H. Putnam
[1968]   Degrees of unsolvability of constructible sets of integers, J. Symbolic Logic 33 (1968),
         497 513.

W. W. Boone
[1954 57] Certain simple unsolvable problems of group theory, I, II, III, IV, V, VI, Nederl.
         Akad. Wetenschappen, Proc. Ser. A 57 (1954), 231 237 and 492 497; ibid. 58 (1955),
         252 256 and 571 577; ibid. 60 (1957), 22 27 and 227 232.

[1959]   The word problem, Ann. of Math. (2) *70* (1959), 207 265.
[1966]   Word problems and recursively enumerable degrees of unsolvability. A sequel on finitely presented groups, Ann. of Math. (2) *84* (1966), 49 84.

W. W. Boone and G. Higman
[1974]   An algebraic characterization of groups with soluble word problem, J. Austral. Math. Soc. *43* (1974), 41 53.

E. Börger
[1985]   Berechenbarkeit, Komplexität, Logik, Vieweg und Sohn, Wiesbaden, 1985.

V. K. Bulitko
[1980]   Reducibility by Zhagalkin-linear tables, Sibirsk. Mat. Zh. *21* (1980), 23 31, 235 (Russian); Siberian Math. J. *21* (1980), 332 339 (English translation).

R. Boyd, G. Hensel, and H. Putnam
[1969]   A recursion-theoretic characterization of the ramified analytical hierarchy, Trans. Amer. Math. Soc. *141* (1969), 37 62.

N. R. Bukharaev
[1979]   On limit enumeration properties of complements of recursively enumerable sets, Veroyatn. Metod. i Kibern. (Probabilistic Meth. and Cybern.) *228* (1979), 40 49 (Russian).
[1981]   On T-degrees of differences of recursively enumerable sets, Izv. Vyssh. Uchebn. Zaved. Mat. *228* No.5 (1981), 40 49 (Russian); Sov. Math. (Iz. VUZ) *25* No. 5 (1981), 40 52 (English translation).

S. Burris and R. McKenzie
[1981]   Decidability and Boolean representations, Memoirs Amer. Math. Soc. *32* No. 246, 1981.

J. Case
[1970]   Enumeration reducibility and partial degrees, Ann. Math. Logic *2* (1970), 419 439.

D. Cenzer, P. Clote, R. Smith, R. I. Soare, and S. S. Wainer
[1986]   Members of countable Π_1^0 classes, Ann. Pure Appl. Logic *31* (1986), 145 163.

C. T. Chong
[ta]   Minimal degrees recursive in 1-generic degrees, to appear.
[ta]   Techniques of Admissible Recursion Theory, Lecture Notes in Mathematics, Springer-Verlag, Berlin, Heidelberg, New York, Tokyo, to appear.

C. T. Chong and C. G. Jockusch, Jr.
[1984]   Minimal degrees and 1-generic sets below **0′**, In: Richter et al. [1984], 63 77.

C. T. Chong and M. J. Wicks
[1983]   Proceedings of the Southeast Asian Conference on Logic (editors), North-Holland, Amsterdam, New York, Oxford, 1983.

A. Church
[1936a]   An unsolvable problem of elementary number theory, Amer. J. Math. *58* (1936), 345 363; reprinted in Davis [1965], 88 107.
[1936b]   A note on the Entscheidungsproblem, J. Symbolic Logic *1* (1936), 40 41 and 101 102; reprinted in Davis [1965], 108 115.

J. P. Cleave
[1961]   Creative functions, Z. Math. Logik Grundlag. Math. *7* (1961), 205 212.

P. Clote
[1985]   Applications of the low-basis theorem in arithmetic, In: Ebbinghaus, Müller, and Sacks [1985], 65 88.

P. F. Cohen
[1975]   Weak Truth-Table Reducibility and the Pointwise Ordering of 1–1 Recursive Func-
         tions, Ph.D. Dissertation, University of Illinois at Urbana-Champaign, 1975.

P. F. Cohen and C. G. Jockusch, Jr.
[1975]   A lattice property of Post's simple set, Illinois J. Math. *19* (1975), 450–453.

P. J. Cohen
[1963]   The independence of the continuum hypothesis I, Proc. Natl. Acad. Sci. USA *50*
         (1963), 1143–1148.
[1966]   Set theory and the continuum hypothesis, W. A. Benjamin, New York, 1966.

S. B. Cooper
[1972a] Minimal upper bounds for sequences of recursively enumerable degrees, J. London
         Math. Soc. *5* (1972), 445–450.
[1972b] Degrees of unsolvability complementary between recursively enumerable degrees,
         Part I, Ann. Math. Logic *4* (1972), 31–73.
[1972c] Jump equivalence of the Δ_2^0 hyperhyperimmune sets, J. Symbolic Logic *37* (1972),
         598–600.
[1972d] Sets recursively enumerable in high degrees, Notices Amer. Math. Soc. *19* (1972),
         A-20.
[1973]   Minimal degrees and the jump operator, J. Symbolic Logic *38* (1973), 249–271.
[1974a] On a theorem of C.E.M. Yates (handwritten notes).
[1974b] Minimal pairs and high recursively enumerable degrees, J. Symbolic Logic *39* (1974),
         655–660.
[1974c] An annotated bibliography for the structure of the degrees below $0'$ with special
         reference to that of the recursively enumerable degrees, Recursive Function Theory
         Newsletter *5* (1974), 1–15.
[1982]   Partial degrees and the density problem, J. Symbolic Logic *47* (1982), 854–859.
[1984]   Partial degrees and the density problem. Part 2: The enumeration degrees of the
         Σ_2 sets are dense, J. Symbolic Logic *49* (1984), 503–513.
[ta]   The strong anti-cupping property for recursively enumerable degrees, to appear.
[ta]   A jump class of non-cappable degrees, J. Symbolic Logic, to appear.
[ta]   Strong minimal covers for recursively enumerable degrees, to appear.
[ta]   Enumeration reducibility using bounded information: counting minimal covers, to
         appear.

S. B. Cooper and R. Epstein
[ta]   Complementing below recursively enumerable degrees, to appear.

W. Craig
[1953]   On axiomatizability within a system, J. Symbolic Logic *18* (1953), 30–32.

J.N. Crossley
[1967]   Sets, Models, and Recursion Theory (editor), Proceedings of the Summer School
         in Mathematical Logic and Logic Colloquium, Leicester, England, 1965, North-
         Holland, Amsterdam, 1967.
[1981]   Aspects of Effective Algebra (editor), Proceedings of a Conference on Aspects of
         Effective Algebra at Monash University, August, 1979, Upside Down A Book Com-
         pany, Yarra Glenn, Victoria, Australia, 1981. (This contains an extensive bibliog-
         raphy of effective algebra by J. N. Crossley and S. Miranda.)

J. N. Crossley and A. Nerode
[1974]   Combinatorial Functors, Ergebnisse der Mathematik und ihrer Grenzgebiete, Band
         81, Springer-Verlag, Berlin, Heidelberg, New York, 1974.

N. J. Cutland
[1980]   Computability: An Introduction to Recursive Function Theory, Cambridge Univer-
         sity Press, Cambridge, 1980.

M. Davis
[1958]   Computability and Unsolvability, McGraw-Hill, New York, 1958; reprinted in 1982
         by Dover Publications, with an appendix which includes Davis [1973].
[1965]   The Undecidable. Basic Papers on Undecidable Propositions, Unsolvable Problems,
         and Computable Functions (editor), Raven Press, Hewlitt, New York, 1965.
[1973]   Hilbert's tenth problem is unsolvable, Amer. Math. Monthly *80* (1973), 233 269.
[1977]   Unsolvable problems, In: Barwise [1977], 567 594.
[1982]   Why Gödel didn't have Church's Thesis, Inform. and Control *54* (1982), 3 24.

M. Davis, Ju. V. Matijasevič, and J. Robinson
[1976]   Hilbert's tenth problem, Diophantine equations: positive aspects of a negative solu-
         tion, In: Mathematical Developments Arising From Hilbert's Problems, Proc. Symp.
         Pure Math. *27*, Amer. Math. Soc., Providence, R.I., 1976, 323 378.

M. Davis, H. Putnam and J. Robinson
[1961]   The decision problem for exponential diophantine equations, Ann. of Math. (2) *74*
         (1961), 425–436.

M. Davis and E. Weyuker
[1983]   Computability, Complexity, and Languages: Fundamentals of Theoretical Computer
         Science, Academic Press, New York, 1983.

A.N. Degtev
[1970]   Remarks on retraceable, regressive, and pointwise decomposable sets, Algebra i
         Logika *9* (1970), 651 660 (Russian); Algebra and Logic *9* (1970), 390 395 (English
         translation).
[1971]   Hypersimple sets with retraceable complements. Algebra i Logika *10* (1971), 235 246
         (Russian); Algebra and Logic *10* (1971), 147 154 (English translation).
[1972a] On *m*-degrees of simple sets, Algebra i Logika *11* (1972), 130 139, 239 (Russian);
         Algebra and Logic *11* (1972), 74 80 (English translation).
[1972b] Hereditary sets and truth table reducibility, Algebra i Logika *11* (1972), 257 279,
         361 (Russian); Algebra and Logic *11* (1972), 145 152 (English translation).
[1973]   tt- and *m*-degrees, Algebra i Logika *12* (1973), 143 161, 243 (Russian); Algebra and
         Logic *12* (1973), 78–89 (English translation).
[1975]   Reducibility of partial recursive functions, Sibirsk. Mat. Zh. *16* No. 5 (1975), 970 988
         (Russian); Siberian Math. J. *16* No. 5 (1975), 741 754 (English translation).
[1976a] Partially ordered sets of 1-degrees contained in recursively enumerable *m*-degrees,
         Algebra i Logika *15* (1976), 249 266, 365 (Russian); Algebra and Logic *15* (1976),
         153 164 (English translation).
[1976b] On minimal 1-degrees and tt-reducibility, Sibirsk. Mat. Zh. *17* (1976), 1014 1022,
         1196 (Russian); Siberian Math. J. *17* (1976), 751 757 (English translation).
[1977]   Reducibility of partial recursive functions II, Sibirsk. Mat. Zh. *18* No. 4 (1977),
         765 774 (Russian); Siberian Math. J. *18* No. 4 (1977), 541 548 (English transla-
         tion).
[1978a] On *m*-degrees of supersets of simple sets, Mat. Zametki *23* (1978), 889 893 (Rus-
         sian); Math Notes *23* (1978), 488 490 (English translation).
[1978b] Three theorems on tt-degrees, Algebra i Logika *17* (1978), 270 281 (Russian); Al-
         gebra and Logic *17* (1978), 187 194 (English translation).
[1978c] Decidability of the ∀∃-theory of some quotient lattice of recursively enumerable sets,
         Algebra i Logika *17* (1978), 134 143, 241 (Russian); Algebra and Logic *17* (1978),
         94 101 (English translation).
[1979a] On truth-table-type reducibilities in the theory of algorithms, Usp. Mat. Nauk *34*
         (1979), 137 168, 248 (Russian); Russ. Math. Surv. *34* (1979), 155 192 (English
         translation).
[1979b] Some results on upper semilattices and *m*-degrees, Algebra i Logika *18* (1979),
         665 679, 754 (Russian); Algebra and Logic *18* (1979), 420 430 (English translation).

[1980]   On reducibilities of numerations, Mat. Sbornik (N.S.) *112(154)* No. 2 (1980),
         207–219 (Russian); Mathematics of the USSR, Sbornik *40* No. 2 (1981), 193–204
         (English translation).
[1981]   Relations between complete sets, Izv. Vyssh. Uchebn. Zaved. Mat. *228* No. 5 (1981),
         50–55 (Russian); Sov. Math. (Iz. VUZ) *25* No. 5 (1981), 53–61 (English translation).
[1982]   Small degrees in ordinary recursion theory, In: Proc. of the Sixth International
         Congress of Logic, Methodology, and the Philosophy of Science, Hannover, 1979,
         North-Holland, Amsterdam, New York, Oxford, 1982, 237–240.

J. C. E. Dekker
[1953]   Two notes on recursively enumerable sets, Proc. Amer. Math. Soc. *4* (1953),
         495–501.
[1954]   A theorem on hypersimple sets, Proc. Amer. Math. Soc. *5* (1954), 791–796.
[1955]   Productive sets, Trans. Amer. Math. Soc. *78* (1955), 129–149.

J. C. E. Dekker and J. Myhill
[1958a] Some theorems on classes of recursively enumerable sets, Trans. Amer. Math. Soc.
         89 (1958), 25–59.
[1958b] Retraceable sets, Canad. J. Math. *10* (1958), 357–373.
[1960]   Recursive equivalence types, University of California Publications in Mathematics,
         New Series *3* (1960), 67–214.

S. D. Denisov
[1970]   On m-degrees of recursively enumerable sets, Algebra i Logika *9* (1970), 422–427
         (Russian); Algebra and Logic *9* (1970), 254–256 (English translation).
[1974]   Three theorems on elementary theories and tt-reducibility, Algebra i Logika *13*
         (1974), 5–8, 120 (Russian); Algebra and Logic *13* (1974), 1–2 (English translation).
[1978]   The structure of the upper semilattice of recursively enumerable m-degrees and
         related questions I, Algebra i Logika *17* (1978), 643–683, 746 (Russian); Algebra
         and Logic *17* (1978), 418–443 (English translation).

R. G. Downey
[1985]   The degrees of r.e. sets without the universal splitting property, Trans. Amer. Math.
         Soc. *291* (1985), 337–351.
[ta]   Localization of a theorem of Ambos-Spies and the strong antisplitting property,
         Arch. Math. Logik Grundlag., to appear.
[ta]   Subsets of hypersimple sets, Pacific J. Math., to appear.
[ta]   Intervals and sublattices in the r.e. wtt-degrees, Ann. Pure Appl. Logic, to appear.
[ta] Δ_2^0 degrees and transfer theorems, Illinois J. Math., to appear.

R. G. Downey and C. G. Jockusch, Jr.
[ta]   T-degrees, jump classes, and strong reducibilities, Trans. Amer. Math. Soc., to ap-
         pear.

R. G. Downey and S. A. Kurtz
[1986]   Recursion theory and ordered groups, Ann. Pure Appl. Logic, *32* (1986), 137–151.

R. G. Downey and M. Stob
[1986]   Structural interactions of the recursively enumerable T- and W-degrees, Ann. Pure
         Appl. Logic *31* (1986), 205–236.
[ta]   Minimal pairs in lower cones, Israel J. Math., to appear.
[ta]   Automorphisms of the lattice of recursively enumerable sets: Orbits, to appear.

R. G. Downey and L. V. Welch
[1986]   Splitting properties of r.e. sets and degrees, J. Symbolic Logic *51* (1986), 88–109.

F. Drake and S. S. Wainer
[1980]   Recursion Theory: Its Generalizations and Applications (editors), Proceedings of
         Logic Colloquium '79, Leeds, August, 1979, London Mathematical Society Lecture
         Notes Series No. 45, Cambridge University Press, Cambridge, U.K., 1980.

H. D. Ebbinghaus, G. H. Müller, and G. E. Sacks
[1985]   Recursion Theory Week (editors), Proceedings of a Conference held in Oberwolfach, West Germany, April 15–21, 1984, Lecture Notes in Mathematics No. 1141, Springer-Verlag, Berlin, Heidelberg, New York, Tokyo, 1985.

S. Eilenberg and C. C. Elgot
[1970]   Recursiveness, Academic Press, New York, London, 1970.

R. L. Epstein
[1975a] Minimal Degrees of Unsolvability and the Full Approximation Construction, Memoirs Amer. Math. Soc. *3* No. 162, Amer. Math. Soc., Providence, R.I., 1975.
[1975b] Review of S. B. Cooper [1972b], J. Symbolic Logic *40* (1975), 86.
[1979]   Degrees of Unsolvability: Structure and Theory, Lecture Notes in Mathematics No. 759, Springer-Verlag, Berlin, Heidelberg, New York, 1979.
[1981]   Initial Segments of Degrees Below **0'**, Memoirs Amer. Math. Soc. *30* No. 241, Amer. Math. Soc., Providence, R.I., 1981.

R. L. Epstein, R. Haas, and R. Kramer
[1981]   Hierarchies of sets and degrees below **0'**, In: Lerman, Schmerl, and Soare [1981], 32–48.

Y. L. Ershov
[1964]   Decidability of the elementary theory of relatively complemented distributive lattices and of the theory of filters, Algebra i Logika *3* (1964), 17–38 (Russian).
[1968a] A hierarchy of sets, Part I, Algebra i Logika *7* (1968), 47–73 (Russian); Algebra and Logic *7* (1968), 24–43 (English translation).
[1968b] A hierarchy of sets, Part II, Algebra i Logika *7* (1968), 15–47 (Russian); Algebra and Logic *7* (1968), 212–232 (English translation).
[1969]   Hyperhypersimple *m*-degrees, Algebra i Logika *8* (1969), 523–552 (Russian); Algebra and Logic *8* (1969), 298–315 (English translation).
[1970a] A hierarchy of sets, Part III, Algebra i Logika *9* (1970), 34–51 (Russian); Algebra and Logic *9* (1970), 20–31 (English translation).
[1970b] On inseparable pairs, Algebra i Logika *9* (1970), 661–666 (Russian); Algebra and Logic *9* (1970), 396–399 (English translation).
[1970c] On index sets, Sibirsk. Mat. Zh. *11* (1970), 326–342 (Russian); Siberian Math. J. *11* (1970), 246–258 (English translation).
[1971]   Positive equivalences, Algebra i Logika *10* (1971), 620–650 (Russian); Algebra and Logic *10* (1971), 378–394 (English translation).
[1973]   The hierarchy of Δ_2^0 sets (in Russian with English abstract), In: Proc. of the Fourth International Congress of Logic, Methodology, and the Philosophy of Science, Bucharest, 1971, North-Holland, Amsterdam, 1973, 69–76.
[1975]   The upper semilattice of numerations of a finite set, Algebra i Logika *14* (1975), 258–284, 368 (Russian); Algebra and Logic *14* (1975), 159–175 (English translation).
[1977]   The Theory of Enumerations, Nauka, Moscow, 1977 (Russian).
[1980]   Decidability problems and constructive models, Nauka, Moscow, 1980 (Russian).

Y. L. Ershov and I. A. Lavrov
[1973]   The uppersemilattice L(γ), Algebra i Logika *12* (1973), 167–189, 243–244 (Russian); Algebra and Logic *12* (1973), 93–106 (English translation).

Y. L. Ershov et al.
[1965]   Elementary theories, Usp. Mat. Nauk *20* (1965), 37–108 (Russian); Russian Math. Surveys *20* (1965), 35–105 (English translation).

S. Feferman
[1957]   Degrees of unsolvability associated with classes of formalized theories, J. Symbolic Logic *22* (1957), 161–175.

L. Feiner
[1967]   Orderings and Boolean algebras not isomorphic to recursive ones, Ph.D. Dissertation, Massachusetts Institute of Technology, Cambridge, Mass., 1967.

P. A. Fejer
[1980] The Structure of Definable Subclasses of the Recursively Enumerable Degrees, Ph.D. Dissertation, University of Chicago, 1980.
[1982] Branching degrees above low degrees, Trans. Amer. Math. Soc. *273* (1982), 157–180.
[1983] The density of the nonbranching degrees, Ann. Pure Appl. Logic *24* (1983), 113–130.

P. A. Fejer and R. A. Shore
[1985] Embeddings and extensions of embeddings in the r.e. tt- and wtt-degrees, In: Ebbinghaus, Müller, and Sacks [1985], 121–140.
[ta]   Infima of recursively enumerable truth table degrees, to appear.
[ta]   Minimal r.e. tt-degrees and minimal wtt-degrees below $\mathbf{0}'$, to appear.

P. A. Fejer and R. I. Soare
[1981] The plus-cupping theorem for the recursively enumerable degrees, In: Lerman, Schmerl, and Soare [1981], 49–62.

J. E. Fenstad
[1980] General Recursion Theory: An Axiomatic Approach, Perspectives in Mathematical Logic, Omega Series, Springer-Verlag, Berlin, Heidelberg, New York, 1980.

J. E. Fenstad, R. O. Gandy, and G. E. Sacks
[1978] Generalized Recursion Theory II (editors), Proceedings of the Second Symposium on Generalized Recursion Theory, Oslo, 1977, North-Holland, Amsterdam, New York, Oxford, 1978.

J. Ferrante and C. W. Rackoff
[1979] Computational complexity of logical theories, Lecture Notes in Mathematics No. 718, Springer-Verlag, Berlin, Heidelberg, New York, 1979.

R. M. Friedberg
[1957a] Two recursively enumerable sets of incomparable degrees of unsolvability, Proc. Natl. Acad. Sci. USA *43* (1957), 236–238.
[1957b] The fine structure of degrees of unsolvability of recursively enumerable sets, In: Summaries of Cornell University Summer Institute for Symbolic Logic, Communications Research Division, Inst. for Def. Anal., Princeton, N. J., 1957, 404–406.
[1957c] A criterion for completeness of degrees of unsolvability, J. Symbolic Logic *22* (1957), 159–160.
[1958a] Three theorems on recursive enumeration: I. Decomposition, II. Maximal Set, III. Enumeration without duplication, J. Symbolic Logic *23* (1958), 309–316.
[1958b] Un contre-exemple relatif aux fonctionnelles récursives, Comptes rendus hebdomadaires des séances de l'Académie des Sciences (Paris) *247* (1958), 852–854.
[1958c] Four quantifier completeness: a Banach-Mazur functional not uniformly partial recursive, Bulletin de l'Académie Polonaise des Sciences, Série des sciences mathématiques, astronomiques et physiques, *6* (1958), 1–5.

R. M. Friedberg and H. Rogers, Jr.
[1959] Reducibility and completeness for sets of integers, Z. Math. Logik Grundlag. Math. *5* (1959), 117–125.

H. Friedman
[1975] 102 Problems in mathematical logic, J. Symbolic Logic *40* (1975), 113–129.

H. Friedman, K. McAloon, and S. G. Simpson
[1982] A finite combinatorial principle which is equivalent to the 1-consistency of predicative analysis, In: Metakides [1982], 197–230.

S. D. Friedman
[1978] Negative solutions to Post's problem I, In: Fenstad, Gandy, and Sacks [1978], 127–133.
[1979] β-Recursion theory, Trans. Amer. Math. Soc. *255* (1979), 173–200.
[1980] Post's problem without admissibility, Adv. Math. *35* (1980), 30–49.

[1981]   Negative solutions to Post's problem II, Ann. Math. (2) *113* (1981), 25–43.

A. Fröhlich and J. C. Shepherdson
[1955]   Effective procedures in field theory, Philos. Trans., Royal Soc. of London, Series A
         248 (1955), 407–432.

J. Gill and P. Morris
[1974]   On subcreative sets and S-reducibility, J. Symbolic Logic *39* (1974), 669–677.

K. Gödel
[1930]   Die Vollständigkeit der Axiome des logischen Funktionenkalküls, Monatsh. Math.
         Phys. *37* (1930), 349–360; translated in Van Heijenoort [1967], 583–591.
[1931]   Über formal unentscheidbare Sätze der Principia Mathematica und verwandter Sys-
         teme I, Monatsh. Math. Phys. *38* (1931), 173–198; translated in Davis [1965], 4–38.
[1934]   On undecidable propositions of formal mathematical systems, Notes by S.C. Kleene
         and Barkley Rosser on lectures at the Institute for Advanced Study, Princeton,
         N. J.; reprinted in Davis [1965], 39–71.
[1946]   Remarks before the Princeton Bicentennial Conference on Problems in Mathemat-
         ics, In: Davis [1965], 84–88.

S. S. Goncharov
[1983]   Universal recursively enumerable Boolean algebras, Sibirsk. Mat. Zh. *24* No. 6
         (1983), 36–43 (Russian); Siberian Math. J. *24* (1983), 852–858 (English transla-
         tion).

J. Grassin
[1974]   Index sets in Ershov's hierarchy, J. Symbolic Logic *39* (1974), 97–104.

A. Grzegorczyk
[1953]   Some classes of recursive functions, Rozprawy matematyczne no. 4, Instytut Matem-
         atyczny Polskiej Akad. Nauk, Warsaw, 1953.

L. Gutteridge
[1971a] Some Results on Enumeration Reducibility, Ph.D. Dissertation, Simon Fraser Uni-
         versity, 1971.
[1971b] The partial degrees are dense, preprint (unpublished).

W. Hanf
[1965]   Model theoretic methods in the study of elementary logic, In: Addison, Henkin, and
         Tarski [1965], 132–145.

L. Harrington
[1974]   Recursively presentable prime models, J. Symbolic Logic *39* (1974), 305–309.
[1976]   On Cooper's proof of a theorem of Yates, Parts I and II (handwritten notes).
[1978]   Plus cupping in the recursively enumerable degrees (handwritten notes).
[1980]   Understanding Lachlan's monster paper (handwritten notes).
[1982]   A gentle approach to priority arguments (handwritten notes for a talk at A.M.S.
         Summer Research Institute in Recursion Theory, Cornell University, July 1982).
[1983]   The undecidability of the lattice of recursively enumerable sets (handwritten notes).

L. Harrington and S. Shelah
[1982a] The undecidability of the recursively enumerable degrees (research announcement),
         Bull. Amer. Math. Soc. (N. S.) *6* No.1 (1982), 79–80.
[1982b] The undecidability of the recursively enumerable degrees (handwritten notes).

L. Harrington and R. A. Shore
[1981]   Definable degrees and automorphisms of *D*, Bull. Amer. Math. Soc. (N. S.) *4* (1981),
         97–100.

L. Harrington and T. A. Slaman
[ta]   Interpreting arithmetic in the Turing degrees of the recursively enumerable sets, to
         appear.

C. A. Haught
[1985]   Turing and truth table degrees of 1-generic and recursively enumerable degrees, Ph.D. dissertation, Cornell University, 1985.
[1986]   The degrees below a 1-generic degree and less than $\mathbf{0}'$, J. Symbolic Logic *51* (1986), 770–777.
[ta]   Lattice embeddings in the r.e. tt-degrees, Trans. Amer. Math. Soc., to appear.

L. Hay
[1973a] The halting problem relativized to complements, Proc. Amer. Math. Soc. *41* (1973), 583–587.
[1973b] The class of recursively enumerable subsets of a recursively enumerable set, Pacific J. Math. *46* (1973), 167–183.
[1975]   Spectra and the halting problem, Z. Math. Logik Grundlag. Math. *21* (1975), 167–176.

L. Hay, A. Manaster, and J. G. Rosenstein
[1975]   Small recursive ordinals, many-one degrees, and the arithmetical difference hierarchy, Ann. Math. Logic *8* (1975), 297–343.
[1977]   Concerning partial recursive similarity transformations of linearly ordered sets, Pacific J. Math. *71* (1977), 57–70.

H. Hermes
[1969]   Enumerability, Decidability, Computability: An Introduction to the Theory of Recursive Functions, Springer-Verlag, Berlin, Heidelberg, New York, 1969.   (This is an English translation of: Aufzählbarkeit, Entscheidbarkeit, Berechenbarkeit, Grundlehren der mathematischen Wissenschaften, Band *109*, Springer-Verlag, Berlin, Heidelberg, New York, 1965.)

E. Herrmann
[1978]   Der Verband der rekursiv aufzählbaren Mengen (Entscheidungsproblem), Seminarbericht Nr. 10, Humboldt-Universität Sektion Mathematik, Berlin, GDR, 1978.
[1981]   Die Verbandseigenschaften der rekursiv aufzählbaren Mengen, Seminarbericht Nr. 36, Humboldt-Universität Sektion Mathematik, Berlin, GDR, 1981.
[1983a] Major subsets of hypersimple sets and ideal families, Dissertation B, Humboldt-Universität Sektion Mathematik, Berlin, GDR, 1983.
[1983b] Orbits of hyperhypersimple sets and the lattice of Σ_3^0 sets, J. Symbolic Logic *48* (1983), 693–699.
[1984a] Classes of simple sets, filter properties and their mutual position, In: Proceedings of the Conference on Model theory, Wittenberg, GDR, 1984, Seminarbericht Nr. 49, 60–72.
[1984b] The undecidability of the elementary theory of the lattice of recursively enumerable sets (abstract), In: Frege Conference 1984, Proceedings of the International Conference at Schwerin, GDR, Akademie-Verlag, Berlin, GDR, 66–72.
[ta]   Definable structures in the lattice of recursively enumerable sets, J. Symbolic Logic, to appear.
[ta]   Definable Boolean pairs in the lattice of recursively enumerable sets, to appear.
[ta]   The structure of the filter of simple sets, In: Proceedings of the Conference at Jadwisin, Poland 1981, School of Mathematics, The University of Leeds, to appear.
[ta]   The index set $\{\, e : W_e \equiv_1 X \,\}$, J. Symbolic Logic, to appear.
[ta]   Automorphisms of the lattice of recursively enumerable sets and hyperhypersimple sets, to appear.

A. Heyting
[1959]   Constructivity in Mathematics (editor), Proceedings of the Colloquium held in Amsterdam 1957, North-Holland, Amsterdam, 1959.

G. Higman
[1961]   Subgroups of finitely presented groups, Proc. Roy. Soc. London, Series A *262* (1961), 455–475.

P. G. Hinman
[1978]   Recursion-Theoretic Hierarchies, Perspectives in Mathematical Logic, Omega Series, Springer-Verlag, Berlin, Heidelberg, New York, 1978.

W. Hodges
[1972]   Conference in Mathematical Logic, London, 1970 (editor), Lecture Notes in Mathematics No. 255, Springer-Verlag, Berlin, Heidelberg, New York, 1972.

S. Homer and G. E. Sacks
[1983]   Inverting the half-jump, Trans. Amer. Math. Soc. *278* (1983), 317–331.

Huang Wen Qi and A. Nerode
[1985]   The application of pure recursion theory to computable analysis, Acta Mathematica Sinica *28* (1985), 625–636.

M. Ingrassia
[1981]   P-Genericity for Recursively Enumerable Sets, Ph.D. Dissertation, University of Illinois at Urbana-Champaign, 1981.
[ta]   P-generic r.e. degrees are dense, to appear.
[ta]   Restricted notions of P-genericity for r.e. sets, to appear.

Sh. T. Ishmuchametov
[1982]   On the classes of recursively enumerable sets, Veroyatn. Metod. i Kibern. (Probabilistic Meth. and Cybern.) *18* (1982), 46–53 (Russian).
[1983]   On the index sets of differences of recursively enumerable sets, Izv. Vyssh. Uchebn. Zaved. Mat. *250* No. 3 (1983), 78–79 (Russian); Sov. Math. (Iz. VUZ) *27* (1983), 99–102 (English translation).
[1985]   On differences of recursively enumerable sets, Izv. Vyssh. Uchebn. Zaved. Mat. *279* No. 8 (1985), 3–12 (Russian).

L. L. Ivanov
[1986]   Algebraic Recursion Theory, Ellis Horwood Publishers, Chichester, U.K., 1986.

C. G. Jockusch, Jr.
[1968a] Semirecursive sets and positive reducibility, Trans. Amer. Math. Soc. *131* (1968), 420–436.
[1968b] Uniformly introreducible sets, J. Symbolic Logic *33* (1968), 521–536.
[1969a] The degrees of hyperhyperimmune sets, J. Symbolic Logic *34* (1969), 489–493.
[1969b] Relationships between reducibilities, Trans. Amer. Math. Soc. *142* (1969), 229–237.
[1969c] The degrees of bi-immune sets, Z. Math. Logik Grundlag. Math. *15* (1969), 135–140.
[1972a] Degrees in which the recursive sets are uniformly recursive, Canad. J. Math. *24* (1972), 1092–1099.
[1972b] Ramsey's theorem and recursion theory, J. Symbolic Logic *37* (1972), 268–280.
[1972c] Upward closure of bi-immune degrees, Z. Math. Logik Grundlag. Math. *18* (1972), 285–287.
[1973a] Review of Lerman [1971a], Math. Reviews *45* (1973), No. 3200.
[1973b] Upward closure and cohesive degrees, Israel J. Math. *15* (1973), 332–335.
[1973c] An application of Σ_4^0 determinacy to the degrees of unsolvability, J. Symbolic Logic *38* (1973), 293–294.
[1974] Π_1^0 classes and Boolean combinations of recursively enumerable sets, J. Symbolic Logic *39* (1974), 95–96.
[1975]   Recursiveness of initial segments of Kleene's $\mathcal{O}$, Fund. Math. *87* (1975), 161–167.
[1977]   Simple proofs of some theorems on high degrees, Canad. J. Math. *29* (1977), 1072–1080.
[1980]   Degrees of generic sets, In: Drake and Wainer [1980], 110–139.
[1981]   Three easy constructions of recursively enumerable sets, In: Lerman, Schmerl, and Soare [1981], 83-91.
[1985]   Genericity for recursively enumerable sets, In: Ebbinghaus, Müller, and Sacks [1985], 203–232.

C. G. Jockusch, Jr. and I. Kalantari
[1984]   Recursively enumerable sets and van der Waerden's theorem on arithmetic progressions, Pacific J. Math. *115* (1984), 143–153.

C. G. Jockusch, Jr., M. Lerman, R. I. Soare and R. M. Solovay
[ta]   Recursively enumerable sets modulo iterated jumps and extensions of Arslanov's completeness criterion, to appear.

C. G. Jockusch, Jr. and T. G. McLaughlin
[1969]   Countable retracing functions and Π_2^0 predicates, Pacific J. Math. *30* (1969), 67–93.

C. G. Jockusch, Jr. and J. Mohrherr
[1985]   Embedding the diamond lattice in the r.e. tt-degrees, Proc. Amer. Math. Soc. *94* (1985), 123–128.

C. G. Jockusch, Jr. and M. Paterson
[1976]   Completely autoreducible degrees, Z. Math. Logik Grundlag. Math. *22* (1976), 571–575.

C. G. Jockusch, Jr. and D. Posner
[1978]   Double jumps of minimal degrees, J. Symbolic Logic *43* (1978), 715–724.
[1981]   Automorphism bases for degrees of unsolvability, Israel J. Math. *40* (1981), 150–164.

C. G. Jockusch, Jr. and R. A. Shore
[1983]   Pseudo jump operators I: The R.E. case, Trans. Amer. Math. Soc. *275* (1983), 599–609.
[1984]   Pseudo jump operators. II: Transfinite iterations, hierarchies, and minimal covers, J. Symbolic Logic *49* (1984), 1205–1236.
[1985]   REA operators, R.E. degrees, and minimal covers, In: Nerode and Shore [1985], 3–11.

C. G. Jockusch, Jr. and S. G. Simpson
[1976]   A degree theoretic definition of the ramified analytical hierarchy, Ann. Math. Logic *10* (1976), 1–32.

C. G. Jockusch, Jr. and R. I. Soare
[1970]   Minimal covers and arithmetical sets, Proc. Amer. Math. Soc. *25* (1970), 856–859.
[1971]   A minimal pair of Π_1^0 classes, J. Symbolic Logic *36* (1971), 66–78.
[1972a] Degrees of members of Π_1^0 classes, Pacific J. Math. *40* (1972), 605–616.
[1972b] Π_1^0 classes and degrees of theories, Trans. Amer. Math. Soc. *173* (1972), 33–56.
[1973a] Post's problem and his hypersimple set, J. Symbolic Logic *38* (1973), 446–452.
[1973b] Encodability of Kleene's $\mathcal{O}$, J. Symbolic Logic *38* (1973), 437–440.

C. G. Jockusch, Jr. and R. M. Solovay
[1977]   Fixed points of jump preserving automorphisms of degrees, Israel J. Math. *26* (1977), 91–94.

I. Kalantari
[1982]   Major subsets in effective topology, In: Metakides [1982], 77–94.

I. Kalantari and J. B. Remmel
[1983]   Degrees of recursively enumerable topological spaces, J. Symbolic Logic *48* (1983), 610–622.

I. Kalantari and G. Weitkamp
[1985]   Effective topological spaces I: A definability theory, Ann. Pure Appl. Logic *29* (1985), 1–27.

S. Kallibekov
[1971a] Index sets of degrees of unsolvability, Algebra i Logika *10* (1971), 316–326 (Russian); Algebra and Logic *10* (1971), 198–204 (English translation).

[1971b] Index sets of m-degrees, Sibirsk. Mat. Zh. *12* (1971), 1292–1300 (Russian); Siberian
        Math. J. *12* (1971), 931–937 (English translation).
[1973a] On degrees of recursively enumerable sets, Sibirsk. Mat. Zh. *14* (1973), 421–426,
        463 (Russian); Siberian Math. J. *14* (1973), 290–293 (English translation).
[1973b] On tt-degrees of recursively enumerable sets, Mat. Zametki *14* (1973), 697–702 (Rus-
        sian); Math. Notes *14* (1973), 958–961 (English translation).

C. F. Kent
[1962] Constructive analogues of the group of permutations of the natural numbers, Trans.
        Amer. Math. Soc. *104* (1962), 347–362.

J. L. Kelley
[1955] General Topology, D. van Nostrand, Inc., Princeton, N. J., 1955.

A. B. Khutoretskii
[1969] Two existence theorems for computable numerations, Algebra i Logika *8* (1969),
        483–492 (Russian); Algebra and Logic *8* (1969), 277–282 (English translation).

E. Kinber
[1977] On T-degrees of Post's hypersimple sets, Latviiskii Mat. Ezhegodnik (Latvian Math-
        ematical Yearbook) *21* (1977), 164–170 (Russian).

S. C. Kleene
[1934] Proof by cases in formal logic, Ann. of Math. (2) *35* (1934), 529-544.
[1935] A theory of positive integers in formal logic, Amer. J. Math. *57* (1935), 153–173,
        219–244.
[1936a] General recursive functions of natural numbers, Math. Ann. *112* (1936), 727–742;
        reprinted in Davis [1965], 236–253.
[1936b] λ-definability and recursiveness, Duke Math. J. *2* (1936), 340–353.
[1938] On notation for ordinal numbers, J. Symbolic Logic *3* (1938), 150-155.
[1943] Recursive predicates and quantifiers, Trans. Amer. Math. Soc. *53* (1943), 41–73;
        reprinted in Davis [1965], 254–287.
[1950] A symmetric form of Gödel's theorem, Neder. Akad. Wetenschappen Proc. Ser. A
        53 (1950), 800–802.
[1952a] Introduction to Metamathematics, Van Nostrand, New York, 1952; Eighth reprint,
        Wolters-Noordhoff Publishing Co., Groningen and North-Holland, Amsterdam, New
        York, Oxford, 1980.
[1952b] Recursive functions and intuitionistic mathematics, Proceedings of the International
        Congress of Mathematicians, Cambridge, Mass., 1950, 679–685, Amer. Math. Soc.,
        Providence, R.I., 1952.
[1955a] Arithmetical predicates and function quantifiers, Trans. Amer. Math. Soc. *79* (1955),
        312–340.
[1955b] On the forms of the predicates in the theory of constructive ordinals (second paper),
        Amer. J. Math. *77* (1955), 405–428.
[1955c] Hierarchies of number-theoretic predicates, Bull. Amer. Math. Soc. *61* (1955),
        193–213.
[1959a] Recursive functionals and quantifiers of finite types I, Trans. Amer. Math. Soc. *91*
        (1959), 1–52.
[1959b] Countable functionals, In: Heyting [1959], 81–100.
[1959c] Quantification of number-theoretic functions, Compositio Math. *14* (1959), 23–40.
[1963] Recursive functionals and quantifiers of finite types II, Trans. Amer. Math. Soc. *108*
        (1963), 106–142.
[1981] Origins of recursive function theory, Ann. Hist. Comput. *3* (1981), 52–67.

S. C. Kleene and E. L. Post
[1954] The upper semi-lattice of degrees of recursive unsolvability, Ann. of Math. (2) *59*
        (1954), 379–407.

J. Knight, A. H. Lachlan, and R. I. Soare
[1984]   Two theorems on degrees of models of true arithmetic, J. Symbolic Logic *49* (1984), 425–436.

G. N. Kobzev
[1973a] On btt-reducibility, Algebra i Logika *12* (1973), 190–204, 244 (Russian); Algebra and Logic *12* (1973), 107–115 (English translation).
[1973b] On btt-reducibility II, Algebra i Logika *12* (1973), 433–444, 492 (Russian); Algebra and Logic *12* (1973), 242–248 (English translation).
[1973c] The complete btt-degree, Algebra i Logika *13* (1973), 22–25, 120 (Russian); Algebra and Logic *13* (1973), 10–12 (English translation).
[1975]   r-separated sets, Issl. Mat. Log. Teor. Algorit. (Studies in Mathematical Logic and the Theory of Algorithms), Tbilisi, 1975, 19–30 (Russian).
[1976]   Relationships between recursively enumerable tt-degrees and w-degrees, Soobshch. Akad. Nauk Gruz. SSR (Bull. Acad. Sci. Georgian SSR) *84* (1976), 585–587 (Russian).
[1977a] Maximal *m*-degrees, Soobshch. Akad. Nauk Gruz. SSR (Bull. Acad. Sci. Georgian SSR) *85* No. 2 (1977), 325–327 (Russian).
[1977b] Recursively enumerable bw-degrees, Mat. Zametki *21* (1977), 839–846 (Russian); Math. Notes *21* (1977), 473–477 (English translation).
[1978a] On tt-degrees of r.e. T-degrees, Mat. Sbornik (N.S.) *106* (1978), 507–514 (Russian); Mathematics of the USSR, Sbornik *35* (1978), 173–180 (English translation).
[1978b] On the upper semilattice of tt-degrees, Soobshch. Akad. Nauk Gruz. SSR (Bull. Acad. Sci. Georgian SSR) *90* No. 2 (1978), 281–283 (Russian, English summary).
[1979]   On the tt-degrees of r.e. T-degrees II, Algebra i Logika *18* (1979), 415–425 (Russian); Algebra and Logic *18* (1979), 252–259 (English translation).

G. Kreisel
[1950]   Note on arithmetic models for consistent formulae of the predicate calculus, Fund. Math. *37* (1950), 265–285.
[1953]   A variant to Hilbert's theory on the foundations of arithmetic, British J. Philos. Sci. *4* (1953), 107–129, 357.

G. Kreisel, D. Lacombe, and J. R. Shoenfield
[1959]   Partial recursive functionals and effective operations, In: Heyting [1959], 290–297.

G. Kreisel and G. E. Sacks
[1965]   Metarecursive sets, J. Symbolic Logic *30* (1965), 318–338.

A. Kučera
[1985]   Measure, Π_1^0-classes and complete extensions of PA, In: Ebbinghaus, Müller, and Sacks [1985], 245–259.
[1986]   An alternative priority-free solution to Post's problem, In: Twelfth Symposium held in Bratislava, Czechoslovakia, August 25–29, 1986, edited by J. Gruska, B. Rovan, and J. Wiederman, Lecture Notes in Computer Science No. 233, Proceedings, Mathematical Foundations of Computer Science '86, Springer-Verlag, Heidelberg, Berlin, New York, Tokyo, 1986.
[ta]   On the role of $\mathbf{0}'$ in recursion theory, to appear.

S. A. Kurtz
[1983]   Notions of weak genericity, J. Symbolic Logic *48* (1983), 764–770.

A. L. Kuznecov
[1950]   On primitive recursive functions of large oscillation, Dokl. Akad. Nauk SSSR, N.S. *71* (1950), 233–236 (Russian).

A. H. Lachlan
[1965a] Some notions of reducibility and productiveness, Z. Math. Logik Grundlag. Math. *11* (1965), 17–44.
[1965b] On a problem of G. E. Sacks, Proc. Amer. Math. Soc. *16* (1965), 972–979.

[1966a] A note on universal sets, J. Symbolic Logic *31* (1966), 573–574.
[1966b] Lower bounds for pairs of recursively enumerable degrees, Proc. London Math. Soc. *16* (1966), 537–569.
[1966c] The impossibility of finding relative complements for recursively enumerable degrees, J. Symbolic Logic *31* (1966), 434–454.
[1967] The priority method I, Z. Math. Logik Grundlag. Math. *13* (1967), 1–10.
[1968a] Degrees of recursively enumerable sets which have no maximal superset, J. Symbolic Logic *33* (1968), 431–443.
[1968b] Distributive initial segments of the degrees of unsolvability, Z. Math. Logik Grundlag. Math. *14* (1968), 457–472.
[1968c] On the lattice of recursively enumerable sets, Trans. Amer. Math. Soc. *130* (1968), 1–37.
[1968d] The elementary theory of recursively enumerable sets, Duke Math. J. *35* (1968), 123–146.
[1968e] Complete recursively enumerable sets, Proc. Amer. Math. Soc. *19* (1968), 99–102.
[1969] Initial Segments of one-one degrees, Pacific J. Math. *29* (1969), 351–366.
[1970a] On some games which are relevant to the theory of recursively enumerable sets, Ann. of Math. (2) *91* (1970), 291–310.
[1970b] Initial segments of many-one degrees, Canad. J. Math. *22* (1970), 75–85.
[1971] Solution to a problem of Spector, Canad. J. Math. *23* (1971), 247–256.
[1972a] Embedding nondistributive lattices in the recursively enumerable degrees, In: Hodges [1972], 149–177.
[1972b] Recursively enumerable many-one degrees, Algebra i Logika *11* (1972), 326–358, 362 (Russian); Algebra and Logic *11* (1972), 186–202 (English translation).
[1972c] Two theorems on many-one degrees of recursively enumerable sets, Algebra i Logika *11* (1972), 216–229 (Russian); Algebra and Logic *11* (1972), 127–132 (English translation).
[1973] The priority method for the construction of recursively enumerable sets, In: Mathias and Rogers [1973], 299–310.
[1975a] A recursively enumerable degree which will not split over all lesser ones, Ann. Math. Logic *9* (1975), 307–365.
[1975b] Uniform enumeration operations, J. Symbolic Logic *40* (1975), 401–409.
[1975c] wtt-complete sets are not necessarily tt-complete, Proc. Amer. Math. Soc. *48* (1975), 429–434.
[1979] Bounding minimal pairs, J. Symbolic Logic *44* (1979), 626–642.
[1980] Decomposition of recursively enumerable degrees, Proc. Amer. Math. Soc. *79* (1980), 629–634.

A. H. Lachlan and R. Lebeuf
[1976] Countable initial segments of the degrees of unsolvability, J. Symbolic Logic *41* (1976), 289–300.

A. H. Lachlan and R. I. Soare
[1980] Not every finite lattice is embeddable in the recursively enumerable degrees, Adv. in Math. *37* (1980), 74–82.

D. Lacombe
[1954] Sur le semi-réseau constitué par les degrés d'indécidabilité récursive, Comptes rendus hebdomadaires des séances de l'Académie des Sciences (Paris) Ser. A-B *239* (1954), 1108–1109.
[1959] Quelques procédés de définition en topologie récursive, In: Heyting [1959], 129–158.
[1960] La théorie des fonctions récursives et ses applications (Exposé d'information générale), Bull. Soc. Math. France *88* (1960), 393–468.

R. E. Ladner
[1973a] Mitotic recursively enumerable sets, J. Symbolic Logic *38* (1973), 199–211.
[1973b] A completely mitotic nonrecursive recursively enumerable degree, Trans. Amer. Math. Soc. *184* (1973), 479–507.

R. E. Ladner and L. P. Sasso
[1975]   The weak truth table degrees of recursively enumerable sets, Ann. Math. Logic *8* (1975), 429–448.

I. A. Lavrov
[1968]   Answer to a question of P. R. Young, Algebra i Logika, *7* (1968), 48–54 (Russian); Algebra and Logic *7* (1968), 98–101 (English translation).
[1974]   Certain properties of retracts of Post numerations, Algebra i Logika *3* (1974), 662–675 (Russian); Algebra and Logic *13* (1974), 379–387 (English translation).

S. Lempp
[1986]   Topics in recursively enumerable sets and degrees, Ph.D. Dissertation, University of Chicago, 1986.
[ta]   A high strongly noncappable degree, J. Symbolic Logic, to appear.
[ta]   Hyperarithmetical index sets in recursion theory, Trans. Amer. Math. Soc., to appear.

S. Lempp and T. A. Slaman
[ta]   A limit on relative genericity in the recursively enumerable sets, to appear.

M. Lerman
[1970a] Recursive functions modulo co-r-maximal sets, Trans. Amer. Math. Soc. *148* (1970), 429–444.
[1970b] Turing degrees and many-one degrees of maximal sets, J. Symbolic Logic *35* (1970), 29–40.
[1971a] Some theorems on r-maximal sets and major subsets of recursively enumerable sets, J. Symbolic Logic *36* (1971), 193–215.
[1971b] Initial segments of the degrees of unsolvability, Ann. of Math. (2) *93* (1971), 365–389.
[1973]   Admissible ordinals and priority arguments, In: Mathias and Rogers [1973], 311–344.
[1976]   Congruence relations, filters, ideals and definability in lattices of alpha-recursively enumerable sets, J. Symbolic Logic *41* (1976), 405–418.
[1977]   Automorphism bases for the semilattice of recursively enumerable degrees, Notices Amer. Math. Soc. *24* (1977), A-251, Abstract #77T-E10.
[1978a] Lattices of α-recursively enumerable sets, In: Fenstad, Gandy, and Sacks [1978], 223–238.
[1978b] On elementary theories of some lattices of α-recursively enumerable sets, Ann. Math. Logic *14* (1978), 227–272.
[1980]   The degrees of unsolvability: Some recent results, In: Drake and Wainer [1980], 140–157.
[1981]   On recursive linear orderings, In: Lerman, Schmerl, and Soare [1981], 132–142.
[1983a] Degrees of Unsolvability, Perspectives in Mathematical Logic, Omega Series, Springer-Verlag, Berlin, Heidelberg, New York, Tokyo, 1983.
[1983b] The structures of recursion theory, In: Chong and Wicks [1983], 77–95.
[1985a] The embedding problem for the recursively enumerable degrees, In: Nerode and Shore [1985], 13–20.
[1985b] On the ordering of classes in high/low hierarchies, In: Ebbinghaus, Müller, and Sacks [1985], 260–270.
[1985c] Upper bounds for the arithmetical degrees, Ann. Pure Appl. Logic *29* (1985), 225–254.
[1986]   Degrees which do not bound minimal degrees, Ann. Pure Appl. Logic *30* (1986), 249–276.

M. Lerman and J. Remmel
[1982]   The universal splitting property I, In: van Dalen, Lascar, and Smiley [1982], 181–208.
[1984]   The universal splitting property II, J. Symbolic Logic. *49* (1984), 137–150.

M. Lerman and J. Rosenstein
[1982]   Recursive linear orderings, In: Metakides [1982], 123–136.

M. Lerman and J. H. Schmerl
[1979]   Theories with recursive models, J. Symbolic Logic *44* (1979), 59–76.

M. Lerman, J. H. Schmerl, and R. I. Soare
[1981]   Logic Year 1979–80: University of Connecticut (editors), Lecture Notes in Mathematics No. 859, Springer-Verlag, Berlin, Heidelberg, Tokyo, New York, 1981.

M. Lerman, R. A. Shore, and R. I. Soare
[1978] r-Maximal major subsets, Israel J. Math. *31* (1978), 1–18.
[1984]   The elementary theory of the recursively enumerable degrees is not $\aleph_0$-categorical, Adv. in Math. *53* (1984), 301–320.

M. Lerman and R. I. Soare
[1980a] d-Simple sets, small sets, and degree classes, Pacific J. Math. *87* (1980), 135–155.
[1980b] A decidable fragment of the elementary theory of the lattice of recursively enumerable sets, Trans. Amer. Math. Soc. *257* (1980), 1–37.

G. Lolli
[1981]   Recursion Theory and Computational Complexity (editor), Proceedings of Centro Internazionale Matematico Estivo (C.I.M.E.), June 14–23, 1979, in Bressanone, Italy, Liguori Editore, Naples, Italy, 1981.

W. Maass
[1981]   A countable basis for Σ_2^1 sets and recursion theory on $\aleph_1$, Proc. Amer. Math. Soc. *82* (1981), 267–270.
[1982]   Recursively enumerable generic sets, J. Symbolic Logic *47* (1982), 809–823.
[1983]   Characterization of recursively enumerable sets with supersets effectively isomorphic to all recursively enumerable sets, Trans. Amer. Math. Soc. *279* (1983), 311–336.
[1984]   On the orbits of hyperhypersimple sets, J. Symbolic Logic *49* (1984), 51–62.
[1985a] Variations on promptly simple sets, J. Symbolic Logic *50* (1985), 138–148.
[1985b] Major subsets and automorphisms of recursively enumerable sets, In: Nerode and Shore [1985], 21–32.
[1986]   Are recursion theoretic arguments useful in complexity theory?, In: Proc. of the Seventh International Congress of Logic, Methodology, and the Philosophy of Science, Salzburg, 1983, North-Holland, Amsterdam, New York, Oxford, 1986, 141–158.

W. Maass and S. Homer
[1983]   Oracle dependent properties of the lattice of NP sets, Theoret. Comput. Sci. *24* (1983), 279–289.

W. Maass, R. A. Shore and M. Stob
[1981]   Splitting properties and jump classes, Israel J. Math. *39* (1981), 210–224.

W. Maass and M. Stob
[1983]   The intervals of the lattice of recursively enumerable sets determined by major subsets, Ann. Pure Appl. Logic *24* (1983), 189–212.

M. Machtey and P. R. Young
[1978]   An introduction to the general theory of algorithms, Elsevier North-Holland, Amsterdam, New York, Oxford, 1978.

D. B. Madan and R. W. Robinson
[1982]   Monotone and 1–1 sets, J. Austral. Math. Soc. (Series A) *33* (1982), 62–75.

A. I. Mal'cev
[1970]   Algorithms and recursive functions (English translation), Wolters-Noordhoff Publishing Co., Groningen, The Netherlands, 1970, 372 pp.

An. A. Mal'cev
[1984]   On the structure of the families of immune, hyperimmune and hyperhyperimmune sets, Mat. Sbornik (N.S.) *124* No. 7 (1984), 307–319 (Russian).

S. S. Marchenkov
[1975]   The existence of recursively enumerable minimal truth-table degrees, Algebra i Logika *14* (1975), 442–429 (Russian); Algebra and Logic *14* (1975), 257–261 (English translation).
[1976a] A class of incomplete sets, Mat. Zametki *20* (1976), 473–478, (Russian); Math. Notes *20* (1976), 823–825 (English translation).
[1976b] On the comparison of the upper semilattices of recursively enumerable m-degrees and truth-table degrees, Mat. Zametki *20* (1976), 19–26 (Russian); Math. Notes *20* (1976), 567–570 (English translation).
[1976c] Truth-table degrees of maximal sets, Mat. Zametki *20* (1976), 373–381 (Russian); Math. Notes *20* (1976), 766–770 (English translation).
[1977]   Recursively enumerable minimal btt-degrees, Mat. Sbornik (N.S.) *103* (*145*) (1977), 550–562, 631 (Russian).

A. A. Markov
[1947]   On the representation of recursive functions, Dokl. Akad. Nauk SSSR, N.S. *58* (1947), 1891–1892 (Russian).
[1951]   The theory of algorithms, Trudy Mat. Inst. Steklov. *38* (1951), 176–189 (Russian); Transl., II Ser., Amer. Math. Soc. *15* (2) (1960), 1–14 (English translation).
[1954]   The theory of algorithms, Trudy Mat. Inst. Steklov. *42* (1954) (Russian); (English translation, 1961, available from the Office of Technical Series, U.S. Dept. of Commerce, Washington, D.C.).
[1960]   Insolubility of the problem of homeomorphy (Russian), In: Proceedings of the International Congress of Mathematicians, 1958, Cambridge University Press, London, 1960, 300–306.

I. Marques
[1973]   Complexity Properties of Recursively Enumerable Sets, Ph.D. Dissertation, University of California, Berkeley, 1973.
[1975]   On degrees of unsolvability and complexity properties, J. Symbolic Logic *40* (1975), 529–540.

D. A. Martin
[1963]   A theorem on hyperhypersimple sets, J. Symbolic Logic *28* (1963), 273–278.
[1966a] Completeness, the recursion theorem, and effectively simple sets, Proc. Amer. Math. Soc. *17* (1966), 838–842.
[1966b] Classes of recursively enumerable sets and degrees of unsolvability, Z. Math. Logik Grundlag. Math. *12* (1966), 295–310.
[1966c] On a question of G. E. Sacks, J. Symbolic Logic *31* (1966), 66–69.
[1966d] The priority method of Sacks (mimeographed notes).
[1967]   Measure, category, and degrees of unsolvability (unpublished manuscript), 16 pp.

D. A. Martin and M. B. Pour-El
[1970]   Axiomatizable theories with few axiomatizable extensions, J. Symbolic Logic *35* (1970), 205–209.

Ju. V. Matijasevič
[1970]   Enumerable sets are diophantine, Dokl. Akad. Nauk. SSSR, N.S. *191* (1970), 279–282 (Russian); Sov. Math. Dokl. *11* (1970), 354–357 (English translation).
[1971]   Diophantine representation of enumerable predicates, Izv. Akad. Nauk SSSR Ser. Mat. *35* (1971), 3–30 (Russian).

A. R. D. Mathias and H. Rogers, Jr.
[1973]   Cambridge Summer School in Mathematical Logic (editors), held in Cambridge, England, August 1–21, 1971, Lecture Notes in Mathematics No. 337, Springer-Verlag, Berlin, Heidelberg, New York, 1973.

K. McEvoy
[ta]   Jumps of quasi-minimal enumeration degrees, to appear.

K. McEvoy and S. B. Cooper,
[1985]   On minimal pairs of enumeration degrees, J. Symbolic Logic *50* (1985), 983–1001.

T. G. McLaughlin
[1965]   On a class of complete simple sets, Canad. Math. Bull. *8* (1965), 33–37.
[1966]   Retraceable sets and recursive permutations, Proc. Amer. Math. Soc. *17* (1966), 427–429.
[1968]   A theorem on intermediate reducibilities, Proc. Amer. Math. Soc. *19* (1968), 87–90.

Yu. T. Medvedev
[1955a] On nonisomorphic recursively enumerable sets, Dokl. Akad. Nauk SSSR, N.S. *102* (1955), 211–214 (Russian).
[1955b] Degrees of difficulty of the mass problem, Dokl. Akad. Nauk SSSR, N.S. *104* (1955), 501–504 (Russian).

G. Metakides
[1982]   Patras Logic Symposion (editor), Proceedings of the Logic Symposium held at Patras, Greece, August 18–22, 1980, North-Holland, Amsterdam, New York, Oxford, 1982.

G. Metakides and A. Nerode
[1975]   Recursion theory and algebra, In: Lecture Notes in Mathematics No. 450, Springer-Verlag, Berlin, Heidelberg, New York, 1975, 209–219.
[1977]   Recursively enumerable vector spaces, Ann. Math. Logic *11* (1977), 141–171.
[1979]   Effective content of field theory, Ann. Math. Logic *17* (1979), 289–320.
[1980]   Recursion theory on fields and abstract dependence, J. Algebra *65* (1980), 36–59.
[1982]   The introduction of non-recursive methods in mathematics, In: Troelstra and van Dalen [1982], 319–335.

G. Metakides, A. Nerode, and R. A. Shore
[1985]   Recursive limits on the Hahn-Banach theorem, Contemp. Math. *39* (1985), 85–91.

T. Millar
[1978]   Foundations of recursive model theory, Ann. Math. Logic *13* (1978), 45–72.
[1985]   Decidable Ehrenfeucht theories, In: Nerode and Shore [1985], 311–321.

C. F. Miller, III
[1971]   On Group-Theoretic Decision Problems and Their Classification, Ann. of Math. Stud. *68*, Princeton University Press, Princeton, N.J., 1971.

D. Miller
[1981a] High recursively enumerable degrees and the anti-cupping property, In: Lerman, Schmerl, and Soare [1981], 230–245.
[1981b] The Relationship Between the Structure and Degrees of Recursively Enumerable Sets, Ph.D. Dissertation, University of Chicago, 1981.

D. Miller and J. B. Remmel
[1984]   Effectively nowhere simple sets, J. Symbolic Logic *49* (1984), 129–136.

W. Miller and D. A. Martin
[1968]   The degree of hyperimmune sets, Z. Math. Logik Grundlag. Math. *14* (1968), 159–166.

M. L. Minsky
[1961]   Recursive unsolvability of Post's problem of "tag" and other topics in the theory of Turing machines, Ann. of Math. (2) *74* (1961), 437–455.

J. Mohrherr
[1982]   Index Sets and Truth-Table Degrees, Ph.D. Dissertation, University of Illinois at Chicago Circle, 1982.

[1983]   Kleene index sets and functional m-degrees, J. Symbolic Logic *48* (1983), 829–840.
[1986]   A refinement of low_n and high_n for the r.e. degrees, Z. Math. Logik Grundlag. Math. *32* (1986), 5–12.

M. D. Morley and R. I. Soare
[1975]   Boolean algebras, splitting theorems and Δ_2^0 sets, Fund. Math. *90* (1975), 45–52.

P. B. Morris
[1974]   Complexity Theoretic Properties of Recursively Enumerable Sets, Ph.D. Dissertation, University of California, Irvine, 1974.

A. Mostowski
[1947]   On definable sets of positive integers, Fund. Math. *34* (1947), 81–112.
[1948]   On a set of integers not definable by means of one-quantifier predicates, Annales de la Société Polonaise de Mathématique *21* (1948), 114–119.

A. A. Muchnik
[1956a] On the unsolvability of the problem of reducibility in the theory of algorithms, Dokl. Akad. Nauk SSSR, N.S. *108* (1956), 194–197 (Russian).
[1956b] On the separability of recursively enumerable sets, Dokl. Akad. Nauk SSSR, N.S. *109* (1956), 29–32 (Russian).
[1958]   Isomorphism of systems of recursively enumerable sets with effective properties, Trudy Moskov. Mat. Obshch. *7* (1958), 407–412 (Russian); Amer. Math. Soc. Translations (Series 2) *23* (1963), 7–13 (English translation).

J. Myhill
[1953]   Three contributions to recursive function theory, In: Actes du XIème congrès international de philosophie, Bruxelles *20–26*, Août 1953 *XIV*, North-Holland, Amsterdam, 1953, 50–59.
[1955]   Creative sets, Z. Math. Logik Grundlag. Math. *1* (1955), 97–108.
[1956]   The lattice of recursively enumerable sets, J. Symbolic Logic *21* (1956), 215, 220 (abstract).
[1959a] Finitely representable functions, In: A. Heyting [1959], 195–207.
[1959b] Recursive digraphs, splinters and cylinders, Math. Ann. *138* (1959), 211–218.
[1961a] A note on degrees of partial functions, Proc. Amer. Math. Soc. *12* (1961), 519–521.
[1961b] Category methods in recursion theory, Pacific J. Math. *11* (1961), 1479–1486.
[1971]   A recursive function, defined on a compact interval and having a continuous derivative that is not recursive, Michigan Math. J. *18* (1971), 97–98.

J. Myhill and J. C. Shepherdson
[1955]   Effective operations on partial recursive functions, Z. Math. Logik Grundlag. Math. *1* (1955), 310–317.

A. Nerode
[1961]   Extensions to Isols, Ann. of Math. (2) *73* (1961), 362–403.

A. Nerode and A. B. Manaster
[1970]   A universal embedding property of the RET's, J. Symbolic Logic *35* (1970), 51–59.

A. Nerode and J. B. Remmel
[1982]   Recursion theory on matroids, In: Metakides [1982], 41–65.
[1985a] Generic objects in recursion theory, In: Ebbinghaus, Müller, and Sacks [1985], 271–314.
[1985b] A survey of lattices of r.e. substructures, In: Nerode and Shore [1985], 323–375.
[1986]   Generic objects in recursion theory II: Operations on recursive approximation spaces, Ann. Pure Appl. Logic *31* (1986), 257–288.

A. Nerode and R. A. Shore
[1980a] Second order logic and first order theories of reducibility orderings, In: Barwise, Keisler, and Kunen [1980], 181–200.

[1980b] Reducibility orderings: Theories, definability and automorphisms, Ann. Math. Logic
         18 (1980), 61–89.
[1985]   Recursion Theory (editors), Proc. Symp. Pure Math. *42*, Proceedings of AMS-ASL
         Summer Institute on Recursion Theory, held at Cornell University, Ithaca, N. Y.,
         June 28 - July 16, 1982, Amer. Math. Soc., Providence, R. I., 1985.

D. Normann
[1980]   Recursion on Countable Functionals, Lecture Notes in Mathematics No. 811,
         Springer-Verlag, Berlin, New York, Heidelberg, 1980.

P. S. Novikov
[1955]   On the algorithmic unsolvability of the word problem in group theory, Trudy Mat.
         Inst. Steklov. *44* (1955) (Russian).

P. Odifreddi
[1981]   Strong reducibilities, Bull. Amer. Math. Soc. (N. S.) *4* (1981), 37–86.
[1983a] Forcing and reducibilities I, J. Symbolic Logic *48* (1983), 288–310.
[1983b] Forcing and reducibilities II: forcing in fragments of analysis, J. Symbolic Logic *48*
         (1983), 724–743.
[1985]   The structure of *m*-degrees, In: Ebbinghaus, Müller, and Sacks [1985], 315–332.
[ta]   Classical Recursion Theory, North-Holland, Amsterdam, New York, Oxford, to ap-
         pear.
[ta]   Recursion-Theoretical Aspects of Complexity Theory, to appear.
[ta]   Church's Thesis: the extent of recursiveness in the universe and man, to appear.

R. Sh. Omanadze
[1976]   On completeness for recursively enumerable sets, Soobshch. Akad. Nauk Gruz. SSR
         (Bull. Acad. Sci. Georgian SSR) *81* (1976), 529–532 (Russian).
[1978]   On reducibilities for the class of recursively enumerable sets, Soobshch. Akad. Nauk
         Gruz. SSR (Bull. Acad. Sci. Georgian SSR) *91* (1978), 549–552 (Russian).

J. C. Owings, Jr.
[1967]   Recursion, metarecursion, and inclusion, J. Symbolic Logic *32* (1967), 173–178.
[1970]   Review of Lachlan [1968c, 1968d] and Robinson [1967a, 1967b], J. Symbolic Logic
         35 (1970), 153–155.
[1973]   Diagonalization and the recursion theorem, Notre Dame J. Formal Logic *14* No. 1
         (1973), 95–99.

J. B. Paris
[1976]   Survey of results on measure and degrees (preprint).
[1977]   Measure and minimal degrees, Ann. Math. Logic *11* (1977), 203–216.

R. Péter
[1967]   Recursive Functions (third revised edition), Academic Press, New York, 1967.

D. Posner
[1977]   High Degrees, Ph.D. Dissertation, University of California, Berkeley, 1977.
[1980]   A survey of non-r.e. degrees ≤ **0′**, In: Drake and Wainer [1980], 52–109.
[1981]   The upper semilattice of degrees ≤ **0′** is complemented, J. Symbolic Logic *46* (1981),
         705–713.

D. Posner and R. Epstein
[1978]   Diagonalization in degree constructions, J. Symbolic Logic *43* (1978), 280–283.

D. Posner and R. W. Robinson
[1981]   Degrees joining to **0′**, J. Symbolic Logic *46* (1981), 714–722.

E. L. Post
[1936]   Finite combinatory processes-formulation I, J. Symbolic Logic *1* (1936), 103–105;
         reprinted in Davis [1965], 288–291.

[1943]   Formal reductions of the general combinatorial decision problem, Amer. J. Math.
         65 (1943), 197–215.
[1944]   Recursively enumerable sets of positive integers and their decision problems, Bull.
         Amer. Math. Soc. *50* (1944), 284–316; reprinted in Davis [1965], 304–337.
[1946]   Note on a conjecture of Skolem, J. Symbolic Logic *11* (1946), 73–74.
[1947]   Recursive unsolvability of a problem of Thue, J. Symbolic Logic *12* (1947), 1–11;
         reprinted in Davis [1965], 292–303.
[1948]   Degrees of recursive unsolvability: preliminary report (abstract), Bull. Amer. Math.
         Soc. *54* (1948), 641–642.

M. B. Pour-El
[1965]   "Recursive isomorphism" and effectively extensible theories, Bull. Amer. Math. Soc.
         71 (1965), 551–555.
[1968a] Effectively extensible theories, J. Symbolic Logic *33* (1968), 56–68.
[1968b] Independent axiomatization and its relation to the hypersimple set, Z. Math. Logik
         Grundlag. Math. *14* (1968), 449–456.
[1974]   Abstract computability and its relation to the general all purpose analogue com-
         puter, Trans. Amer. Math. Soc. *199* (1974), 1–28.

M. B. Pour-El and S. Kripke
[1967a] Deduction-preserving "recursive isomorphisms" between theories, Bull. Amer.
         Math. Soc. *73* (1967), 145–148.
[1967b] Deduction-preserving "recursive isomorphisms" between theories, Fund. Math. *61*
         (1967), 141–163.

M. B. Pour-El and I. Richards
[1979]   A computable ordinary differential equation which possesses no computable solu-
         tion, Ann. Math. Logic *17* (1979), 61–90.
[1981]   The wave equation with computable initial data such that its unique solution is not
         computable, Adv. in Math. *39* (1981), 215–239.
[1983a] Computability and noncomputability in classical analysis, Trans. Amer. Math. Soc.
         275 (1983), 539–560.
[1983b] Noncomputability in analysis and physics: a complete determination of the class of
         noncomputable linear operators, Adv. in Math. *48* (1983), 44–74.

M. O. Rabin
[1960]   Computable algebra, general theory and theory of computable fields, Trans. Amer.
         Math. Soc. *95* (1960), 341–360.

J. B. Remmel
[1978]   Recursively enumerable Boolean algebras, Ann. Math. Logic *14* (1978), 75–107.
[1980]   Recursion theory on algebraic structures with independent sets, Ann. Math. Logic
         18 (1980), 153–191.
[ta]   Recursive Boolean Algebras, Handbook of Boolean Algebras, to appear.

H. G. Rice
[1953]   Classes of recursively enumerable sets and their decision problems, Trans. Amer.
         Math. Soc. *74* (1953), 358–366.
[1954]   Recursive real numbers, Proc. Amer. Math. Soc. *5* (1954), 784–791.
[1956a] Recursive and recursively enumerable orders, Trans. Amer. Math. Soc. *83* (1956),
         277–300.
[1956b] On completely recursively enumerable classes and their key arrays, J. Symbolic
         Logic *21* (1956), 304–308.

M. M. Richter, E. Börger, W. Oberschelp, B. Schinzel and W. Thomas
[1984]   Computation and Proof Theory (editors), Proceedings of the Logic Colloquium
         held in Aachen, July 18–13, 1983, Part II, Lecture Notes in Mathematics No. 1104,
         Springer-Verlag, Berlin, Heidelberg, New York, Tokyo, 1984.

J. Robinson

[1949]   Definability and decision problems in arithmetic, J. Symbolic Logic *14* (1949), 98–114.

[1968]   Recursive functions of one variable, Proc. Amer. Math. Soc. *19* (1968), 815–820.

R. W. Robinson

[1966a] The Inclusion Lattice and Degrees of Unsolvability of the Recursively Enumerable Sets, Ph.D. Dissertation, Cornell University, 1966.

[1966b] Recursively enumerable sets not contained in any maximal set, Notices Amer. Math. Soc. *13* (1966), 325. Abstract #632–4.

[1967a] Simplicity of recursively enumerable sets, J. Symbolic Logic *32* (1967), 162–172.

[1967b] Two theorems on hyperhypersimple sets, Trans. Amer. Math. Soc. *128* (1967), 531–538.

[1968]   A dichotomy of the recursively enumerable sets, Z. Math. Logik Grundlag. Math. *14* (1968), 339–356.

[1969]   Review of Sacks [1964a], J. Symbolic Logic *34* (1969), 294–295.

[1971a] Interpolation and embedding in the recursively enumerable degrees, Ann. of Math. (2) *93* (1971), 285–314.

[1971b] Jump restricted interpolation in the recursively enumerable degrees, Ann. of Math. (2) *93* (1971), 586–596.

H. Rogers, Jr.

[1959]   Computing degrees of unsolvability, Math. Ann. *138* (1959), 125–140.

[1967a] Theory of Recursive Functions and Effective Computability, McGraw-Hill, New York, 1967.

[1967b] Some problems of definability in recursive function theory, In: Crossley [1967], 183–201.

G. F. Rose and J. S. Ullian

[1963]   Approximations of functions on the integers, Pacific J. Math. *13* (1963), 693–701.

J. B. Rosser

[1936]   Extensions of some theorems of Gödel and Church, J. Symbolic Logic *1* (1936), 87–91; reprinted in Davis [1965], 230–235.

M. Rubin

[1976]   The theory of Boolean algebras with a distinguished subalgebra is undecidable, Ann. Sci. Univ. Clermont-Ferrand II Math. *13* (1976), 129–134.

G. E. Sacks

[1961a] A minimal degree less than **0**′, Bull. Amer. Math. Soc. *67* (1961), 416–419.

[1961b] On suborderings of degrees of recursive unsolvability, Z. Math. Logik Grundlag. Math. *7* (1961), 46–56.

[1963a] Degrees of unsolvability, Ann. of Math. Stud. *55*, Princeton University Press, Princeton, N.J., 1963 (see revised edition, 1966).

[1963b] On the degrees less than **0**′, Ann. of Math. (2) *77* (1963), 211–231.

[1963c] Recursive enumerability and the jump operator, Trans. Amer. Math. Soc. *108* (1963), 223–239.

[1964a] The recursively enumerable degrees are dense, Ann. of Math. (2) *80* (1964), 300–312.

[1964b] A maximal set which is not complete, Michigan Math. J. *11* (1964), 193–205.

[1964c] A simple set which is not effectively simple, Proc. Amer. Math. Soc. *15* (1964), 51–55.

[1966a] Degrees of unsolvability, rev. ed., Ann. of Math. Studies *55*, Princeton University Press, Princeton, N. J., 1966.

[1966b] Post's problem, admissible ordinals and regularity, Trans. Amer. Math. Soc. *124* (1966), 1–23.

[1967]   On a theorem of Lachlan and Martin, Proc. Amer. Math. Soc. *18* (1967), 140–141.

[1969]   Measure theoretic uniformity in recursion theory and set theory, Trans. Amer. Math. Soc. *142* (1969), 381–420.

[1971] Forcing with perfect closed sets, In: Scott [1971], 331–355.
[1976] Countable admissible ordinals and hyperdegrees, Adv. in Math. *19* (1976), 213–262.
[1977a] The k-section of a type n-object, Amer. J. Math. *99* (1977), 901–917.
[1977b] R.e. sets higher up, In: Logic, Foundations of Mathematics and Computability Theory, ed. D. Reidel, Dordrecht, Holland, 1977, 173–194.
[1982] Post's problem in E-recursion, In: Nerode and Shore [1985], 177–193.
[1985] Some open questions in recursion theory, In: Ebbinghaus, Müller, and Sacks [1985], 333–342.
[1986] The limits of E-recursive enumerability, Ann. Pure Appl. Logic *31* (1986), 87–120.
[ta] On Kučera's solution to Post's problem, to appear.

G. E. Sacks and S. Simpson
[1972] The α-finite injury method, Ann. Math. Logic *4* (1972), 343–367.

L. P. Sasso, Jr.
[1970] A cornucopia of minimal degrees, J. Symbolic Logic *35* (1970), 383–388.
[1971] Degrees of Unsolvability of Partial Functions, Ph.D. Dissertation, University of California, Berkeley, 1971.
[1973] A minimal partial degree $\leq \mathbf{0}'$, Proc. Amer. Math. Soc. *38* (1973), 388–392.
[1974a] Deficiency sets and bounded information reducibilities, Trans. Amer. Math. Soc. *200* (1974), 267–290.
[1974b] A minimal degree not realizing least possible jump, J. Symbolic Logic *39* (1974), 571–574.
[1975] A survey of partial degrees, J. Symbolic Logic *40* (1975), 130–140.

S. Schwarz
[1982] Index Sets of Recursively Enumerable Sets, Quotient Lattices, and Recursive Linear Orderings, Ph.D. Dissertation, University of Chicago, 1982.
[1984a] The quotient semilattice of the recursively enumerable degrees modulo the cappable degrees, Trans. Amer. Math. Soc. *283* (1984), 315–328.
[1984b] Recursive automorphisms of recursive linear orderings, Ann. Pure Appl. Logic *26* (1984), 69–73.
[ta] Index sets related to prompt simplicity, Ann. Pure Appl. Logic, to appear.
[ta] Index sets related to the high-low hierarchy, Israel J. Math., to appear.

D. Scott
[1962] Algebras of sets binumerable in complete extensions of arithmetic, In: Proc. Symp. Pure Math. *5*, Providence, R.I., 1962, 117–121.
[1971] Axiomatic Set Theory I (editor), Proc. Symp. Pure Math. *13*, Los Angeles, 1967, Amer. Math. Soc., Providence, R.I., 1971.

V. L. Selivanov
[1978a] On index sets of computable classes of finite sets, Algoritmy i Avtomaty (Algorithms and Automata) *10* (1978) (Kazan. Gos. Univ., Kazan), 95–99 (Russian).
[1978b] Some remarks about classes of r.e. sets, Sibirsk. Mat. Zh. *19* (1978), 153–160, 238 (Russian); Siberian Math. J. *19* (1978), 109–113 (English translation).
[1979] On the structure of degrees of unsolvability of index sets, Algebra i Logika *18* (1979), 463–480, 508–509 (Russian); Algebra and Logic *18* (1979), 286–299 (English translation).
[1982a] On index sets in the Kleene-Mostowski hierarchy, In: Mathematical Logic and Theory of Algorithms, Akad. Nauk SSSR, Sibirsk. Otdel. *2*, 1982, Nauka, Novosibirsk, USSR, 135–158 (Russian).
[1982b] On one class of reducibilities in the theory of recursive functions, Veroyatn. Metod. i Kibern. (Probabilistic Meth. and Cybern.) *18* (1982), 83–100 (Russian).
[1984a] On a hierarchy of limiting computations, Sibirsk. Mat. Zh. *25* (1984), 146–156 (Russian).
[1984b] Index sets in the hyperarithmetical hierarchy, Sibirsk. Mat. Zh. *25* (1984), 164–181 (Russian); Siberian Math. J. *25* (1984), 474–488 (English translation).

[1985]   On Ershov's hierarchy, Sibirsk. Mat. Zh. *26* (1985), 134 149 (Russian).

J. C. Shepherdson and H. E. Sturgis
[1963]   Computability of recursive functions, J. Assoc. Comput. Mach. *10* (1963), 217 255.

J. R. Shoenfield
[1957]   Quasicreative sets, Proc. Amer. Math. Soc. *8* (1957), 964 967.
[1958a] The class of recursive functions, Proc. Amer. Math. Soc. *9* (1958), 690 692.
[1958b] Degrees of formal systems, J. Symbolic Logic *23* (1958), 389 392.
[1959]   On degrees of unsolvability, Ann. of Math. (2) *69* (1959), 644 653.
[1960a] Degrees of models, J. Symbolic Logic *25* (1960), 233 237.
[1960b] An uncountable set of incomparable degrees, Proc. Amer. Math. Soc. *11* (1960), 61 62.
[1961]   Undecidable and creative theories, Fund. Math. *49* (1961), 171 179.
[1965]   Application of model theory to degrees of unsolvability, In: Addison, Henkin, and Tarski [1965], 359 363.
[1966]   A theorem on minimal degrees, J. Symbolic Logic *31* (1966), 539 544.
[1967]   Mathematical Logic, Addison-Wesley, Reading, Mass. (1967), 344 pp.
[1971]   Degrees of unsolvability, Mathematics Studies *2*, North-Holland, Amsterdam, 1971.
[1975]   The decision problem for recursively enumerable degrees, Bull. Amer. Math. Soc. *81* (1975), 973 977.
[1976]   Degrees of classes of r.e. sets, J. Symbolic Logic *41* (1976), 695 696.
[1984]   Strategies in the r.e. degrees (handwritten notes).

J. R. Shoenfield and R. I. Soare
[1978]   The generalized diamond theorem, Recursive Function Theory Newsletter *19* (1978), #219 (abstract).

R. A. Shore
[1977a] Determining automorphisms of the recursively enumerable sets, Proc. Amer. Math. Soc. *65* (1977), 318 325.
[1977b] α-recursion theory, In: Barwise [1977], 653 680.
[1978a] Nowhere simple sets and the lattice of recursively enumerable sets, J. Symbolic Logic *43* (1978), 322 330.
[1978b] On the ∀∃-sentences of α-recursion theory, In: Fenstad, Gandy, and Sacks [1978], 331 354.
[1979]   The homogeneity conjecture, Proc. Natl. Acad. Sci. U.S.A. *76* (1979), 4218 4219.
[1980a] $\mathcal{L}^*(K)$ and other lattices of the recursively enumerable sets, Proc. Amer. Math. Soc. *80* (1980), 143 146.
[1980b] Some constructions in α-recursion theory, In: Drake and Wainer [1980], 158 170.
[1981a] The theory of the degrees below **0**′, J. London Math. Soc. Ser. II *24* (1981), 1 14.
[1981b] The degrees of unsolvability: global results, In: Lerman, Schmerl, and Soare [1981], 283 301.
[1982a] Finitely generated codings and the degrees r.e. in a degree **d**, Proc. Amer. Math. Soc. *84* (1982), 256 263.
[1982b] On homogeneity and definability in the first order theory of the Turing degrees, J. Symbolic Logic *47* (1982), 8 16.
[1982c] The theories of the truth-table and Turing degrees are not elementarily equivalent, In: van Dalen, Lascar, and Smiley [1982], 231 237.
[1984a] The degrees of unsolvability: the ordering of functions by relative computability, Proc. International Congress of Mathematicians, Warsaw, 1983, North-Holland, New York, 337 345.
[1984b] The arithmetical degrees are not elementarily equivalent to the Turing degrees, Arch. Math. Logik Grundlag. *24* (1984), 137 139.
[1985]   The structure of the degrees of unsolvability, In: Nerode and Shore [1985], 33 51.
[ta]   A non-inversion theorem for the jump operator, Ann. Pure Appl. Logic, to appear.

S. G. Simpson

[1977a] First-order theory of the degrees of recursive unsolvability, Ann. of Math. (2) *105* (1977), 121–139.

[1977b] Degrees of unsolvability: A survey of results, In: Barwise [1977], 631–652.

[1978] Sets which do not have subsets of every higher degree, J. Symbolic Logic *43* (1978), 135–138.

[1985] Recursion theoretic aspects of the dual Ramsey theorem, In: Ebbinghaus, Müller, and Sacks [1985], 357–371.

M. Simpson

[1985] Arithmetical Degrees: Initial Segments, ω-REA Sets and the ω-Jump Operator, Ph.D. Dissertation, Cornell University, 1985.

T. A. Slaman

[1985] Reflection and the priority method in E-recursion theory, In: Ebbinghaus, Müller, and Sacks [1985], 372–404.

[ta] The density of infima in the recursively enumerable degrees, to appear.

[ta] A recursively enumerable degree that is not the top of a diamond in the Turing degrees, to appear.

[ta] The recursively enumerable degrees as a substructure of the Δ_2^0 degrees, to appear.

T. A. Slaman and J. R. Steel

[ta] Complementation in the Turing degrees, to appear.

[ta] Definable functions on degrees, to appear.

T. A. Slaman and W. H. Woodin

[ta] Definability in the Turing degrees, Illinois J. Math., to appear.

R. M. Smullyan

[1964] Effectively simple sets, Proc. Amer. Math. Soc. *15* (1964), 893–895.

R. I. Soare

[1969a] Recursion theory and Dedekind cuts, Trans. Amer. Math. Soc. *140* (1969), 271–294.

[1969b] A note on degrees of subsets, J. Symbolic Logic *34* (1969), 256.

[1969c] Cohesive sets and recursively enumerable Dedekind cuts, Pacific J. Math. *31* (1969), 215–231.

[1972] The Friedberg-Muchnik theorem re-examined, Canad. J. Math. *24* (1972), 1070–1078.

[1974a] Automorphisms of the lattice of recursively enumerable sets, Bull. Amer. Math. Soc. *80* (1974), 53–58.

[1974b] Automorphisms of the lattice of recursively enumerable sets, Part I: Maximal sets, Ann. of Math. (2) *100* (1974), 80–120.

[1976] The infinite injury priority method, J. Symbolic Logic *41* (1976), 513–530.

[1977] Computational complexity, speedable and levelable sets, J. Symbolic Logic *42* (1977), 545–563.

[1978] Recursively enumerable sets and degrees, Bull. Amer. Math. Soc. *84* (1978), 1149–1181.

[1980a] Recursive enumerability, In: Proceedings of the International Congress of Mathematicians, Helsinki, August 15–23, 1978, Academia Scientiarum Fennica, Hungary, 1980, 275–280.

[1980b] Fundamental methods for constructing recursively enumerable degrees, In: Drake and Wainer [1980], 1–51.

[1981] Constructions in the recursively enumerable degrees, In: Lolli [1981], 172–225.

[1982a] Automorphisms of the lattice of recursively enumerable sets, Part II: Low sets, Ann. Math. Logic *22* (1982), 69–107.

[1982b] Computational complexity of recursively enumerable sets, Inform. and Control *52* No. 1 (1982), 8–18.

[1985] Tree arguments in recursion theory and the $0'''$-priority method, In: Nerode and Shore [1985], 53–106.

R. I. Soare and M. Stob
[1982]   Relative recursive enumerability, In: Proceedings of the Herbrand Symposium Logic
         Colloquium '81, ed. J. Stern, North-Holland, Amsterdam, New York, Oxford, 1982,
         299–324.

V. D. Solov'ev
[1974]   Q-reducibility and hyperhypersimple sets, Veroyatn. Metod. i Kibern. (Probabilistic
         Meth. and Cybern.) 10–11 (1974), 121–128 (Russian).

C. Spector
[1955]   Recursive well-orderings, J. Symbolic Logic 20 (1955), 151-163.
[1956]   On degrees of recursive unsolvability, Ann. of Math. (2) 64 (1956), 581–592.
[1961]   Inductively defined sets of natural numbers, In: Infinitistic Methods, Proceedings of
         the Symposium on Foundations of Mathematics, Warsaw, 1959, Pergamon Press,
         Oxford, 1961, 97–102.

J. Steel
[1975]   Descending sequences of degrees, J. Symbolic Logic 40 (1975), 59–61.
[1982]   A Classification of Jump Operators, J. Symbolic Logic 47 (1982), 347–358.

M. Stob
[1979]   The Structure and Elementary Theory of the Recursively Enumerable Sets, Ph.D.
         Dissertation, University of Chicago, 1979.
[1982a] Invariance of properties under automorphisms of the lattice of recursively enumer-
         able sets, Pacific J. Math. 100 (1982), 445–471.
[1982b] Index sets and degrees of unsolvability, J. Symbolic Logic 47 (1982), 241–248.
[1983]   wtt-degrees and T-degrees of recursively enumerable sets, J. Symbolic Logic 48
         (1983), 921–930.
[1985]   Major subsets and the lattice of recursively enumerable sets, In: Nerode and Shore
         [1985], 107–116.

V. Stoltenberg-Hansen and J. V. Tucker
[ta]   Computable Algebra: An Introduction to the Theory of Computable Rings and
         Fields, to appear.

A. Tarski, A. Mostowski, and R. M. Robinson
[1953]   Undecidable theories, North-Holland, Amsterdam, 1953.

S. Tennenbaum
[1961]   Degrees of unsolvability and the rate of growth of functions, Notices Amer. Math.
         Soc. 8 (1961), 608.
[1962]   Degrees of unsolvability and the rate of growth of functions, In: Proc. Sympos.
         Math. Theory of Automata, Microwave Res. Inst. Sympos. Ser. 12, Polytechnic
         Press, Brooklyn, New York, 1962, 71–73.

S. K. Thomason
[1971]   Sublattices of the recursively enumerable degrees, Z. Math. Logik Grundlag. Math.
         17 (1971), 273–280.

A. S. Troelstra and D. van Dalen
[1982]   The L. E. J. Brouwer Centenary Symposium, Proceedings of the Conference held in
         Noordwijkerhout, 8–13 June 1981, North-Holland, New York, Amsterdam, Oxford,
         1982.

R. E. Tulloss
[1971]   Some Complexities of Simplicity: Concerning the Grades of Simplicity of Recursively
         Enumerable Sets, Ph.D. Dissertation, University of California, Berkeley, 1971.

A. M. Turing
[1936–37] On computable numbers, with an application to the Entscheidungsproblem, Proc.
         London Math. Soc. 42 (1936), 230–265; A correction, ibid. 43 (1937), 544–546;
         reprinted in Davis [1965], 115-154.

[1937] Computability and λ-definability, J. Symbolic Logic *2* (1937), 153–163.
[1939] Systems of logic based on ordinals, Proc. London Math. Soc. *45* (1939), 161–228; reprinted in Davis [1965], 154–222.

J. S. Ullian
[1960] Splinters of recursive functions, J. Symbolic Logic *25* (1960), 33–38.
[1961] A theorem on maximal sets, Notre Dame J. Formal Logic *2* (1961), 222–223.

V. A. Uspenskii
[1955a] On computable operations, Dokl. Akad. Nauk SSSR, N.S. *103* (1955), 773–776 (Russian).
[1955b] Systems of denumerable sets and their enumeration, Dokl. Akad. Nauk SSSR, N.S. *105* (1955), 1155–1158 (Russian).
[1957] Some notes on recursively enumerable sets, Z. Math. Logik Grundlag. Math. *3* (1957), 157–170 (Russian); Amer. Math. Soc. Translations *23* (1957), 89–101 (English translation).
[1960] Lectures on computable functions, Gosudarstvennoye Izdat. Fiz.-Mat. Lit., Moscow, 1960 (Russian).

F. I. Validov
[1984] Recursively enumerable sets and discrete families of total recursive functions, Izv. Vyssh. Uchebn. Zaved. Mat. *4* (1984), 6–11 (Russian); Sov. Math. (Iz. VUZ) *28* (1984), 6–11 (English translation).

D. van Dalen, D. Lascar, and J. Smiley
[1982] Proceedings of the Logic Colloquium '80: Papers intended for the European Summer Meeting of the Association of Symbolic Logic planned for Prague, August 24–30, 1980, but never held, (editors), North-Holland, Amsterdam, New York, Oxford, 1982.

J. van Heijenoort
[1967] From Frege to Gödel: A Source Book in Mathematical Logic, 1879–1931, Harvard University Press, Cambridge, Mass., 1967.

V. V. V'yugin
[1974] Segments of recursively enumerable *m*-degrees, Algebra i Logika *13* (1974), 635–654, 719 (Russian); Algebra and Logic *13* (1974), 361–373 (English translation).

S. S. Wainer
[1985] Subrecursive ordinals, In: Ebbinghaus, Müller, and Sacks [1985], 405–418.

L. V. Welch
[1981] A Hierarchy of Families of Recursively Enumerable Degrees and a Theorem on Bounding Minimal Pairs, Ph.D. Dissertation, University of Illinois at Urbana-Champaign, 1981.
[1984] A hierarchy of families of recursively enumerable degrees, J. Symbolic Logic *49* (1984), 1160–1170.

A. Yasuhara
[1971] Recursive Function Theory and Logic, Academic Press, New York, 1971.

C. E. M. Yates
[1962] Recursively enumerable sets and retracing functions, Z. Math. Logik Grundlag. Math. *8* (1962), 331–345.
[1965] Three theorems on the degree of recursively enumerable sets, Duke Math. J. *32* (1965), 461–468.
[1966a] A minimal pair of recursively enumerable degrees, J. Symbolic Logic *31* (1966), 159–168.
[1966b] On the degrees of index sets, Trans. Amer. Math. Soc. *121* (1966), 309–328.
[1967] Recursively enumerable degrees and the degrees less than $\mathbf{0'}$, in Crossley [1967], 264–271.

418   References

[1969]   On the degrees of index sets II, Trans. Amer. Math. Soc. *135* (1969), 249–266.
[1970a] Initial segments of the degrees of unsolvability, Part I: A survey, In: Bar-Hillel [1970],
         63–83.
[1970b] Initial segments of the degrees of unsolvability, Part II: Minimal degrees, J. Symbolic
         Logic *35* (1970), 243–266.
[1972]   Initial segments and implications for the structure of degrees, In: Hodges [1972],
         305–335.
[1974]   Prioric games and minimal degrees below $\mathbf{0}'$, Fund. Math. *82* (1974), 217–237.
[1976]   Banach-Mazur games, comeager sets, and degrees of unsolvability, Math. Proc. Cam-
         bridge Philos. Soc. *79* (1976), 195–220.

P. R. Young
[1963]   On the Structure of Recursively Enumerable Sets, Ph.D. Dissertation, Massachu-
         setts Institute of Technology, Cambridge, Mass., 1963.
[1966]   Linear orderings under one-one reducibility, J. Symbolic Logic *31* (1966), 70–85.
[1985]   Gödel theorems, exponential difficulty and undecidability of arithmetic theories: an
         exposition, In: Nerode and Shore [1985], 503–522.

Notation Index

Introduction

Chapter I

Chapter IV

Chapter V

$G_{e,\infty}$ permanent residents of gate G_e 148

Chapter IX

$D(\vec{x})$ diagram   151
$\diamond$ diamond lattice   151
$l(e,s)$ length function (for Theorem IX.1.2)   153
$m(e,s)$ maximum length function (for Theorem IX.1.2)   153
$r(e,s)$ restraint function (for Theorem IX.1.2)   154
$x \prec y$ stronger priority (for Theorem IX.2.1)   158
$\mathbf{M}$ $\{\,\mathbf{a} : \mathbf{a} = 0 \text{ or } \mathbf{a} \text{ is r.e. and is one half of a minimal pair}\,\}$ 167, 174
$\mathbf{M}^+$ ideal of $\mathbf{R}$ generated by $\mathbf{M}$ 167
$\mathbf{R}(< \mathbf{a})$ $\{\,\mathbf{x} : \mathbf{x} \in \mathbf{R} \ \& \ \mathbf{x} < \mathbf{a}\,\}$ 169
$\mathbf{R}(> \mathbf{a})$ $\{\,\mathbf{x} : \mathbf{x} \in \mathbf{R} \ \& \ \mathbf{x} > \mathbf{a}\,\}$ 169
$\mathbf{R}(\leq \mathbf{a})$ $\{\,\mathbf{x} : \mathbf{x} \in \mathbf{R} \ \& \ \mathbf{x} \leq \mathbf{a}\,\}$ 173
$\mathbf{R}(\geq \mathbf{a})$ $\{\,\mathbf{x} : \mathbf{x} \in \mathbf{R} \ \& \ \mathbf{x} \geq \mathbf{a}\,\}$ 173
$\mathbf{R}^1$ the first Cantor-Bendixson derivative of $\mathbf{R}$ 173
$\mathbf{NC}$ class of noncappable r.e. degrees   174
$\mathbf{ENC}$ class of effectively noncappable (r.e.) degrees   175

Chapter X

$\mathcal{I}(a)$ $\{\,b \in \mathcal{L} : b \leq a\,\}$ 178
$\mathcal{D}(a)$ $\{\,b \in \mathcal{L} : b \geq a\,\}$ 178
$\mathcal{L}^*$ quotient lattice $\mathcal{L}/\mathcal{F}$ 179, 343
$\mathcal{N}^*$ quotient lattice $\mathcal{N}/\mathcal{F}$ 179
$\mathcal{E}^*$ quotient lattice $\mathcal{E}/\mathcal{F}$ 179
$\mathcal{R}^*$ quotient lattice $\mathcal{R}/\mathcal{F}$ 179
A^* $\{\,B : A =^* B\,\}$ 179
$\mathrm{Rec}(x)$ $(\exists y)\,[x \vee y = 1 \ \& \ x \wedge y = 0]$ 179
$\mathrm{Fin}(x)$ $(\forall y)\,[y \leq x \implies \mathrm{Rec}(y)]$ 179
$\mathrm{Sim}(x)$ $x < 1 \ \& \ (\forall y)\,[y > 0 \implies x \wedge y > 0]$ 180
$\mathrm{Max}(x)$ $x < 1 \ \& \ (\forall y)\,[x < y \implies y = 1]$ 180
$\mathcal{E}(S)$ lattice $\{\,W \cap S : W \text{ r.e.}\,\}$ 182
W_S $W \cap S$ in $\mathcal{E}(S)$ 182
$\mathcal{L}(A)$ principal filter $\{\,B : A \subseteq B \ \& \ B \in \mathcal{E}\,\}$ 184
$\sigma(e, x, s)$ e-state of x at stage s 187
$\sigma > \tau$ σ is a higher (stronger) e-state than τ 187, 235
$\sigma(-1, x, s)$ $\emptyset$, for all x and s 187
η-maximal (see page 190)
$A \subset_m B$ A a major subset of B 191
$A \subset_s B$ A small in B 193
$A \subset_{sm} B$ $A \subset_s B$ and $A \subset_m B$ 194
Σ_3 Boolean algebra (see page 203)

Chapter XI

$c_A(x)$ computation function, $(\mu s)\,[A_s \upharpoonright x = A \upharpoonright x]$ 214, 223
$\deg(\mathcal{C})$ $\{\,\deg(W) : W \in \mathcal{C}\,\}$ 217
$\mathcal{M}$ class of maximal sets   217, 230
$\widetilde{\Phi}_s(x)$ $\widetilde{\Phi}(A_{i,s} \cup C_s; e, x, s)$ (for Theorem XI.3.2)   225
$\cong^{\mathrm{eff}}$ effectively isomorphic (automorphic)   230, 344

Chapter XV

Chapter XVI

Subject Index